Modern Concrete Materials: Binders, Additions and Admixtures

Proceedings of the International Conference
held at the University of Dundee, Scotland, UK
on 8-10 September 1999

Edited by

Ravindra K. Dhir
Director, Concrete Technology Unit
University of Dundee

and

Thomas D. Dyer
Research/Teaching Fellow, Concrete Technology Unit
University of Dundee

Thomas Telford

Published by Thomas Telford Publishing, Thomas Telford Limited, 1 Heron Quay, London E14 4JD.

URL: http://www.t-telford.co.uk

Distributors for Thomas Telford books are
USA: ASCE Press, 1801 Alexander Bell Drive, Reston, VA 20191-4400, USA
Japan: Maruzen Co. Ltd, Book Department, 3–10 Nihonbashi 2-chome, Chuo-ku, Tokyo 103
Australia: DA Books and Journals, 648 Whitehorse Road, Mitcham 3132, Victoria

First published 1999

The full list of titles from the 1999 International Congress 'Creating with Concrete' and available from Thomas Telford is as follows

- *Creating with concrete*
- *Radical design and concrete practices*
- *Role of interfaces in concrete*
- *Controlling concrete degradation*
- *Extending performance of concrete structures*
- *Exploiting wastes in concrete*
- *Modern concrete materials: binders, additions and admixtures*
- *Utilizing ready-mixed concrete and mortar*
- *Innovation in concrete structures: design and construction*
- *Specialist techniques and materials in concrete construction*
- *Concrete durability and repair technology*

A catalogue record for this book is available from the British Library

ISBN: 0 7277 2822 9

Printed and bound in Great Britain by MPG Books, Bodmin, Cornwall

PREFACE

Concrete is the key material for Mankind to create the built environment, the requirements for which are both demanding in terms of technical performance and economy and yet greatly varied from architectural masterpieces to the simplest of utilities. This presents the greatest challenge and the question is how best to advance concrete and create imaginatively.

In response, the Concrete Technology Unit (CTU) of the University of Dundee organised this Congress following on from its established series of events, namely, Concrete in the Service of Mankind in 1996, Concrete 2000: Economic and Durable Concrete Construction Through Excellence in 1993 and Protection of Concrete in 1990.

Under the theme of Creating with Concrete, the Congress consisted of five Seminars: (i) Radical Design and Concrete Practices, (ii) Role of Interfaces in Concrete, (iii) Controlling Concrete Degradation, (iv) Extending Performance of Concrete Structures and (v) Exploiting Wastes in Concrete, and five Conferences: (i) Modern Concrete Materials: Binders, Additions and Admixtures, (ii) Utilising Ready-Mixed Concrete and Mortar, (iii) Innovation in Concrete Structures: Design and Construction, (iv) Specialist Techniques and Materials for Concrete and Construction and (v) Concrete Durability and Repair Technology. In all, a total of 421 papers were presented from 67 countries.

The Opening Addresses were given by Mr Henry McLeish, MP, MSP, Minister for Enterprise and Lifelong Learning, Scotland, Dr Ian Graham-Bryce, Principal and Vice Chancellor of Dundee University, Mrs Helen Wright, Lord Provost, City of Dundee and Professor Peter Hewlett, President of the Concrete Society. This was followed by four Opening Papers by leading international experts; Dr Bryant Mather, US Army Corps of Engineers, Professor Charles F Hendriks, Delft University of Technology, Netherlands, Dr Bjørn Jensen, Danish Technological Institute, Dr Oliver Kornadt, Philipp Holzmann AG, Germany, Professor Jurek Tolloczko, Concrete Society, UK, Mr Michael Téménidès, CIMBÉTON, France and Professor Yves Malier, Ecole Normale Superieure de Cachan, France. The Closing Address was given by Professor John Morris, University of the Witwatersrand, South Africa.

The support of 20 International Professional Institutions and 31 sponsors was a major contribution to the success of the Congress. An extensive Trade Fair, participated in by 50 organisations, formed an integral part of the Congress. The work of the Congress was an immense undertaking and all of those involved are gratefully acknowledged, in particular, the members of the Organising Committee for managing the event from start to finish; members of the International Advisory and National Technical Committees for advising on the selection and reviewing of papers; the Authors and the Chairmen of Technical Sessions for their invaluable contributions to the proceedings.

All of the proceedings have been prepared directly from the camera-ready manuscripts submitted by the authors and editing has been restricted to minor changes where it was considered absolutely necessary.

Dundee
September 1999

Ravindra K Dhir
Chairman, Congress Organising Committee

INTRODUCTION

Most practitioners in the field of concrete technology would agree that binders, and the manner in which they are used, are the most critical factors in determining the performance of the final product. Although Portland cement will continue as the major binder, the need to move towards sustainable practices in construction, through improved durability, recycling, reduced energy consumption and reduced pollutant emissions, has lead to both the emergence of a diversity of other binder materials, as well as changes in the way Portland cement itself is manufactured.

With these, a better understanding of the role of binders, which is helping to optimise their use in the concrete industry has developed. New methods of combining materials to extract maximum benefit, both by exploiting physical characteristics and harnessing inherent chemical energy, are emerging.

Refinement of molecular function and synthesis is yielding admixtures which outperform, radically, their predecessors. The widespread use of computers is permitting greater meaning to be extracted from experimental data than ever before, as well as allowing many of the complex processes relating to binder systems to be modelled practically.

Notwithstanding this, many questions regarding the mechanisms of reaction, the structures of some hydration products and the role of hydration product microstructure in concrete have yet to be fully resolved.

The Proceedings of this Conference; *'Modern Concrete Materials: Binders, Additions and Admixtures'* dealt with all of these subject areas and the issues thus raised, under six clearly identified themes: (i) Basic Portland Cements, (ii) Portland Cement Based Binders, (iii) Non-Portland Cement-Based Binders, (iv) Water-Reducing / Set-Controlling Admixtures, (v) Durability Enhancing Admixtures, (vi) Designer Mix Methodology. Each theme started with a Leader Paper presented by the foremost exponents in their respective fields. There were a total of 64 papers presented during the International Conference which are compiled into these Proceedings.

Dundee
September 1999

Ravindra K Dhir
Thomas D Dyer

ORGANISING COMMITTEE
Concrete Technology Unit

Professor R K Dhir, OBE (Chairman)

Dr M R Jones (Secretary)

Mr M D Newlands (Joint Secretary)

Professor P C Hewlett
British Board of Agrément

Dr N A Henderson
Mott MacDonald Ltd

Professor V K Rigopoulou
National Technical University of Athens, Greece

Dr S Y N Chan
Hong Kong Polytechnic University

Dr N Y Ho
L & M Structural Systems, Singapore

Dr M J McCarthy

Dr M C Limbachiya

Dr T D Dyer

Dr K A Paine

Dr T G Jappy

Mr P A J Tittle

Mr J C Knights

Mr S R Scott (Unit Assistant)

Miss A M Duncan (Unit Secretary)

INTERNATIONAL ADVISORY COMMITTEE

INTERNATIONAL ADVISORY COMMITTEE (CONTINUED)

Professor R Park
Professor of Civil Engineering
University of Canterbury, New Zealand

Mr S A Reddi
Managing Director
Gammon India Limited, India

Professor H-W Reinhardt
Head of Construction Materials Institute
University of Stuttgart, Germany

Professor R Rivera-Villarreal
Chief of Concrete Technology Department
Ciudad Universitaria, Mexico

Professor A Samarin
Consultant
Sustainable Development Technological Sciences and Engineering, Australia

Professor A E Sarja
Research Professor
Technical Research Centre of Finland, Finland

Professor S P Shah
Walter P Murphy Professor or Civil Engineering
Northwestern University, USA

Professor H Sommer
Head of Research Institute
VÖZ, Austria

Professor I Soroka
Professor of Civil Engineering
National Building Research Institute, Israel

Professor M Tang
Research Professor
Nanjing University of Chemical Technology, China

Professor T Tassios
Professor
National Technical University of Athens, Greece

Professor K Tuutti
Vice President
Skanska Technik AB, Sweden

Professor T Vogel
Professor of Structural Engineering
Swiss Federal Institute of Technology ETH , Switzerland

Professor F H Wittmann
Head of Building Materials Laboratory
Swiss Federal Institute of Technology ETH , Switzerland

Professor A V Zabegayev
Head of Department RC Structures
Moscow State University of Civil Engineering, Russia

SUPPORTING INSTITUTIONS

American Concrete Institute, USA

American Society of Civil Engineers, USA

Australian Concrete Institute

Concrete Association of Finland

Concrete Society of Southern Africa

Concrete Society, UK

Danish Concrete Society, Denmark

Fédération de l'Industrie du Beton, France

German Concrete Association (DBV)

Hong Kong Institution of Engineers

Indian Concrete Institute

Institute of Concrete Technology, UK

Institution of Civil Engineers, UK

Instituto Brasileiro Do Concreto, Brazil

Japan Concrete Institute

Netherlands Concrete Society

New Zealand Concrete Society

Norwegian Concrete Association

Singapore Concrete Institute

Spanish Association for Structural Concrete

Swedish Concrete Association

SPONSORING ORGANISATIONS WITH EXHIBITION

AMEC Civil Engineering Ltd.

Babtie Group Ltd.

Bardon Aggregates

Blue Circle Cement

Blyth & Blyth

British Board of Agrément

British Cement Association

Building Research Establishment

Castle Cement Ltd.

Cementitious Slag Makers Association

CIMBÉTON, France

Du Pont de Nemours International S.A., Switzerland

ECC International Ltd.

Elkem Ltd. (Materials)

Fosroc International Ltd.

Grace Construction Products

HERACLES General Cement Co., Greece

John Doyle Group

Lafarge Aluminates

L M Scofield Europe Ltd.

Minelco Ltd.

Mott MacDonald Ltd.

O'Rourke Group

Ove Arup and Partners

SPONSORING ORGANISATIONS WITH EXHIBITION (CONTINUED)

Readymix (UK) Ltd.

Rugby Cement

Scottish Enterprise Tayside

Sika Ltd.

SKW - MBT Construction Chemicals

Thomas Telford Publishing Ltd.

United Kingdom Quality Ash Association

W A Fairhurst & Partners

ADDITIONAL EXHIBITORS

Christison Scientific Equipment Ltd.

CMS Pozament Limited

The Concrete Society

David Ball Group plc.

E & FN Spon

Flexcrete Ltd.

Germann Instruments A/S, Denmark

Natural Cement Distribution Limited

Palladian Publications Ltd.

Quality Scheme for Ready Mixed Concrete

UK Certification Authority for Reinforcing Steel

Wacker-Chemie GmbH, Germany

Wexham Developments

CONTENTS

LEADER PAPER

CONCRETE BINDERS, MINERAL ADDITIONS AND CHEMICAL ADMIXTURES: STATE OF THE ART AND CHALLENGES FOR THE 21ST CENTURY

J S Damtoft

D Herfort

Aalborg Portland

E Yde

Nordisk Bygge Kemi A/S

Denmark

ABSTRACT. Although a wide range of concrete binders is available for specialist applications, by far the greatest volume is based on Portland cement. Mineral additions to Portland cement, particularly of artificial pozzolans such as fly ash, silica fume and blast furnace slag, and the use of more chemically specific admixtures continue to gain acceptance in many parts of the world. Although the hydrate mineralogy of the Portland cement systems is not radically altered by the presence of pozzolans, the content of C-S-H gel is generally increased at the expense of $Ca(OH)_2$ due to the reactive SiO_2 contents of the pozzolans, giving lower capillary porosities and therefore higher strengths and durability. Limestone fillers are also gaining acceptance and in some cases improve strengths and durability even at levels of replacement of up to 20%. For concretes where early strengths are less important than long term strengths and durability, mineral additions can be used to reduce CO_2 emissions and lower production costs. Systems containing two or more mineral additions combined with the optimum use of admixtures are creating opportunities for the development of high strength, highly durable concretes without compromising workability.

Keywords: Binders, Portland cement, Pozzolans, Admixtures, CO_2 emission.

Mr Jesper S Damtoft is Executive Manager, Research Development & Testing at Aalborg Portland, Denmark. His research interests include: Development of new cement and clinker types, recycling of waste and by-products as raw materials or fuels in the cement and concrete industry, novel cementitious materials and life cycle analyses.

Mr Elo Yde is Technical Manager of Nordisk Bygge Kemi A/S, Denmark, a member of SKW-MBT group. His responsibilities include research & development of admixtures for concrete and Quality Management.

Mr Duncan Herfort is Chief Geologist at Aalborg Portland. His responsibilities include all mineralogical aspects associated with raw material assessment, process and quality control, and concrete analysis.

INTRODUCTION

Concrete is the world's most important construction material. Modern construction in industry and of residential housing and infrastructure would be unthinkable without concrete. The volume produced is huge with approximately 5 km^3 produced annually [1]. Cement binders, additions and admixtures comprise a comparatively minor portion of modern concrete; typically 10-15%. Nevertheless, it is these minor constituents which are mainly responsible for the properties of fresh and hardened concrete, how much it costs to produce, and its impact on the environment. The main challenges faced by the concrete industry as it enters the 21st century include the need to

- reduce the impact on the environment
- improve durability
- compete with alternative building materials

against a background of increasing demand in the developing world, and little or no growth in the industrialised west.

The materials at the industry's disposal in order to meet these challenges are the binder, which is the topic of this paper, and the aggregate. The vast bulk of concrete binders consist of Portland cement with or without mineral additions and admixtures.

BINDERS

The term "binder" is normally used to define a material which, after mixing in a plastic state, generally with an aggregate, sets and hardens to form a solid mass. An hydraulic binder such as Portland cement is a finely ground substance which achieves a plastic and workable state on mixing with water and then reacts to form hydrated minerals. A wide range of binders exists for specialised applications such as Plaster of Paris for the gypsum product industry, Sorel cement for industrial flooring, phosphate cement and water glass for rapid set and acid resistance, and many more. In terms of volume, the great majority of binders used in the construction industry are calcareous which, for the purpose of this paper, are defined as cements whose compositions fall within the CaO-rich region of the C-S-A ternary phase diagram shown in Figure 1.

Lime and hydraulic lime cements (as well as lime cements strengthened by additions of Portland cement as for example specified by ENV 459-1 [2]) are today mainly restricted to masonry applications, although they had wider applications in the past. They were the forerunners of Portland cement leading up to the Industrial Revolution, with a wide range of applications including some large scale constructions such as the famous Eddystone Lighthouse [3]. Ancient pozzolanic cements had a similar compositional range to that given in Figure 1 for the hydraulic limes. High aluminate cements (HAC) are used for high temperature and chemical resistant applications, or where particularly rapid hardening is required [4]. The anhydrous and hydrate mineralogy of HAC is markedly different from the other calcareous cement types. The remainder of this paper will focus on Portland cement and its various derivatives.

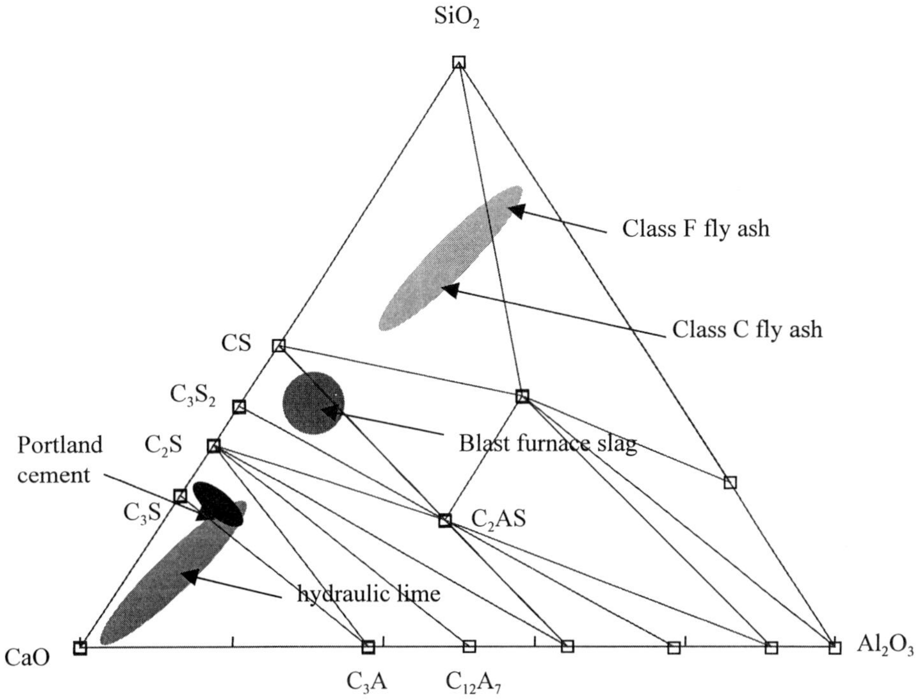

Figure 1 C-S-A phase diagram showing compositions of common calcareous cements

Portland Cement

In terms of chemical composition, modern Portland cement differs from the pozzolanic and hydraulic lime cements. The early Victorian Portland cements (termed by Blezard [3] as proto- and meso-Portland cements), had narrower and more consistent chemical compositions, and higher sulfate contents generally obtained by the controlled addition of gypsum for set regulation.

In terms of mineralogical composition, modern Portland cements have a much more well defined mineral assemblage, i.e. alite (C_3S), belite (C_2S), the aluminate phase (C_3A), the ferrite phase (C_4AF), in addition to approximately 1 % free lime. This is in contrast to the early Portland cements, hydraulic limes (which contained many non-equilibrium phases such as mayenite ($C_{12}A_7$) and gehlenite (C_2AS) due to inadequate blending of the raw feed and low burning temperatures), and lime-pozzolan cements in which the mineralogy is dependent on the raw materials (e.g. zeolites and volcanic glass in pozzolans of volcanic origin).

A feature common to all the cements plotted in Figure 1 is that they, at least in principle, form the same hydrate phase assemblage in the absence of carbonation, i.e. C-S-H, CH, AFt (usually ettringite) and AFm (usually monosulfate), an iron hydrate phase (probably an iron hydroxide gel containing Ca [5]) and the pore solution. The rate at which these hydrate phases are formed, and therefore the rate at which strength is gained through the reduction in capillary porosity, is dependent on the reactivity of the anhydrous minerals. The clinker phases in Portland cement are generally much more reactive than the minerals present in both natural and the artificial pozzolans described below. That is why mineral additions until now have generally not been included in rapid hardening cements.

The final hydrate assemblage is dependent on the combined chemical composition of all the hydraulic phases, and not on their mineralogy or reactivity. Portland pozzolan cements, which are generally richer in SiO_2 than pure Portland cements, result in a final hydrate assemblage with more C-S-H and less CH which, because of the comparative lower density of the former, results in lower capillary porosity at the same initial water/cement ratio. Greater durability due to lower permeability is also obtained in addition to higher ultimate strengths.

Apart from the capillary porosity which is determined by the chemical composition (as noted above) and the initial water/cement ratio, durability is also to a large extent dependent on the resistance of the individual hydrate phases to certain types of chemical attack. One of the best known examples of this is the inferior resistance of the aluminate-bearing phases, i.e. AFm and unreacted C_3A in the clinker, to sulfate attack. Because cements with a high content of Al_2O_3 can form more potentially expansive ettringite on reaction with sulphates from an external source, such as brine-rich groundwater, limits are placed on the C_3A content in cements specified as sulfate resisting (e.g. max. 5% C_3A for ASTM Type V and 8% for ASTM Type II) [6].

ADDITIONS

Blast Furnace Slag, Fly Ash, Silica Fume and Other Pozzolans

Mineral additions to Portland cement, added at the cement works or the concrete plant, have been common practice throughout most of this century with blast furnace slag in the heavily industrialised regions perhaps having the longest and most consistent track record. Over recent decades Portland cements containing fly ash from coal-fired power stations and silica fume from ferrosilicon production have gained increasing acceptance, whilst Portland cements containing natural pozzolans, rice husk ash and burnt oil shale are common in regions where these raw materials are readily available (Table 1). Replacement levels of Portland cement by blast furnace slag vary considerably with contents of well over 50 % by weight common in some regions. Fly ash typically replaces 10-30 % of the Portland cement although levels of 50-60 % have been advocated [7]. Where silica fume is added, it commonly comprises 5-10 % of the binder. ASTM standards exist for the use of natural pozzolans, fly ash, silica fume and blast furnace slag in concrete [8,9,10]. The only comparable European EN standard is for fly ash [11], although a standard for silica fume is currently in the course of preparation [12]. There are no plans for any EN standards for the use of blast furnace slag and natural pozzolans in concrete.

Table 1 Estimated availability of mineral additions worldwide [1, 13, 16, 17]

ADDITION	QUANTITY (MILLION. T/YEAR)
Fly ash	370
Granulated blast furnace slag	35
Silica fume	2
Rice husk ash	20
Portland cement (for comparison)	1,400

As shown in Table 1, fly ash is clearly the most abundant artificial pozzolan. Even so only approximately one third of it is used in the construction industry [13]. A large portion is produced in China and India, most of which could conceivably be used to advantage as cement and concrete production continues to increase in these countries [14]. The supply of fly ash in the industrialised countries, however, is expected to decline in the future, as coal-fired power stations are replaced by plants powered by renewable forms of energy [15].

Limestone Additions

Perhaps the most significant development in recent years has been the increased use of finely divided limestone, particularly in France, parts of Germany and in Denmark where advantage is taken of the European prestandard ENV 197 [18] allowing up to 35% $CaCO_3$. Many other countries permit the inclusion of up to 5% limestone as a minor addition.

ADMIXTURES

Admixtures are used to modify the physical properties of fresh concrete, the concrete's setting behaviour or the properties of the hardened concrete. The properties of the fresh concrete are modified by increasing its workability which can be used to reduce the water/cement ratio, reduce the cement content or increase the pumpability of the concrete. Other properties include the freezing temperature which can be lowered when using cold materials.

The setting and hardening of the concrete can be retarded or accelerated depending on the desired pot life, early strength development, etc. The properties of the hardened concrete can be modified by air entrainment which minimises the risk of freeze/thaw damage, by making it water-proof or more resistant to corrosion. Until recently, admixtures used in the concrete industry were usually bi-products of unrelated industries. Specific modifications for optimum use in the concrete industry were rarely possible.

Water-reducing admixtures were first introduced in 1935 as plasticizers based on ligno- and naphthalene sulfonates. In the early 1960's the first high range, or "super"-plasticizer, based on melamine and naphthalene formaldehyde, was introduced [19]. The dispersion mechanism of these superplasticizers in cement paste is best explained in terms of electrostatic repulsion between cement particles which are negatively charged by the adsorption of the polymer molecule onto the cement surface.

Table 2 Development of Admixtures in Concrete since 1885 [19]

1885	Patent of Calcium chloride
1925	Water-tightness admixtures
1930	Alumina powder
1932	Naphthalene-formaldehyde-con.
1938	Air entraining admixtures
1938	Accelerator and retarder
1939	Ligno-sulphonates
1950	De-foaming agent
1955	Freezing point lowering admixtures
1960	Melamine-formaldehyde condensate
1993	Polyacrylate
1997	Polycarboxlic ether

Over the past 10 years a new generation of polycarboxylate-based superplasticizers has been developed specifically for the concrete industry allowing far better control of key parameters such as strength, w/c ratio and slump retention. Structurally the new plasticizers are made up of a main flexible backbone, containing negatively charged carboxylic groups, and a large number of side chains (Figure 2).

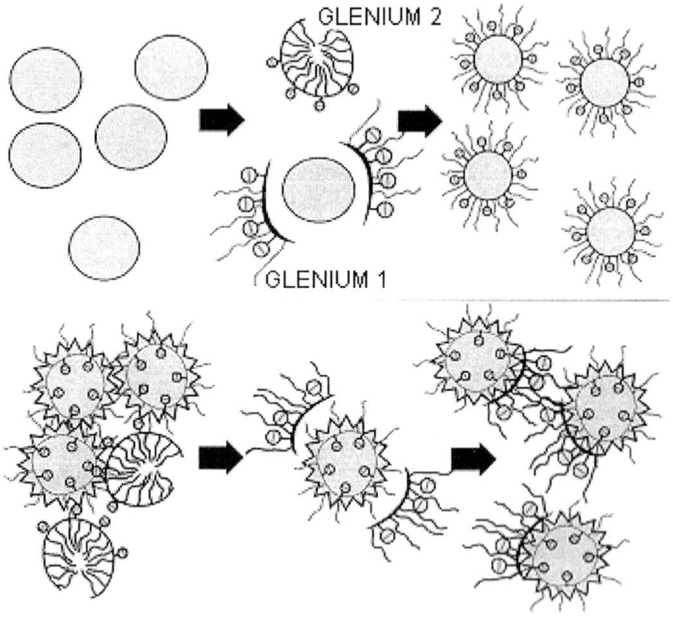

Figure 2 Mechanism of the action of polycarboxylic ether-based superplasticizisers, molecule 1 is activated immediately on mixing, whereas molecule 2, which requires a sufficiently high pH is usually activated after about one hour

Table 3 Water reduction from various water reducing admixtures [19]

Ligno-sulfonates	5-15 %
Melamine sulfonates	5-25 %
Naphtalene sulfonates	15-25%
Polyacrylics	20-30%
Polycarboxylic ether	25-40%

The dispersion mechanism involves two different types of repulsion forces between cement particles, i.e. electrostatic repulsion due to the presence of the negative charge of the carboxylic groups, and steric repulsion of the long side chains.

ADVANTAGES IN THE USE OF ADMIXTURES AND MINERAL ADDITIONS

Three main factors are currently driving trends in the use of admixtures and mineral additions in cement and concrete production:

1. Legislation and government incentives to reduce the impact on the environment, particularly CO_2 emissions.
2. Customer demand for more durable concrete.
3. Increased competition.

CO_2 Emissions

The most important political development over recent years has been the increasing commitment by industrialised countries to reduce CO_2 emissions. For the cement manufacturer, this is adding greater incentive to lower the energy input per unit clinker by improving kiln and grinding efficiency, to replace fossil fuel with CO_2-neutral waste materials such as used tires or plastic derived-fuels, and finally, to lower the clinker content of cement by replacement with pozzolans or limestone filler. In the final analysis, of course, it is not the amount of CO_2 emission per unit cement supplied by the cement manufacturer that counts, but the amount of CO_2 emission per unit concrete.

The CO_2 emission during the life cycle of four different types of concrete is shown in Figure 3. The contribution from the cement (which is calculated for ordinary Portland cement without mineral addition) is solely dependent on the cement content of the concrete, with the edge beam containing the highest content and the lightweight concrete containing the lowest. As noted in the previous section, the replacement level for mineral additions to Portland cement on a global basis is just over 10% compared to a potential replacement of approximately 30% based on present day availability. In other words, there is a potential for reducing the contribution made by the cement on CO_2 emission in concrete production by approximately 20%. However, whilst a weight for weight replacement of Portland cement by fly ash or slag may result in comparable or even improved long term strengths and durability, early strengths are reduced unless the overall content of the Portland cement is increased accordingly or the rate of clinker hydration enhanced by finer grinding.

For certain sections of the concrete industry such the precast and concrete product industries where high early strengths are required to minimised production costs, savings in Portland cement contents are probably minimal. For the ready mix producers, however, savings on Portland cement (and CO_2 emissions) are comparable to the level of mineral replacement and are therefore considerable.

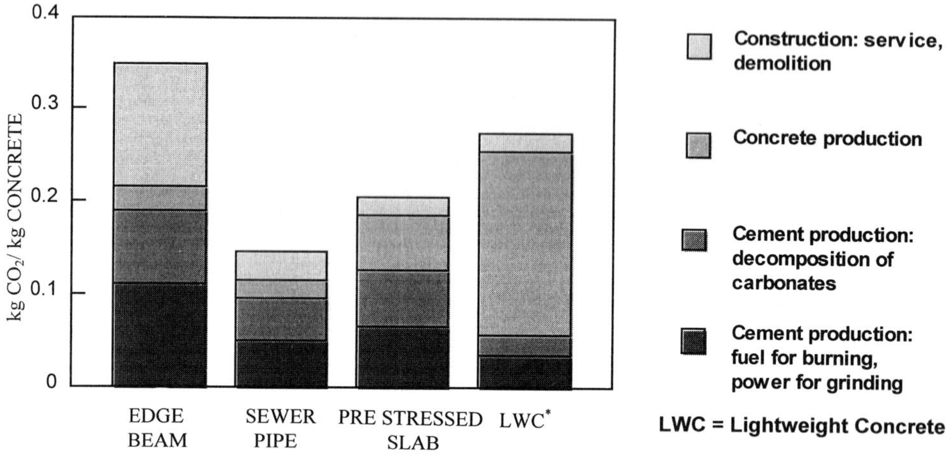

Figure 3 CO_2 emission during the life cycle of concrete [20]

Small additions of finely divided limestone of up to approximately 5% can result in comparable or improved strengths. Since the energy required to grind limestone is considerably less than that involved in the production of Portland cements, some reduction in CO_2 emissions is possible at these levels of replacement. At greater levels of limestone replacement, however, the limestone does not normally react and total cement contents in the concrete must be increased by an equivalent amount in order to maintain strengths, thus outweighing the environmental benefit. With special types of clinker, however, cements containing 20% limestone are possible without reducing strengths and porosity. This is described in more detail in the section of this paper on developments in Scandinavia and in a separate paper in these proceedings [21].

Durability

There is a general consensus nowadays that whilst prescriptive specifications on the composition of Portland cements that have been in place for many decades, such as the limit of 5% C_3A for ASTM type V cement, remain valid, the most effective method of improving durability is by reducing the w/c+m ratio (where m stands for mineral addition) [22,23]. The replacement of cement by artificial or natural pozzolans which are almost invariably richer in SiO_2 than Portland cement inevitably leads to lower capillary porosity at the same w/c+m ratio once most or all of the pozzolan has reacted.

Although this may take months or years, the end result is lower permeability and improved durability. It can be argued that the higher Al_2O_3 contents of fly ash and blast furnace slag compared to Portland cement may reduce resistance to sulfate attack [24].

The fact that limestone filler participates in the hydration reactions has been reported by a number of workers over recent years [25,26,27] despite other reports that it essentially acts as an inert filler [28]. It reacts mainly by forming monocarbonate ($C_3A.CaCO_3.12H_2O$), in addition to the hydrate phases already present. At replacements of approximately 5 wt% (depending on the aluminate content which is the main constraint on the amount of $CaCO_3$ that can react) capillary porosity is reduced, and from that point of view durability must be improved. Comparable or improved durability is confirmed by numerous performance tests [28,29,30]. Improved sulfate resistance from a mineralogical perspective has also been indicated [31], although concerns have been expressed that cool and moist conditions may lead to thaumasite formation [32]. As already noted in the previous section, except where special types of clinker are concerned, limestone replacement levels of more than 5-10% do not generally react and capillary porosities are therefore increased.

Water reducing agents such as plasticizers and superplasticizers are mainly used to reduce the water/cement ratio resulting in lower capillary porosity and therefore increased strength, lower permeability and improved durability. Improved workability may also indirectly improve durability by minimising the formation of placement related defects such as entrapped air. Air entraining agents generate air voids which, if sufficiently spaced, can relieve expansive pressures in the paste by filling up with the pore solution. They are primarily used to minimise the risk of freeze/thaw damage, but other expansive stresses such as those resulting from sulfate attack and alkali aggregate reaction may also be alleviated by air entrainment [33,34,35].

Increased Competition

The globalisation of the cement market, regional over capacity and the reduction in the cost of transporting bulk materials by sea have led to increased competition even in times of economic boom. This, and the increased harmonisation of cement standards has increased the perception of cement as an international commodity with nothing much to choose between Portland cements complying to the same specification apart from price. The replacement of Portland cement by pozzolans for applications where late strengths and durability are of primary importance, the use of limestone fillers, improved production processes, and the use of alternative fuels and raw materials in the production of cement increases competitiveness through a direct reduction in costs and improved environmental performance.

DEVELOPMENTS IN SCANDINAVIA

Reduction in Portland Cement Clinker Contents in Denmark

By replacements of fly ash and silica fume between 1977 and 1990

From the late 1970's until 1990, demand for higher concrete durability led to a general increase in average concrete strengths in Denmark.

This was achieved by replacing on average approximately 20% of the clinker with fly ash and silica fume, without increasing the binder content of the concrete [36]. Although the pozzolanic reaction of the fly ash and silica fume was largely responsible for this, improvements in other practices listed in Table 4 were also significant.

Table 4 Factors reducing cement content of Danish concrete 1977-90 [36]

Higher and more uniform cement strength	2-4%
Substitution of cement with fly ash	8-10%
Substitution of cement with silica fume	3-5%
Use of water reducing admixtures	1-3%
Improved process control at concrete plants	1-3%
Total estimated reduction in cement content	20%

By replacing with limestone filler between 1990 and 1999

Since 1990, further reductions in the clinker content in concrete have been achieved by the successful introduction of a Portland limestone cement based on mineralised clinker. The filler content of this cement is 14%. Since this cement enjoys a market share of approximately half of all the Portland cement sold in Denmark, the average reduction in the clinker content of concrete amounts to approximately 7%. That this has proved possible is due to the synergetic effect between clinker mineralised by CaF_2 and SO_3, and finely divided limestone filler, in which strengths are much higher than would normally be expected [21,25,37]. The new Portland limestone cement has significantly higher early as well as late strengths (EN 196 mortar 1 and 28 day strengths of 27 and 66 MPa respectively). This enables it to be used in applications, such as in the precast industry where high early strengths are a requirement, without needing to increase the cement content.

Since fly ash and silica fume additions are already common as indicated in Table 4, with the permitted levels of replacement shown in Table 5, it is quite common for the binder in concrete designed for all but the most aggressive exposure conditions to consist of four mineral constituents, i.e. limestone, fly ash, silica fume and as little as 50% Portland cement.

100 year guaranteed service life

An example of the way in which mineral additions and admixtures have been used to produce highly durable concrete is the recently completed Great Belt bridge and tunnel connection between the Danish islands of Fyn and Sjælland. Among other things, the contractor was required to guarantee a 100 year service life. The mix design for the bridge is shown in Table 6.

Further specifications include a slump of 180 mm (std. < 30 mm), a slump loss of < 50 mm after 1 1/2 hours, an air void content of 6.5% (std < 1.5%) with a spacing factor of < 0.20 mm, and a specified limits for autogenous and thermal cracks.

Considerable effort was employed in finding a compatible combination of admixtures in order to meet these requirements. Among other things, a water reducer/retarder was needed that did not adversely affect air entrainment. Once the correct combination was developed it proved possible to maintain sufficient workability for the specified period, although in some cases it was necessary to re-dose superplasticizer during use.

Table 5 Fly ash and silica fume according to the Danish concrete pre-standard [38]

EXPOSURE CLASS	(FA+SF)/(FA+SF+C)	SF/(FA+SF+C)
Passive	no requirements	no requirements
Moderate	maximum 35%	maximum 10%
Aggressive	maximum 25%	maximum 10%
Extra aggressive	maximum 25%	maximum 10%

FA: Fly ash, MS: Micro silica, C: Cement

Exposure classes

Passive:	Dry or subterranean exposure with no risk of damage due to freeze/thaw, chemical attack or corrosion of reinforcement
Moderate:	Outdoor exposure, but with no risk of freeze/thaw combined with water saturation, and with no significant external source of alkalis, sulfates, chlorides or other potentially deleterious components.
Aggressive:	Outdoor exposure, with significant risk of freeze/thaw combined with water saturation, or a significant risk of attack by of alkalis, sulfates, chlorides or other potentially deleterious components.
Severe:	Option available in which additional control measures may be required where exposure is judged to particularly aggressive.

Table 6 Mix design for the Great Belt Bridge project (kg/m^3) [39]

Cement (low heat, sulfate resisting, low alkali)	325
Fly ash (class F)	40
Silica fume	40
Air entraining agent	1.0
Plasticizer/retarder	1.7
Superplasticizer (melamine)	7.3
Maximum water content	130
Maximum water/cement ratio	0.33

The low heat, low alkali, sulfate resisting cement was chosen to minimise the risk of thermal and autogenous cracking, alkali aggregate reaction and sulfate attack, respectively. Apart from improving workability, the silica fume and fly ash were designed to increase ultimate strength and durability, mainly by reducing the ultimate capillary porosity, although the Al_2O_3 content of the fly ash is also expected to reduce chloride diffusion.

The technology applied in the Great Belt project was further developed for the Sound Link Project - a major tunnel-bridge construction linking Denmark and Sweden, which is currently under construction. Some of the specifications for the Sound Link concrete are listed in Table 7. In order to obtain a concrete with a relatively high degree of workability and stability, the Sound Link Tunnel Contractors I/S chose a superplasticizer based on melamine, in combination with a stabiliser (Melstab 21) and an additional retarder. Initial slump was typically 220 - 250 mm, being reduced to approximately 170 - 200 mm after pumping. Typical mix designs for the Sound Link concretes are shown in table 8.

Table 7 Specifications for the Sound Link concrete, Types A and B [39]

CONCRETE DEMANDS	TYPE A: TUNNEL SEGMENTS	TYPE B: RAMPS
General condition	- High degree of workability - Adjustable setting time - Precise density - Quick strength development	- Normal workability - Frost resistant
Water/cement ratio	< 0.40	<0.45
Cement content	> 275 kg/m^3	
Fly ash content	< 15 % of total powder	
Silica fume content	< 5 % of total powder	
Total powder	< 340 kg/m^3	
Compressive strength	40 MPa	35 MPa
Air content	no demands	> 3.0 vol. %
Frost resistance SS137244	no demands	< 1.0 kg/m^2

Table 8 Mix design for Sound Project Concrete types A and B (kg/m^3) [40]

	TYPE A	TYPE B
Cement	324	295
Fly ash	52	50
Silica fume	24	30
Air entrainment	-	1.0
Retarder	0.4	-
Plasticizer	-	0.9
Superplasticizer	10.5	5.4

The superplasticizers used for for types A and B are Melstab 21 and Melment L1504x respectively.

Self-compacting Concrete

Self-compacting concrete is a relatively new type of high performance concrete with very high workability necessary for placement in complicated structures such as tightly spaced reinforcement. This is made possible by a new generation of superplasticizers allowing much lower water/cement ratios than was previously possible. This technology has already been successfully applied in several European constructions.

As an example, three bridges were constructed in Sweden in 1998 using self-compacting concrete with a slump of 600 - 700 mm, and a 28 day compressive strength of 70 MPa. Part of the mix design is shown in table 9. Performance testing showed the concretes to be freeze/thaw resistant. Visual analysis including thin section microscopy revealed a dense homogeneous structure with none of the flaws characteristic of poor compacting and placement

Table 9 Part of mix design for Swedish bridge unit using self-compacting concrete [41]

MIX CONSTITUENT	kg/m^3
Cement	415
Limestone Filler, max 0.5mm	180
Superplasticizer, Glenium 51	3.3
Air Entrainer	0.04

w/c ratio = 0.40

Interest in self-compacting concrete is not restricted to major constructions. There is also considerable interest in the precast industry, where self-compacting concrete can reduce or even completely eliminate the need to vibrate. This has obvious benefits in reducing costs and improving workmanship as well as working conditions. Table 10 shows some typical mix designs for self-compacting, precast concrete.

Table 10 Mix design for self-compacting concrete used in pre-cast concrete [42]

ADMIXTURES (% OF POWDER)	CEMENT (kg/m^3)	FLY ASH (kg/m^3)	W/C RATIO	SLUMP FLOW	COMPRESSIVE STRENGTH (MPa)	
					1 Day	28 Days
0.6	242	72	0.50	500	-	53
0.9	330	80	0.39	600	19	55
1.0	320	60	0.41	670	21	62

The admixtures used in these mix designs are all second generation superplasticizers

The advantages of using mineral additions such as fly ash and silica fume described in earlier sections also apply to self-compacting concrete further improvements in performance.

CONCLUDING REMARKS

Concrete is a cost-effective and relatively environmentally friendly construction material. However, it does not exist in a vacuum and must remain competitive if it is to maintain its dominance as the preferred construction material into the 21st century.

The challenge faced by the concrete industry is to reduce the environmental impact of concrete throughout its life cycle, from the quarry to the end of the concrete's service life, whilst at the same time improving durability and cost efficiency.

The concrete binder based on Portland cement, mineral additions and chemical admixtures will remain the most effective way of meeting this challenge for the foreseeable future. The finite availability of artificial pozzolans, such as fly ash and blast furnace slag, however, is likely to pave the way for alternative replacement materials such as limestone filler.

REFERENCES

1. JAHREN, P. Har bransjen tatt problemet på alvor. BETON Gindustrien, No. 4, 1998, pp 26-30. (In norwegian)

2. ENV 459-1:1994:E, Building lime - Part 1: Definitions, specifications and conformity criteria, 1994.

3. BLEZARD, R G. "The history of Calcareous cements" in Lea's Chemistry of Cement and Concrete, 4 th edition Ed. P.C.Hewlett, Pub. Arnold, 1998, pp 1-23.

4. TAYLOR, H F W. Cement Chemistry, Pub. Thomas Telford, 1997, pp. 295-313.

5. EMANUELSON, A, AND HANSEN, S. Distribution of Iron Among Ferrite Hydrates. Cem. and Conc. Res, Vol. 27, No. 1, 1997, pp 1167-1177.

6. ASTM C 150-97, Standard Specification for Portland Cement.

7. BILODEAU, A, and MALHOTRA, V M. High-Volume Fly Ash System: The Concrete Solution for Sustainable Development. In Compilation of Papers, Three-Day CANMET/ACI Symposium on Sustainable Development of the Cement and Concrete Industry, October 21-23 1998, Ottawa, pp. 193-214

8. ASTM C 612-97, Standard Specification for Coal Fly Ash and Raw or Calcined Natural Pozzolan for Use as a Mineral Admixture in Concrete, 1997.

9. ASTM C 989-95, Standard Specification for Blast-Furnace Slag for Use in Concrete and Mortars, 1995.

10. ASTM C 1240-97, Standard Specification for Silica Fume for Use as a Mineral Admixture in Hydraulic-Cement Concrete, Mortar and Grout, 1997.

11. EN 450:1994E, Fly Ash for Concrete - Definitions, Requirements and Quality Control, 1994.

12. pre EN 13263:1998, Silica Fume for Concrete - Definitions, Requirements and Quality Control, 1998.

13. MANZ, O E. Worldwide Production of Coal Ash and Utilisation in Concrete and Other Products. Fuel, Vol. 76, No. 8, pp. 691-696, 1997.

14. MEHTA, P K. Concrete technology for sustainable development - An overview of essential principles. In Compilation of Papers, Three-Day CANMET/ACI Symposium on Sustainable Development of the Cement and Concrete Industry, October 21-23 1998, Ottawa, pp. 1-14.

15. DAMTOFT, J S. Use of Fly Ash and Other Waste Materials as Raw Feed and Energy Source in the Danish Cement Industry. In Compilation of Papers, Three-Day CANMET/ACI Symposium on Sustainable Development of the Cement and Concrete Industry, October 21-23 1998, Ottawa, pp. 95-105.

16. World Cement Annual Review 1997. World Cement, Vol. 28, No. 7, July 1997, p. 4.

17. MORANVILLE-REGOURD, M. Cements Made From Blastfurnace Slag. In HEWLETT, P C (ed.) Lea's Chemistry of Cement and Concrete. Fourth Edition. Arnold, London, pp. 633-674.

18. ENV 197-1, Cement - Composition, specifications and conformity criteria - Part 1: Common cements, October 1992.

19. KJÆR, U. Admixtures for Concrete, a Handbook. BKI Report 1, 1982.

20. Industry Analysis, Concrete - Cleaner Technology in Concrete Production. Environmental report No. 350, Danish Environmental Protection Agency, 1997. (In danish).

21. RASMUSSEN, S, AND HERFORT, D. Mineralised Cement, Separate peper in these proceedings, Conference 1: Modern Concrete Materials: Binders, Additions and Admixtures, Basic Portland Cement.

22. SKALNY, J, AND PIERCE, J S. Sulfate Attack Issues, Proceedings of the "Quebec Seminar on Sulfate", October 5 & 6, 1998, to be published as "Materials Science of Concrete, Special Volume: Sulfate Attack Mechanisms (J. Marchand & J. Skalny, Eds.) by the American Ceramic Society.

23. Guide to Durable Concrete (1992), section 2.2, ACI 201.2R-92.

24. LOCHER, F W. The problem of Sulfate resistance of slag cements (in German), Zement-Kalk-Gips, Vol. 55, No. 9, 1966, pp. 395-401.

25. KLEMM, W A, AND ADAMS, L D. An Investigation of the Formation of Carboaluminates. in Carbonate Additions to Cement, Klieger/Hooton, Eds. ASTM STP 1064, 1990, pp. 60-72.

26. BORGHOLM, H E, HERFORT, D, AND RASMUSSEN, S. A new blended cement based on mineralised clinker, World Cement vol. 8, pp. 27-33, (1995).

27. INGRAM, K D, AND DAUGHERTY, K E. A review of Limestone Additions to Portland Cement and Concrete. Cement and Concrete Composites, Vol. 13, No. 3, 1991, pp. 165-170.

28. SIEBEL, E, AND SPRUNG, S. Influence of Limestone in Portland Limestone Cement on the Durability of Concrete (in German). Beton, Vol. 41, No. 3, 1991, pp. 113-117.

29. RANC, R, MORANVILLE-REGOURD, M, COCHET, G, AND CHAUDOUARD, G. Durability of Cements with fillers, 2nd- CANMET/ACI International Conference on Durability of Concrete, Eds V.M.Malhotra, 1991, Vol. II, pp. 1239-1257.

30. SCHMIDT, B. Cement with interground additives - Capabilities and Environmental Relief, Part 2, Zement-Kalk-Gips, Vol. 45, No. 6, June 1992, pp. 296-301.

31. SOROKA, I, AND SETTER, N. Effect of Mineral Fillers on Sulfate Resistance of Portland Cement Mortars. Durability of Building materials and Components, ASTM STP 691, P.J. Sereda and G.G. Litvan, Eds., American Ceramic Society for Testing and Materials, 1980, pp. 326-335.

32. BENSTED, J. Scientific Background to Thaumasite Formation in Concrete. World Cement, November, 1998, pp. 102-105.

33. YANG, Q, WU, X, AND HUANG, S. concrete deterioration due to physical attack by salt crystallisation. Proc. 10th ICCC, Gothenburg, 1997, 4iv032 5p.

34. STARK, J, AND BOLLMANN, K. Ettringite Formation - A durability Problem of Concrete Pavements, Proc. 10th ICCC, Gothenburg, 1997, 4iv062 8p.

35. JENSEN, A D, CHATTERJI, S, CHRISTIANSEN, P, AND THAULOW, N. Studies of Alkali-Silica Reaction – Part II; Effect of Air Entrainment on Expansion, Cem. and Conc. Res, Vol. 14, No. 3, 1984, pp 311-314.

36. Cementindustrien og CO_2-afgiften. Aalborg Portland, 1990, 20 p.

37. BORGHOLM H E, AND DAMTOFT, J S. European Patent, EP 0640062B1, 1996.

38. DS 481:1998, Danish pre-standard for composition of concrete and concrete materials, 1998.

39. BAEKMARK, HANSEN AND REICHERT, "Stabilized Air Void System in Superplasticized Concrete", ACI SP-144, Concrete Technology Past, Present, and Future, pp.177-190, 1994.

40. LUNDBERG L. "Øresundstunnel - Performance and Demands on Concrete" (in Danish), Dansk Betondag 1997, Concrete Society of Denmark.

41. PETERSSON, Ö. Applications of Self-Compacting Concrete for Bridge Castings. Swedish Cement and Concrete Research Institute, 1998.

42. Self Compacting Concrete with Glenion 51, NBK unpublished internal report (in Danish), 1998.

THEME ONE:

BASIC PORTLAND CEMENTS

Keynote Paper

ECO-CEMENT: A NEW PORTLAND CEMENT TO SOLVE MUNICIPAL AND INDUSTRIAL WASTE PROBLEMS

T Shimoda

S Yokoyama

Taiheiyo Cement Corporation

Japan

ABSTRACT. Eco-cement is a new type of Portland cement being developed not only to solve the municipal and industrial waste problem caused by limited availability of landfill sites, but also to contribute to the protection of the environment by providing a complete recycling system of wastes that would otherwise be dumped. This new cement is designed to use municipal waste incinerator ash as up to 50% of raw materials. Combustible wastes like waste oil, plastics, RDF(refuse-derived fuel) also can be used as fuel. Eco-cement consists of the same main components as normal Portland cement (NPC); alite(C_3S), belite(C_2S), calcium aluminate (C_3A), calcium alumino-ferrite (C_4AF) and calcium sulfate. The manufacturing process of Eco-cement is also mostly the same as NPC. Incinerator ash generally contains substances undesirable for cement, such as C1 and a small amount of heavy metals. The metals vaporize in the form of chlorides through the sintering process and are caught as kiln dust in the bag filter. The heavy metals, then, are extracted from the dust through the metal recovery process and delivered to a smelter for refining. This makes the Eco-cement process a complete recycle system for municipal and industrial wastes. Another type of Eco-cement featuring rapid hardening in one hour was also designed. This cement contains 20% of calcium chloroaluminate ($C_{11}A_7 \cdot CaCl_2$), instead of C_3A. Other main components of this type are C_3S, C_2S, C_4AF and calcium sulfate. This cement is suitable for concrete blocks, wood-fiber board and other non-rebar concrete products.

Keywords: Eco-cement, Portland cement, Waste, Garbage, Incinerator ash, Sewage sludge, RDF(refuse-derived fuel), Recycle, Heavy metal, Chloride, Calcium-chloro-aluminate.

Mr Takashi Shimoda is General Manager of the Central Research Laboratory, Sakura, Taiheiyo Cement Corporation, Japan. His career includes being a cement chemist, cement manufacturing engineer, and the manager of the company's cement/concrete operations in the United States of America.

Mr Shigerij Yokoyama is a Certified Professional Engineer and currently the leader of the Eco-cement project at the Central Research Laboratory, Sakura, Taiheiyo Cement Corporation. He specializes in the application of cement chemistry to concrete technology.

INTRODUCTION

Through the 20th century, the cement industry has greatly contributed to society, providing basic materials for construction of infrastructures, modern cities, durable houses and so forth. Now, toward the 21st century, it is an emerging fact that the cement industry will become a greater contributor to society, creating a new role for the industry, that is, a provider of a solution to municipal and industrial waste problems, in addition to being a provider of basic construction materials as it has been [1].

Rapid increase in municipal and industrial waste has been causing several serious environmental problems in both industrialized and developing countries.

In Japan, municipal waste (e.g. garbage) reached 50 million tons in 1997. The opening of a new landfill site anywhere faces fierce opposition from the local residents while the life expectancy of existing sites has been rapidly decreasing. For years, municipalities have chosen the use of garbage incinerators so that the life of the existing disposal sites can be extended by reducing the volume of dumping materials. The incinerator, however' does not give a final solution as it produces ash which has to be dumped, keeping the municipalities still looking for the next site. Moreover, the incinerator ash often contains toxic substances such as dioxins and heavy metals which call for very expensive care and treatment.

It appears that this problem of decreasing availability of landfill sites and the toxicity of the ash is not limited to Japan but exists worldwide. Eco-cement has been developed to provide a solution to this problem [2][3].

The fundamental targets in the development of Eco-cement are:

1. As much as 50% of the raw materials has to be replaced by incinerator ash or other waste materials such as sewage sludge.

2. The cement has to have general wide use.

3. Both the manufacturing process and the products have to be environment-friendly.

4. The entire process has to be a complete recycling system.

Incinerator ash generally contains a high concentration of chlorides as well as a very small amount of toxic substances such as dioxins and heavy metals; therefore, decomposition, removal, or enclosure of these substances is the key for success of this project. Concerning prevention of environmental pollution, the Eco-cement manufacturing plant has to meet strict standards for emission, eliminating NOx, SOx, HCl, dioxins and any other toxic substances. The product, Eco-cement, has to be safe when it is used, for instance, safe regarding leach-out of toxic substances from concrete. Anv secondary pollution must not be caused.

The development of Eco-cement was a collaboration project between a foundation funded by the Japanese government and three private companies including Chichibu Onoda Cement (now Taiheiyo Cement) as the leader of the project (see the Note).

Taibeiyo Cement operates a 50 ton-cement/d test plant in Japan and two big local governments have approved the construction of commercial plants that produce 200,000 ton-cement/year and 95,000 ton-cement/year (using 130,000 t/y and 90,000 t/y of incinerator ash, respectively), which are now at the designing stage.

This paper discusses every aspect of Eco-cement; raw materials, manufacturing processes, clinker minerals, characteristics and properties of the cement, applications and metal recovery.

Note: The project was implemented by the New Energy Development Organization(NEDO), an extra-governmental organization of the Ministry of International Trade and Industry, and the research is entrusted to a foundation named Clean Japan Center with which the three private companies, Chichibu Onoda Cement, Ebara and Aso Cement, formed a collaboration.

CHEMICAL DESIGN OF ECO-CEMENT

The typical chemical composition of incinerator ash and sewage sludge in Japan are shown in Table 1. Also, a typical mix design of raw materials is shown in Table 2. Using incinerator ash close to 50% of raw materials generally leads to an increase Cl, Na_2O, K_2O and P_20_5 as well as $A1_2O_3$ in cement. In consideration of this, two types of Eco-cement, (1) Portland cement type and (2) rapid-hardening type, are designed:

1. Portland cement type
   This cement consists of the same clinker minerals as normal Portland cement (NPC); C_3S, C_2S, C_3A and C_4AF (Table 3) [4]. Chlorides vaporize in the sintering process, mostly combining with alkalis; thus, chloride ion in cement is reduced down to below 0.1%. The C_3A content is generally higher than NPC because of use of incinerator ash which contains 10-20% of $A1_2O_3$ In order to maintain the setting time and workability within a preferable range, C_3A content is controlled by adjusting Fe_2O_3 content to get C_4AF formed.

   As a whole, the physical properties of this type are similar to NPC.

2. Rapid-hardening type
   This type of cement is designed to take advantage of chlorides in the incinerator ash to make rapid-hardening cement. C1 is taken into calcium aluminate, forming calcium chloro-aluminate ($C_{11}A_7 \cdot CaCl_2$), eliminating C_3A. $C_{11}A_7 \cdot CaCl_2$ contributes to rapid hardening.

   In the sintering process, chlorides firstly vaporize with alkalis, the same as in the case of the Portland cement type; therefore, in order to reserve Cl for the formation of $C_{11}A_7 \cdot CaCl_2$, Cl in the raw meal has to be in excess relative to alkalis and the excess has to be balanced accordingly to form chloro-aluminate. The Cl content in the final product is approximately 1%. Performance of this cement is similar to the rapid-hardening Jet cement which develops that performance because of $C_{11}A_7 \cdot CaF_2$.

Table 1 Chemical composition of incinerator ash

		LOI %	SiO_2 %	Al_2O_3 %	Fe_2O_3 %	CaO %	MgO %	SO_3 %	Na_2O %	K_2O %	Cl %	TiO_2 %
Incinerator ash	Max.	4	13.7	16.4	4	24.3	2.2	1.4	2.4	2	5.02	0.87
	|	|	|	|	|	|	|	|	|	|	|	|
	Min	15	29.1	24.9	6.8	39.8	12.1	2.4	3.9	2.9	10.6	1
	Average	11	22.9	19.7	5.6	30.4	4.8	2.1	3.3	2.6	8.5	0.9
Sewage sludge*		79.8	5.2	4.6	1.9	1.6	0.5		2 (R_2O = 0.3)		0.07	
		P_2O_5 %	ZnO %	CuO %	Cr mg/kg	As mg/kg	Cd mg/kg	Hg ppb	Pb mg/kg	F mg/kg	CN mg/kg	PCB mg/kg
Incinerator ash	Max.	0.88	0.55	0.12	22	1	1	0.1	100	-	-	-
	|	|	|	|	|	|	|	|	|	|	|	|
	Min	2.3	0.66	1.29	2000	90	30	70	9200	-	-	-
	Average	1.8	0.6	0.6	438	55	11	3.5	311	120	ND	ND
Sewage sludge**		16.8	-	-	153.3	3.9	0.72	0.8	13.4	-	0.7	0.07

Note *: An example of dewatered cake of sewage sludge with polymeric flocculant

**: Average of dewatered cake of sewage sludge with same amount of lime

Table 2 Typical mix design of raw meal

Type of cement	Incinerator ash	Limestone	Clay	Ferro- materials	Alumina
PC	58.2	40	1.3	0.5	-
RH	52.2	45	2.2	0.3	0.3

PC:Portland cement type
RH:Rapid-hardening

Table 3 Chemical and mineral composition in Eco-cement

Type of cement	Chemical compositions (%)									
	ig.loss	SiO_2	Al_2O_3	Fe_2O_3	CaO	MgO	SO_3	Na_2O	K_2O	Cl
PC	0.6	19.1	8.1	4.5	62.7	1.4	3.7	0.05	0.00	0.04
RH	0.8	15.5	11.0	1.9	58.5	1.4	8.8	0.6	0	1
NPC	0.6	22.2	5.1	3.0	63.8	1.4	2.0	0.3	0.2	0

Type of cement	Mineral compositions (%)					
	C_3S	C_2S	C_3A	$C_{11}A_7CaCl_2$	C_4AF	$CaSO_4$
PC	50	10	15	-	14	7.7
RH	50	7	-	20	6	15
NPC	46	29	8	-	10	3.4

PC:Portland cement type
RH:Rapid-hardening

The two types of Eco-cement can be produced from the same production line by controlling the amount of Cl in cement clinker. Typical chemical and mineral compositions of Eco-cement are shown in Table 3 in comparison with NPC.

MANUFACTURING PROCESS

The manufacturing process of Eco-cement is shown in Figure 1. The whole process basically consists of the same unit processes used for NPC production: raw meal preparation, sintering and finish processes.

Incinerator ash generally contains heavy metals such as Pb, Zn, Cu, Cr and As. To recover these metals from the kiln dust, a metal recovery process (Figure 2) is designed and connected to the cement production line. The recovered Cu, Pb and other accompanying minor components are sent to a smelter for refining, making the entire Eco-cement process a complete recycling system.

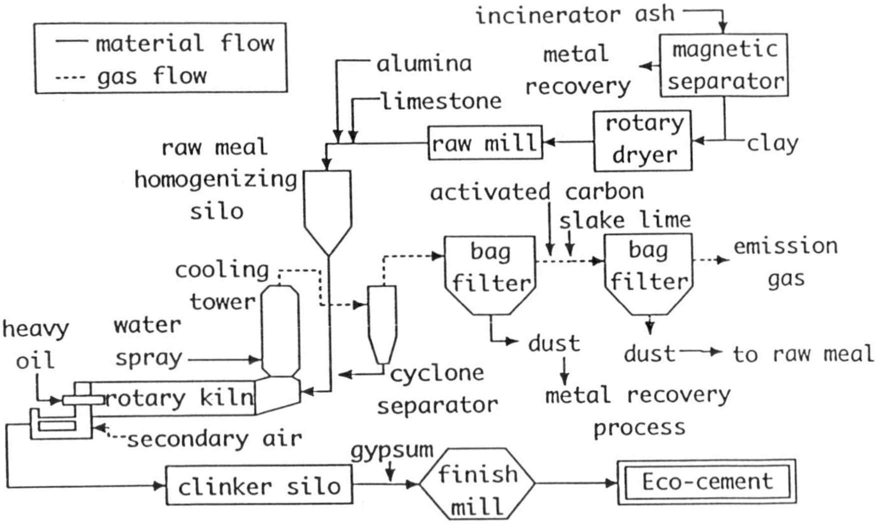

Figure 1 Manufacturing process of Eco-cement

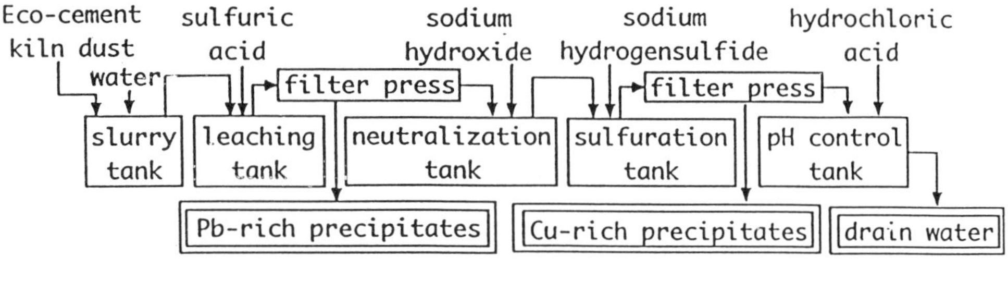

Figure 2 Metal recovery process (BY DOWA MINING CO.LTD)

Raw Meal Preparation

As the Eco-cement manufacturing uses waste materials, additional care has to be taken, compared with the NPC manufacturing which uses natural resources.

1. Chemical compositions of incinerator ash generally show substantial differences depending on their municipality of origin. Consequently, to get the raw meal homogenized with acceptable fluctuation, the raw preparation process has to be equipped with ash blending vessels, X-ray fluorescence analyzer and feeders of limestone and clay for fine tuning.

2. Alkalis (Na_2O, K_2O) and other metals (such as Pb, Zn, Cu, Cd, As, Hg) vaporize in the form of chloride in the sintering process. Therefore, the amounts of alkalis and C1 in the raw meal have to be appropriately designed, taking into consideration the vaporization of alkalis and metals through the sintering process.

For the Portland cement type, the amount of Cl in the raw meal is designed to be equivalent to the amount of alkalis and metal components. If alkalis in the raw meal is insufficient relative to Cl, Na_2CO_3 is added to make the excess Cl vaporize.

For the rapid-hardening type, first, most of the chlorides in the raw meal vaporizes with alkali in the sintering process; then, the excess Cl forms calcium-chloro-aluminate ($C_{11}A_7 \cdot CaCl_2$). It is important to maintain a constant amount of $A1_20_3$ in the raw meal so that the amount of $C_{11}A_7 \cdot CaCl_2$ in the clinker also becomes constant. This is essential to control the quality of cement. If Cl is insufficient relative to alkalis, $CaCl_2$ is added as a supplement.

Table 4 shows the chemical composition of the raw meal for both types.

Sintering Process

Both types of Eco-cement are heated to 1350°C, that is, 100 degrees lower than NPC which calls for 1450^{0}C at the highest, owing to a mineralizer-effect of Cl in the raw meal. Dioxins, which tend to come with the ash, is completely decomposed above 800°C. The exhaust gas from the kiln, however, has to be quickly cooled down below 250°C using the cooling tower to prevent dioxins from recomposing at 250-350°C.

Heavy metals such as Pb, Zn, Cu, Cd as well as alkalis (Na_2O, K_2O) vaporize in the form of chloride at temperatures between 800-1300°C, then condense and precipitate under 1000°C. Therefore, the suspension preheater is not recommended for the Eco-cement process as the chloride condensation may cause clogging in the cyclones.

The chlorides in the exhaust gas are cooled through the cooling tower and the separator; then they are collected by the bag filter. The collected dust is sent to the metal recovery process.

The mass balance of the entire Eco-cement process is shown in Figure 3.

Table 4 Chemical composition of raw meals for both types of Eco-cement

Type of cement	ig.loss	SiO_2	Al_2O_3	Fe_2O_3	CaO	MgO	SO_3	Na_2O	K_2O	Cl	Total
PC	30.1	13	5.8	6.4	42.5	1.1	0.4	2.0	0.7	2.1	99.6
RH	29.4	12.3	7.7	1.8	41.9	1.5	0.8	1.3	0.9	2.3	99.5

PC:Portland cement type
RH:Rapid-hardening

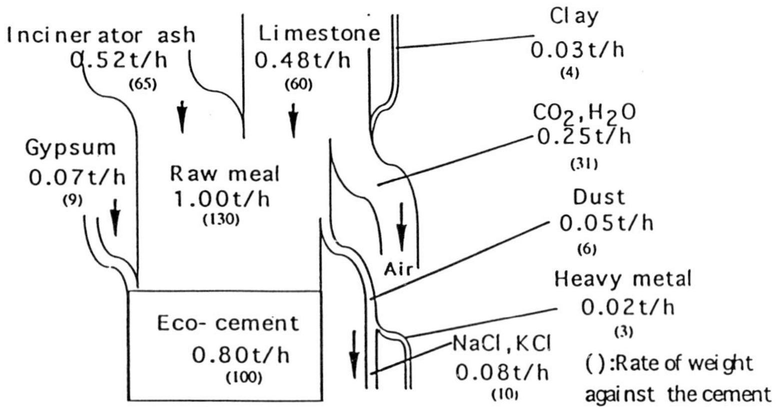

Figure 3 Mass balance of the Eco-cement process

Finish Process

Following the sintering process, Eco-cement clinker is sent to the finish process.

For the Portland cement type, the clinker, with the addition of gypsum, is ground up to 4,000 cm^2/g. The gypsum is dehydrated to hemi-hydrate during grinding and helps control the setting time of the C_3A-rich cement. SO_3 in Eco-cement has to be 3.5-4.0%, a little higher than NPC because of the higher C_3A.

For the rapid-hardening type, the clinker, with the addition of 1% N&4 and without gypsum, is ground up to 4,500 cm^2/g. Then anhydrite powder of about 7,000 cm^2/g is mixed, making the molar ratio of $A1_2O_3$ to SO_3 in cement between 1.0-1.2.

Na_2SO_4 controls the setting time of the rapid-hardening cement while the anhydrite helps $C_{11}A_7 \cdot CaCl_2$ rapidly form ettringite and maintain stable development of strength.

Metal Recovery Process

Municipal waste incinerator ash generally contains a small amount of heavy metals as shown in Table 1. Among them, most of Cr and Zn are taken into the clinker minerals through the sintering process, while Pb, Cu, Cd and some other heavy metals vaporize and are caught in the kiln dust as chlorides. The mass balance of heavy metals in the Eco-cement process is shown in Table 5.

The kiln dust, collected in the bag filter and containing chlorides of heavy metals and alkalis is processed in the metal recovery process as shown in Figure 2.

In this process, first, the dust is mixed with water. Then the Pb-rich sulfate solid precipitates out of the solution under acid conditions produced by sulfuric acid. After solid-liquid separation, sodium hydroxide and sodium hydrogensulfide are added to the filtrate, controlling pH and oxidation-reduction potential. Then the Cu-rich sulfide solid precipitates and is separated from the liquid. Both of the precipitates, the Pb-rich and the Cu-rich, are sent to a smelter to recover metals. Chemical compositions of the precipitates are shown in Table 6. The final metal products are comparable in quality to those from natural ores[5].

Table 5 Mass balance of heavy metals in the Eco-cement process

		Ratio of Production	(unit)	Pb	Zn	Cu	Cr	Cd	Hg	As	F	CN	PCB
Raw meal	Concentration	-	(mg/Kg)	490	1340	1250	70	2.5	0.8	12.3	26.8	ND	ND
(Input)	Distribution	100	(%)	100	100	100	100	100	100	100	100	-	-
Clinker	Concentration	-	(mg/Kg)	14	1838	821	100	0.07	0.02	17.5	26.8	ND	ND
(Output)	Distribution	70	(%)	2	96	46	100	2	2	100	70	-	-
Kiln dust	Concentration	-	(mg/Kg)	14552	1624	20455	0	72.9	23.2	0	243	ND	ND
(Output)	Distribution	3.3	(%)	98	4	54	0	98	98	0	30	-	-

ND:Not ditected

Table 6 Chemical compositions of the Pb-rich and Cu-rich products

	Ratio of weight	Pb (%)	Zn (%)	Cu (%)	Cd (mg/Kg)	Hg (mg/kg)	As (mg/Kg)	Fe (%)
Kiln dust	100	1.53	0.08	1.91	98	12.18	<1	0.01
(Distribution)		(100)	(100)	(100)	(100)	(100)	(100)	(100)
Pb-rich product	4.7	32.13	0.02	0.52	3	2.66	2	0.26
(Distribution)		(97.9)	(1.2)	(1.5)	(0.1)	(99.3)	(100)	89.0)
Cu-rich product	3.6	0.9	2.21	52.15	2730	2.3	<1	0.04
(Distribution)		(2.1)	(98.2)	(98.5)	(99.8)	(0.7))	(0.0)	(10.6)

ND:Not ditected

CHARACTERISTICS AND PROPERTIES OF ECO-CEMENT

Eco-cement Clinker

Figure 4 shows back-scattered electron images of polished sections of Eco-cement clinker in comparison with NPC clinker.

1. Portland cement type
   The clinker matrix consists of C_3S, C_2S, C_3A and C_4AF. The elite of this type of Eco-cement is a monoclinic phase and shows the automorphic form, the same as NPC clinker. Its average size is approximately 30 micron-m. The belite is type I with cross striations, and its average size is 15 micron-m. The amount of the interstitial phase is larger than NPC. C_3A is in the cubic form with low alkali content, which is the result of alkali vaporization while forming chloride in the sintering process. C_4AF is the same as in NPC.

Cl in the raw meal works as a mineralizer, accelerating the elite formation and the reduction of free-lime

2. Rapid-hardening type
   The clinker matrix of this type of Eco-cement consists of C_3S, C_2S, $C_{11}A_7 \cdot CaCl_2$ and C_4AF. The elite shows the automorphic form. Its average size is about 30 micron-m. The belite is observed as type I and its average size is 15 micron-m. Cl forms $C_{11}A_7 \cdot CaCl_2$ and works as a mineralizer of alite[6][7].

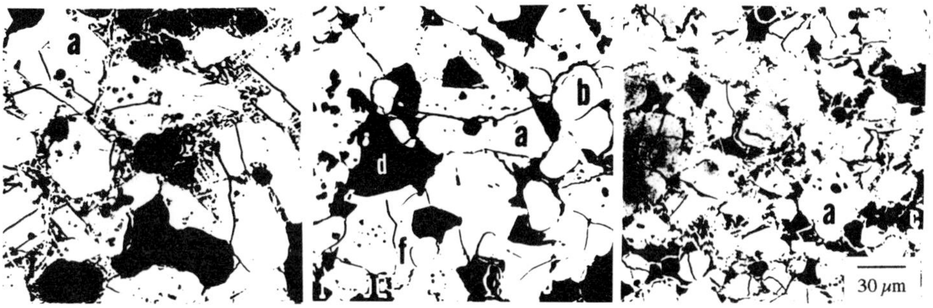

Figure 4 Back-scattered electron images of the clinker of each type of cement

Hydration of Eco-cement

1. Portland cement type
   As this type of Eco-cement contains a larger amount of C_3A than NPC, heat generation at initial contact with water is larger than NPC. Immediately after the cement is mixed with water, C_3A combines with hemi-hydrate gypsum and forms ettringite ($C_3A \cdot 3CaSO_4 \cdot 32H_2O$). Then the hydration enters into the induction period, the same as seen with NPC. Gypsum is then rapidly consumed, forming calcium-aluminate monosulfate hydrate ($C_3A \cdot CaSO_4 \cdot 12H_2O$), as the amount of gypsum is not sufficient to form ettringite with C_3A. In 2 hours after mixing with water, elite begins active hydration in the same manner as NPC.

   In the end a dense structure of hardened cement paste is formed with calcium-silicate hydrates(C-S-H), calcium hydroxide and monosulfate hydrate, the same as in the case of NPC

2. Rapid-hardening type
   $C_{11}A_7 \cdot CaCl_2$ begins to form ettringite with anhydrite immediately after the cement is mixed with water. At the same time, the chloride ion reacts to form.a small amount of Friedel's salt. In one hour, elite begins active hydration. After that, the matrix of hydrates increases its density as calcium-silicate hydrates(C-S-H), formed by hydration of elite and belite, fill up pore spaces between needle-shaped ettringite crystals of 1 micron-m.

In the end, hardened cement is stabilized with hydrates such as C-S-H, $Ca(OH)_2$, ettringite and calcium aluminate monosulfate hydrate.

Physical Properties

Physical properties of both types of Eco-cement are shown in Table 7 and Figure 5.

1. Portland cement type
   This type of Eco-cement with 3.5-4.0% SO_3 develops nearly the same physical properties as NPC

2. Rapid-hardening type
   The most distinguishing character of this type of Eco-cement is the very short setting time and high early-strength development as observed in the test ages of 3 hours, 1 day and 3 days, because calcium chloro-aluminate actively hydrates and rapidly forms ettringite. This is in the category of rapid hardening cement. The use of a retarder such as citric acid or other appropriate additives like blast furnace slag extends the setting time.

Table 7 Physical properties and heat generation of Eco-cement

Type of cement	Specific gravity (-)	Fineness Specific surface (cm²/g)	Time of setting Initial (h-min)	Final (h-min)	Compressive strength (N/mm²) 3hr	6hr	1day	3days	7days	28days
PC	3.19	4500	2-0	4-30	-	-	9	22	37	53
RH	3.13	4600	0-9	0-13	10	16	23	30	38	46
NPC	3.17	3220	2-22	3-20	-	-	11	27	43	59

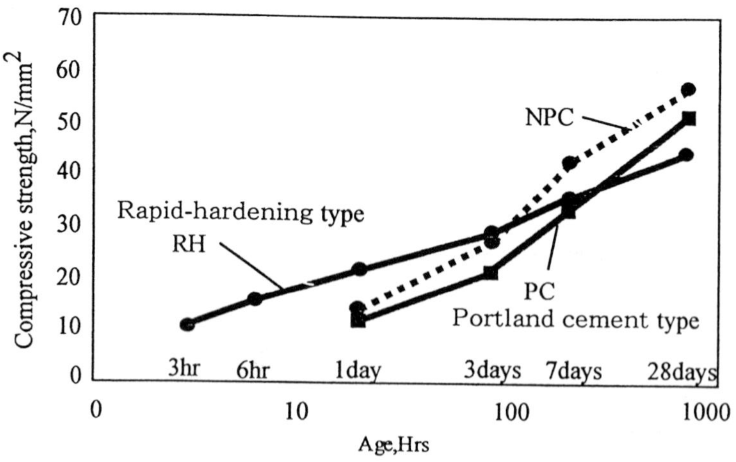

Figure 5 Strength growth of Eco-cement

CONTRIBUTION TO PROTECTION OF ENVIRONMENT

Recovery of Heavy Metals

The kiln dust containing chlorides of heavy metals and alkalis is processed in the metal recovery process as shown in Figure 2.

The drain water from this process contains NaCI and KCI of the concentration close to the level of sea water. Heavy metals in the drain water also meet the Japanese Clean Water Act regulations (Table 8). Therefore, the drain can be discharged to the sea or sewage plant.

Gas Emission from the Eco-cement Plant

The emission gas from the Eco-cement plant has to be clean and meet every standard to protect the environment. Incinerator ash often contains very small amounts of dioxins in addition to heavy metals and chlorides. The dioxins in the raw meal, however, are decomposed when they are heated up to 1350°C in the rotary kiln. To prevent recomposition, the gas has to be cooled down quickly below 250°C.

SOx and HCl are mostly combined with CaO during the clinker formation, taken into clinker, and discharged from the process. Thus, sulfur oxide and hydrogen chloride are not discharged in the emission gas.

Both an activated carbon filter and a calcium hydroxide filter are equipped in preparation for an emergency breakdown of the process[8].

Analysis of the emission gas from the Eco-cement test plant is shown in Table 9. The results for dioxins and other toxic substances meet the Japanese Clean Air Act.

Table 8 Heavy metal content in waste water

	Pb (mg/l)	Zn (mg/l)	Cu (mg/l)	Cd (mg/l)	Hg (ppb)	As (mg/l)
Waste water	0.01	0.48	0.38	0.01	<0.5	<0.01
Emission standerd	<0.1	< 5	< 3	<0.1	<0.5	<0.1

ND:Not ditected

Table 9 Analysis of emission gas from Eco-cement process

	NO_X (ppm)	SO_X (Nm^3/h)	HCl (mg/m^3N)	Dioxins ($ng\text{-}TEQ/m^3N$)	Dust (g/m^3N)
Analytical result	<100	<0.23 (<20ppm)	<32 (20ppm)	0.05	0.001
Emission standard	250	4.76	700	0.1	0.08

Leach-out of Heavy Metals from Hardened Eco-cement

Leach-out of heavy metals from hardened Eco-cement mortar is determined in accordance with the procedures set by the Japanese Environment Protection Agency(JEPA). The results are shown in Table 10. Leach-out of all the heavy metals listed in the JEPA regulations is lower than detectable limits

Concerning the rapid-hardening type Eco-cement, a small amount of chloride ion leaches out; however, the concentration is still within the regulations for drinking water in Japan. The mass balance of heavy metals is summarized in Table 5.

Table 10 Leaching of heavy metals from hardened Eco-cement

	Cd mg/l	Pb mg/l	As mg/l	CN mg/l	Cu mg/l	Hg mg/l	Cr^{6+} mg/l	Se mg/l	Cl mg/!
1d	ND	ND	ND	ND	ND	ND	ND	ND	41
3d	ND	ND	ND	ND	ND	ND	ND	ND	39
7d	ND	ND	ND	ND	ND	ND	ND	ND	38
28d	ND	ND	ND	ND	ND	ND	ND	ND	34
Emission standard	<0.01	<0.05	<0.01	<0.01	<1	<0.0005	<0.05	<0.01	<200

ND:Not detected

APPLICATION OF ECO-CEMENT

Concrete

The Portland cement type of Eco-cement can be used just as NPC; thus, it is applicable for very wide use.

The rapid-hardening type contains a high chloride ion content of 1°/43herefore, non-rebar concrete products are a major area of application; for instance, will/blocks, interlocking blocks, wood fiber board, organic fiber board, etc. The setting time can be regulated by a retarder or by mixing with slag.

Soil Improvement

Both types of Eco-cement can be used as solidifying agent for wet land, swamps and other soft ground as shown in Figure 6 [9][10].

CONCLUSIONS

1. Municipal waste incinerator ash and sewage sludge has long been dumped at landfills. Eco-cement has made it possible to turn the ash and the sludge into industrial resources.

2. The Eco-cement process not only produces quality cement but also contributes to the protection of the environment by recycling heavy metals.

3. Currently, two Eco-cement plant projects totaling 295,000 tons-cement/year are proceeding in Japan

Figure 6 Using Eco-cement for wet land

REFERENCES

1. UCHIKAWA, H.Management strategy of cement industry in technology toward the next century, Proc.Engineering Foundation Conference, Advances in Cement and Concrete edited by GRUTESK,W. and SARKAR,S.L., Amer.Soc.Civ.Eng., 1994, pp.524-56 1.

2. OBANA, H, ANZAI, T and FUKUNAGA, T. Ecocement recycled urban garbage and waste materials. Proceedings of International Symposium on Environmental Issues of Ceramics, 1995, pp.63-67.

3. UCHIKAWA, H and OBANA, H. Ecocement - frontier of recycling of urban composite waste. World cement. NOVEMBER 1995. pp.33-36.

4. TAYLOR, H F W. The Chemistry of Cements, Academic press, London and New York, 1964, pp. 1-24.

5. T.ANZAI, H, OBANA, C, IZUMIYA and ABUMIYA, M. Ecocement which is used Waste and the Recycling of the Heavy Metals, The Mining and Materials Processing Institute of Japan, 1997, pp. 126- 130.

6. TAKUMA, Y, TSUCHIDA, Y and UCHIDA, S. Characteristics and hydration of cement produced from ash from incinerated urban garbage. Proceedings of the 10th International Congress on the Chemistry of Cement, Gothenburg, Sweden, June 2-6, 1997, 8pp.

7. NAKANO, T, TSUCHIDA, Y, UWABO, T and UCHIDA, S. Fundamental Research on Manufacture of Cement from Urban Composite Waste, The Third International Conference Ecomaterials, September, 1997, Tsukuba, Japan, pp.87-90.

8. YAMAMOTA, Y and MICHIBATA, Y, Behavior of harmful components in Ecocement plant system, The Third International Conference on ECOMATERIALS, September, 1997, Tsukuba, Japan, pp 87-90.

9. OHMORI, M, MARUTA, T, MORIYA, Y and ISODA, H. Decreasing the Amount of Chloride Ion Elution in the Ecocement made from Municipal Waste Ash, Inorganic Materials, Vol.4, Mar., 1997, pp. 119-125.

10. MARIJTA, T, OHMORI, T, MORIYA, M, UCHIDA, K and ISODA, H. Properties of the Soil Improved by the Special Cement Made from Municipal Solid Waste Incineration Ashes, Inorganic Materials, Vol.4, Mar., 1997, pp.152-155.

EARLY PREDICTION OF CEMENT MORTAR STRENGTH FROM RAW MIX CONTAINING METALLIC PARTICLES

J O Odigure

Federal University of Technology

Nigeria

ABSTRACT. The use of metallic particles containing byproducts in cement raw mix is associated with some difficulties. Apart from the technological problems associated with the sintering, it is very difficult to forecast the properties of the clinker produced.

The major objective of this work is to develop an optimization procedure for the formulation of raw mix components containing metallic particles. Result showed that for raw mix containing less than 1 mass % metallic particles currently used formulae can be used. For raw mix containing more than 1 mass % Fe and FeO, the expected Fe_2O_3 content can be predicted using the formula

$$Fe_2O_3 = (1.43F + 1.111FeO + Fe_2O_{3initial})(1.3 - 1.4)$$

Optimization of raw mix components should be complimented with the prediction of the chemical composition of the raw mix. This will enhance early determination of the suitability of the raw mix for clinkerization and consequently prognosis of the physio - chemical and mechanical properties of the clinker.

Keywords: Prediction, Metallic particles, Oxidation, Optimization, Clinker, Chemical composition.

Dr J O Odigure is a senior lecturer in the Chemical Engineering department, Federal University of Technology, Minna, Niger State, Nigeria. He specializes in chemical technology of silicates and refractory non-metallic substances. His present interest include durability and protection of building material in aggressive environments. Dr Odigure has published widely. He is the Head of Chemical Engineering Department and Chairman, Equipment maintenance Centre of his University and Technical secretary, Nigerian Society of Engineers, Minna Branch.

INTRODUCTION

The quality of cement clinker and consequently that of the cement produced can be prognosed using the following characteristics: chemical composition of the individual oxides; the modules values; the clinker microstructure, the size and configuration of the clinker minerals. The first two characteristics are widely used in the preparatory stage to prognose the burnability of raw mix and the expected chemical and mineral composition of the clinker to be produced. The morphology of the clinker minerals not only depends on the quantity and chemical composition of the liquid phase [1], but also on the homogeneity and fineness of raw mix grains [2].

Consequently the quality of clinker produced is determined by a combination of technological factors and the morphology of the raw mix components. Prediction of cement quality can therefore be performed based on the analysis of the physical state and chemical composition of raw mix components, prevailing technological condition during preparation of raw mix, sintering and cooling. In the traditional raw mix, all the components exist in the oxide from and the physical state (hardness) is dependent on the genealogy. Continuous depletion of available traditional deposits had necessitated the use of industrial byproducts - slag, flyash, abrasive slurry, tyres, etc., as cement raw mix components [3,4]. Unlike raw materials of natural origin, some industrial byproducts the likes of metallurgy slag and abrasive slurry - a byproduct of machine building industry, contain metallic particles.

Analyses of the effect of metallic particles on the kinetics of cement production [5], microstructure [6] and hydration of cement paste [7] had shown that optimization of raw mix containing such byproducts can not be performed using known methods [8,9,10] without modifications.

The major objective of this paper is to develop a characteristic model for formulating raw mix component containing metallic particle and its oxide.

EXPERIMENT

The raw materials used for the experiments were limestone, sand, abrasive slurry and open hearth slag. The compositions of the raw materials are presented in Table 1. The granulometric composition of abrasive slurry is, mass %: 0 - 0.12mm 48 - 80, 0.2 - 1mm 20 - 29 and above 1mm the remainder. Part of the abrasive slurry was burnt at 500°C and the chemical composition was also determined. The Fe, FeO and Fe_2O_3 contents were determined as in [11]. Quantitative analyses of Fe, FeO and Fe_2O_3 in the abrasive slurry were determined using the methodologies that prevent oxidation of Fe and FeO. The determination of Fe_2O_3 in clinker allows for the oxidation of FeO.

Using the simplex - lattice diagram Composition - Properties [12], the compositions of the raw mixes were determined (Table 2 and 3). The dried raw materials mixed in the required proportion were ground in a ceramic lined ball mill to a fineness of 6 - 8 mass % residue on a 80:m size mesh. They were then watered and manually granulated. Their granules were dried and burnt in an electric furnace using silicon heating elements.

The burning temperature was gradually raised to 1400°C, with a retention time of 30min. at the maximum temperature. The entire mass of the produced clinkers were ground to a surface area of 3200cm^3/kg.

The mechanical strength of the clinkers, determined in accordance with British standard (vibrated mortar) are presented in Table 2 and 3. The optimization parameters were 7, 28 and 180 days compressive strength. All the samples were chemically analyzed and the results are presented in Table 6 and 9. The calculated chemical compositions of the raw mixes are presented in Table 4 and 7.

Table 1 Chemical composition of raw materials

MATERIAL	OXIDE CONTENT, MASS %								
	SiO_2	Al_2O_3	Fe_2O_3	FeO	Fe	CaO	Cr_2O_3	MnO	LOI[1]
Limestone	7.66	1.19	0.62	-	-	49.50	-	-	39.65
Sand	76.71	6.60	2.81	-	-	6.64	-	-	5.61
Slag [2]	37.17	11.05	14.56	-	-	20.72	3.57	9.32	-
Abrasive Slurry	3.02	27.00	4.34	6.72	54.0	0.84	1.82	0.66	-
Burnt Slurry [3]	1.10	22.54	74.24	-	1.10	-	-	-	-
Burnt Slurry [4]	1.32	16.42	60.52	9.5	10.3	-	-	1.31	-

1 Loss on Ignition
3 Expected chemical composition of abrasive slurry at 1400°C

2 Open Hearth Slag
4 Abrasive slurry burnt at 500°C

Table 2 Composition of raw mix and compressive strength of cement

	COMPOSITION OF RAW MIX, MASS %				COMPRESSIVE STRENGTH, MPa		
N	Limestone	Open Hearth Slag	Abrasive Slurry	Sand	7 days	28 days	180 days
1	82.00	5.00	4.00	9.00	37.0	40.1	52.5
2	79.00	7.00	5.30	9.00	25.0	37.7	46.7
3	80.00	8.00	3.00	9.00	28.5	34.5	52.5
4	80.33	6.34	4.33	9.00	26.5	36.4	46.0
5	79.67	6.66	4.67	9.00	34.0	43.5	52.5
6	79.25	7.30	4.40	9.00	19.0	35.0	49.1
7	79.53	7.80	3.67	9.00	34.0	48.2	61.9
8	81.03	6.67	3.30	9.00	40.0	34.4	59.9
9	79.73	7.60	3.67	9.00	32.0	42.6	53.7
10	80.03	6.67	7.30	9.00	23.0	37.7	49.1

Table 3 Composition of cement raw mix and compressive strength

	COMPOSITION OF RAW MIX, MASS %				COMPRESSIVE STRENGTH, MPa	
N	Limestone	Open Hearth Slag	Abrasive Slurry [1]	Sand	7 days	28 days
1	82.00	7.00	2.00	9.00	33.6	52.5
2	80.00	8.00	3.00	9.00	32.5	57.5
3	81.00	5.00	4.00	9.00	34.0	50.2
4	81.66	5.68	2.66	9.00	38.0	56.2
5	81.34	6.33	3.33	9.00	37.5	53.7
6	80.68	6.66	3.66	9.00	39.1	54.8
7	80.33	7.33	3.33	9.00	41.2	56.3
8	80.66	7.68	2.66	9.00	35.0	55.3
9	81.33	7.33	2.33	9.00	36.0	52.5
10	81.00	7.00	3.00	9.00	41.7	63.5

[1] Burnt at 500°C

RESULTS AND DISCUSSION

Analyses of the simplex - lattice diagrams taking maximum compressive strength as the optimization criteria is presented in [10]. The regression equation approximating the relation between the components, at sand content = 9 mass % are:

For Table 2:

$$y_{180} = 52.5X_1 + 46.7X_2 + 52.5X_3 +\text{-}1.68X_1X_3 +19.55X_1X_3 + 26.43X_2X_3 +\text{-}172.41X_1X_2X_3 +56.81X_1X_2(X_1\text{-}X_2) + 41.78X_1X_3(X_1\text{-}X_3) +\text{-}\ 73.19X_2X_3(X_2\text{-}X_3)$$

For Table 3:

$$y_{28} = 52.5X_1 + 57.5X_2 + 50.2X_3 + 0.11X_2 + 11.47X_1X_2 + 7.44X_2X_3 + 216.27X_1X_2X_3 + 27.78X_1X_2(X_1\text{-}X_2) + 13.72X_1X_3(X_1\text{-}X_3) +\text{-}\ 25.80X_2X_3(X_2\text{-}X_3)$$

From the above it is clear that the simplex - composition is effective as an optimization tool for a defined composition of raw mix. The calculated chemical composition of the raw mixes (Table 2) using results of analyses in Table 1 are presented in Table 4. It determine the region of optimum composition of the raw mix, but not the chemical composition of the raw mix. Therefore the composition of the raw mix for sintering can not be predicted.

Table 4 Calculated chemical composition of raw mix

N	OXIDE CONTENT, MASS %								
	CaO	Fe_2O_3	Al_2O_3	SiO_2	Fe	FeO	Cr_2O_3	MnO	Total
1	64.56	2.54	4.89	23.16	3.30	0.41	0.38	0.75	99.99
2	61.85	4.67	3.79	23.58	4.07	0.50	0.51	1.03	100.00
3	63.47	3.10	4.91	24.08	2.45	0.30	0.52	1.16	99.99
4	63.21	2.82	5.18	23.25	3.54	0.44	0.46	0.94	99.84
5	62.45	2.89	5.34	23.54	3.80	0.47	0.49	0.98	99.96
6	62.22	3.11	5.29	23.78	3.57	0.44	0.51	1.07	99.99
7	62.69	3.08	5.11	24.11	2.98	0.37	0.52	1.13	99.99
8	63.99	2.85	4.84	23.84	2.71	0.34	0.45	0.98	100.00
9	62.93	3.04	5.10	24.08	2.99	0.37	0.36	1.11	99.98
10	62.95	2.89	5.08	23.66	3.51	0.44	0.48	0.99	100.00

During heating in air, the weight of metallic particles is expected to increase by 1.8 - 2.1 times depending on the extent of oxidation (13). From the reaction stoichiometry the increase in weight of FeO after oxidation to Fe_2O_3 and Fe to Fe_2O_3 are 1.111 and 1.43 respectively. Using the above stoichiometric coefficients the composition of the raw mixes were computed assuming that the Fe and FeO present are converted to Fe_2O_3. The expected chemical composition of the clinkers is presented in Table 5.

Table 5 Calculated chemical composition of clinker from table 2

N	OXIDE CONTENT, MASS %			
	CaO	Fe_2O_{3c}	Al_2O_3	SiO_2
1	63.85	7.63	4.84	22.90
2	60.76	10.53	3.62	22.58
3	62.78	6.87	4.86	23.82
4	62.23	8.26	5.11	22.92
5	61.44	8.71	5.25	23.16
6	61.24	8.58	5.20	23.41
7	61.87	7.66	5.04	23.79
8	63.22	7.00	4.79	23.56

Note: $Fe_2O_{3\,c} = 1.43F + 1.111FeO + Fe_2O_3$

Table 6 Chemical composition of clinker

	OXIDE CONTENT, MASS %							CLINKER MODULE		
N	CaO	Fe_2O_3	Al_2O_3	Cf*	SiO_2	MnO	Cr_2O_3	LCF	SR	Fe_2O_{3E}
1	65.00	9.80	3.40	1.4	20.20	0.60	0.30	0.99	1.53	28.1
2	62.30	13.00	3.40	0.2	20.10	0.68	0.34	0.92	1.22	23.5
3	64.50	9.50	3.70	1.0	20.90	0.76	0.34	0.94	1.58	38.3
4	64.50	10.40	3.00	0.8	21.30	0.62	0.30	0.94	1.54	25.6
5	62.70	11.30	3.90	0.6	20.40	0.60	0.32	0.91	1.32	29.7
6	62.70	11.20	3.60	0.6	21.10	0.70	0.34	0.89	1.42	30.5
7	63.30	10.70	3.20	0.4	20.80	0.75	0.34	0.93	1.50	39.7
8	64.50	9.60	3.60	1.1	21.30	0.64	0.30	0.92	1.61	37.1
9	63.00	11.00	3.60	0.4	21.00	0.73	0.34	0.90	1.44	-
10	64.30	10.40	3.10	0.7	20.40	0.65	0.31	0.97	1.57	-

Cf* - CaO_{free}

Analysis of Table 5 and 6 showed that the $Fe_2O_{3\ E}$ content in the clinker produced far exceeded the expected amount of $Fe_2O_{3\ C}$. The extent of disparity between the calculated and the experimentally obtained values varies considerably and could be as high as 40% (Table 6).

Table 3 shows the composition of raw mixes containing burnt abrasive slurry with calculated chemical composition is in Table 7. Analysis of the expected clinker is shown in Table 8 and the obtained experimental values are given in Table 9. Analyses of Table 7, 8 and 9 shows that the Fe_2O_3 contents in Table 7 and 8 are similar. Reduced Fe and FeO contents in the raw mix have little contribution to the final Fe_2O_3 value in clinker. From the above analysis the presence of less than 1.0 mass % of Fe and FeO may not alter the chemical composition of clinker to be produced. Consequently, the existing formulae can be used without any modifications to determine the suitability of raw mix containing metallic particles.

Table 7 Calculated chemical composition of raw mix

	OXIDE CONTENT, MASS %								
N	CaO	Fe_2O_3	Al_2O_3	SiO_2	Fe	FeO	Cr_2O_3	MnO	Total
1	65.22	4.08	4.57	24.16	0.32	0.29	0.43	0.99	100.06
2	63.24	4.40	5.62	24.23	0.47	0.42	0.48	1.13	99.99
3	64.31	4.25	6.00	23.13	0.63	0.53	0.35	0.72	99.92
4	65.32	4.07	4.95	23.67	0.42	0.38	0.35	0.82	99.98
5	64.19	4.28	5.60	23.63	0.50	0.48	0.40	0.89	99.97
6	63.07	4.33	5.91	23.43	0.57	0.52	0.37	0.93	99.13

Table 8 Calculated chemical composition of clinker

N	OXIDE CONTENT, MASS %			
	CaO	Fe_2O_3	Al_2O_3	SiO_2
1	65.07	5.34	4.07	24.10
2	63.09	6.73	4.48	24.17
3	64.00	7.67	4.23	23.02
4	65.18	5.96	4.06	23.62
5	64.02	6.86	4.26	23.56
6	62.88	4.32	7.29	23.40

Table 9 Chemical composition of clinker

N	OXIDE CONTENT, MASS %			CLINKER MODULE		
	CaO	Fe_2O_3	Al_2O_3	LCF	SR	AR
1	66.10	4.40	4.40	23.1	0.91	1.00
2	64.90	5.60	4.70	23.4	0.87	0.84
3	64.70	5.90	4.60	21.8	0.92	0.78
4	65.60	4.80	4.40	22.2	0.93	0.89
5	65.60	5.60	4.60	22.6	0.91	0.82
6	65.20	5.90	4.80	22.7	0.89	0.81
7	65.10	5.80	4.80	23.0	0.88	0.82
8	65.50	5.30	4.70	22.3	0.92	0.88
9	65.90	4.90	4.50	23.2	0.90	0.92
10	65.40	5.40	4.80	23.0	0.89	0.89

Note: No free CaO present

At high Fe and FeO content the influence of the oxidation process must be considered as used during calculation of Fe_2O_3 in Table 5. Comparative analysis .of $Fe_2O_{3\ D}$ contents in Table 6 showed that the difference between the calculated values using the above formula and the experimental ones was about 23 - 40%. According to (11) the mass of metallic particles is expected to increase by 1.8 - 2.1 times. Therefore the approximate value of Fe_2O_3 produced from raw mix containing metallic particles can be determined by multiplying by 1.3 - 1.4.

$$Fe_2O_3 = (1.43F + 1.111FeO + Fe_2O_{3initial})(1.3 \text{ to } 1.4)$$

Early prognosis of the Fe_2O_3 content in the raw mix could greatly influence the optimization process and the determination of the composition suitability for clinkerization.

The ability to effectively control the composition of the raw mix from byproducts containing metallic particles will not only greatly enhance their utilization but also the production of a more reactive cement at lower temperature.

CONCLUSIONS

Currently used characteristic formulae for optimization of PCC raw mix can be applied for cement raw mix component containing <1% by mass metallic particles. For raw mix containing >1% by mass Fe and FeO, the expected Fe_2O_3 content can be predicted using the formula

$$Fe_2O_3 = (1.43F + 1.111FeO + Fe_2O_{3initial})(1.3 - 1.4)$$

Optimization of raw mix components should be complimented with prediction of the chemical composition of the raw mix. this will enhance early determination of the suitability of the raw mix for clinkerization and consequently prognosis of the physio-mechanical properties.

REFERENCES

1. NIKOLAEV, M M, BUTT Y M, TIMASHEV V V. Sintering of alite – aluminoferrite clinker. Investigation into chemistry and Technology of Silicate. Moscow Chemico-Technology Inst. Moscow, pp. 25 - 29 (1964).
2. TIMASHEV V V, BUTT Y M. Crystal formation in portland cement raw mixes during sintering. Proc. All Union Conf. on chem. and Tech. of cement. Moscow, pp.52-87 (1967).
3. ZHOU, P Y. 9th Intern. Congress on the Chemistry of Cement. New Delhi. Vol. II. pp.271 - 277 (1992).
4. RIGANTI V, FIAMARA A, ODOBEZ G B. Waste Management and Res. Vol. 6, No 4, pp. 293 - 302 (1986).
5. ODIGURE J O. Kinetic modelling of cement raw mix containing iron particles and clinker microstructure. Cem. and Concr. Res. Vol. 26, N. 9. pp. 1435 - 1442, (1996).
6. ODIGURE J O. Mineral composition and microstructure of clinker from raw mix containing metallic particles. Cem and Concr. Res. Vol. 26, N 8, pp. 1171 - 1178 (1996).
7. ODIGURE J O. Hydration of cement paste and concrete from raw mix containing metallic particles. Cem and Concr. Res. Vol. 24, N 8, pp. 1549 - 1557 (1994).
8. BUTT Yu M, CICHEV, M M, TIMASHEV, V V. Chemical Technology of binding materials. Vishaya Shola Publication, Moscow, 472p (1980).
9. CHATTERJEE A K. Chemico- Physio- Mineralogical characteristics of Raw Materials. Advances in Cement Technology. Edited by SN Ghosh, Pergamon Press, pp. 39-68 (1983).
10. ODIGURE J O. Preparation ofcement raw mix containing metallic particles. Cem and Concr. Res. Vol. 27, N 11, pp. 1641 - 1648 (1997).
11. NECRASOV B V. Basic General Chemistry. T 2. Ximiya Publication, Moscow, T. 2, pp.262, 391 - 398 T. 3 416p (1970).
12. MONTGOMERY D C. Design and Analysis of Experiments. John Wiley and Sons Publication, pp 551 - 558, (1991).

GRAPHICAL ESTIMATION OF THE CONCRETE COMPRESSIVE STRENGTH AT OLDER AGES (>7 DAYS) ON THE BASIS OF THE CONCRETE STRENGTH VALUES MEASURED AT THE EARLY AGES OF HYDRATION

K K Sideris

Democritus University of Thrace

Greece

ABSTRACT. A graphical method was proposed in order to plot the straight line for the value pairs of mortar and concrete strength (Mortar versus concrete strength) using only two concrete strength values and the respective mortar strengths, determined at early ages of hydration. Based on this straight line the concrete strength was estimated graphically at the age of 3, 7, 14, 21, 28, 90, 365, 730 , and 5475 days. From the results of the analysis which has being made on three different cements, it was clearly seen that the variation between the graphically estimated values and the accurate reference values (calculated on the basis of the concrete compressive strength equation) is less than 2%.

Keywords: Cement hydration equation, Mortar versus concrete strength, Two points proportionality between mortar and concrete strength.

Dr K K Sideris is a Lecturer at Democritus University of Thrace. His main research interests include the use of pozzolanas in the chemistry of cement and concrete, with particular reference to the application of the cement hydration equation in the chemistry of cement and concrete.

INTRODUCTION

It is of practical importance to be able to predict the concrete compressive strength at older ages (> 7 days) without having rely concrete strength tests. For this purpose the cement hydration equation [1-5] was applied in order to plot the straight line of the linear correlation between the mortar and concrete strength [1, 2]. On the basis of this straight line (mortar versus concrete strength) it is possible to estimate the concrete compressive strength at any hydration age up to the end of hydration.

According to the cement hydration equation proposed by Professor K. Sideris [1, 2] the study of the linear proportionality between mortar and concrete strength is possible (mortar versus concrete strength) (Figure 1). The hydration criteria of mortar and concrete are regarded as established hydration criteria [2] and therefore their measured value pairs at any hydration age lay on a straight line [1, 2].

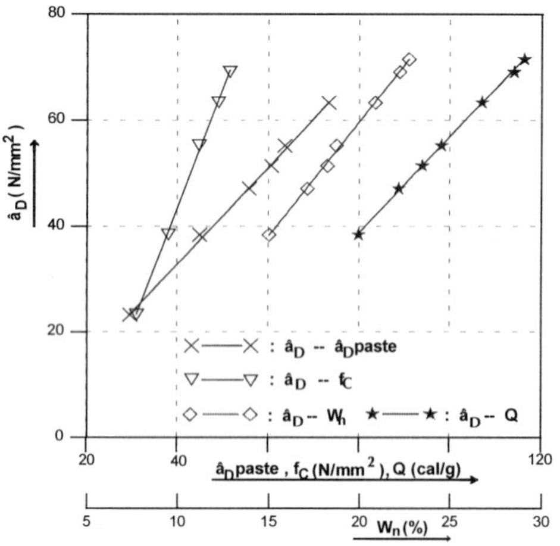

Figure 1 Correlation between the mortar compressive strength ($\hat{a}_D$) and other hydration criteria [1]

The first and second phase of the hydration, characterized by a hydration criterion K in the co-ordinate system with K on the ordinate and $(1/t)^p$ on the abscissa, follow a straight line path (Figure 2). The quantitative relationship, which is valid for the two phases by the same value of p (Figure 2), is given by the following equation [1, 2]:

$$\mathbf{K = K_{\infty} \pm b{*}t^{-p}} \qquad (1)$$

where:
- K: is the hydration criterion examined (Table 1).
- K_{∞}: is a constant, which represents the intersection of the straight line of Equation 1 with the ordinate
- b: is the slope of the straight line of Equation 1.
- t: is the hydration time expressed in days (t>>0).
- p: is the hydration number, depending only on the chemical and mineralogical composition of the cement type used.

The value of the hydration number p remains the same as long as the chemical and mineralogical composition of the cement type used also remain the same [1, 2].

The value of the hydration number p does not depend on:

1. the hydration criterion used
2. the curing temperature [5-60°]
3. The fineness of the cement
4. the ratio W/C
5. the aggregates' grading curve and the aggregates' quality when the hydration criterion of the compressive strength of mortar or concrete is examined.

Table 1 Hydration criteria [5]

CEMENT HYDRATION CRITERIA		EQUATIONS
1. Heat of hydration (cal/g)	Q:	$Q=Q_{\infty}-bt^{-p}$
2. Non evaporable water (%)	W_n:	$W_n=W_{n\infty} - t^{-p}$
3. $Ca(OH)_2$ elimination (%)	CaO:	$CaO=CaO_{\infty}-bt^{-p}$
4. Density (g/cm^2)	d:	$d=d_{\infty}+bt^{-p}$
5. Total porosity (%)	P:	$P=P_{\infty}+bt^{-p}$
6. Porosity with D > 75 Å (cm^3/g, dry paste)	P:	$P=P_{\infty}+bt^{-p}$
7. Gas permeability (pPm)	G_p	$G_p=G_{p\infty}+bt^{-p}$
8. Specific surface (m^2/g)	O_s:	$O_s=O_{s\infty}-bt^{-p}$
9. Vapor sorption (%)	V_s:	$V_s=V_{s\infty}-bt^{-p}$
10. Water sorption (%)	W_s:	$W_s=W_{s\infty}+bt^{-p}$
11. Compr. strength of cement paste (MPa)	$\hat{a}_D$:	$\hat{a}_D=\hat{a}_{D\infty}-bt^{-p}$
12. Compressive strength of mortar (MPa)	$\hat{a}_D$:	$\hat{a}_D=\hat{a}_{D\infty}-bt^{-p}$
13. Compressive strength of concrete (MPa)	f_c:	$f_c=f_{c\infty}-bt^{-p}$
14. Flexural strength of mortar (MPa)	$\hat{a}_B$:	$\hat{a}_B = \hat{a}_{B\infty}-bt^{-p}$
15. Pulse velocity (km/sec)	V_t:	$V_t=V_{\infty}-bt^{-p}$
16. Rebound hammer units	R_t:	$R_t=R_{\infty}-bt^{-p}$
CLINKER PHASES' HYDRATION CRITERIA		**EQUATIONS**
17. Compressive strength of C_3S (MPa)	$\hat{a}_D$:	$\hat{a}_D=\hat{a}_{D\infty}-bt^{-p}$
18. Compressive strength of C_2S (MPa)	$\hat{a}_D$:	$\hat{a}_D=\hat{a}_{D\infty}-bt^{-p}$
19. Compressive strength of C_3A (MPa)	$\hat{a}_D$:	$\hat{a}_D=\hat{a}_{D\infty}-bt^{-p}$
20. Compressive strength of C_4AF (MPa)	$\hat{a}_D$:	$\hat{a}_D=\hat{a}_{D\infty}-bt^{-p}$
21. $Ca(OH)_2$ elimination of C_3S paste (%)	CaO:	$CaO=CaO_{\infty}-bt^{-p}$
22. Non evapor. water of C_3S paste (%)	W_n:	$W_n=W_{n\infty} - t^{-p}$
23. Non evapor. water of C_2S paste (%)	W_n:	$W_n=W_{n\infty} - t^{-p}$
24. Non evapor. water of C_3A paste (%)	W_n:	$W_n=W_{n\infty} - t^{-p}$
25. Non evapor. water of C_4AF paste (%)	W_n:	$W_n=W_{n\infty} - t^{-p}$
26. Compressive strength of C_3S and C_2S mixture paste	$\hat{a}_D$:	$\hat{a}_D=\hat{a}_{D\infty}-bt^{-p}$
27. Non evaporable water of C_3S and C_2S mixture paste	W_n:	$W_n=W_{n\infty} - t^{-p}$
28. Compressive strength of C_3S, C_2S and gypsum mixture paste (MPa)	$\hat{a}_D$:	$\hat{a}_D=\hat{a}_{D\infty}-bt^{-p}$
29. Non evaporable water of C_3S, C_2S and gypsum mixture paste (%)	W_n:	$W_n=W_{n\infty} - t^{-p}$

When the hydration criterion of the mortar' s and concrete' s compressive strength is examined, Equation 1 takes the following form:

$$\hat{a}_D = \hat{a}_{D\infty} - b * t^{-p} \qquad (2), \qquad\qquad f_C = f_{C\infty} - b * t^{-p} \qquad (3)$$

The procedure of the evaluation of the hydration number p, for defining the equations of the compressive strength and the straight line for the value pairs $\hat{a}_D$ - f_C (linear correlation between two hydration criteria) are discussed in previous works [1-7]. In Tables 2-4 is given the procedure for the evaluation of the hydration number p and for the definition of the compressive strength equations. Figures 3-5 show the shapes of the diagrams of the concrete' s compressive strength equations, defined on the basis of measured values taken from the literature [8-10].

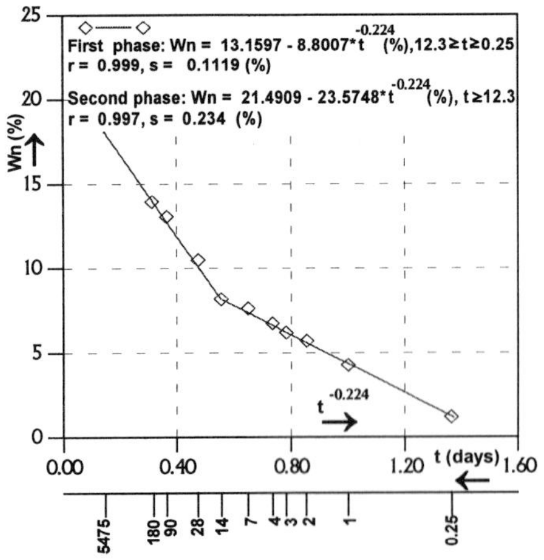

Figure 2 Non-evaporable water of cement. (meas. values taken from Copeland et all [8])

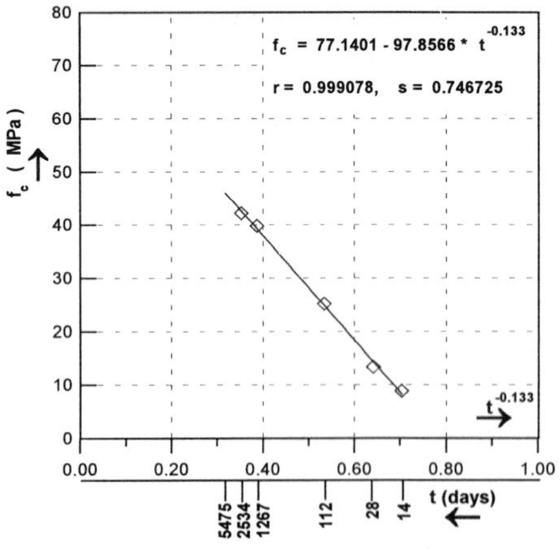

Figure 3 Compressive strength of concrete (meas.values taken from Hansen [9])

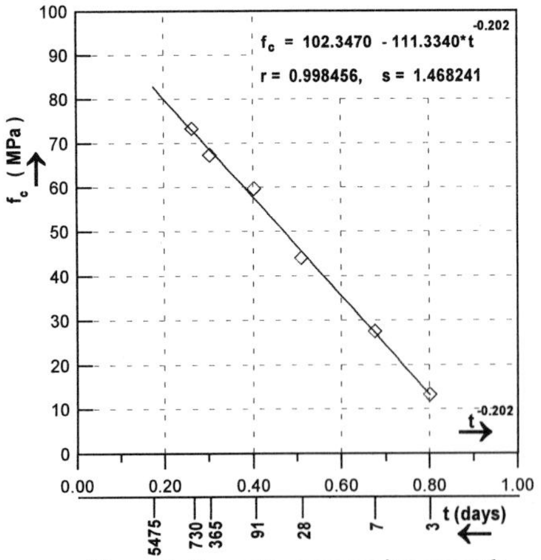

Figure 4 Concrete compressive strength (meas. values taken from Sivasundaram et al, [10])

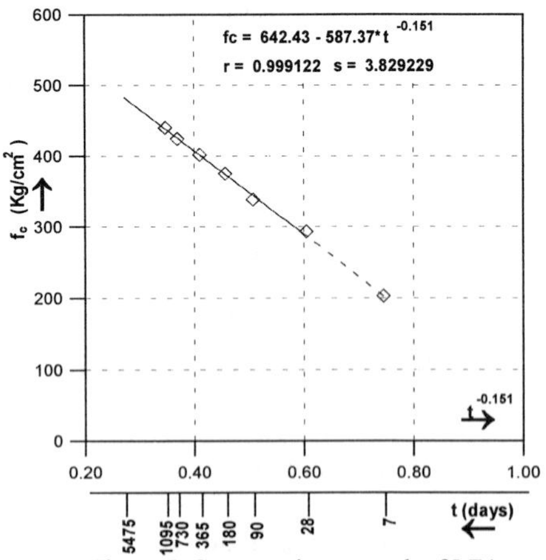

Figure 5 Compressive strength of LFA concrete with 40% Fly Ash (meas. values taken from Papayanni [11])

Based on the measured strength values of mortar and concrete which have been prepared with the same cement, the compressive strength equations for mortar and concrete can be defined. Further, on the basis of these equations the accurate values [2] of the compressive strength of mortar ($\hat{a}_{Dcalc}$) and concrete (f_{Ccalc}) are calculated at any hydration age up to the end of hydration. Using the calculated value pairs ($\hat{a}_{Dcalc}$ - f_{Ccalc}) at any hydration ages (e.g., at 3, 7, 28, 90, 365, 730 and 5475 (15 years)) the respective equation for the straight line is defined (linear proportionality between the value pairs $\hat{a}_{Di}$ - f_{Ci}) [1,2].

In the present work a graphical method is proposed in order to plot the straight line for the value pairs of $\hat{a}_D$-f_C using only two concrete strength values (determined at early ages, e.g., at 3 and 7 days) and the respective values of mortar strength ($\hat{a}_{Dcalc}$). Based on this straight line the concrete strength was estimated by graphical means at any hydration age up to the end of hydration. Further, the accuracy of the estimated values (f_{Cgraph}) was checked at the ages of 28, 90, 365 and 5475 days on the basis of the comparison with the respective calculated values f_{Ccalc}.

MIXTURES

For the purposes of this paper three mortars and three concrete mixtures were prepared using three different types of cements. All mixtures were tested at 3, 7, 14, 21, 28 and 92-98 days (Tables 2-4). The cements used were an OPC and two blended cements. The blended cements were prepared by the replacement of OPC by two different pozzolanas; one natural pozzolana (10% Skydra earth: SE cement) and one lignite fly ash (20% Ptolemaida fly ash: PFA cement).

All mortars were prepared, cured and tested according to DIN 1164. The concrete mixtures were prepared with limestone aggregates, 300 Kg cement per m^3 and W/C=0.55. At each test age the concrete compressive strength was determined on three cubes (15x15x15 in cm). At the ages of 3 and 7 days, the concrete compressive strength was determined on 12 (Tables 2 and 4) and on 6 cubes (Table 3). The cubes were cured at 20±2 °C in a room with a relative humidity of 95-98 %.

PROCEDURE OF THE PROPOSED METHOD

The proposed method requires a set of mortar compressive strength tests and two concrete compressive strengths (e.g., strength at 3 and 7 days). Great accuracy can be achieved if the mortar testing is carried out at several test ages (more than 6 different ages) and the two compressive strength values come out as the average test value of 12 (or 6) specimens (cubes or cylinders).

The proposed method includes the following steps:

(a) Estimation of the hydration number p for each type cement used, according to the procedure described in previous works [1-5] and given in Tables 2-4 on the basis of the mortar hydration criterion.

(b) Setting up the compressive strength equation for the mortar used, using the measured values at 3, 7, 10, 14, 21, 28 (or at 3, 7, 14, 21, 28, 60 (or 90)) days.

(c) Calculating the accurate compressive strength values of mortar ($\hat{a}_{Dcalc}$) at 3, 7, 28, 90, 365 and 5475 days with the aid of the as above defined mortar strength equation.

(d) Plotting the straight line of the correlation between mortar and concrete strength for the value pairs $\hat{a}_{Dcalc}$ - f_C (f_C=averaged value of 6 or 12 cubes) at the age of 3 and 7 days (two points proportionality) on a millimeter paper. (Figure 8).

(e) Using the calculated values of the mortar strength ($\hat{a}_{Dcalc}$) at the age of 28, 90, 365 and 5475 days (or at any other age) the respective concrete strengths are estimated according to the procedure shown in Figure 8.

The accuracy of the concrete strength values (f_{Cgraph}) estimated by the as above plotted straight line (Figure 8), was checked by comparing these values with the respective calculated values f_{Ccalc} given in Tables 2-4.

In Figure 6 the shapes of the hydration diagrams of mortar and concrete strength are given as a function of hydration time (Table 2). In Figure 7 is shown the straight line of the linear correlation for the calculated values of the mortar and concrete strength.

Table 2 Compressive strength equations of mortar and concrete (OPC)

MORTAR ACCORDING TO DIN 1164 (MPa)											
Evaluation of p:	t (d)	3	7	14	21	28	90	98	365	730	5475
Use of PC program for	$\hat{a}_{Dmeas}$	23.22	38.38	47.1	51.4	55.2	---	63.4	---	---	---
the value pairs			p=0.375			r=0.9996			s=0.4334 (MPa)		
$\hat{a}_{Dmeas}$ - t			$\hat{a}_D = 78.39 - 83.30 * (1/t)^{0.375}$					(MPa)		t ≥ 3	
$\hat{a}_{Dcalc.}$		23.2	38.2	47.4	51.8	54.5	63.0	63.5	69.3	71.4	75.1
CONCRETE (MPa) CEMENT 300 Kg/m³, W/C = 0.55											
Linear regre-	t (t)	3	7	14	21	28	90	92	365	730	5475
ssion for the	f_{Cmeas}	31.2	37.5	43.4	44.6	44.9	---	50.7	---	---	---
value pairs	$t^{-0.375}$	0.662	0.482	0.372	0.319	0.287	0.185	0.184	0.109	0.084	0.04
f_{Cmeas} - $(1/t)^{0.375}$			p = 0.375			r = 0.9942			s=0.81081 (MPa)		
			$f_C = 57.40 - 39.97 * (1/t)^{0.375}$					(MPa)			t ≥ 3
$f_{Ccalc.}$		30.9	38.1	42.6	44.6	46.0	50.0	50.1	53.0	54.0	55.8
CHECKING THE ACCURACY OF THE STRENGTHS ESTIMATED BY THE GRAPHICAL METHOD. THE STRAIGHT LINE WAS PLOTTED ON THE BASIS OF THE VALUE PAIRS $\hat{a}_{Dcalc}$ - f_{Cmeas} AT 3 AND 7 DAYS. $f_{Cmeas(3)}$ =31.1 (MPa), $f_{Cmeas(7)}$ =37.9 (MPa). f_{Cmeas} AVERAGED OF 12 CUBES.											
f_{Ccalc}		30.9	38.1	42.6	44.6	46.0.	50.0	---	53.0	54.0	55.8
$f_{Cgraph.}$		31.1	37.9	42.1	44.0	45.3	49.15	---	52.1	53.0	54.7

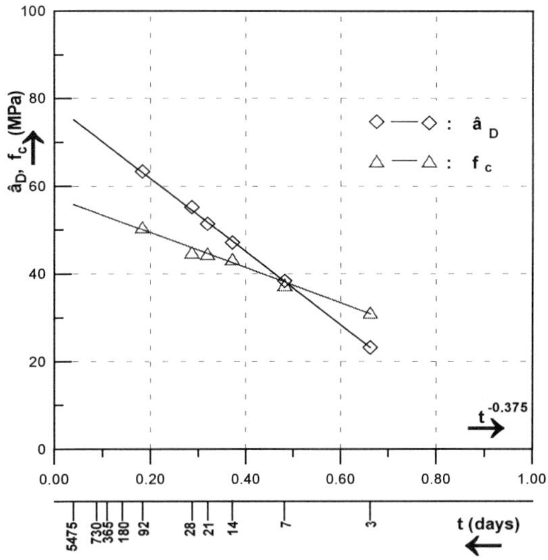

Figure 6 Compressive strength of mortar and concrete as a function of hydration time [values taken from Table 2]

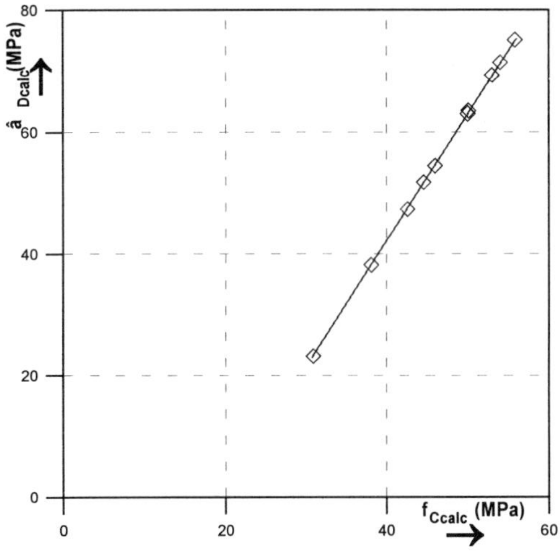

Figure 7 Linear correlation between mortar and concrete strength $â_{Dcalc}$ - f_{Ccalc} [values taken from Table 2]

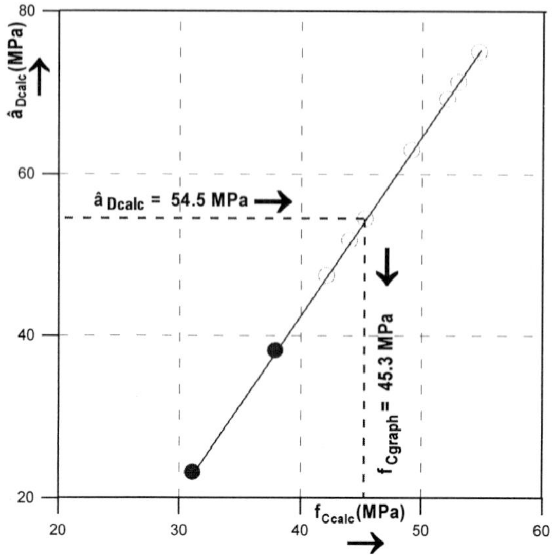

Figure 8 Mortar versus concrete strength. Two points proportionality [values taken from Table 2]

Table 3 Compressive strength equations of mortar and concrete (SE cement)

MORTAR ACCORDING TO DIN 1164 (MPa)											
Evaluation of p:	t (d)	3	7	14	21	28	90	92	365	730	5475
Use of PC program for	$\hat{a}_{Dmeas}$	18.54	33.85	44.06	49.58	52.70	---	65.83	---	---	---
the value pairs			P = 0.212			r = 0.9999		s = 0.5290		(MPa)	
$\hat{a}_{Dmeas}$ - t			$\hat{A}_D = 109.88 - 115.17 * (1/t)^{0.212}$					(MPa) $t \geq 3$			
$\hat{a}_{Dcalc.}$		18.6	33.6	44.1	49.5	53.1	65.6	65.7	76.9	81.4	91.3
CONCRETE STRENGTH (MPa). CEMENT 300 Kg/m³, W/C = 0.55											
Linear regression	t (d)	3	7	14	21	28	90	92	365	730	5475
for the	f_{Cmeas}	29.20	36.53	41.50	42.90	45.63	---	53.04	---	---	---
value pairs	$t^{-0.212}$	0.7922	0.6619	0.5715	0.5244	0.4934	0.3852	0.3834	0.2863	0.2471	0.1612
f_{Cmeas} - $(1/t)^{0.212}$			p = 0.212			r = 0.9962		s = 0.7932		(MPa)	
			$f_C = 73.97 - 56.90 * (1/t)^{0.212}$					(MPa) $t \geq 3$			
$f_{Ccalc.}$		28.9	36.3	41.5	44.1	45.9	52.1	52.2	57.7	59.9	64.8
CHECKING THE ACCURACY OF THE STRENGTHS ESTIMATED BY THE GRAPHICAL METHOD. THE STRAIGHT LINE WAS PLOTTED ON THE BASIS OF THE VALUE PAIRS $\hat{a}_{Dcalc}$ - f_{Cmeas} AT 3 AND 7 DAYS. $f_{Cmeas(3)}$ =28.7 AND $f_{Cmeas(7)}$ =36.5 (MPa). f_{Cmeas} AVERAGED OF 6 CUBES.											
f_{Ccalc}		28.9	36.3	41.5	44.1	45.9	52.1	52.2	57.7	59.9	64.8
f_{Cgraph}		28.7	36.5	41.8	44.6	46.4	52.6	---	58.2	60.6	65.6

Table 4 Compressive strength equations of mortar and concrete (PFA cement)

MORTAR ACCORDING TO DIN 1164 (ÌPa)										
Evaluation of p:	t (d)	3	7	14	21	28	92	365	730	5475
Use of PC program for	$â_{Dmeas}$	14.69	27.40	36.98	41.78	43.96	52.90	---	---	---
the value pairs		p = 0.068		r = 0.998972			s = 0.779976			(MPa)
$â_{Dmeas}$ - t		$â_D = 229.66 - 231.33 * (1/t)^{0.068}$					(MPa) $t \geq 3$			
$â_{Dcalc.}$		15.0	27.0	36.3	41.6	45.2	59.6	74.8	81.9	100.8
CONCRETE (MPa): CEMENT 300 Kg/m³, W/C = 0.55										
Linear regression	t (d)	3	7	14	21	28	92	365	730	5475
for the	f_{Cmeas}	22.16	32.80	37.93	40.08	43.20	49.30	---	---	---
value pairs	$t^{-0.068}$	0.9280	0.8761	0.8357	0.8130	0.7972	0.7353	0.6695	0.6387	0.5569
$f_{Cmeas} - (1/t)^{0.068}$		p = 0.068		r = 0.9964			s = 0.6013			(MPa)
		$f_C = 136.29 - 117.84 * (1/t)^{0.068}$					(MPa) $t \geq 7$			
f_{Ccalc}		---	33.0	37.8	40.5	42.3	49.6	57.4	61.0	70.7
CHECKING THE ACCURACY OF THE STRENGTHS ESTIMATED BY THE GRAPHICAL METHOD. THE STRAIGHT LINE WAS PLOTTED ON THE BASIS OF THE VALUE PAIRS $â_{Dcalc} - f_{Cmeas}$ AT 7 AND 14 DAYS. $f_{Cmeas(7)}$ =32.72 AND $f_{Cmeas(14)}$ =38.1. f_{Cmeas} AVERAGED OF 12 CUBES										
f_{Ccalc}		---	33.0	37.8	40.5	42.3	49.6	57.4	61.0	70.7
f_{Cgraph}		---	32.72	38.1	40.1	41.9	49.1	56.8	59.9	69.8

DISCUSSION

From the results of the foregoing analysis it is clearly seen that the deviation between the graphically estimated values and the reference values (calculated on the basis of the concrete compressive strength equation) is less than 2%.

CONCLUSIONS

The proposed method of the two points proportionality leads to reliable results and is useful because only simple measurements are needed. For this reason it can be used in each usual concrete laboratory without the aid of special equipment and special tests.

ACKNOWLEDGMENTS

The author is grateful for the financial support provided by the Technical Chamber of Greece.

REFERENCES

1. SIDERIS K.: The cement hydration equation, Zement-Kalk-Gips, 12 (1993), Edition B, pp.E337-E344.

2. SIDERIS K., SIDERIS K. K.: The Cement Hydration Equation and its Application to several Hydration Criteria according to the Literature, Proc. of the 10th International Congress on the Chemistry of Cement, Göthenburg, Sweden, June 2-6, 1997, Volume 2, paper 2ii061.

3. SIDERIS K. K., KONSTA M. S., KARAYANNIS C. G.: Application of the Cement Hydration Equation to Concrete, Proc. of the International Congress on Concrete in the Service of Mankind, Dundee, Scotland, 24-28 June 1996 pp. 343-349.

4. SIDERIS K. K., SAVVA A. E., BALTZOPOULOU K. D., ECONOMOU C. M., SIDERIS K.: Influence of Silica and Limestone Aggregates on the Final Compressive Strength of Blended Cement Concretes prepared with the Use of Three Different Pozzolanas, Proc. of the 10th International Congress on the Chemistry of Cement, Göthenburg, Sweden, June 2-6, 1997, Volume 4, paper 4iv052.

5. SIDERIS K. K., KONSTA-GDOUTOS M.: The cement Hydration Equation and its Application to the Strength of Concrete according to the Literature, Proc. of the First Hellenic Conference on Composite Materials and Structures, Xanthi, Greece, July 2-5, 1997, pp. 789-802.

6. KARAYANNIS C.A.: Determination of the influence of W/C ratio on the final compressive strength of concrete, according to the cement hydration equation, Proc. of 11th Greek Concrete Conference, Corfu May 1994, Vol. I, pp 54-63 (in Greek).

7. SIDERIS K. K.: Influence of Natural Pozzolans on the Compressive Strength and the Porosity of Cement Paste and Concrete, Ph. D. Thesis, Xanthi 1996, Greece.

8. COPELAND L. E ., KANTRO D. L., VERBECK G.: Chemistry of hydration of portland cement, Proc. of the Fourth Intenational Symposium on the Chemistry of Cement, Washington 1960, pp. 429-465.

9. HANSEN T. C.: Long-term strength of high fly ash concretes, Cement and Concrete Research, vol. 20, 1990, pp. 193-196

10. SIVASUNDARAM V., GARETTE G. G., MALHOTRA V. M.: Mechanical properties creep and resistance to diffusion of cloride of concretes incorporating high volumes of ASTM Class F Fly Ashes from seven different sources, ACI Materias Journal, vol. 88, pp. 407-416, July-August 1991.

11. PAPAGIANNI I.: Research for the use of Ptolemaida Fly Ash on Concrete production, Ph. D. Thesis, Thessaloniki 1981, Greece.

FACTORS INFLUENCING THE RATIO OF MORTAR TO CONCRETE STRENGTHS

G K Moir

Blue Circle Industries PLC

United Kingdom

ABSTRACT. The factors which influence the ratio of EN 196-1 mortar to BS 4550 concrete strengths have been examined using data derived from the testing of random audit samples taken as part of the BSI Kitemark Scheme. Selected cements with a range of chemical and physical characteristics were also tested in EN 196-1 mortar, ASTM C 109 mortar and a range of concretes prepared using BS 4550 aggregates. It is concluded that no single test, whether in mortar or concrete, can be expected to reliably predict concrete strength over a range of cement contents. Nevertheless for a given cement source a significant change in mortar strength is likely to be reflected in a significant change in concrete strength.

Keywords: EN 196-1 mortar, ASTM C 109 mortar, BS 4550 concrete, Alkalis, Cement content, Portland cement.

Dr G K Moir is Chief Chemist at the Blue Circle Industries PLC Technical Centre at Greenhithe in Kent. His main responsibilities and interests concern cement production quality control and the factors which influence both the short and long term performance of cements in use. He is currently chairman of CEN TC51/WG10 (masonry cement) and of CEN TC51/WG12/TG1 (sulfate resistance test method).

INTRODUCTION

All national cement standards have minimum strength requirements which are determined utilising a standardised test procedure. Cement type may also be classified according to strength and the current British cement standards, in common with the proposed European Standard for Common Cement (pr EN 197-1) has several strength classes with well defined upper, as well as lower strength limits.

In the world today most national standards utilise one of 4 test procedures whose main characteristics are summarised below:

Table 1 Standardised strength tests for cement used internationally

DESIGNATION	MIX PROPORTIONS CA:FA:C:W*	SPECIMEN DIMENSIONS mm	CURING TEMP °C
BS 4550 concrete‡	3.5 : 2.5 : 1 : 0.60	100 x 100 x 100	20
EN 196-1/ISO 679 mortar	0 : 3 : 1 : 0.50	40 x 40 x 160	20
ASTM C109 mortar	0 : 2.75 : 1 : 0.485	50 x 50 x 50	23
BS 4550 mortar	0 : 3 : 1 : 0.40	70.7 x 70.7 x 70.7	20

* CA = Coarse aggregate, FA = Fine Aggregate, C = Cement, W = Water
‡ Now withdrawn but maintained as BS 1881 : Part 131 and as the UK Cement Industry Test Procedure.

In the UK the vibrated mortar cube test was introduced in 1940 but was never popular with either cement users or manufacturers. The cement industry developed its own 4 inch cube (~ 100 mm) concrete method and this was standardised in 1978 with the publication of BS 4550 : Part 3.4. This method formed the basis of quality control testing within the industry until the introduction of BS 12 : 1991 which required strength testing using the EN 196-1 mortar prism method. The adoption of the new testing procedure meant that customers were supplied with test certificate data based on mortar rather than concrete, and an ongoing debate has continued regarding the significance of these results to concrete production quality control data.

Schramli [1] has investigated the relationship between standard mortar test results and standard concrete results with the results of ready-mixed concrete strength testing. It was found that the standardised concrete test was no better than the standard mortar test for predicting the strength of production concrete. This was attributed to the relatively low variability of the cement used in the investigation, and the poor reproducibility of the concrete test.

Gaynor [2] has analysed ASTM C 109 mortar and field concrete data and concluded that provided reliance is not placed on the results of single tests, mortar results do give a meaningful guide to concrete strengths. The relationships improve markedly if cement variability is increased, and/or repeat testing and average results are used.
De Hayes [3] investigated the relationship between ASTM C 109 mortar strengths and concrete cylinder strengths. Although a good correlation was found between 1 day mortar and concrete strengths, correlations at later ages were poor.

Bloomer [4] has investigated the factors which affect the ratio of EN 196-1 mortar to concrete strengths. It was concluded that cement properties other than strength have little effect on this ratio.

This paper compares strength data obtained when cements from different sources were tested in mortar (ASTM C 109 and EN 196-1) and concrete. Factors responsible for differences in the ratio of mortar to concrete strength are examined.

SOURCES OF DATA

In the preparation of this paper, data have mainly been drawn from 3 sources:

- Cement plant routine quality control data
- Audit sample testing at Blue Circle Industries PLC central laboratory
- An experimental programme utilising 17 cements from different plants in the UK and overseas.

REPRODUCIBILITY OF TEST METHOD

If a test method is to yield meaningful results, it is essential that the inherent variability of the method is relatively low in relation to cement variability. The reproducibility of the EN 196-1 and ASTM C 109 methods can be assessed from the annual round-robin testing exercises organised respectively by the French Cement industry association (ATILH) and the US Cement and Concrete Research Laboratory (CCRL).Typical mean values in recent years for all participating laboratories are given in Table 2

Table 2 Reproducibility of EN 196-1 and ASTM C 109

METHOD	COEFFICIENT OF REPRODUCIBILITY
EN 196-1	3%
ASTM C 109	7%

Within the cement industry itself, a higher level of reproducibility and repeatability is normally achieved, and within laboratory coefficients of variation of 2% and 4% should be achievable for the EN and ASTM methods respectively.

Thus, the European method has a level of inherent variability, which is approximately half that of the ASTM method. This is borne out by quality control data for US plants, where despite similar levels of control of cement chemistry, fineness and mineralogy, the coefficients of variation of the mortar test results are approximately twice those of UK plants which employ the EN 196-1 mortar test procedure. These differences in apparent strength variability are illustrated in Figure 1.

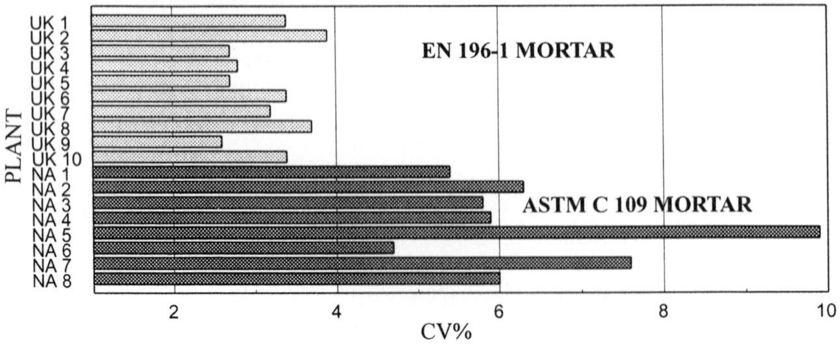

Figure 1 Coefficient of variation of 28 day strength results for 1997

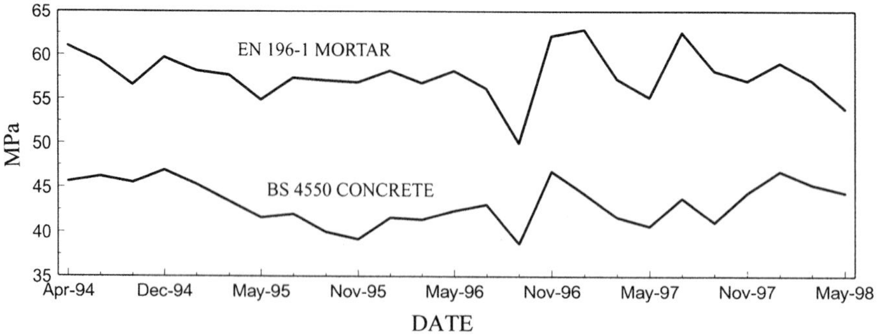

Figure 2 Works A audit samples tested at Central Laboratory

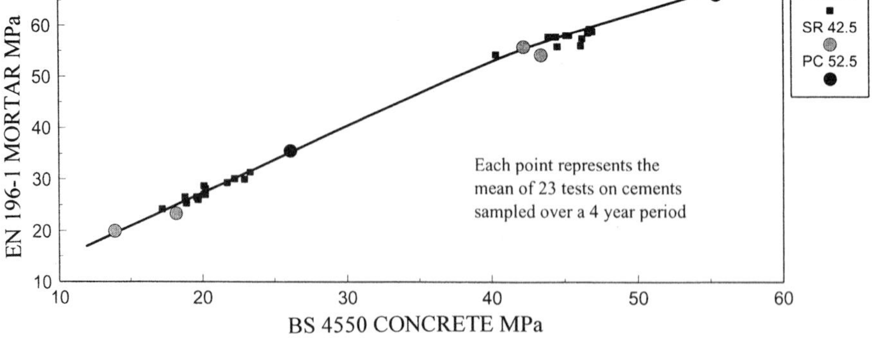

Figure 3 Relationship between EN 196-1 mortar and BS 4550 concrete strengths

The BS 4550 concrete testing procedure has been withdrawn, but continues within the UK cement industry as the Cement Industry Test Procedure and in other laboratories as BS 1881 : Part 131. The method uses 10 mm crushed granite aggregate from a single source and selected quartz sands. In a competent laboratory the coefficient of variation of 28 day strength determined using this method is as good as that for the EN 196-1 mortar method, that is ~2%.

CORRELATION BETWEEN EN 196-1 MORTAR STRENGTHS AND BS 4550 CONCRETE STRENGTHS

Parallel testing of cement samples in mortar and concrete has demonstrated that a significant change in mortar strengths will always be reflected in concrete strengths, and vice versa. This relationship is illustrated in Figure 2 where data from the testing of audit samples from a cement source which shows higher than normal variability are plotted.

Since March 1994 the 6 random despatch audit samples taken each year for 16 cements, (13 PC 42.5, 2 sulfate resisting PC 42.5 and 1 PC 52.5), as part of the requirements of the BSI Kitemark Scheme have been tested in concrete, as well as mortar, at the Blue Circle UKAS accredited central laboratory. The sand used during the period March 1994 – December 1997 was Normensand from Germany and the method of compaction used was vibration. Figure 3 illustrates the relationship between mortar and concrete 2 day and 28 day strengths. Each point represents the average result obtained for 23 samples from each cement source/type. The relationship appears satisfactory, but when the data are examined, it can be seen that there are systematic differences in mortar:concrete (m/c) ratio for different products. At 28 days the m/c ratios range from 1.19 (PC 52.5) to 1.35 (relatively coarse PC 42.5). Figure 4 illustrates that there is clearly a relationship between m/c ratio and cement surface area. The surface area of PC 42.5 cement reflects the fineness of grinding required to achieve a mean 28 day EN 196-1 mortar strength of ~59 Mpa. The fineness of grinding required is influenced by

- the cement milling equipment (eg. open or closed circuit)
- the clinker chemistry (silicates content, alkali level, etc.).
- clinker mineralogy

and tends to be characteristic of a given plant.

The reduction in m/c ratio with increasing fineness is related to the influence of cement fineness on the degree of compaction achieved during specimen preparation. Prism densities were determined by weighing the mortar specimens in air and water immediately following demoulding at an age of 24 hours.

In Figure 5 it can be seen that prism density decreases with increasing cement surface area. This suggests that mortar compaction and the elimination of entrapped air is inhibited with finer cements, at least with the vibration method of compaction.

Figure 6 confirms that m/c ratio increases with increasing mortar density. It can be deduced that mortar compaction is influenced by cement fineness to a greater extent than is the case with concrete compacted using the BS 4550 hand tamping procedure.

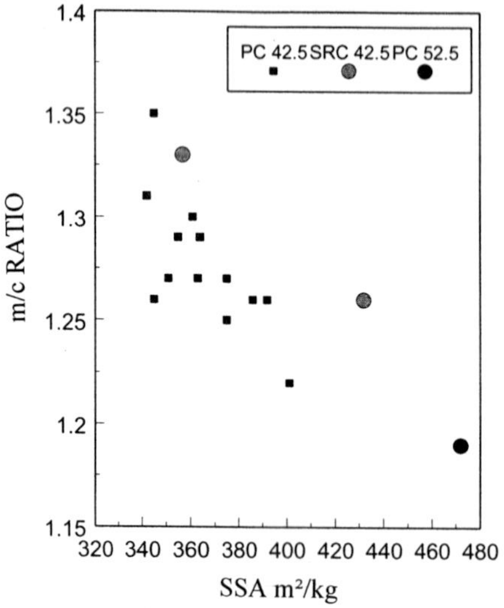

Figure 4 Relationship between m/c strength ratio and cement SSA

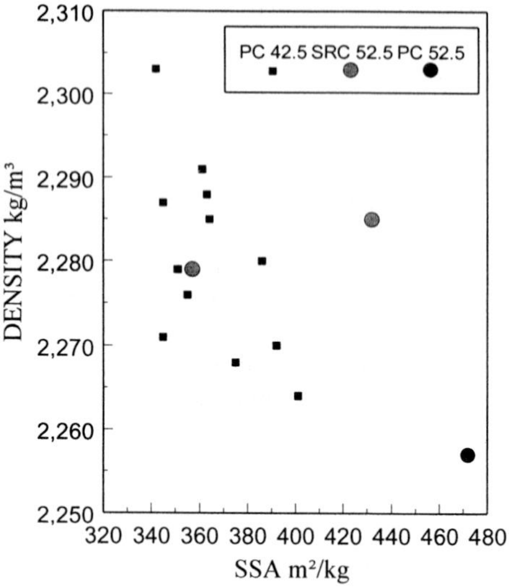

Figure 5 Relationship between prism density and cement SSA

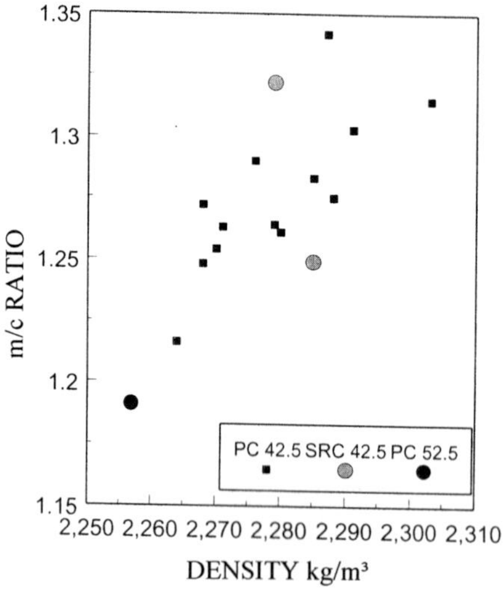

Figure 6 Relationship between m/c strength density and mortar density

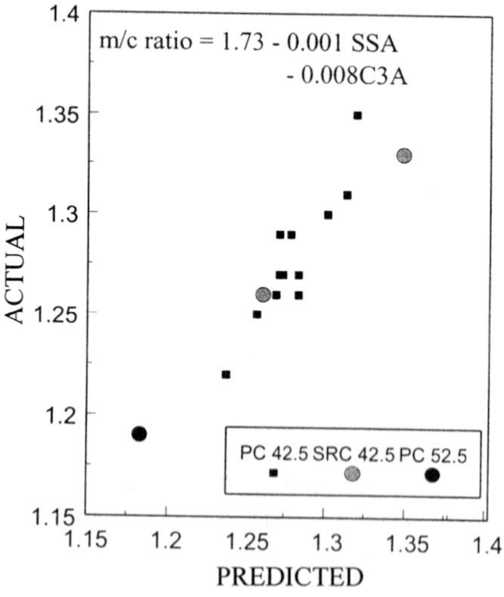

Figure 7 Actual v predicted m/c Strength ratio

A secondary influence identified using multiple linear regression analysis is cement C_3A content. Figure 7 illustrates that m/c ratio can be satisfactorily predicted using the expression

$$m/c = 1.73 - 0.001\ SSA - 0.008\ C_3A.$$

Thus the 28 day m/c ratio is not the same for all cements, but unless significant changes are made to cement fineness or clinker chemistry the ratio remains a characteristic of that particular cement.

INFLUENCE OF MORTAR TEST PROCEDURE AND CONCRETE CEMENT CONTENT

Whilst it has been shown that m/c ratio is not the same for all cements, the implications for the prediction of concrete strength need to be assessed taking into consideration the other important influences on concrete strength, such as:

- Cement content
- w/c ratio
- Aggregate type

In order to investigate the influence of mortar test method and concrete cement content, 19 different cements were obtained from the UK, US, Australia and South Africa. This selection enabled a much wider range of cement characteristics to be investigated, than would have been the case had only UK class 42.5 cements been included in the programme.

All of the cements were tested in concretes prepared from BS 4550 aggregates, with nominal cement contents of 200, 260, 310, 360, 430 and 550 kg/m^3. The concretes were gauged to a constant slump of 60 mm at w/c ratios ranging from 1.10 to 0.36. In addition, 17 of the cements were tested in BS 4550 concrete (w/c 0.60) and using the EN 196-1 and ASTM C 109 mortar methods.

Figures 8 and 9 respectively illustrate the relationship between EN 196-1 and ASTM C 109 mortar strengths, and those given by the BS 4550 procedure. Note that these are the results of single pairs of tests and thus the scatter is considerably greater than in Figure 3.

Figure 10 illustrates that if the cements are ranked according to BS 4550 28 day strength the overall agreement with both ASTM and EN 196-1 strengths is relatively poor. However, it can also be seen that the agreement of BS 4550 concrete strengths with those of concretes having different cement contents is also poor.

Several factors affect the response of different cements to changes in cement content and thus w/c ratio. One of the most important is cement alkali content. Figure 11 illustrates that in the rich 550 kg/m^3 mix the cement alkalis have the expected negative influence on 28 day strength. However, in the lean 200 kg/m^3 mix, where the w/c was ~1.1, the cement alkalis had a positive influence on 28 day strength. Thus, no single test method, whether based on mortar or concrete, can be expected to predict performance in a range of production concretes. It is likely that the closer the w/c of the concrete to that used in the standardised test, the better will be the relationship.

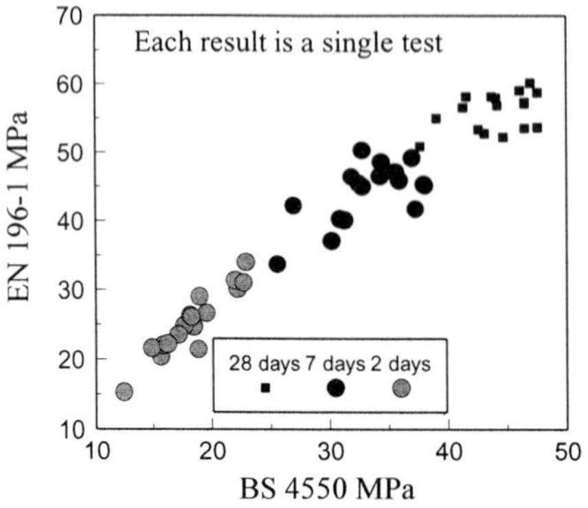

Figure 8 Relationship between EN196-1 mortar and BS 4550 concrete

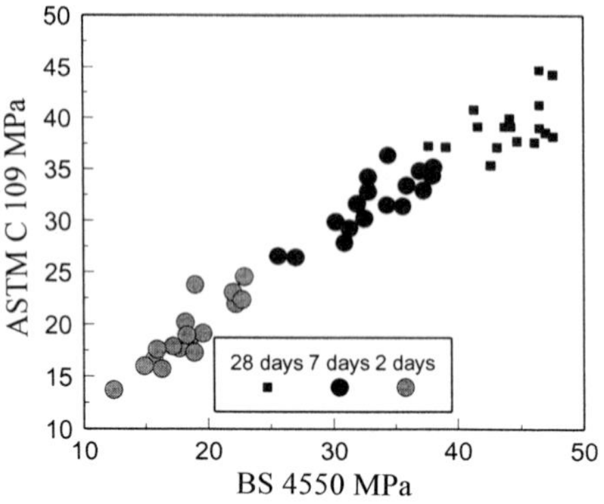

Figure 9 Relationship between ASTM C 109 mortar and BS 4550 concrete

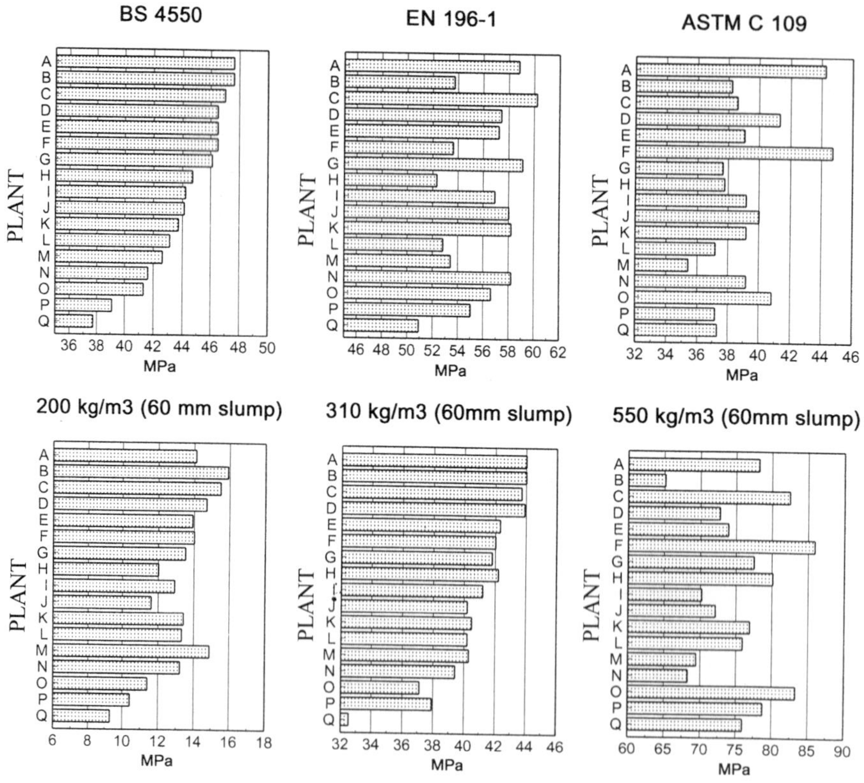

Figure 10 Influence of test procedure and cement content on 28 day strength

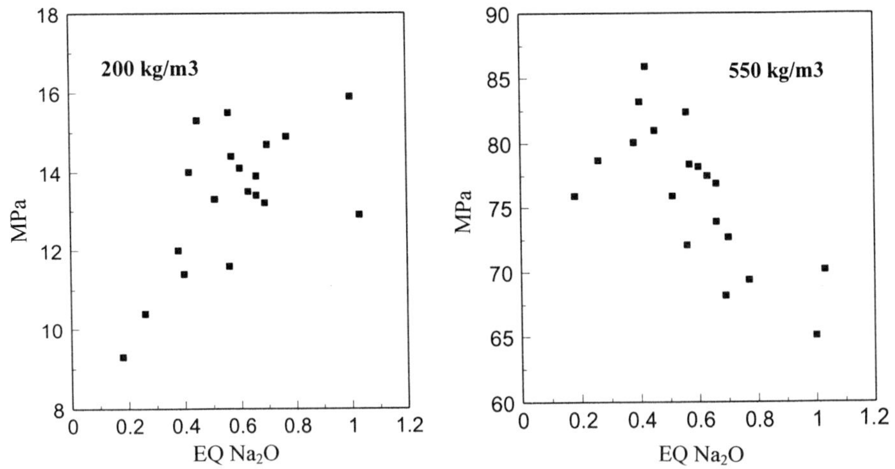

Figure 11 Influence of cement alkali content on concrete 28 day strength

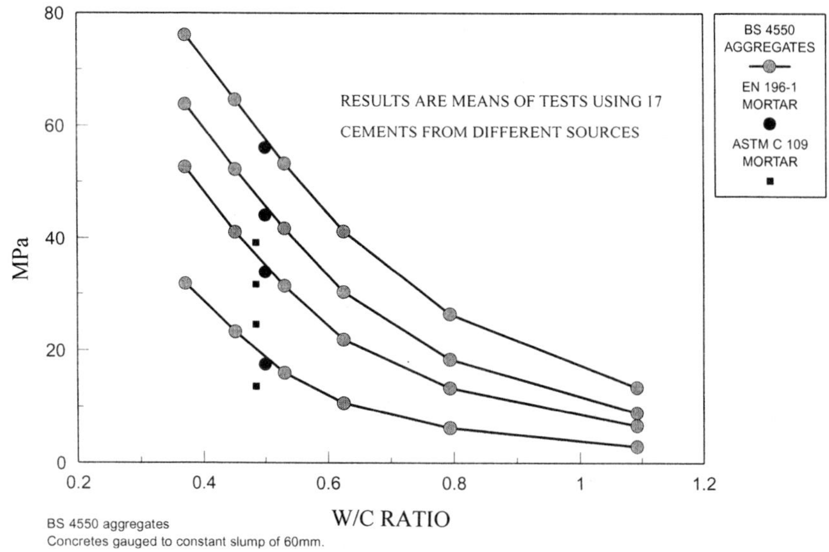

Figure 12 Relationship between cement content and 28 day strength

Apart from its lower inherent variability a further advantage of the EN 196-1 procedure, compared to the ASTM C 109 procedure, is that the strengths obtained are similar to those given by a BS 4550 type concrete mix, proportioned to give a w/c of 0.50. The ASTM test gives much lower strengths, despite the lower w/c of 0.485. This difference is illustrated in Figure 12 using average data for the 17 cements. The main reason for the much lower strengths given by the ASTM method is the much higher level of air in the mortar. The ASTM C 109 mortars have typical air contents of 10%, whilst in EN 196 mortar the level should be less than 4%, and in BS 4550 concretes less than 2%.

CONCLUSIONS

1. The ratio of EN 196-1 mortar to BS 4550 concrete strength is mainly influenced by cement surface area. A secondary influence is cement C_3A content.

2. The ratio of EN 196-1 mortar to BS 4550 concrete at 28 days strengths can be estimated using the expression

$$m/c = 1.73 - 0.001\ SSA - 0.008\ C_3A$$

3. The decrease in m/c ratio with increasing cement surface area is associated with reduced mortar prism density.

4. For any given cement source a significant change in EN 196 mortar strength is likely to be reflected in a significant change in concrete strength.

5. It is not possible to predict concrete strengths over a range of cement contents from any single strength test either based on standard mortar or concrete.

6. The EN 196 mortar strength is more satisfactory for quality control purposes than the ASTM C 109 test procedure, on account of its lower inherent variability, and the much better agreement between strengths obtained and those expected for a concrete of similar w/c ratio.

REFERENCES

1. SCHRAMLI, W, and WOLTER, H. Determining the quality of cement by means of standard mortar or standard concrete testing – an investigation. Zement Kalk Gips, Vol 41, No 10, 1988, pp 403-408.

2. GAYNOR, R D. Cement strength and concrete strength – an apparition or a dichotomy. Cement Concrete and Aggregates, Vol 15, No 2, 1993, pp 135-144.

3. De HAYES, S M. C 109 vs concrete strengths: a preliminary investigation. Proceedings of the 12th International conference on cement microscopy, Vancouver, 1990, pp 246-263.

4. BLOOMER, S J. Effect of cement properties on the strength performance of various cements in commercial concrete. Advanced Concrete Technology Project 95/4. Institute of Concrete Technology, 1995.

MINERALISED CLINKER AND CEMENT

S Rasmussen

D Herfort

Aalborg Portland

Denmark

ABSTRACT. Mineralised clinkers and the potential benefits of using cements based on mineralised clinkers have been studied for many years. During the past two decades attention has mainly focused on the benefits arising from the combined use of fluorine and SO_3. Until recently, however, no producers of grey cement, have utilised this technology. The Danish producer, Aalborg Portland has, since 1994, produced a cement based on mineralised clinker as one of its main products for the domestic market. The clinker contains approximately 2% SO_3 and 0.2% F. The combined mineralising action lowers burning temperatures in the kiln by approximately 100°C, giving considerable reductions in emissions of NO_x in particular. Cements ground from this clinker are more reactive with both higher early and 28 day strengths than for otherwise identical non-mineralised cements. Where rapid strength development is not desirable this technology can be used to partially replace the clinker by a material such as limestone. In Denmark this is achieved by replacing 14% of the clinker with a limestone rich dust giving a further reduction in NO_x emission in addition to a considerable reduction in CO_2 emission by weight of the cement produced.

Keywords: Mineralised cement, Clinker, Fluorine, Sulphate, Limestone filler.

Mr Søren Rasmussen is Manager of the Project Service Department at Aalborg Portland. His main interests include the properties of mineralised and composite cements, and the effects of admixtures.

Mr Duncan Herfort is chief geologist at Aalborg Portland. His responsibilities include all mineralogical aspects associated with raw material assessment, process and quality control, and concrete analysis.

INTRODUCTION

During recent decades the cement industry has had to face growing demands as a result of increasing fuel costs, stricter emission limits, higher expectations regarding cement quality, and in many cases increased competition. Reduced fuel consumption has mainly been achieved by investing in more efficient kilns. Modern precalciner kilns, for example, use only half the amount of fuel required by their wet process predecessors to produce a tonne of clinker, and at much higher outputs. Further reductions in fuel consumption have been achieved by partially replacing the clinker with pulverised fly ash, ground blast furnace slag, natural pozzolans, limestone dust etc. The wide range of such composite cements produced in Europe is exemplified by the long list of cement types specified in the European pre-standard, ENV 197.

Another way of reducing fuel consumption as well as emissions of CO_2 and NO_x is by using a mineraliser where this is not already present in the raw materials. The mineralisers which have received most attention, particularly since the work done by Blue Circle in the 1970's and early 1980's [1,2], are CaF_2 and $CaSO_4$, because of the marked improvements in clinker combinability and strength development of the cement when these components are used in combination. This technology has been successfully applied at Aalborg Portland since 1994, with considerable savings in fuel consumption and improvement in cement quality [3].

MINERALISED CLINKER

Although at times used loosely, the term mineralisation in clinker production is defined in the same way as it is in geology, i.e. it is the process of promoting mineral formation. A mineraliser is, therefore, the agent by which this is achieved, for example by lowering the temperature of formation of a mineral through solid solution, or by increasing the content of the melt, or changing the properties of the melt. Strictly speaking, then, a fluxing agent is also a mineraliser if it promotes mineral formation. Not all fluxing agents are mineralisers, however, since some are known to inhibit the formation of clinker minerals. The term is conventionally used to describe the minor components which promote the formation of alite. Al_2O_3 and Fe_2O_3 are, therefore, not regarded as mineralisers in the production of Portland cement clinker, because they occur as major components. For a comprehensive review of mineralisers in Portland cement clinker, the reader is referred to Moir & Glasser [4]. Some of the major trends are briefly described below.

The effects that transition metals have on the formation of alite varies considerably with no clear relationship with their effect on the properties of the melt. Of the P-block elements, S and P are known to inhibit the formation of alite, except where this is offset by sufficiently high contents of other minor components, particularly F. The presence of halogens, notably F and Cl, are known to promote the formation of alite (or alinite in the case of Cl). The effect of the alkalis is strongly dependent on the degree of sulphurisation.

The above minor components are always present in cement but, apart from the alkalis and sulphates, are usually at too low concentrations to have any significant effect. A combination of sulphate and fluoride is known to have a particularly positive effect, although it is unusual for the desired combination to occur naturally in the raw materials and/or fuel.

A small number of producers, including Aalborg Portland have utilised this technology through the controlled addition of $CaSO_4$ and CaF_2 to the raw mix used to produce grey clinker.

Two types of mineralised clinker are currently produced at Aalborg, both in its 5500 t/d semi-wet, 2-stage separate line precalciner kiln, the details of which are published elsewhere [5,6]. The composition of the more normal clinker is given in Table 1, with SO_3 and F contents of 1.9% and 0.18%, respectively. A high C_3S clinker is also produced in which the F content is increased to 0.23% as shown in Table 2. Combination to between 1 and 2% free lime is achieved at approximately 1350°C which is about 100°C below that for the non-mineralised clinker. The bulk of the sulphate melt forms between 1000°C and 1200°C. Most of the alite forms fairly rapidly over a narrow temperature interval between 1150°C and 1200°C before any significant oxide melt formation begins. The rapid, early formation of alite is partly due to its greater stability at lower temperatures, and partly to the high rate of diffusion of Ca^{2+} through the low viscosity sulphate melt. About 2/3 of the sulphate occurs in a readily water soluble form, mainly as Ca-langbeinite, with the remainder incorporated in the silicate phases, with the highest concentrations occurring in the belite phase which significantly increases its hydraulic reactivity.

Low temperatures for clinker formation and relatively high partial pressures of O_2 prevent excessive volatilisation of the sulphates. Volatilisation of F is negligible. Ring formation and other volatile related kiln build-ups are therefore limited. Direct fuel savings are marginal at approximately 5%. Formation of NO_x in the kiln is reduced by up to 50%. Since the total NO_x emission from this type of kiln is also dependent on fuel derived NO_x formed in the calciners, the overall reduction in emission of NO_x can vary from 10 to 30% depending on the nitrogen content of the fuel.

MINERALISED CEMENT

The properties of cement based on all known mineralisers is beyond the scope of this paper and only mineralised cement prepared from clinker containing controlled levels of SO_3 and F will be discussed. The properties of mineralised cement are no different from those of normal Portland cement in terms of the dependency on the relative content of clinker minerals, fineness, ratio of soluble sulphate to C_3A, etc. The only difference lies in the modified reactivity of the silicate phases which gives longer setting times and faster strength development.

Setting Times

Setting times are increased with increased contents of F in the alite phase. Work by Renichi Kondo *et al* [7] indicates that this is due to the precipitation of CaF_2 on the surface of the alite crystals during initial hydration. The relationship between setting time and the fluorine content of the mineralised cement produced at Aalborg shown in Figure 1 corresponds to 250 minutes/%F. The setting times were determined from pastes prepared to normal consistency, according to EN-196, and do not necessarily correspond to the setting times of concretes which often contain water reducing and air entraining additives.

Studies at Aalborg Portland have shown that the difference in concrete setting times is approximately 30 minutes for the cements shown in Table 1 regardless of the type and dosage of the admixture. In other words, even though the admixture may increase the setting time of the concrete by up to 5 times that of the paste, the difference between concretes containing the mineralised and non-mineralised cements remains constant at approximately 30 minutes. In relative terms, then, the effect on concrete is not as great as would appear from the EN-196 setting times alone.

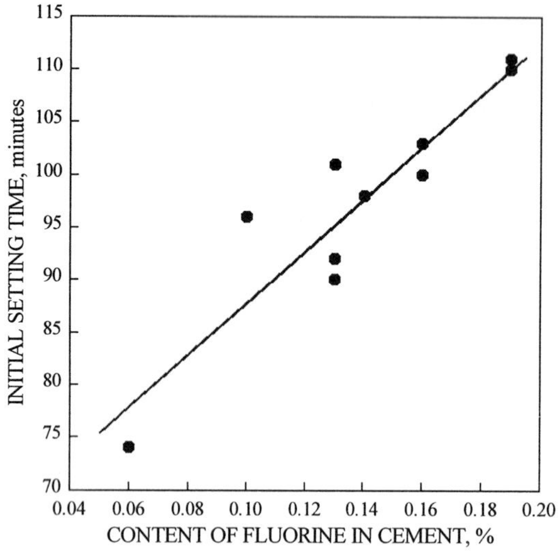

Figure 1 Relationship between initial setting time (EN-196) and content of fluorine in cement

Compressive Strengths

Apart from the benefits on clinker production discussed above, the main advantage of mineralisation lies in the increased reactivity of the cement resulting in a more rapid strength development. The increased reactivity is the result of increased solid solution of the mineralisers, i.e. S and F in the silicate phases, along with an additional incorporation of Al which this facilitates. An example of the enhanced strength development of mineralised cement compared to a similar, but non-mineralised cement, is shown in Figure 2. The strengths plotted are the average of two, and sometimes three separate EN-196 determinations. The chemical and physical characteristics of these cements are shown in Table 1.

The higher 7 and 28 day strengths are mainly due to the increased reactivity of the belite phase in the mineralised clinker. With the possible exception of the ferrite phase, which makes little contribution to strength development, belite is normally the least reactive phase, making relatively little contribution to 28 days strengths.

In mineralised clinker, however, a far higher proportion of the belite has reacted at 28 days, to the extent that increasing the ratio of C_2S to C_3S in the clinker (essentially lowering the LSF) results in comparable or often higher 28 days strength.

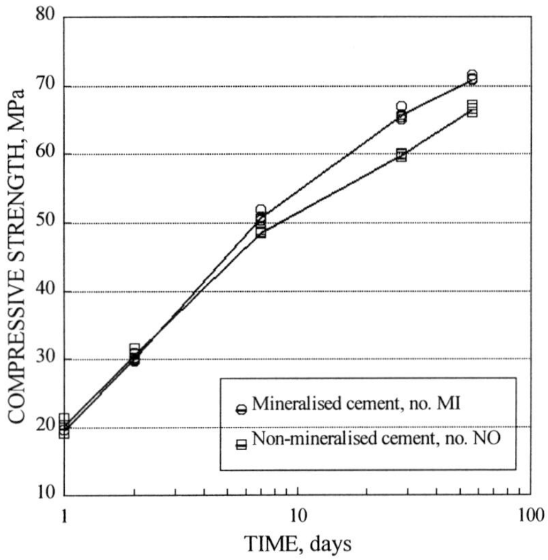

Figure 2 Compressive strength of mortars with mineralised cement no MI and non-mineralised cement no NO

Table 1 Clinker and cement data

		NON-MINERALISED CEMENT, no. NO	MINERALISED CEMENT, no. MI
CLINKER:	%	52.9	54.1
C_3S[1)]	%	20.1	20.0
C_2S[1)]	%	6.1	6.2
C_3A	%	11.6	10.8
C_4AF	%	1.36	0.98
Free lime	%	0.06	0.18
Fluorine	%	0.90	1.90
SO_3 in clinker	%	0.59	0.56
Na_2O eqv. acid sol.			
CEMENT:	m^2/kg	426	412
Fineness (Blaine)	%	2.4	2.5
45μm residue	%	2.94	2.91
SO_3 in cement	%	4.0	4.0
Limestone extender			

1) Bogue composition corrected for SO_3 and free lime

Although the early strengths, i.e. 1 and 2 day strengths, are essentially the same for the two cements shown in Figure 2, higher early strengths can be achieved by increasing the LSF of the mineralised clinker which is readily accomplished owing to its easier burnability. Strength development for this type of cement is shown in Figure 3 for the cements labelled MH420 and MH251 (ground to specific surface areas of 420 and 251 m^2/kg), with relevant chemical and physical data shown in Table 2. The results show, among other things, that very high early strengths can be achieved without grinding to excessive fineness. This allows the addition of a relatively inactive filler such as limestone [8]. The strength development obtained with a replacement of the clinker by 14% calcareous filler (cements MH420L and MH251L) are also plotted in figure 3. The cement labelled MH420L in Table 2 has been produced in Denmark since 1994, where it is the preferred cement for the pre-cast and concrete product producers. It is formally designated by ENV 197-1 as a PC 52.5 R, Portland limestone cement (type II/A-L).

For the cements to comply with the upper limit of 62 MPa for ENV 197, class 42.5 cements, the clinker fineness must be kept below 250 m^2/kg as shown by the strength plots in Figure 3. The presence of a relatively high content of fine filler (MH251L) eliminates the risk of bleeding which would normally be a problem at such a low fineness of the clinker. Compared with conventional PC 42.5 cement, savings achieved on grinding would be at least 20%.

For both filler cements reductions in fuel consumption are more or less the same as the level of clinker replacement, i.e. 14% in the examples shown. Reductions in NOx emissions are estimated at slightly more than 30%.

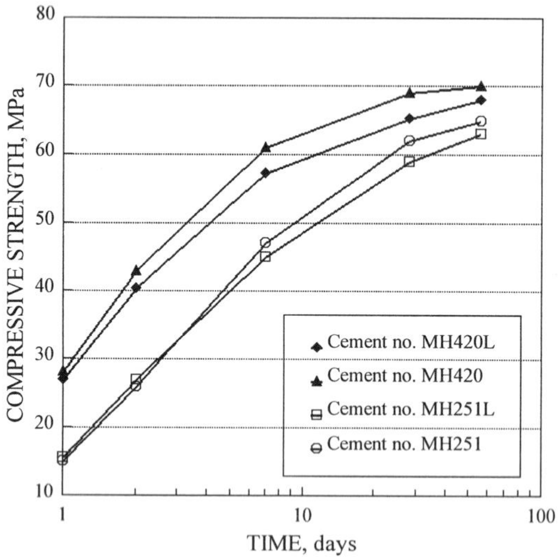

Figure 3 Compressive strength of mineralised cement with relatively high content of C_3S and with varied fineness and content of limestone filler, mortar strength determined according to EN-196

Table 2 Clinker and cement data.

		MINERALISED CEMENT			
		MH251	MH251L	MH420	MH420L
CLINKER	%	66.4			
C_3S[1)]	%	6.7			
C_2S[1)]	%	7.9			
C_3A	%	9.7			
C_4AF	%	1.80			
Free lime	%	0.23			
Fluorine	%	1.90			
SO_3 in clinker	%	0.59			
Na_2O eqv. acid sol.					
CEMENT	%	3.01	2.92	3.04	2.95
SO_3 in cement	m^2/kg	251	251	420	420
Fineness of clinker (Blaine)	%	19.0	17.3	2.1	2.0
45μm residue	%	0.0	14.0	0.0	14.0
Limestone filler					

1) Bogue composition corrected for SO_3 and free lime

Durability

Experience in Denmark to date has been that concrete produced from mineralised cement is equally durable to concrete produced from conventional Portland cements. This is documented in a report by the Danish Institute of Testing of Building Materials (DTI) [9], and in publications by the present authors [10,11].

Despite having many properties in common with the other halogens, including chlorine, fluorine does not cause corrosion of steel reinforcement due to the low solubility of CaF_2 in pore solutions found in concrete [12].

Another concern has been that clinker containing high sulphate contents may give rise to a type of secondary ettringite formation which may lead to deleterious expansion in much the same way as delayed ettringite formation (DEF). As noted in the section on clinker, however, most of the sulphate present in the clinker is in a readily soluble form which contributes to regulation of the initial set.

All but a negligible fraction of the remaining sulphate, which occurs in the unreacted silicates, is available for any further ettringite formation, and even here an invariable excess of Al_2O_3 precludes any ettringite formation whatsoever. Numerous experimental studies [10,13,14] have recently shown that expansion is unaffected by the level of sulphate in the clinker.

CONCLUSIONS

The production of mineralised clinker using controlled additions of fluorine and sulphates lowers burning zone temperatures by approximately 100°C. Although this results in fuel savings of only 5%, NO_x formed in the kiln is reduced by up to half, which in a precalciner kiln typically equates to an overall reduction in NO_x emission of approximately 20%.

In terms of fuel consumption and CO_2 emission the main advantage of producing mineralised cement lies in its greater hydraulic reactivity which allows the replacement of clinker by CO_2 neutral additions such as limestone.

REFERENCES

1. MOIR, G, K. Mineralised High Alite Cements. World Cement, 1982, Vol. 13, No. 10, pp 374-382.

2. MOIR, G, K. Improvements in the early strength properties of Portland cement. Phil Trans. Lond., 1983, A310, pp 127-138.

3. BORGHOLM, H, E, HERFORT, D, AND RASMUSSEN, S. A New Blended Cement Based on Mineralised Clinker. World Cement, 1995, Vol. 8, pp 27-33.

4. MOIR, G, K. Effect of the Use of Mineralisers, Modifiers and Activators. 9th ICCC, New Delhi, 1998, Vol. 1, pp 125-152.

5. BORGHOLM, H, E, AND NIELSEN, P,B. Ein neues Halbtrocken-Ofen-System fur 4000t/d im Aalborg Portland-Zementwerk. ZKG Inernatioanl, 1988, Vol. 41, No. 12, pp 595-600.

6. BORGHOLM, H, E. Comissioning the world's largest semi-dry process kiln system. World Cement, 1989, Vol. 20, No. 3, pp 72-79.

7. KONDO, R, DAIMON, M, SAKAI, E AND USHIYAMA, H. The Influence of Inorganic Salts on the Hydration of Tricalcium Silicate. J. appl. Chem. Biotechnol. 1977, 27, pp 191-197.

8. BORGHOLM, H,E, AND DAMTOFT, J, S. EUROPEAN PATENT, EP 0640 062 B1, 1993.

9. PETERSEN, E, J, AND HAUGAARD, M. New Cement Types: Evaluation of the Suitability of Basis-Cement (= Portland Limestone Cement Based on Mineralised Clinker) in Concrete designed for Moderate and Aggressive Environments. DTI Byggeteknisk Institut, March 21, 1994, In Danish.

10. HERFORT, D, RASMUSSEN, S, JØNS, E, AND OSBÆCK, B. Mineralogy and Performance of Cement Based on High Sulphate Clinker. ASTM Symposium on Internal Sulphate Attack on Cementitious Systems: Implications for Standards. In press, Cement, Concrete and Aggregates.

11. HERFORT, D, SØRENSEN, J AND COULTHARD, E. Mineralogy of Sulphate Rich Clinker and the Potential for Internal Sulphate Attack. World Cement, 1997, Vol. 28, No. 5, pp 77-85.

12. ESCUDERO, M, L, AND MACIAS, A. Corrosion of Reinforcing Steel in Mortar of Cement with CaF_2 as a Minor Component. Cement and Concrete Research, 1995, Vol. 25, No. 2, pp 376-386.

13. KELHAM, S. Effects of Cement Composition and Hydration on Volume Stability of Mortar. 10th ICCC, Gothenburg, 1997, Vol. 4, paper iv060.

14. TENNIS, S, BHATTACHARJA, S. AND KLEMM, W. Ambient Temperature Delayed Ettringite: A Reapraisal. ASTM Symposium on Internal Sulphate Attack on Cementitious Systems: Implications for Standards. In press, Cement, Concrete and Aggregates.

INVESTIGATIONS INTO THE INFLUENCE OF LIMESTONE ADDITIONS TO PORTLAND CEMENT CLINKER PHASES ON THE EARLY PHASE OF HYDRATION

J Stark

E Freyburg

K Löhmer

Bauhaus-Universitat Weimar

Germany

ABSTRACT. This paper is concerned with the influence of limestone powder on the hydration of cement. Differences in dilation and shrinkage of concretes made from cements with limestone additives prompted an investigation into the effects of the use of limestone in cement on the process and the products of hydration of the four major cement clinker phases. In order to do so, separately produced cement clinker phases were hydrated one by one with a variety of four different limestone powders together with sulfatic additives. By means of X-ray diffractometry for phase analysis monocarbonate (MC) was detected, proving the participation of the calcium carbonate added in the chemical reaction with both aluminatic phases C_3A and C_4AF. It showed that calcium carbonate has an influence on the heat of reaction and the conversion of the aluminatic phases as well as C_3S during the early age of hydration. These effects were noticeably increased solely by extending the specific surface of the limestone powder.

Keywords: β-C_2S, C_3A, C_4AF, C_3S, Cement clinker, Conduction calorimetry, Dolomite powder, Limestone powder, Monocarbonate, Thaumasite, X-ray diffractometry.

Professor J Stark is Director of the F.A.Finger-Institute of Building Material Sciences at the Bauhaus-Universität Weimar. The main fields of research are the use of binders, the durability of concrete, structural light-weight concrete, high-performance concrete, and historic mortars.

Dr E Freyburg is a Research/Teaching Fellow at the F.A.Finger-Institute of Building Material Sciences at the Bauhaus-Universität Weimar. His main research interests include the use of binders, the durability of concrete and especially the assessment of damaged concrete structures.

K Löhmer is a research assistant at the F.A.Finger-Institute of Building Material Sciences at the Bauhaus-Universität Weimar. She specialises on the use of binders, in particular on the improvement of the properties of cement.

INTRODUCTION

By German and European standards, except for CEM II/A-L and CEM II/B-L, the addition of limestone powder is permissible as "subsidiary component" in cements and as "inorganic substance" in plaster and mortar binders. In these cases limestone powder does not have to meet any requirements concerning its chemical reactivity.

From various investigations [1,4,5,6] it is known that the addition of even relatively small amounts -still ranging as subsidiary component- changes the processing features of cement. It had to be asked whether or not the use of limestone in quantities classified as " subsidiary components" according to ENV 197 part 1, i.e. about 5 %, accounts for any chemical reactions besides the purely physical effect of a filler. Therefore it was investigated if in the presence of very small amounts of limestone detectable chemical reactions do occur and how they might influence the process of hydration (heat and phases of hydration).

In order to be able to classify possible effects and to identify converted compounds the complex matter "cement" was abstracted as singular clinker phases.

BASIC MATERIALS AND COMPOSITION OF MIXTURES

The cement clinker phases C_3S, C_2S, C_3A and C_4AF were produced, processed and tested separately. In order to maintain comparable conditions it was defined that each cement clinker comprises of 100 % of the cement clinker phase investigated. So, the actual proportions of clinker and additives (or the stoichiometrically precise ratio) were not considered. Common to all mixtures was the addition of 6 % sulfatic substance, corresponding to an SO_3-content of 3.4 %. The sulfatic substance contained 50 % natural anhydrite and 50 % synthetically produced hemihydrate.

The reference sample contained nothing but one cement clinker phase each and sulfatic substance.

To all other samples 3 and 6 % limestone powder were added to respectively. As additives were used: two different types of limestone powder (A) and (B), type (B) of different fineness (Ba) and (Bb), and dolomite powder (C).

Table 1 shows the specific surface and the calcite/dolomite content of each powder used.

Table 1 Fineness and chemical purity of the stonepowders used

TYPE	SPECIFIC SURFACE [cm²/g]	CALCITE CONTENT
A	13,700	98 %
Ba	6,800	85 %
Bb	10,500	85 %
C	11,500	98 % (dolomite)

The value for the specific surface of the cement clinker phases ranged between 3,300 and 4,200 cm²/g.

By means of calorimetric testing only reference samples and type (A) mixtures containing 3 and 6 % limestone powder were analysed.

All mixtures were produced with a water-to-binder ratio of 0.5.

METHODS OF INVESTIGATION

Calorimetric measurements were run by means of a conduction calorimeter (DCA-06, Toni-Technik). Only the rate of heat evolution was taken under consideration. Each measurement lasted over a period of 72 hours.

The influence of the stone powder content was determined concerning the time and peak value at which a maximum in heat evolution is reached. Furthermore, these maxima determined served as an indicative to limit the hydration period prior to X-ray analysis.

Before being analysed by means of XRD, the samples have been mixed for 90 seconds and were then stored quasi-adiabaticly. The hydration was stopped by grinding the sample in isopropanol and subsequent rinsing with acetone. The samples were dried at 40 °C; to examine them a X-ray diffractometer was used ("D 5000", Siemens). The results were evaluated qualitatively and semi-quantitatively, the latter by measuring and comparing peak values of certain reflections.

The d-values of the reflections are listed in Table 2.

Table 2 d-values of certain reflections

COMPOUND	REFLECTIONS [nm] (d-value)
Cement clinker mineral	C_3S: 0.276; 0.218; β-C_2S: 0.219; 0.287;
	C_3A: 0.191; C_4AF: 0.181; 0.205; 0.725
Portlandite	0.49
Anhydrite	0.35
Calcite	0.209; 0.228; 0.303
Dolomite	0.289
Gypsum	0.761; 0.761; 0.428
Thaumasite	0.966
Ettringite / AFt-phase	0.973 / 0.56; 0.973
Monosulfate (MS) / AFm-phase	0.223; 0.893
Monocarbonate (MC)	0.378; 0.756 / 0.757
C_3AH_6 / $C_3(A,F)H_6$	0.23; 0.513 / 0.515
$C_xA_yH_z$ / $C_x(A,F)_yH_z$	0.287 - 0.289; 0.796 - 0.834

RESULTS AND INTERPRETATION

Conduction Calorimetry

With the C_3S samples an exothermal effect of 5 J/g·h was indicated within minutes after the samples were exposed to water which has to be explained with the moistening of the solid material. A second exothermal effect of about 20 J/g·h was observed after 10 to 12 hours. If limestone powder had been added the exothermal maximum occurred up to one hour earlier, as illustrated in Figure 1.

During the period of observation the β-C_2S samples did not show any exothermal effect besides the one due to moistening.

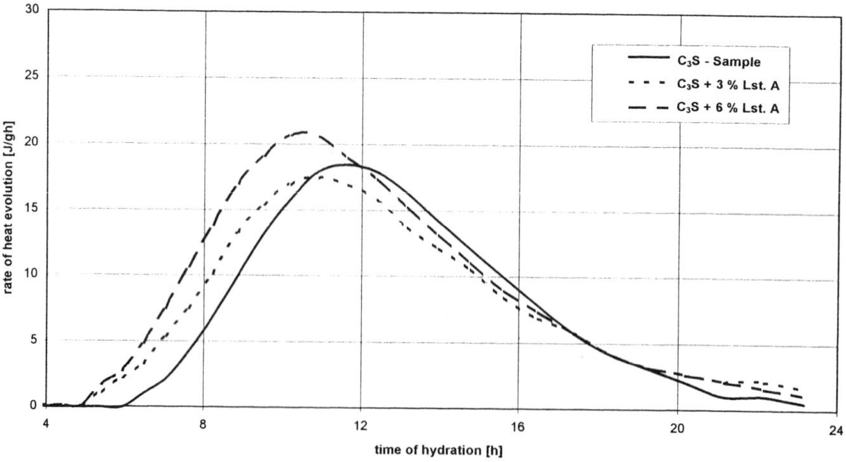

Figure 1 Course of heat emission of C_3S samples

All C_3A samples are of very high reactivity during the first minutes of hydration. The heat release of this effect was found to be reduced if limestone powder had been added. A slight acceleration is suggested.

Figure 2 illustrates the course of heat emission of C_3A samples during the first minutes of hydration.

The reduction in heat release is evident. This could be attributed to a "thinning" of the reactive substance i.e. the clinker phase with "inert" filler as well as to a formation of compounds involving calcium carbonate. The latter is supported by the results of the X-ray analysis (see next chapter).

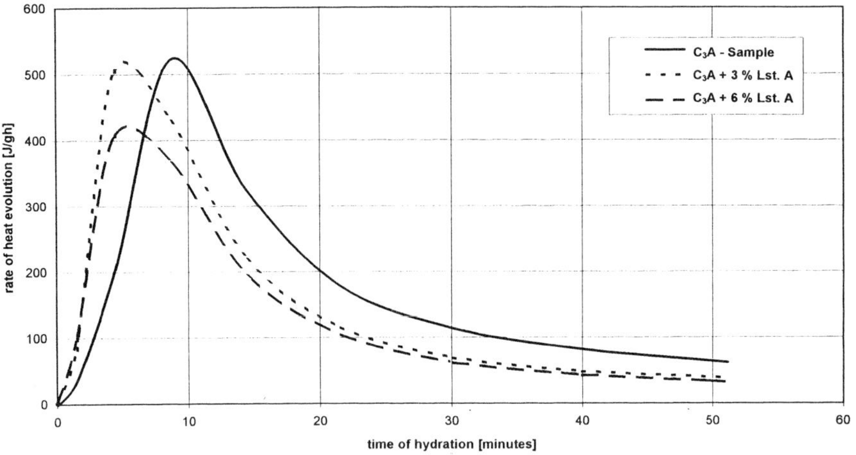

Figure 2 Course of heat emission of C_3A samples

Regarding C_4AF samples three exothermal effects were detected. The first one corresponds to the C_3A effect, but it occurs later, lasts longer and is less intense than the one of the iron-free calcium aluminate. A second and third effect was observed during the following five hours.

Figure 3 illustrates the course of heat emission of C_4AF samples containing no limestone and 3% of stonepowder respectively.

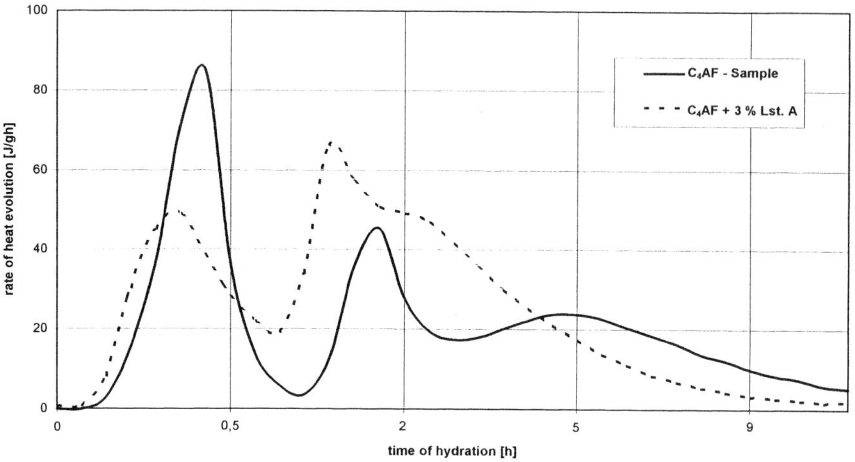

Figure 3 Course of heat emission of C_4AF samples

The addition of limestone powder prompted an acceleration of the reaction, so that all three effects occurred earlier. The first effect then was, as with the C_3A samples, less intense but in the following two effects the samples containing limestone tended towards a higher rate of heat release.

X-ray Diffractometry

The influence of limestone additives on the hydration of C_3S, as described in other publications, could be explained mineralogically by means of X-ray diffractometry. In the presence of limestone (3 and 6%) an intensified conversion of C_3S was observed, especially in the beginning.

A related acceleration in the formation of portlandite could not be established.

Figure 4 illustrates the different conversion of C_3S in samples containing no and 6% limestone powder respectively after two and 24 hours, taking the reflections of 2-θ=32.07 and 41.1 (d-value=0.276 and 0.218) as examples.

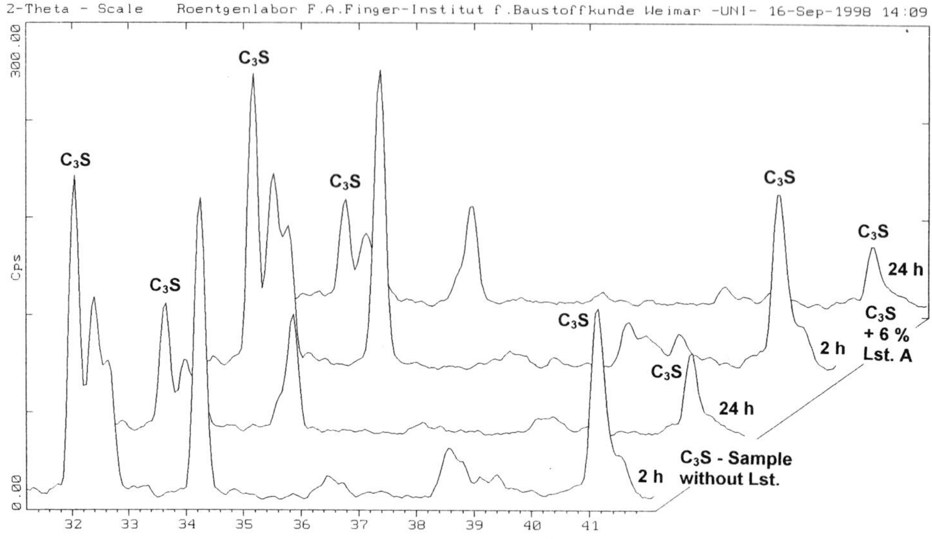

Figure 4 X-ray diffraction pattern of C_3S samples after two and 24 hours of hydration

Occasionally, some thaumasite was found; its presence was proven through analysis by means of SEM of samples hydrated for five days.

This suggests at least the possibility of a chemical reaction of siliceous and sulfatic phases with calcium carbonate, as discussed by some. Yet, it can not be excluded that atmospheric carbon dioxide penetrated into the fresh samples (small volume, large surface) and participated in a chemical reaction.

Because of the low reactivity of β-C_2S no influence of the limestone additive could be observed.

Besides the slight increase in the conversion of C_3A when limestone was added, the reaction of calcium carbonate with both aluminatic phases was established. Calcium aluminate monocarbonate hydrate as well as calcium ferrite monocarbonate hydrate was detected, its amount increasing in relation with the supply of more calcite.

Figure 5 is an X-ray diffraction pattern depicting the different formation of monocarbonate in varying C_3A mixtures after four days of hydration.

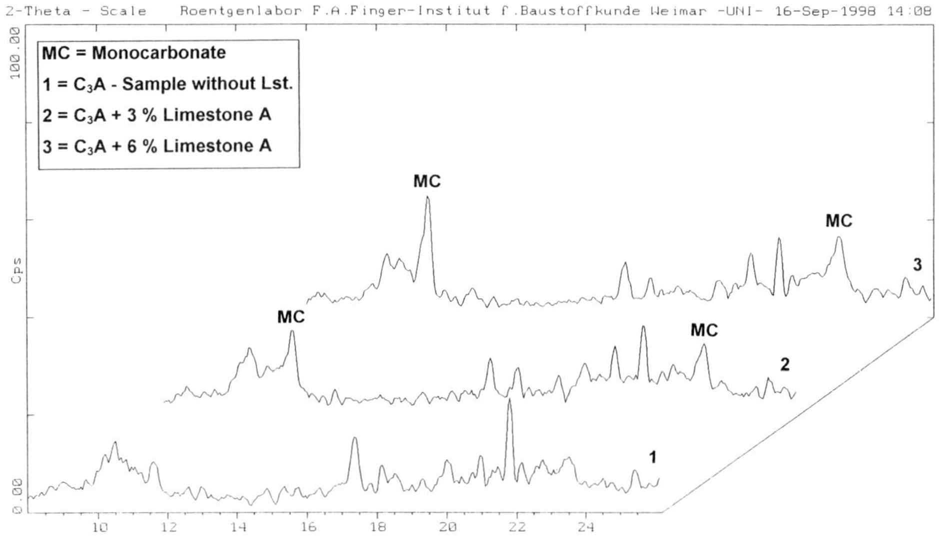

Figure 5 X-ray diffraction pattern of monocarbonate in C_3A samples after four days of hydration

There were no apparent differences between the two types (A) and (B) of limestone used. When dolomite powder (C) or the more coarse type (Ba) was used less monocarbonate was formed, especially in C_4AF samples, due to a lower reactivity of the additives.

Figure 6 illustrates that fact with the help of the monocarbonate reflections at 2-θ = 11.68 and 23.5 (d-value = 0.757 and 0.378).

The detection of monocarbonate in samples with dolomite stonepowder added was possible using the same d-value. In that, the peak changed into the shape of a "shoulder" at d = 0.757nm, as depicted in Figure 7.

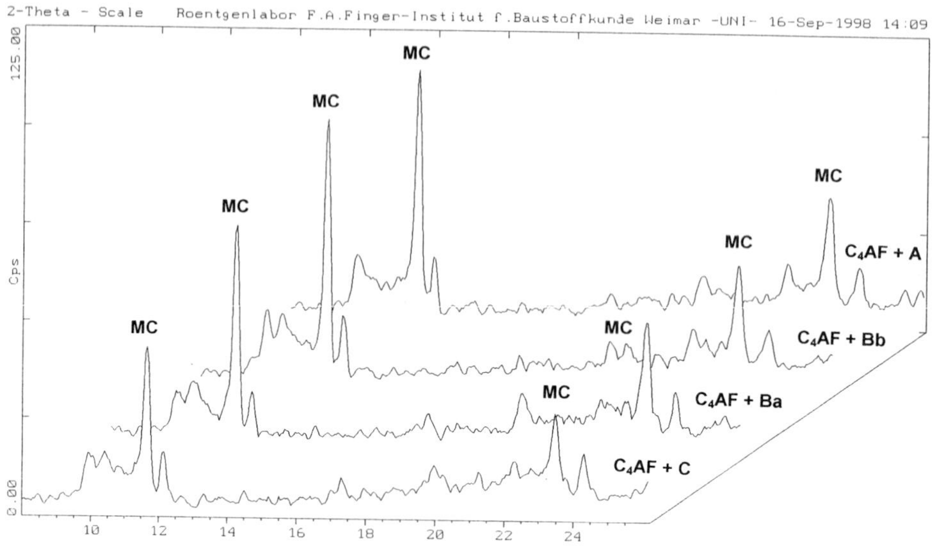

Figure 6 X-ray diffraction pattern of C_4AF samples with 6% stonepowder each after four days of hydration

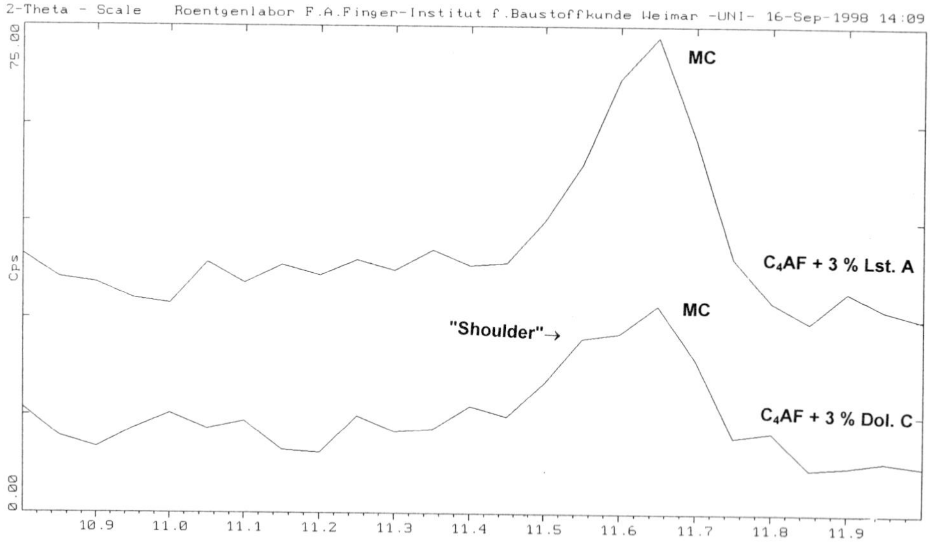

Figure 7 X-ray diffraction pattern depicting monocarbonate in C_4AF samples with limestone (A) and dolomite (C) powder

The above mentioned influence of limestone powder on the second and third exothermal effect during the hydration of C_4AF samples could not be investigated by means of phase analysis. Detection was possibly hindered by the formation of minor amounts of calcium aluminate ferrite sulfate hydrate phases as well as by not yet appropriately gentle enough ways of preparation.

CONCLUSION

It was established that the addition of 3 and 6% limestone powder indeed influences the early phase of hydration of C_3S, C_3A and C_4AF.

The finding that thaumasite is formed in an early stage by reaction of C_3S with the sulfatic substance and possibly with limestone powder (at least, the C_3S reaction is accelerated in the presence of limestone powder) may be exploited by purposefully producing this "rising" mineral to control the shrinkage of concrete from early on comparable to ettringite.

The conjecture that the thaumasite found originates from the reaction of the source materials and is not formed during carbonation is currently being reviewed.

If existing unspecified concrete has to be identified with respect to its original binding agent, these results indicate that monocarbonate may have been formed primarily, not only by carbonation of ettringite.

REFERENCES

1. FREYBURG, E, STARK, J, SPRINGENSCHMID, R AND MANGOLD, M. Einfluß der Zementeigenschaften auf die Hydratationswärmeentwicklung und die Reißneigung des Beton. Wissenschaftliche Zeitschrift der HAB Weimar, 1993, Vol. 39, No. 3, pp 149-155.

2. INGRAM, K, D, AND DAUGHERTY, K, E. Limestone additions to Portland Cement : Uptake, Chemistry and Effects. 9. International Congress on the Chemistry of Cement (ICCC), 1992.

3. PERLT-SCHUSCHNIG, E. Einfluß von Kalksteinmehl auf die Reißneigung von jungem Beton. HAB Weimar, Master Thesis 1993.

4. SCHIEL, T. Eigenschaften von jungem Beton bei Einsatz von Kalksteinmehl, Silicastaub und unterschiedlichen Zuschlägen. HAB Weimar, Master Thesis 1994.

CEMENT SULFATE RESISTANCE EVALUATION BY THE DRAFT EUROPEAN STANDARD OF DIRECT TESTING METHOD

S Peukert

A Garbacik

S Chladzyñski

IMMB Cement Department

Poland

ABSTRACT. In the paper the problems of determination of the cement resistance to attack by sulfates have been considered. A method of testing proposed by the draft European Standard prENV 196-10 was examined. For testing cements covered in the new Polish Standard PN-B-19701 elaborated according to the European Standard ENV 197-1 were used. Various Portland cements CEM I, Portland blended cements CEM II, blastfurnace cements CEM III and pozzolanic cements CEM IV were examined. It was stated that the test method proposed by prENV 196-10 can be applicable for estimation of sulfate resistance of cements. Diversified expansion ratios for different cements as a result of attack by sulfate solution were obtained. Very high sulfate resistance of CEM I containing 3,6% C_3A, CEM III/B and CEM IV/B was confirmed.

Keywords: Cement, Standardization, Corrosion, Sulfates.

Professor Dr Eng Stanislaw Peukert is Scientific Director of Institute for Mineral Building Materials Cracow Division. Specialist in technology of common and special cements. Author of many patents and applications of new technologies in Polish cement plants. He is the author of over 100 publications on properties and application of cement, including the co-author of MPz class cement.

Dr Eng Albin Garbacik is Manager of Cement Department at IMMB. Specialist in technology of cement and concrete. Author of tens of publications on cement and concrete and their applications into the building materials industry and applications of new technologies in Polish cement plants.

MSc Eng Slawomir Chladzyñski is a research worker in Cement Department at IMMB. He is a candidate for doctor´s degree, specialing in cement and concrete technology, especially sulfate corrosion in concrete and is author of several publications.

INTRODUCTION

The rules for elaboration of European Standards are defined by the Directives of European Commission. Product signed by the CE has to meet specified technical, safety, health and ecological requirements determined by suitable standards and instructions. The principles of standardization for building materials are determined by Construction Products Directive (CPD) 89/106/EEC.

Standard requirements for cement and concrete belong to the group of so called material standards. The main aim of every material standard is to characterise material in the most precise and possible to control way. The characteristics of material is given by the group of descriptive features determining properties of material composition and its production conditions. The supplement of these features are the performance criteria determining possibilities of material application fields. In case of cement standards instance, the performance criteria are connected with proper durability of concrete.

According to the mentioned Directive 89/106/EEC European Standards for cement and concrete should be elaborated taking into consideration, first of all, the requirements concerning performance criteria. This is the main condition of creating the harmonized standards hEN for these materials. This condition is the base for the program of elaboration of testing methods for additional performances criteria for cement and concrete not covered in the existing standards.

It is especially important from the point of view of requirements of the prestandard 197-1 and final harmonized standard hEN 197-1 for common cements [1]. This standard takes into account very big variety of kind of cements and qualities and quantities of components used for their production. The standard makes the users afraid of the worse durability of concrete made from some cements or their unfavourable action in concrete on environment. So, the necessity of more detailed assessment of new kind of cements on account of durability features of concrete is especially stressed.

For the present CEN elaborates the standardization procedures for testing methods for the following additional performance criteria of cement and concrete:

- sulfate resistance,
- heat of hydration,
- resistance to frost,
- resistance to de-icing agents,
- concrete reinforcement corrosion,
- resistance to carbonation,
- resistance to alkali-aggregate reaction,
- leaching of heavy metals, aluminium and calcium,
- radioactivity and radon emission.

From the mentioned above properties the evaluation of cement resistance to sulfate attack is of especially significant meaning. An evaluation in this subject by the direct method of testing can be applied to assess the durability and resistance of concrete to corrosive action of environment.

At present there is a lack of direct standardized method of testing cements to attack by sulfates. Long-term methods are unpractical taking into account long measuring periods. On the other hand, accelerated methods are not verifiable due to lack of repeatability and reproducibility of results. The actions co-ordinated by CEN are intended to elaborate of a new uniform method of cement testing of sulfate resistance.

In the paper the results of testing of cement resistance to sulfate attack according to the procedure given in the standard prENV 196-10 [2] were presented. Cements covered in the new Polish Standard PN-B-19701 [3] elaborated according to the European Standard ENV 197-1 were investigated. Cements with potentially diversified resistance to sulfates were investigated.

EXPERIMENTAL DETAILS

Description of Testing Method

A project of the European Standard prENV 196-10 covering methodology of testing of cement resistance to sulfates elaborated by CEN is based on the French Standard P 18-837 [4]. The standard gives a procedure of testing of cement mortar resistance to deterioration caused by the attack of sodium sulfate or sea water solution. The procedure is based on the determination of changes of length for mortar samples sized 20×20×160 mm. The samples are moulded from the standard mortar prepared according to EN 196-1. The mortar is compacted using standard jolt-moulding machine, but with the number of jolts decreased to 10. Mortar is compacted in the mould according to EN 196-1 adapted to simultaneous moulding of 6 prisms with studs to measure the length changes in a comparator. The sum of the masses of jolt-moulding machine should meet the requirements of the standard EN 196-1. Samples in moulds are cured in a moisture atmosphere for 24 hours and then demoulded and stored in distilled water. After 28 days the length of the samples is measured and then 3 prisms are placed in Na_2SO_4 solution or artificial sea water and 3 control prisms are placed in the distilled water. Length changes of samples are measured every 4 weeks up to one year.

The sulfate solution should have a concentration of 16±0,5 g SO_4 per litre and should be prepared by adding of pure Na_2SO_4 or $Na_2SO_4 \cdot 10H_2O$ to the distilled water. Sulfate or sea water solution should be changed after every 28 days. During storing of control samples the distilled water should not be changed, and a constant level of water in container is to be kept.

The results of measurements of prism length changes in aggressive solution should be given in relation to control samples stored in distilled water. The object of evaluation of expansion is the average change of length calculated for the difference between corresponding pairs of measurements for samples stored in the sulfate or sea water solution and for samples stored in distilled water.

Materials

Cements according to the standard PN-B-19701 with potentially different sulfate resistance were used for investigations. A criterion to predict high sulfate resistance of cements was the low concentration of C_3A in Portland cements CEM I, high slag concentration in blastfurnace cements CEM III and high fly ash content in pozzolanic cements CEM IV. According to the Polish Standard PN-B-19705 [5] regarding the sulfate resistant cements, the following cements were investigated:

- Portland cement CEM I with C_3A=3,6% content,
- blastfurnace cement CEM III/B with slag content of 75%,
- pozzolanic cement CEM IV/B with fly ash content of 38%.

The comparison materials for investigation were cements with potentially lower resistance to sulfates; so that, Portland cements CEM I with higher C_3A content; Portland-composite cements CEM II with the fly ash and/or slag content up to 33% and blastfurnace cement CEM III/A containing 56% of slag.

Characteristics of the investigated cements are presented in the Table 1.

Table 1 Characteristics of investigated cements

CEMENT TYPE	PHASE COMPOSITION OF CEMENT CLINKER, %				BLAINE SPECIFIC SURFACE	WATER DEMAND	CEMENT COMPOSITION weight %				COMPRESSIVE STRENGTH MPa		
	C_3S	C_2S	C_3A	C_4AF	cm^2/g	%	Clinker K	Slag S	Fly Ash V	Gypsum	2d	7d	28d
CEM I 42,5	60,2	16,1	3,7	14,1	2760	23,0	96,0	—	—	4,0*	13,7	32,7	49,4
CEM I 52,5 R	66,7	13,4	9,9	6,9	4530	27,5	95,0	—	—	5,0	33,9	50,8	56,4
CEM I 42,5					2760	26,0	95,0	—	—	5,0	21,3	35,2	44,5
CEM II/A-S 42,5					3250	25,5	76,0	19,0	—	5,0	19,8	35,3	46,1
CEM II/B-S 32,5R					3580	24,0	61,8	33,2	—	5,0	13,6	30,4	44,3
CEM II/A-SV 42,5	58,8	16,4	13,7	8,2	2980	26,5	76,0	9,5	9,5	5,0	18,5	35,0	46,6
CEM II/B-SV 32,5R					2980	26,0	61,8	16,6	16,6	5,0	13,6	23,9	38,9
CEM III/A 42,5					4500	27,0	37,6	56,4	—	6,0	14,9	34,8	53,5
CEM III/B 32,5					5020	26,5	18,8	75,2	—	6,0	5,0	25,6	42,9
CEM I 42,5					2760	25,0	95,0	—	—	5,0	20,9	36,0	46,2
CEM II/A-V 42,5	63,8	15,2	9,3	7,0	3410	26,5	76,0	—	19,0	5,0	21,8	35,0	48,1
CEM II/B-V 32,5R					3600	27,0	61,8	—	33,2	5,0	13,8	24,6	37,2
CEM IV/B 32,5					5220	28,5	57,0	—	38,0	5,0	12,3	21,7	37,3

*Anhydrite

RESULTS

The cements characterised in the Table 1 were investigated according to the methodology previously described by using of sodium sulfate solution as the aggressive medium. The results are presented in Figures 1-4.

The length changes in time for the individual group of cements are shown graphically and given as the absolute values. The characteristics of the following group of cements were compared:

- cements CEM I with various C_3A content - Figure 1,
- cements CEM I, CEM II, CEM III with increasing slag content - Figure 2,
- cements CEM I, CEM II, CEM III with increasing fly ash content - Figure 3,
- cements CEM I, CEM II with increasing slag and fly ash content - Figure 4.

CURING TIME (WEEKS)	CEM I 42.5 (3,6 % C3A)	CEM I 42.5 (8,8 % C3A)	CEM I 52.5 R (9,4 % C3A)	CEM I 42.5 (13,0 % C3A)
0	0,000	0,000	0,000	0,000
4	0,006	0,010	0,002	0,023
8	0,008	0,023	0,015	0,323
12	0,006	0,038	0,048	destr.
16	0,023	0,119	0,210	
20	0,044	0,298	0,581	
24	0,077	0,648	1,356	
28	0,133	1,033	destr.	
32	0,235	1,379		
36	0,375	1,781		
40	0,542	2,22		
44	0,721	destr.		
48	0,881			
52	1,040			

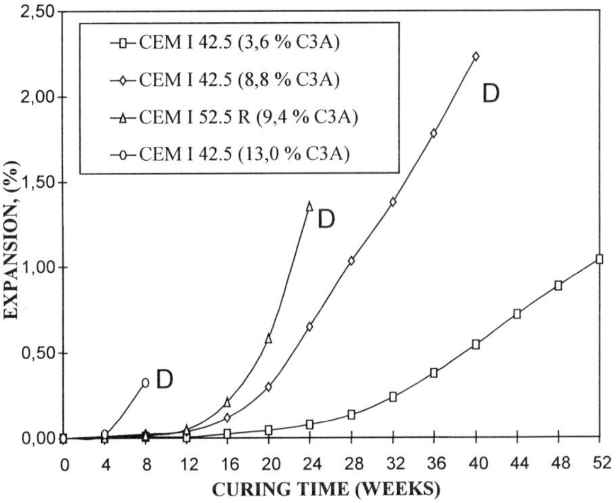

Figure 1 Expansion of Portland cements

CURING TIME (WEEKS)	CEM I 42.5 (13 % C_3A)	CEM II/A-S 42.5	CEM II/B-S 32.5 R	CEM III/A 42.5	CEM III/B 32.5
0	0,000	0,000	0,000	0,000	0,000
4	0,023	0,013	0,008	0,006	0,000
8	0,323	0,079	0,048	0,008	0,000
12	destr.	0,427	0,173	0,015	0,000
16		2,265	0,448	0,019	0,000
20		destr.	0,885	0,031	0,004
24			1,406	0,038	0,004
28			2,140	0,060	0,004
32			destr.	0,096	0,004
36				0,142	0,008
40				0,198	0,004
44				0,246	0,015
48				0,310	0,008
52				0,417	0,006

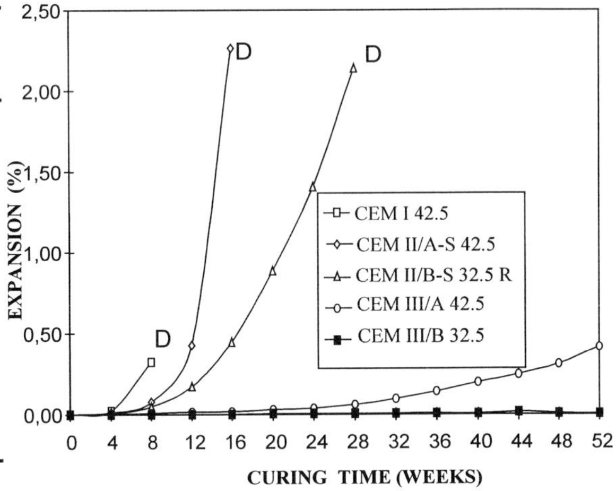

Figure 2 Expansion of slag cements

CURING TIME (WEEKS)	CEM I 42.5 (8,8 % C_3A)	CEM II/A-V 42.5	CEM II/B-V 32.5 R	CEM IV/B 32.5
0	0,000	0,000	0,000	0,000
4	0,010	0,000	0,000	0,000
8	0,023	0,002	0,002	0,002
12	0,038	0,008	0,006	0,002
16	0,119	0,017	0,010	0,002
20	0,298	0,021	0,013	0,004
24	0,648	0,025	0,017	0,004
28	1,033	0,027	0,019	0,006
32	1,379	0,031	0,023	0,006
36	1,781	0,035	0,025	0,004
40	2,229	0,042	0,027	0,004
44	destr.	0,050	0,029	0,004
48		0,058	0,029	0,004
52		0,067	0,029	0,006

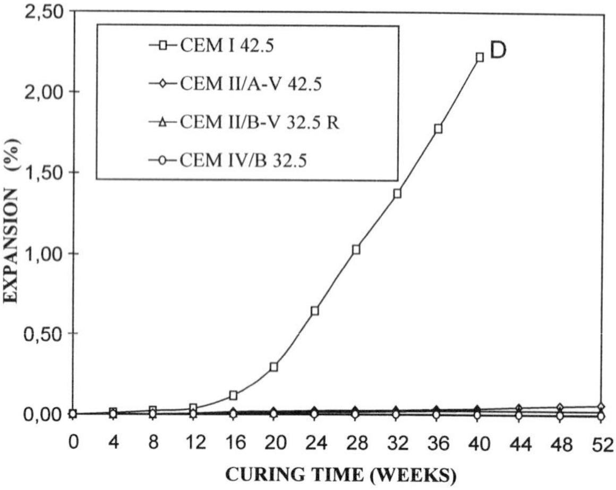

Figure 3 Expansion of fly ash cements

CURING TIME (WEEKS)	CEM I 42.5 (13 % C_3A)	CEM II/A-S 42.5	CEM II/B-S 32.5 R	CEM II/A-SV 42.5	CEM II/B-SV 32.5 R
0	0,000	0,000	0,000	0,000	0,000
4	0,023	0,013	0,008	0,017	0,008
8	0,323	0,079	0,048	0,058	0,013
12	destr.	0,427	0,173	0,250	0,017
16		2,265	0,448	1,402	0,021
20		destr.	0,885	destr.	0,027
24			1,406		0,040
28			2,140		0,058
32			destr.		0,081
36					0,110
40					0,148
44					0,185
48					0,235
52					0,302

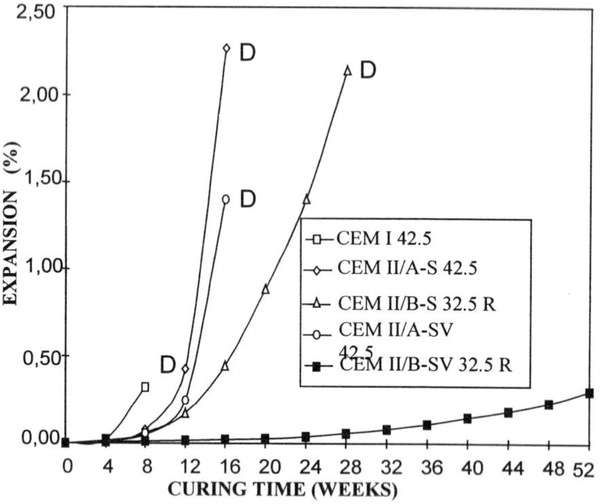

Figure 4 Expansion of Portland blended cements

DISCUSSION

Investigations of cement resistance to sulfate attack were carried out according to the methodology given in the European prestandard prENV 196-10 [2]. The method is based on the estimation of corrosive action of sulfates on the cement paste by measurements of expansion of cement mortar samples stored in Na_2SO_4 solution. The methodology takes into account the basic criteria of deterioration processes of cements in a sulfate solution. Expansion and destruction processes are caused by deteriorating influences on the hardened cement paste of reaction products of SO_4^{2-} ions with cement paste constituents. In Na_2SO_4 solution the main reactions concerning expansion are:

$$3CaO{\cdot}Al_2O_3+3Na_2SO_4+3Ca(OH)_2+32H_2O=3CaO{\cdot}Al_2O_3{\cdot}3CaSO_4{\cdot}32H_2O+6NaOH \quad (1)$$
$$Ca(OH)_2+Na_2SO_4+2H_2O=CaSO_4{\cdot}2H_2O+2NaOH \quad (2)$$

The mentioned above reactions proceed with large increase of volume on account of forming of expansive gypsum and ettringite. Potentially, increased process of destruction will have place in paste from cements rich in C_3A and high-basic cements generating large quantities of $Ca(OH)_2$ during hydration process.

Assuming a diffusion character of sulfate corrosion, the increased rate of process is established by high concentration of SO_4^{2-} ions, low degree of compacting of samples and high surface to volume ratio of the prisms. Therefore, a quantity evaluation of corrosion process is possible in the period of time not exceeding one year. Procedures of the test method should provide the suitable degree of repeatability and reproducibility of results.

The methodology of investigations of cement sulfate resistance according to the standard prENV 196-10 given in the paper confirms the correctness of assumptions of testing procedures of method and evaluation conditions of results. The results obtained for the tested cements are conforming with the potential sulfate resistance of cements predicting from their phase composition.

Portland cements CEM I (Figure 1) show very clear dependence of expansion rate on C_3A content in cement. Cements with maximum C_3A content occurring in Poland (about 14%) become destroyed already after 4 months of storing in the aggressive solution.

Correspondingly, cement CEM I with C_3A=3,6% content produced in Poland as the so called „bridge cement" with high sulfate resistance does not become destroyed even after 13 months. Therefore, the criterion of cement CEM I durability related to the restricted C_3A content in cements with high sulfate resistance was confirmed during the direct method evaluation.

Assumed high diversification of sulfate resistance was confirmed for cements with different slag and/or fly ash addition, too. Cements with increasing slag content show lower expansion (Figure 2). The expansion of cement based on high C_3A clinker and owing 19 to 76% of slag additions is lowered. For the blastfurnace cement CEM III/A containing 56% of slag and especially for cement CEM III/B containing 76% of slag the slight linear changes of samples were obtained, which allows for considering them to the group with high sulfate resistance. Such testing results find confirmation in the destruction mechanism of slag containing cements. A high slag content in the cement lowers the C_3A content and the expansion effect connected with ettringite formation. A similar influence on the expansion is exerted by the limited $Ca(OH)_2$ content formed from the clinker phases hydration and bonded partially by slag into the calcium silicate hydrates.

The property of increased sulfate resistance described for slag cements was confirmed for fly ash cements (Figure 3). The effect of positive influence of fly ash in cement on the sulfate resistance clear. The pozzolanic cement CEM IV/B containing 38% of fly ash shows a negligible expansion close to blastfurnace cement CEM III/B containing 75% slag.

A higher effectiveness of fly ash compared to slag on the limiting of sulfate expansion and destruction of cements clear is confirmed by the data compiled in Figure 4. It can be interpreted by the more effective bonding of $Ca(OH)_2$ during the pozzolanic reaction in fly ash cement pastes and by a more advantageous, from the sulfate resistance point of view, formation of the microstructure of paste owing to the generation of C-S-H gel .

CONCLUSIONS

1. The method of testing of cement sulfate resistance presented in the paper, being the object of the European Standard project prENV 196-10 is based on the evaluation of linear changes of cement mortar samples stored in SO_4^{2-} solution.

2. The results of testing presented in the paper confirm the usability of applied method for the direct evaluation of cement sulfate resistance. According to descriptive features of cements with different phase composition and differential quantity of additives a high differentiation of the sulfate resistance for individual cements was obtained.

3. The results of determinations using the method according to the prENV 196-10 confirmed the high resistance to sulfate attack for Portland cement with low C_3A content, blastfurnace and pozzolanic cements with correspondingly high slag and fly ash content.

REFERENCES

1. ENV 197-1 „Cement - composition, specifications and conformity criteria. Part 1: Common cements."

2. prENV 196-10 „Determination of the resistance of cements to attack by sulfate solution or by sea water."

3. PN-B-19701 „Cement. Cementy powszechnego u¿ytku. Sk³ad, wymagania i ocena zgodnoœci."

4. P 18-837 „Essai de tenue a l'eau de mer et/ou a l'eau a haute teneur en sulfates."

5. PN-B-19705 „Cementy specjalne. Cement portlandzki siarczanoodporny."

THEME TWO:

PORTLAND CEMENT-BASED BINDERS

Keynote Paper

PORTLAND CEMENT-BASED BINDERS - CEMENTS FOR THE NEXT MILLENNIUM

M Moranville-Regourd

Ecole Normale Supérieure de Cachan

France

ABSTRACT Cements for the next millenium will be largely composite cements which are first energy-saving materials. Blended cements include either hydraulic products like blastfurnace slag and calcareous fly ash, or pozzolanic materials as siliceous fly ash, burnt clay, metakaolin, silica fume or limestone filler. Properties of Portland cement-based materials are presented through their chemical composition, rate of hydration, hardened paste microstructure, pore distribution, durability, and uses.

Beside these usual and normalised products, industrial modified cements appeared in the last decades. MDF, DSP, CBC RPC resulting from the chemistry and microstructure monitoring have started to be used in new design construction like high towers, long bridges and tunnels.

Keywords: Composite cements, Modified Portland cements, Characterisation, Hydration, Durability.

Professor Micheline Moranville-Regourd is Honorary Professor of Materials Science and member of the Civil Engineering and Environment Research Group in the Laboratory of Mechanics and Technology at Ecole Normale Supérieure de Cachan, France and Guest Professor at the University of Sherbrooke, Canada. Her main fields of research are materials characterisation, cement hydration and durability of concrete.

INTRODUCTION

The predicted cement consumption for the year 2000 is 1.625 million tons extended to 1.835 million tons in 2005 [1]. Portland cement was the main hydraulic binder of the 20th century because it offered fast setting and hardening more appropriate to modern construction speeds. However one ton of Portland cement needs 1.5 tons of raw material and 3530 kJ, and releases one ton of CO_2 beside SO_2 and NO_x emissions in the atmosphere. New supplementary cementing materials were used in blended cements during the last two decades. Also novel cements and cement composites were elaborated [2, 3, 4] responding to the Earth Summit in Rio de Janiero in 1992 which stated a sustainable economic development with respect to the earth's ecosystems. These new products occurred as results of the chemistry and microstructure monitoring. They improve concrete properties and are potential for innovative applications.

This paper will present Portland-based materials and novel cement composites for the next millenium.

PORTLAND CEMENT-BASED MATERIALS

Usual and normalised Portland cement-based materials in Europe [5] are categorised as CEM II/A and CEM II/B containing respectively 6-20 and 21-35 % by mass of mineral additions. Natural or artificial fillers able to improve the workability can be used and up to 5% by mass. Calcium sulfate is added to the clinker at its optimum determined by the cement producer. All the mineral additions are either interground with the clinker or mixed with the Portland cement, at the cement plant.

There are three strength classes CEM II 32.5, 42.5, and 52.5 which correspond to the minimum characteristic strength (in N/mm² or MPa) at 28 days, the average strength being respectively 45, 55, and 65 MPa. A subclass R includes an earlier strength i.e. 12 MPa for 32.5 R and 18 MPa for 42.5 R, at 2 days. Nine cements are CEM II. They contain either blast furnace slag S, silica fume D, natural pozzolan P, industrial pozzolan Q, siliceous fly ash V, calcareous fly ash W, burnt shale T, limestone filler L, or a mixture of them. The CEM II/A (D) 52.5 excepted, all Portland cement-based materials are CEM II 32.5 or 42.5 or 52.5.

Portland blast-furnace slag cements

Slags are by-products of the metallurgical industry. Iron blast-furnace slags (BFS) have latent hydraulic properties when they have been rapidly cooled from a molten stream at 1400-1500°C. Granulated slags quenched with water or pelletized slags quenched with air, appear as angular grains in a glassy state mainly. The glassy part of a slag is characterised as a large halo in X-ray diffraction patterns peaking at around 0.3 nm. Crystalline phases are dendritic merwinite and melilite, quartz and calcite. The slag composition can vary from one steel manufacture to another one due to the metallurgical requirements. Iron slags contain on average 40% CaO, 35 % SiO_2, 12% Al_2O_3, 8% MgO, 1% S^{2-} and 1-4% Fe_2O_3-FeO, MnO, TiO_2, Na_2O and K, to be compared to Portland cement clinker with on average 65% CaO, 21% SiO_2, 5% Al_2O_3, 3% Fe_2O_3, 1.5% MgO, 0.6% K_2O, 0.3% Na_2O and 0.3% TiO_2. So slags are more siliceous than Portland cement clinker.

Acid-base considerations have been used to correlate the chemical composition with hydraulic reactivity. The reactivity of slags has also been related to the amount and structure of glass, minor elements as Na, K, S, P, Ti, microheterogeneities and sub-micro-crystals of melilite and merwinite. Slags to be used in CEM II have to follow physical and chemical specifications [5] concerning the amount of glass (>2/3 by mass), the CaO + MgO + SiO_2 composition as at least 2/3 of the sum C + M + S by mass and the ratio (C + M)/ S>1. Slags hydrate slowly. In Portland slag cements, slags are activated by the high alkalinity of the solution [6] due to the hydration of calcium silicate giving $Ca(OH)_2$, and by the heat liberated by C_3S and C_3A hydration. Other chemical activators are soda, sodium carbonate, sodium silicate and calcium chloride and calcium sulfate as gypsum. A thermal treatment also accelerates the slag hydration. Slags first dissolve incongruently as shown by ESCA, Electron Spectroscopic Chemical Analysis. Their hydrates are the same as Portland cement hydrates i.e. C-S-H and sulfoaluminates, ettringite AFt and monosulfoaluminate AFm. However the slag C-S-H is less crystalline but richer in Mg and Al than the Portland C-S-H [7]. The pseudomorphic hydrate surrounding slag grains contains a zoned texture with an outer layer composed of a granular C-S-H, a lamellar layer of a hydrotalcite-like composition, $Mg_6\ Al_2\ CO_3\ (OH)_{16}\ 4\ H_2O$, and a foil like inner structure observed under TEM (Figure 1) [8].

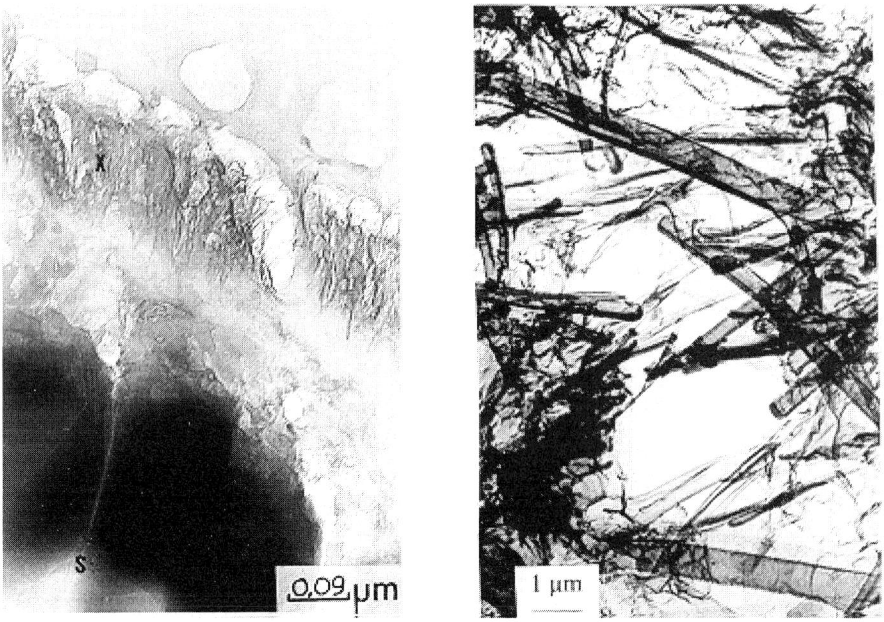

(a) (b)

Figure 1 Transmission electron micrographs

(a) foil-like C-S-H given by a glassy slag. Courtesy of Z. Andriambololona [8]

(b) fibrillar C-S-H given by C_3S of a Portland cement

Two processes appear to be involved (1) nuclear growth which concerns calcium silicate C-S-H and carbonate and (2) restructuring of a residual Si-Al hydrated glass by fixation of Ca and Mg. NMR spectra of the foil-like inner structure shows that the alumino-silicate chains get longer when the proportion of slag increases. The change from fibrillar Portland C-S-H to the foil-like slag C-S-H results in a less interconnected capillary pore structure and consequently a lower permeability [9]. The slag C-S-H contains less calcium than the Portland cement C-S-H. The C/S ratio is 1.7 in a Portland cement CEM I and 1.5 in a CEM II with 30% slag.

In a Portland slag cement, the early strength is mainly due to the fineness of clinker and later strengths to the slag fraction. Due to their different grindabilities, the intergrinding of the clinker and the slag to a high degree of fineness gives a clinker which is very finely ground and a slag which is still coarse. Separate grinding of the slag and the clinker will provide a much finer slag. Slags have been ground at fineness as high as 600 to 1200 m^2 per kg. The early hydration of these fine slags has given nearly amorphous C-S-H and ettringite in thin needles (Figure 2). Granulated slags separately ground increased early strengths.

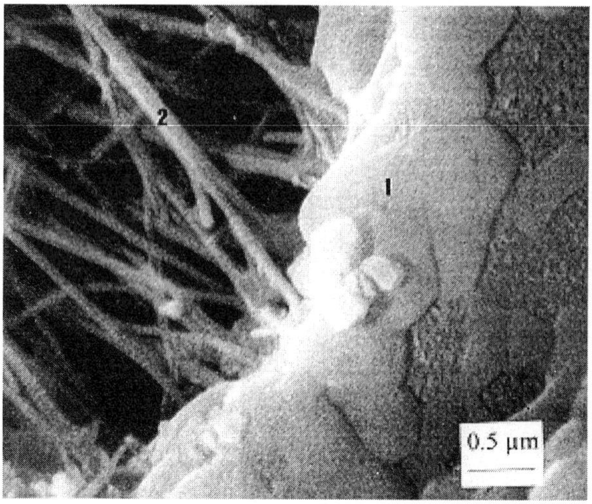

Figure 2 Scanning electron micrograph of a fine slag grain hydrated 7 days (1) like-amorphous C-S-H, (2) thin needles of ettringite

Portland silica fume cement

Silica fume (D) also called condensed silica fume or microsilica is a by-product of the electrometallurgy of silicon alloys. It occurs as spherical particles of 0.1 – 0.2 µm on average size (Figure 3) with a large BET surface area (20 000 m^2/kg) and a very low density (200 kg/m^3). The chemical composition depends on the composition of the raw material and type of furnace. However it is a highly siliceous product i.e. more than 90% in Si and Fe-Si alloys. As shown by X-ray diffraction, silica fume is an amorphous material characterised by a halo peaking at 0.4 nm (Figure 4).

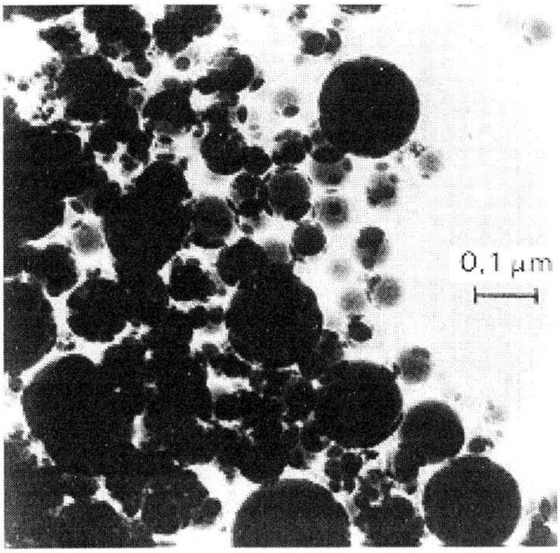

Figure 3 Transmission electron micrograph of silica fume particles

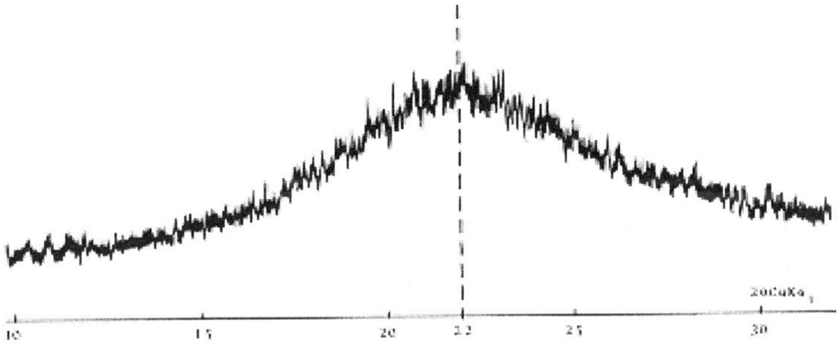

Figure 4 X-ray diffraction pattern of a totally amorphous Ca-Si silica fume
Halo at $2\theta = 22°$ (Cu K_{α_1} radiation)

The European standardisation [5] limits the amount of silica fume D to 10% by mass in the CEM II/A with the following requirements : amorphous silica $\geq$ 85% by mass, loss on ignition $\leq$ 4% by mass and specific BET $\geq$ 15 m^2/g. For an easier handling silica fume can be densified or transformed into a slurry [10]. However the effective particle size of commercial silica fumes is not always about 0.1 – 0.2 µm and undispersed or larger particles can persist [11].

In silica fume cements, silica microspheres first play a physical role of filler between Portland clinker grains of 30-80 µm size. During the cement hydration, silica particles are sites of nucleation for hydrates and then react as an active pozzolan giving homogeneous and like amorphous C-S-H under scanning electron microscope (Figure 5). The pozzolanic reaction consuming $Ca(OH)_2$ is easily followed by X-ray diffraction. The C/S ratio of C-S-H at 28 days is 1.3 compared to 1.7 for the Portland cement used [12].

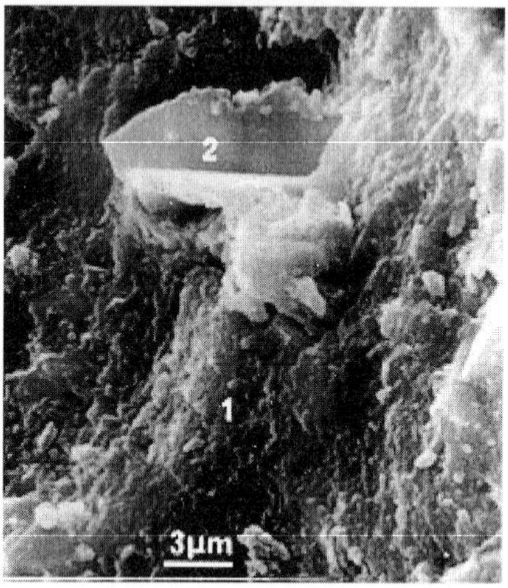

Figure 5 Scanning electron micrograph of a silica fume cement paste showing (1) homogeneous C-S-H and (2) $Ca(OH)_2$ partially dissolved by the pozzolanic reaction

The percolation or connectivity of hydrates like CH and C-S-H as a function of hydration has been determined by a three-dimensional digital image-based simulation [13]. A critical amount of silica fume to be added to Portland cement resulting in capillary pore discontinuity is a function of both w/c and α degree of hydration. Lower w/c ratios require less silica fume at a constant degree of hydration. With 10% D, w/c = 0.5 asks for α = 72%. At 60% hydration the amount of D is 10% for w/c = 0.45.

Portland pozzolana cement

Natural pozzolanas (Z) have been largely used in Italy [14]. Pyroclastic rocks of volcanic origin mostly glassy rhyolites, strass and tuf, andesite, zeolites, diatomaceous, montmorillonite clay, gaize, shales. They all are siliceous materials containing on average 60% SiO_2, 15% Al_2O_3, 6% Fe_2O_3, 6% CaO, 3% MgO, 2% Na_2O, and 5% K_2O. Their reactivity with lime is related to their content of glassy phases or crystalline zeolites, and their surface area.

Industrial pozzolanas (Q) are calcined clays, rice husks, and air cooled non-ferrous slags. Metakaolins have been obtained by thermal treatment of kaolins at 700-800°C and recently from paper sludge at 600-700°C [15]. Rice husks when burnt under controlled conditions at 600-650°C are mostly siliceous [16]. The silica is amorphous as shown by X-ray diffraction.

A typical chemical analysis of rice husk ash gives 92% SiO_2, 3% K_2O, traces of Al_2O_3, Fe_2O_3, CaO, MgO. The ignition loss, 3%, is due to the amount of carbon. The pozzolanic activity of rice husk ashes is related to their surface area and the amount of amorphous silica, balanced against their water demand.

Portland fly ash cements

Fly ashes called pfa, pulverised fly ash, are largely world wide produced in coal power stations. The annual production is estimated to be around 650 million tons [1]. Two types of fly ash have been distinguished on the basis of their chemical composition [17]:

- Siliceous fly ash (V) or low-lime pfa which in Europe contains under 10% CaO would be classified as Class fly ash in USA. These siliceous pfa are produced from anthracite and bituminous coals. Low lime pfa are pozzolanic materials.
- Calcareous fly ash (W) or high-lime pfa produced from lignite and sub-bituminous coals, called Class C in USA contains SiO_2 + Al_2O_3 +Fe_2O_3 higher than 70%. High-lime pfa can be hydraulic.

Fly ashes are mainly glassy. Crystalline quartz, mullite, hematite and magnetite are minor phases only. The glass composition is siliceous in low-lime pfa i.e. 50% SiO_2, 28% Al_2O_3, 9 Fe_2O_3, 3% CaO, 4% K_2O, 1% MgO, and 1% SO_3. Pulverized-fuel ashes which occur as spherical and often hollow particles (Figure 6) are also characterised by their fineness, density, loss on ignition, moisture content, water requirement and pozzolanicity. They improve the concrete workability due to their morphology, compared to Portland cement [18].

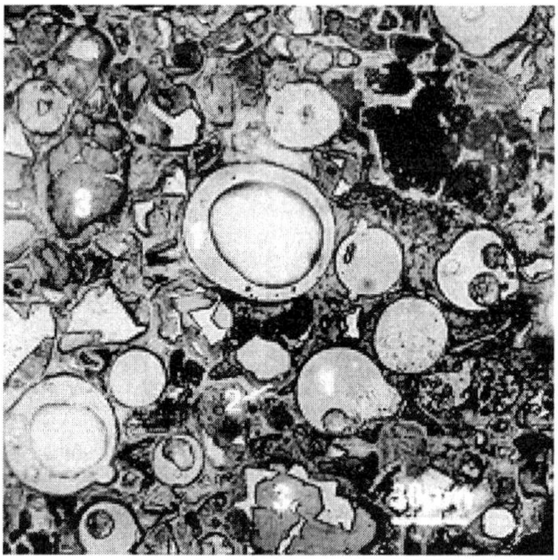

Figure 6 Light microscopy micrograph of a CEM II/B fly ash V cement at 7 days, (1) anhydrous fly ash, (2) C-S-H from C_3S, (3) anhydrous C_3S in a clinker grain

For being used in CEM II the composition of fly ash V has to follow ENV 197-1 [5] requirements on loss on ignition : <5%, reactive CaO : <5% and reactive SiO_2 : >25%. In fly ash W the reactive CaO is higher than 5%. If the amount of reactive CaO is between 5 and 15%, the reactive SiO_2 is higher than 25%. Calcareous fly ash with more than 15% reactive CaO is hydraulic if its compressive strength is higher than 10 MPa at 28 days and its expansion with 70% cement is lower than 10 mm.

Pozzolanic materials hydrate slowly. They give C-S-H with a lower C/S ratio than in CEM I. Depending on the pozzolanic reactivity this ratio is between 0.7 and 1 in CEM II/B. Beside AFt and AFm can coexist hydrogarnet C_2ASH_6 or strätlingite C_2ASH_8. Fly ash cements can be activated by thermal treatment.

Portland burnt shale cements

Burnt shales (T) produced at 800°C from oil shales are hydraulic as they contain clinker phases like dicalcium silicate C_2S but also calcium aluminate CA. Requirements for their use in cements are the same as calcareous fly ash W with however a compressive strength of 25 MPa at 28 days when finely ground. Some siliceous pozzolanic materials can also be present.

Portland limestone cements

Limestone filler (L) in CEM II cements is largely used in France since 1979 [19]. The carbonate is mainly calcite with grindability and distorted crystal structure different from one origin to another one. A limestone filler will be chosen with the following requirements [5]: $CaCO_3$>75%, clay content $\leq$1.2% determined by methylene blue adsorption and organic materials $\leq$0.5% by mass. These characteristics do not modify the rheology and water demand of mortar and concrete.

In limestone filler cements, the clinker-filler compatibility is previously optimised by the cement producer. Limestone is interground with the clinker but depending on its softness it can be finer than clinker i.e. for a mixture of clinker (75%) and limestone (25%) ground at 420 m²/kg, the filler was ground at 737 m²/kg and the clinker at 319 m²/kg. Filler and clinker can be ground separately and their fineness adjusted for a given compressive strength. Compared to Portland cement CEM I at the same 28 day-strength, a Portland limestone filler CEM II will have a higher early strength due to its fineness and the higher fineness of clinker. As an example the compressive strength of ISO mortars is at one day, 11 MPa for a CEM I 42.5 at 255 m²/kg and 15 MPa for the CEM II/B at 397 m²/kg with 27% L, or 15 MPa for a CEM I 52.5 at 334 m²/kg and 18 MPa for the CEM II/B at 397 m²/kg with 15% L. This is an important point. The quality of clinker now produced has been improved. So a CEM I 42.5 would require a fineness of less than 250 m²/kg with coarse grains which will not hydrate completely [20].

Limestone fillers play both physical role by their fineness and chemical role by their reactivity [21]. Hydration of clinker constituents, C_3S and C_3A, is accelerated in the presence of fillers. This is clearly shown by the increased appearance of $Ca(OH)_2$ liberated by C_3S and the formation of carboaluminates in reaction with C_3A, carboaluminates which coexist with ettringite.

There is also a good bond between cement and calcite with a transgranular fracture (7). Optimised CEM II (L) are as durable as CEM I of the same strength class [22, 23].

Figure 7 Scanning electron micrograph of a CEM II/B (L), transgranular fracture at 75 days, (1) cement matrix, (2) calcite grain

Composite Portland cement

Composite Portland cements CEM II/A and CEM II/B contain mixed supplementary cementing materials M like slag + fly ash + silica fume + limestone filler. The addition of 5% of silica fume largely improves strength and compactness of CEM II/B containing slowly reactive minerals like some natural and industrial pozzolans or crystalline slag [24].

Properties of Portland cement-based materials

Supplementary cementing materials added to Portland cement not only save energy but also improve the concrete compactness due to the filler effect and enhance the durability due to a chemical effect related to a lower amount of $Ca(OH)_2$ able to be leached and a reduced interconnected porosity of hydrates like C-S-H.

Portland cement-based materials CEM II/A - II/B 32.5 and 32.5 R are used for common works like foundations and ground stabilisation, all in non aggressive environments, masonry, and pavements. In open air adequate curing is needed. In 1995, 47% of the cement production in France corresponded to CEM II A - B, 32.5 and 32.5 R. There was no CEM I 32.5.

Portland CEM II/A - B, 42.5 and 42.5 R, are used for precast concrete on site, reinforced concrete, prestressed concrete, thermal treated concrete. In open air an adequate curing is needed.

In aggressive environments, cements have to be checked regarding their resistance to sulfates or seawater even if their resistance to aggressive environment is higher than that of CEM I. These CEM II/A - B have the same uses as CEM I of the same strength class. In 1995, 5.8% of the cement production in France corresponded to CEM II/A - B, 42.5 and 42.5 R, and 5.3% for CEM I/A - B, 42.5 and 42.5 R.

Portland CEM II/A - B, 52.5 and 52.5 R are used for precast concrete on site, reinforced concrete, prestressed concrete with no special treatment at ordinary temperatures higher than 5°C, high long term strength concrete, and high performance concrete. They have to be adequately cured and sulfate resisting or marine clinkers will be used in aggressive environments. They behave in the same way as CEM I, 52 and 52.5 R. A special mention is due to the superplasticized CEM II/A (D) with high strength and high durability. In 1995, 1.4% of the cement production in France corresponded to CEM II/A - B, 52.5 and 52.5 R, beside 29.8% of CEM I, 52.5 and 52.5 R.

HIGH PERFORMANCE CONCRETE AND NOVEL CEMENTS

First improvements in concrete characteristics in the 1980's were related to the comparison with high compressive and flexure strengths of materials like steel or sintered alumina. The weakness of concrete has been mostly connected to the capillary porosity and to the excess of water needed for the workability of the fresh paste. Reducing the w/c ratio by use of superplasticizers or extracting the excess of water by pressure or filling in capillary pores by polymer impregnation are processes that are able to densify the cement paste and the internal transition zone between matrix and aggregate in concrete. These processes were used in :

DSP - Densified Systems containing homogeneously arranged ultrafine Particules [25] with compressive strengths higher than 100 MPa, due to silica fume and superplasticizer. They resulted in HSC, High Strength Concretes. However, the strength is not always a characteristic for a concrete structure. As an example, the Channel Tunnel lining was designed for a specified compressive strength of 45 MPa, and for structural reasons. However on the French side, chalk formations permeable to seawater were considered as an aggressive environment. Concrete had to resist to the diffusion of chloride ions and durability requirements resulted in a higher strength, in fact 63 MPa [26]. So HSC became HPC, High Performance Concrete. This new material called for new structural design like in high-rise buildings, bridges, offshore structures, tunnels [27],

MDF - Macro Defect Free cements [28], develop flexure strengths as high as 150 MPa. The absence of large capillary pores and cracks is related to the addition of a polymer which first disperses and lubricates cement grains, then is highly bound to cement grains. MDF processed with a polyvinyl alcohol form an interphase between cement and PVA [29]. MDF can be used in acoustics, cryogenics, tooling, road overlays, panels, claddings,

CBC - Chemically Bonded Ceramics [2] in which there is a combination of the advantages of rapid strength development by alkali activation and high ultimate strength given by slowly reactive materials like slag or fly ash, in a close particle packing. They have potential applications in the nuclear and waste management and rapid repairs,

SAC and **FAC** - respectively belite sulfo-aluminate cement and belite ferro-aluminate cement [30], are energy-saving materials, since synthetised at lower temperature than Portland cement i.e.1200 instead of 1450°C and possibly from by-products as slag or fly ash, also friable, so easy to grind. Their main constituents are yeelimite or Klein's compound 4 CaO. 3 Al_2O_3. SO_3, α' C_2S, C_4AF, and $CaSO_4$. Early strengths are due to the formation of ettringite. These cements can be rapid-hardening cement, expansive cement, self-stressing cement, low-alkalinity cement, and high strength cement (80 MPa for a cast-on-site concrete). They have been used in China for more than twenty years, in building construction even at winter time, underground structures, and marine construction,

RPC - Reactive Powder Concrete [4] invented for ultra-high strength and high ductility by eliminating coarse aggregates replaced by finely ground quartz (4 μm) and silica fume, optimising the granular distribution (sand of 400 μm), applying a pressure during setting, controlling the microstructure by post-set heat-treatment which enhances the silica fume pozzolanicity, and improving the ductility by incorporation of small-sized steel fibres (Figure 8). The microstructure of these composites with a low w/c due to superplasticizers is very dense. Clinker grains and silica fume spheres are still visible but well embedded in a CEM I at w/c = 0.2 and for α = 0.45 (Figure 9). There is no oriented $Ca(OH)_2$ crystals.

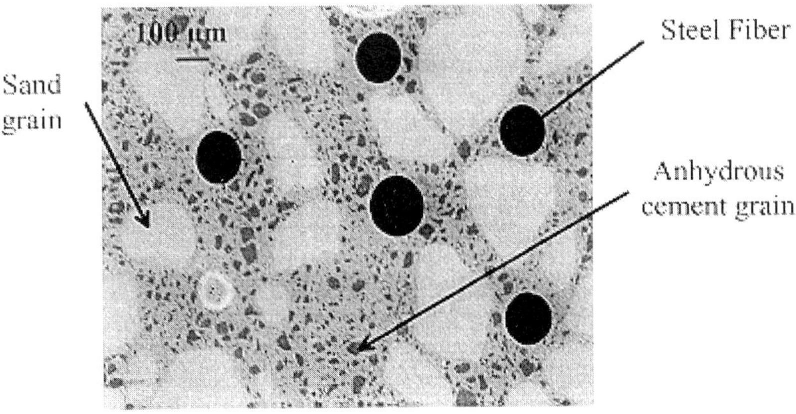

Figure 8 Scanning backscattered electron image of an RPC of 216 MPa

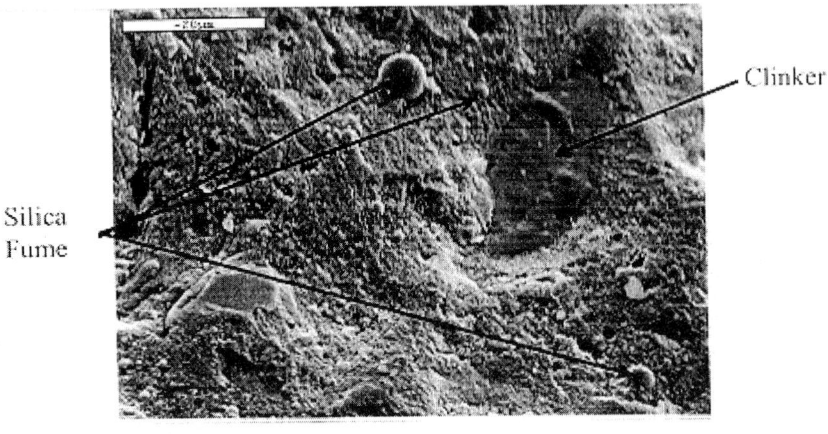

Figure 9 Scanning secondary electron micrograh of the RPC of Figure 8

Compressive strengths can be 230 MPa after a treatment at 90°C but 680 MPa for the same RPC treated at 90°C and pressed under 500 at. When steel fibres are replaced by stainless steel micro-fibres and quartz sand by steel powder, a treatment at 250°C results in a compressive strength of 800 MPa. Such products have already been used in the production of protective panels against impact from projectiles. Their durability is studied in the expected use for radioactive waste storage through a leaching process [31]. First results show that the tritium diffusion, chosen for its inactivity towards cement and assuming that it reproduces the transfer of the radio-elements through RPC, is still lower than that of granite after a three-month-leaching in a severe leaching process at pH=7. The durability factor has been found equal to 100 for different mixes performing in freezing and thawing cycles. The chloride ion permeability measured by the ASHTO test translated a quasi-impermeability of the material.

CONCLUSIONS

Portland cement-based materials CEM II represented 41% of the European production in 1993. The production of Portland cement CEM I was 44.5%.

Supplementary cementing materials, used up to 35% in composite cements optimised at the cement plant, are either hydraulic like blastfurnace slag and calcareous fly ash or pozzolanic as volcanic pozzolan, burnt clay, siliceous fly ash, rice husk ash, silica fume. When finer than Portland cement clinker, mineral additions play first a physical filler effect, then a chemical effect enhancing the rate of hydration of Portland cement constituents, C_3S and C_3A more particularly. They produce hydrates different from those of Portland cement as C-S-H with lower Ca/Si ratios and a refined pore structure, decreasing the interconnectivity of the capillary porosity.

In the case of limestone filler a composite cement with the same compressive strength at 28 days as Portland cement develops a higher early strength due to a greater fineness of the clinker. Carbonates are able to react in the cement paste forming carboaluminates which coexist with ettringite. They are as durable as CEM I of the same strength class.

Novel cements benefited from chemical and microstructure engineering through particle close packing, heat treatment, pressing, and incorporation of organic polymers. These new materials are MDF, CBC, DSP. They result in high performance concretes and new design in building construction.

The next millenium will use more by-products, saving energy, protecting the nature, and meeting the cement demand of 500 millions tons in the year 2000.

REFERENCES

1. MEHTA, P K. Role of pozzolanic and cementitious materials in sustainable development of the concrete industry. Proc.6th CANMET/ACI Int. Conf. Fly Ash, Silica Fume, Slag & Natural Pozzolans in Concrete, Ed. V M Malhotra, ACI SP-178, 1998, Vol 1, 1-20.

2. ROY, D M. and SILSBEE, M R. Novel cements and cement products for applications in the 21th century., Proc. Mohan Malhotra Symp., Concrete Technology, Past, Present, and Future, Ed. P.K. Mehta, ACI SP-144, 1994, 349-374.

3. TAYLOR, H F W. Composite cements. Cement Chemistry, Chapter 9, Thomas Telford, 1997, 261-294.

4. RICHARD, P. and CHEYREZY, M H. Reactive powder concretes with high ductility and 200-800 MPa compressive strength, Ibid Ref. 2, 1994, 507-518.

5. ENV 197-1. Cements – composition, specifications and conformity criteria. Part 1. Common cements, 1992.

6. GLASSER, F P. Chemical, mineralogical and microstructure changes occurring in hydrated slag-cements blends, Materials Science of Concrete II, Ed. J.P. Skalny and S. Mindess, The American Ceramic Society, 1991, 41-82.

7. MORANVILLE-REGOURD, M. Cements made from blastfurnace slag, Lea's Chemistry of Cement and Concrete, 4th Edition, Ed. P.C. Hewlett, ARNOLD, 1998, 633-674.

8. ROJAOKARIVONY-ANDRIAMBOLOLONA, Z, BAILLIF, P, THOMASSIN, J H and TOURAY, J C. Mécanismes d'hydratation d'un verre de laitier de haut fourneau dans l'eau et en milieu alcalin, C.R. Acad. Sci. Série II, 1988, 347-354.

9. RICHARDSON, I G. The nature of hydration products in hardened cement pastes, J.G. Cabrera Symp. Durability of Concrete Materials, Ed. R. N. Swamy, Ibid. Ref. 1, 1998, 13-37.

10. AITCIN, P-C. Physico-chemical characteristics of condensed silica fume, Condensed Silica Fume, Ed. P-C Aïtcin, Les Editions de l'université de Sherbrooke, 1983, 16-19.

11. DIAMOND, S. Silica fume agglomerates and alkali silica reaction, Proc; Int. RILEM Conf. Concrete from Material to Structure, Ed. J.P. Bournazel et Y. Malier 1998, 98-106.

12. REGOURD, M, MORTUREUX, B and GAUTIER, E,. Hydraulic reactivity of various pozzolanas, Fifth Int. Symp. Concrete Technology, Universidad Autonoma de Nuevo Leon, 1981, 1-14.

13. BENTZ, D P and GARBOCZI, E J. Percolation of phases in a three-dimensional cement paste microstructural model, Cem. Concr. Res., 1991, Vol. 21, N° 2/3, 325-344.

14. MASSAZZA, F. Pozzolana and pozzolanic cements, Ibid Ref. 7, 1998, 471-631.

15. PERA, J and AMBROISE, J. Pozzolanic properties of metakaolin obtained from paper sludge, Ibid Ref.1., 1998, Vol. 2, 1007-1020.

16. COOK, D. J. Rice husk ash, Concrete Technology & Design. Cement Replacement Materials, Ed. R. N. Swamy, Surrey University Press, 1986, Vol. 3, 171-198.

17. DHIR, R K. Pulverized –fuel ash, Ibid Ref.17, 1986, 197-255.

18. UCHIKAWA, H. Effect of blending components on hydration and structure formation, Proc. 8th Congress Chemistry of Cement, 1986, Vol. 1, Principal Report 3.2, 249-280.

19. REGOURD, M. Caractéristiques et activation des produits d'addition, Ibid Ref. 19, 1986, Vol. 1, Principal Report 3.1, 199-229.

20. GUYOT, R, RANC, R and CARIOU, B. Utilisation de la potentialité hydraulique du clinker dans les ciments Portland avec ajout de filler calcaire, Ibid Ref. 20, 1986, Vol. 6, 198-202.

21. MORTUREUX, B, HORNAIN, H and REGOURD M. Liaison pâte de ciment-filler dans les ciments composés, Colloque int. Liaisons Pâte de Ciment-Matériaux Associés, INSA Toulouse-RILEM, 1982, A 64-72.

22. RANC, R, M. MORANVILLE-REGOURD, COCHET, C and CHAUDOUARD, G. Durability of cements with fillers, 2nd Int. Conf. Durability of Concrete, 1991, ACI SP-126, Vol. II, 1239-1258.

23. MATTHEWS, J D. Performance of limestone cement concrete, Euro-cements : Impact of ENV 197 on Concrete Construction, Ed. R. K. Dhir & M. R. Jones, E & FN Spon Publisher, 1994, 113-147.

24. REGOURD, M, MORTUREUX, B and HORNAIN, H. Use of condensed silica fume as filler in blended cements, 1rst CANMET/ACI Int. Conf. The Use of Fly Ash, Silica Fume, Slag and Other Mineral By-Products in Concrete, 1986, ACI SP-79, Vol. 2, 847-865.

25. BACHE, H H. Densified cement/ultra-fine particle-based materials, 2nd Int. Conf. Superplasticizers in Concrete, 1981, 35P.

26. BROCHERIEUX, J M. High performance concrete in tunnel linings, High Performance Concrete-From Material to Structure, Ed. Y. Malier, E & FN SPON, 1992, 347-362.

27. MALIER, Y. Introduction, Ibid Ref. 28, 1992, xiii-xxiv.

28. KENDALL, K, HOWARD, A J and BIRCHALL, J D. The relation between porosity, microstructure and strength, and the approach to advanced cement-based materials, Phil. Trans. R. Soc. Lond. A 310, 1983, 139-153.

29. YOUNG, J F. Microstructural development in very high strength cement pastes, Int. RILEM Workshop on Hydration and Setting of Cements, Dijon, France, Ed. A. Nonat and J.C. Mutin, E & FN SPON, 1991, 381-392.

30. SU, M, WANG, Y, ZHANG, L and LI, D. Preliminary study on the durability of sulfo/ferro-aluminate cements, 10th Int. Cong. Chemistry of Cement, 1997, Vol. 4, Paper 4iv029, 12pp.

31. MATTE, V and MORANVILLE, M. The leaching of the reactive powder concretes : results on transfer properties. Nondestructive characterization of materials in aging systems, Materials Research Society, Vol. 503, 1998, 145-150.

A LOGARITMIC MODEL FOR THE PREDICTION OF STRENGTH DEVELOPMENT OF POZZOLAN CEMENT

G Ch Kostogloudis

C Ftikos **V Kasselouri**

National Technical University of Athens

Th Philippou

Hellenic Cement Research Centre Ltd

Greece

ABSTRACT. A logarithmic model was developed for the prediction of the strength development of Portland Pozzolan Cement (PPC) containing Santorin Earth as an admixture. Two multiple regression models were constructed for the determination of the coefficients A and B of the logarithmic model. The independent variables used in the multiple regression models were: (1) the compressive strength at the age of 1 day, (2) the compressive strength at the age of 2 days, (3) the CaO_f of the Portland clinker, (4) the compressive strength after autoclave hydrothermal treatment at 300 psi and 214°C for 3 h, (5) the coefficient A, (6) the insoluble residue of the cement, and (7) the C_3S/C_2S ratio of the clinker. The evaluation of the proposed model was performed by various statistical tests, which were all successful. There was very good agreement between the strength predicted by the logarithmic model and experimental results.

Keywords: Blended cement, Portland pozzolan cement (PPC), Santorin Earth, Compressive strength development, Logarithmic model.

George Ch Kostogloudis is a PhD Postgraduate Student in the Department of Chemical Engineering, National Technical University of Athens (NTUA), Greece. He specializes on electroceramic materials, energy-saving cements and reinforcement corrosion.

Professor Christos Ftikos is Director of the Lab. of Inorganic Materials Technology, Dept. of Chemical Engineering, NTUA. He specializes on cement and concrete technology, pozzolans and durability of their products, as well as on electroceramics. Professor Ftikos has published widely and serves on many Technical Committees.

Dr Vassilia Kasselouri is an Associate Professor in High Temperature Chemistry and Technology, Dept. of Chemical Engineering, NTUA. She specializes in the chemistry and technology of cement, glass and ceramics.

Dr Th Philippou is Director of the Hellenic Cement Research Center Ltd. He specializes on cement and concrete science and technology

INTRODUCTION

Many efforts have been directed towards the development of models with the ability to predict the compressive strength of cement [1-5]. These models relay upon empirical relationships between various physicochemical and mineralogical parameters of cement clinker and the strength of cement [1]. A carefully designed model can serve as a valuable tool during the production of cement and the control of its quality. However, the use of such models is restricted to the specific applications for which they are designed. In the case of blended cements, which have drawn much attention over the past two decades mainly due to energy saving reasons, little information is available concerning strength prediction [6].

In the present work, a logarithmic model (Y=A+BlnX) was developed for the prediction of the strength development up to the age of 28 days of a Portland Pozzolan Cement (PPC) containing Santorin Earth as an admixture. The compressive strength of 17 industrial PPC samples was measured at the ages of 1, 1.5, 2, 7 and 28 days, and for each sample a logarithmic model was developed by linear regression analysis. Two multiple regression models where then developed for the prediction of the coefficients A and B, respectively, of the logarithmic model for the prediction of strength development of PPC. An evaluation of the multiple regression models was performed by various statistical tests.

MATERIALS AND METHODS

17 industrial PPC samples were used for the development of the model. The natural pozzolan material used as an admixture for the production of PPC was Santorin Earth. The sampling was random, and it was performed during a period of three months at the place of production. Chemical analysis and mineralogical characterization was performed for all Portland clinker samples. The fineness (Blaine), R90μm, R32μm, SO_3 content, insoluble residue (IR) and loss on ignition (LOI) of all PPC samples were also measured. It should be noted that the IR of the PPC cement refers to the pozzolan content of the cement, and should not exceed 20%, according to Greek specifications.

The compressive strength of the PPC samples was measured at the ages of 1, 1.5, 2, 7 and 28 days, according to EN specifications. The compressive strength of PPC mortar specimens was measured after hydrothermal treatment for 3h at 300 psi and 214°C.

DEVELOPMENT OF THE LOGARITHMIC MODEL

General

The logarithmic model for the prediction of the strength development of PPC is described by the following equation:

$$S = A + B \ln t \tag{1}$$

where,

S	=	compressive strength of PPC
A and B	=	coefficients
t	=	time

Two multiple regression models where developed for the prediction of the coefficients A and B, respectively. The procedure was as follows: For each of the PPC samples a logarithmic model was developed by linear regression analysis using the measured compressive strengths up to the age of 28 days. The coefficients A and B of these models, as well as the R-squared values are shown in Table 1. These sets of 17 A and B values were used as the observed values during the development of the two multiple regression models. After the models were constructed, they were evaluated using various statistical tests. In order to test the predicting ability of the multiple regression models, the predicted A and B values were compared with the observed A and B values of Table 1. The determination of the coefficients of the multiple regression models and all statistical tests were performed using the EXECUSTAT statistical computer package [7].

Table 1 Coefficients A and B for each PPC sample, used for the development of the two multiple regression models

	1	2	3	4	5	6	7	8	9
A	6.062	5.150	6.005	5.525	3.428	4.808	4.840	3.929	5.156
B	9.093	9.144	8.521	9.663	10.011	9.146	9.190	9.065	9.659
R-squared (%)	99.47	99.57	99.77	99.47	99.39	99.68	99.87	99.91	99.97
	10	11	12	13	14	15	15	17	
A	5.341	7.084	5.859	5.484	5.741	6.446	5.624	4.662	
B	9.037	9.508	8.588	9.293	8.624	8.373	7.673	8.213	
R-squared (%)	99.90	99.98	99.77	99.98	99.97	99.92	99.68	99.80	

Multiple Regression Model for the Prediction of Coefficient A

Construction of the model

The multiple regression model for the prediction of coefficient A of the logarithmic model of Equation (1) had 3 independent variables, and it is described by the equation:

$$A = a_0 + a_1 S1 + a_2 S2 + a_3 CaO_f \tag{2}$$

where,

a_0, a_1, a_2, a_3	=	coefficients
S1	=	the compressive strength at the age of 1 day
S2	=	the compressive strength at the age of 2 days
CaO_f	=	the free lime content of the Portland clinker

Multiple correlation

Before the determination of the coefficients of the model it is necessary to ensure that the independent variables do not correlate with each other. Table 2 shows the correlation coefficients between pairs of variables, and the P values, which are used to test whether the coefficients are significantly different from zero. As can be seen, P>0.05 for all pairs, indicating that all variables are independent from one another.

Table 2 Correlation coefficients for the model of Equation (2). Values in parentheses are P values for the hypothesis of zero correlation

	S1	S2	CaO_f
S1		0.270 (0.30)	-0.343 (0.18)
S2	0.270 (0.30)		0.263 (0.31)
CaO_f	-0.343 (0.18)	0.263 (0.31)	

Estimation of coefficients, test of significance and confidence intervals

The estimated coefficients for the model are shown in Table 3. All coefficients have $P<0.05$, indicating statistically significant nonzero coefficients. Since none of the 95% confidence intervals includes 0, there is an actual linear relationship between the dependent variable and each of the independent variables, and so, all variables should be included in the model.

Table 3 Estimated coefficients for the model of Equation (2) and their confidence intervals

	ESTIMATE	STANDARD ERROR	T VALUE	P VALUE	95% CONFIDENCE LIMITS	
					Lower	Upper
Constant	-1.6072	0.4881	-3.29	0.0058	-2.6617	-0.5528
S1	0.8434	0.0479	17.59	0.0000	0.7398	0.9470
S2	0.1448	0.0429	3.38	0.0050	0.0522	0.2374
CaO_f	0.5147	0.1588	3.24	0.0064	0.1716	0.8579

R-squared = 97.14%
Standard error of estimation = 0.1658
Durbin-Watson statistic = 2.0927

Test for outliers and unusual residuals

Figure. 1 shows the Studentized residuals as a function of the predicted A values. As can be seen, the plot appears reasonably random, and no outliers (i.e. points that do not follow the same pattern as the others) are noticed. Only one unusual residual exists, since its Studentized value is greater than 2. However, none of the Studentized residuals is outside the ±3 limit.

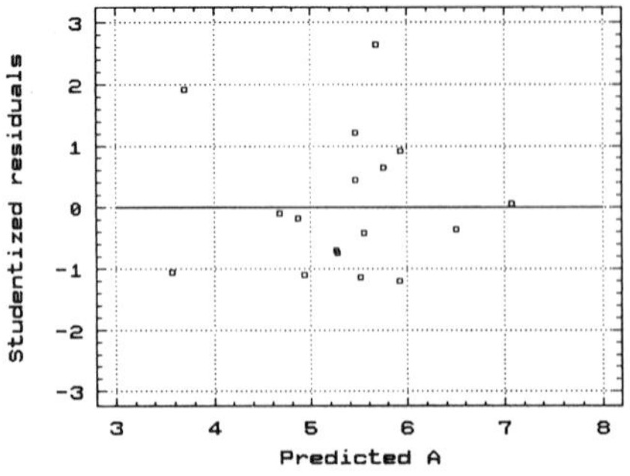

Figure 1 Plot of Studentized residuals versus predicted A values

Analysis of variance

Table 4 shows the analysis of variance (ANOVA) table. The R-squared statistic, based upon the ratio of the model sum of squares divided by the total (corrected) sum of squares, indicates that the model accounts for 97.14% of the variation of A. The mean squared error estimates the variance of the deviations around the model to be equal to 0.0275. Since the P value corresponding to the F-ratio is less than 0.05, the model as a whole is statistically significant.

Table 4 Analysis of variance for the model of Equation (2)

SOURCE	SUM OF SQUARES	D.F.	MEAN SQUARE	F-RATIO	P VALUE
Model	12.1385	3	4.0462	147.29	0.0000
Error	0.3571	13	0.0275		
Total (corr.)	12.4956	16			

Multiple Regression Model for the Prediction of Coefficient B

Construction of the model

The multiple regression model for the prediction of coefficient B of the logarithmic model of Equation (1) had 5 independent variables, and it is described by the equation:

$$B = b_0 + b_1 S2 + b_2 Sa + b_3 A + b_4 IR + b_5 C_3S/C_2S \qquad (3)$$

where,

$b_0, b_1, b_2, b_3, b_4, b_5$	=	coefficients
S2	=	the compressive strength at the age of 2 days
Sa	=	the compressive strength after autoclave hydrothermal treatment at 300 psi and 214°C for 3 h
A	=	the coefficient A determined by the model of (2)
IR	=	the insoluble residue of the cement
C_3S/C_2S	=	the C_3S/C_2S ratio of the Portland clinker

Multiple correlation

In Table 5, the correlation coefficients between pairs of variables of the model of Equation (3) are shown. Since P>0.05 in all cases, the variables are independent from one another.

Table 5 Correlation coefficients for the model of Equation (3), values in parentheses are P valuesfor the hypothesis of zero correlation

	S2	Sa	A	IR	C_3S/C_2S
S2		-0.107 (0.68)	0.483 (0.05)	0.098 (0.71)	0.195 (0.45)
Sa	-0.107 (0.68)		-0.380 (0.13)	0.353 (0.16)	0.352 (0.17)
A	0.483 (0.05)	-0.380 (0.13)		-0.481 (0.05)	0.099 (0.70)
IR	0.098 (0.71)	0.353 (0.16)	-0.481 (0.05)		0.270 (0.29)
C_3S/C_2S	0.195 (0.45)	0.352 (0.17)	0.099 (0.70)	0.270 (0.29)	

Estimation of coefficients, test of significance and confidence intervals

The estimated coefficients for the model of Equation (3) are shown in Table 6. All coefficients are statistically significant, since they have P values below 0.05. None of the 95% confidence intervals includes 0, and so, a linear relationship exists between the dependent variable and each of the independent variables.

Table 6 Estimated coefficients for the model of Equation (3) and their confidence intervals

	ESTIMATE	STANDARD ERROR	T VALUE	P VALUE	95% CONFIDENCE LIMITS	
					Lower	Upper
Constant	3.6571	1.0814	3.38	0.0061	1.2771	6.0372
S2	0.5480	0.0442	12.38	0.0000	0.4506	0.6454
Sa	0.1334	0.0189	7.05	0.0000	0.0917	0.1751
A	-0.5429	0.0683	-7.94	0.0000	-0.6933	-0.3925
IR	-0.2457	0.0560	-4.39	0.0011	-0.3689	-0.1225
C_3S/C_2S	0.3111	0.1230	2.53	0.0280	0.0403	0.5819

R-squared = 95.88%
Standard error of estimation = 0.1454
Durbin-Watson statistic = 1.7287

Test for outliers and unusual residuals

The Studentized residuals versus predicted B plot (Figure. 2) appears to be random. As can be noticed, one unusual residual exists, which, however, is located very close to the -3 limit.

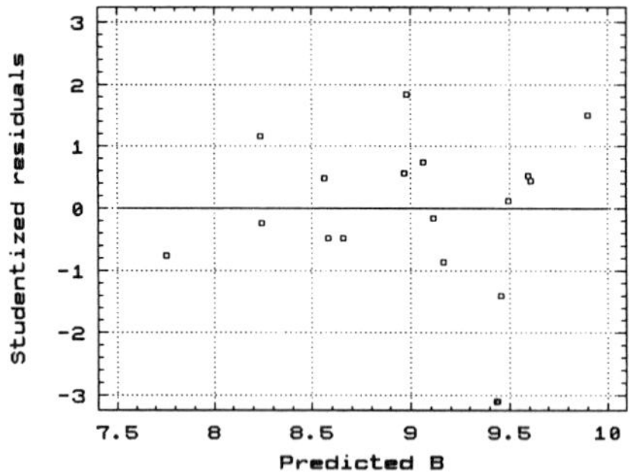

Figure 2 Plot of Studentized residuals versus predicted B values

Analysis of variance

The ANOVA table for the model of Equation (3) is shown in Table 7. The R-squared statistic indicates that the model accounts for 95.88% of the variation of B. The mean squared error estimates the variance of the deviations around the model to be equal to 0.0211. Since the P value, corresponding to the F-ratio, is less than 0.05, the model as a whole is statistically significant.

Table 7 Analysis of variance for the model of Equation (3)

SOURCE	SUM OF SQUARES	D.F.	MEAN SQUARE	F-RATIO	P VALUE
Model	5.4114	5	1.0823	51.21	0.0000
Error	0.2325	11	0.0211		
Total (corr.)	5.6439	16			

DISCUSSION

The two multiple regression models that were developed in the present work for the determination of the coefficients A and B, respectively, of the logarithmic model that predicts the strength development of PPC, were shown to satisfy all the statistical tests used for their evaluation. Figures 3a and 3b show the plots of the observed versus predicted values for A and B, respectively. A line with slope equal to 1 is also included. The points of both plots fell very close to this line, while they show approximately the same scatter around the line everywhere. Therefore, the proposed models have a very good predicting ability.

The choice of the independent variables of the multiple regression models was based on statistical as well as physical considerations. A large number of variables with physical importance was considered, and those satisfying certain statistical criteria were actually selected. The variables S1 and S2 are very significant, and their presence is essential. The lack of a satisfactory mechanism for the strength development of blended cements [8], does not yet permit the substitution of S1 and S2 with other physicochemical parameters. The variable Sa is also of high significance. This was expected, since the strength development under conditions of accelerated hardening provides a good measure of the actual hydraulic behavior of the cement. The IR is an important variable, since it is directly related to the concentration of pozzolan in the blended cement. The negative dependence of B on IR can be explained by the fact that an increase of the pozzolan content is known to reduce the compressive strength of blended cements. The C_3S/C_2S ratio is also an important variable, which has also been used in other works [2]. It incorporates trends arising from the variation of other important parameters that affect the C_3S and C_2S contents.

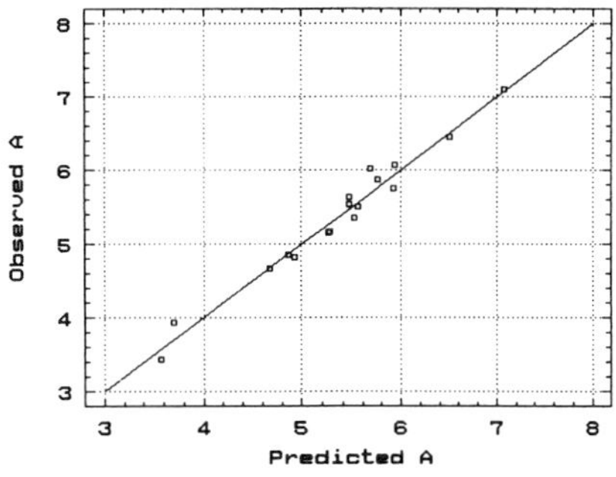

(a)

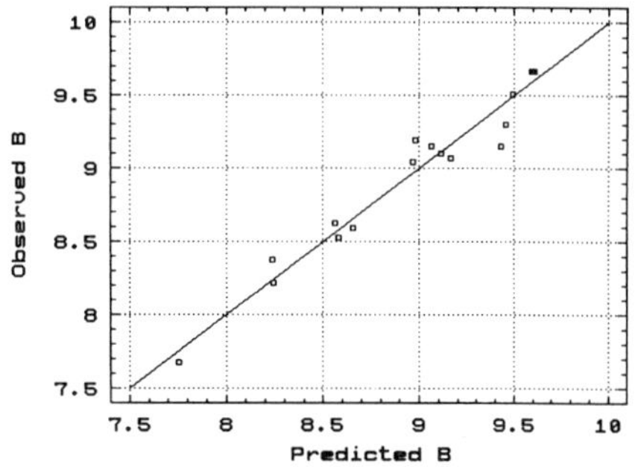

(b)

Figure 3 Plots of observed versus predicted values for the coefficients (a) A and (b) B of the logarithmic model of Equation (1)

These parameters include the CaO_f content, the alkalis occurring in solid solution with the above calcium silicate phases, and the degree of burning of the clinker. The CaO_f content itself, was used as a variable for the model that predicts the coefficient A.

CONCLUSIONS

The developed multiple regression model was shown to be able to predict with adequate accuracy the compressive strength development up to the age of 28 days of Portland pozzolan cement (PPC) containing Santorin Earth as an admixture. The model was evaluated by various statistical tests, which were all successful. The usefulness of the proposed model to the cement industry is based on its ability to make predictions on the strength development of PPC, only 2 days after its production. This gives the opportunity to cement engineers to make the necessary adjustments to improve the quality of the product in due time.

ACKNOWLEDGMENTS

The authors wish to thank HERACLES General Cement Company for providing the industrial cement samples used in this work.

REFERENCES

1. BRUGGERMANN, H AND BRENTUP, L. Relationship between cement strength and the chemico-mineralogical parameters of the clinker. Translation ZKG, 1/90, pp 30-33.

2. ALEXANDER, K, TAPLIN, J AND WARDLAW, J. Correlation of strength and hydration with composition of Portland cement. Proceedings of the 5th International Congress on the Chemistry of Cement, Tokyo, 1968, pp III-152-165.

3. POPOVICS, S. Model for the quantitative description of the kinetics of hardening of Portland cement. Cem. Concr. Res., Vol. 17, 1987, pp 821-838.

4. RELIS, M, LEDBETTER, W AND HARRIS, P. Prediction of mortar-cude strength from cement characteristics. Cem. Concr. Res., Vol. 18, 1988, pp 674-686.

5. PETRASINOVIC, L AND DURIC, M. Predicting the compressive strength of Portland cement and optimizing its raw mixture composition. Cem. Concr. Res., Vol. 20, 1990, pp 484-492.

6. FTIKOS, CH. Study of the action of Santorin Earth during the hydration of cements, PhD Thesis, 1977.

7. EXECUSTAT, PWS-Kent, Boston, 1991.

8. BARAGANO CORONAS, J AND VASQUEZ DE LA TORRE, P Y. Difficulties in the manufacture of blended cements. Translation ZKG, 11/86, pp 617-621.

EFFECT OF VOLCANIC ASH ON CEMENT BASED BINDER IN CONCRETE PRODUCTION

K M A Hossain

Papua New Guinea University of Technology

Papua New Guinea

ABSTRACT. The suitability of using volcanic ash (VA) in the manufacture of cement based binder and concrete is reported based on experimental investigation. Tests were conducted on cement, mortar and concrete replacing cement by VA from 0 to 50%. The physical and chemical properties of VA are critically reviewed to evaluate the possible influence and investigations included testing on both fresh and hardened state of cement and concrete. The standard tests on different Portland cement -VA mixes showed potential of manufacturing blended Portland volcanic ash cement (PVAC) using up to 20% VA. The properties of volcanic ash concrete (VAC) using different percentages of VA are also evaluated. Typical design charts are developed to evaluate mix design parameters of VAC.

Keywords: Volcanic ash, Cement, Mortar, Concrete, Compressive strength, Workability, Physical, Chemical, Mix-design, Setting.

Dr Khandaker M Anwar Hossain is a Lecturer in the Department of Civil engineering, The Papua New Guinea University of Technology. His research interests include the properties of concrete, small-scale modelling of structures, thin walled composite construction and non-linear finite element analysis of reinforced and composite structures.

INTRODUCTION

The search for alternative binders or cement replacement materials had been continued for the last decades. Cement placement materials play an undisputed role in the construction industry. Comprehensive research had been carried out in the past [1,2,3,4,5] on the use of fly ash (FA), pulverized-fuel ash (PFA), blast furnace slag, rice husk ash, silica fume etc. as cement replacement material. The volcanic ash (VA), pumicite, PFA and FA are pozzolanic materials, because of their reaction with lime liberated during the hydration of cement. Amorphous silica present in the pozzolanic materials combines with lime (Calcium hydroxide) and forms cementitious materials. They impart not only technical advantages to the resulting concrete but also enable larger quantities of cement replacement to be achieved. These materials can also improve the durability of concrete and the rate of gain in strength and can also reduce the rate of liberation of heat which is beneficial for mass concrete. This study was undertaken to investigate the effect of VA on cement based binder and to explore its potential use in the manufacture of blended cement [6], mortar and concrete.

RESEARCH SIGNIFICANCE IN THE CONTEXT OF PAPUA NEW GUINEA

Volcanic activities are a common phenomena for a country like Papua New Guinea and due to frequent volcanic eruption, volcanic debris such as : volcanic ash and pumice are found abundantly. The 1994 volcanic eruption that occurred in the East New Britain province was the second destructive one in history, which devastated the province and created an environmental disaster. This research is put forward with assistance from local cement and construction industries in an attempt to explore the possible utilization of volcanic debris in concrete production which can not only provide low cost concrete but also helps to decrease environmental hazard.

EXPERIMENTAL INVESTIGATIONS

The volcanic ash (VA) used in this investigation was collected from the Rabaul area in the East New Britain province of Papua New Guinea and the source is a volcano called Mount Tavurvur. The cement used was locally manufactured Portland cement called 'Paradise'. 20 and 10 mm maximum size crushed gravel and local river sand were used aggregate.

Chemical And Physical Analysis of VA

Chemical and physical properties of VA are compared with those from Paradise cement in Table 1. Chemical analysis indicated that the VA is principally composed of silica (about 60%) while the main components of cement is calcium oxide (maximum 70%). However, VA has cementitious compounds like calcium oxide, alumina and iron oxide (total about 31%). The amount of oxides of sodium and potassium known as 'alkalis' is found to be higher in VA (5.83%) than that in cement (2.6% maximum). Higher alkali presence in the VA may have deleterious effects leading to disintegration of concrete due to reaction with some aggregate and affect the rate of gain in strength of cement. Dry sieve analysis was carried out according to AS (Australian Standard) 1289.C6.1-1977 and results show that the

VA is composed mainly of sandy fraction of about 78% and the remaining 22% goes to silty fraction. VA is found to be much coarser than cement which may lead to the increase of setting time. The particle size distributions for aggregates and VA are presented in Figure 1.

Table 1 Comparative study of chemical and physical properties

CHEMICAL COMPOUND (%)			PHYSICAL PROPERTIES		
	VA	PC		VA	PC
Calcium oxide (CaO)	6.10	60-67	Fineness, m^2/kg	242	320
Silica (SiO2)	59.32	17-25	Residue on 150µm	42%	---
Alumina (Al_2O_3)	17.54	3-8	75µm sieve	78%	0.1-1.5%
Iron oxide (Fe_2O_3)	7.06	0.5-6.0	Sp. Gravity	---	3.5
Sulphur trioxide (SO_3)	0.71	1-3	Unit mass, kg/m^3	---	3150
Magnesia (MgO)	2.55	0.1-4.0	Bulk density ; kg/m^3	2450*	---
Sodium oxide (Na_2O)	3.80	0.5-1.3	Water absorption	0.41%	
Potassium oxide (K_2O)	2.03	0.5-1.3			
Loss on ignition	1.03	1.22			
Chloride	<0.01	--	VA: Volcanic Ash; PC: Paradise Cement		
Sulphate	0.08- 0.68	--	* oven dry basis		
pH	5.4	--			

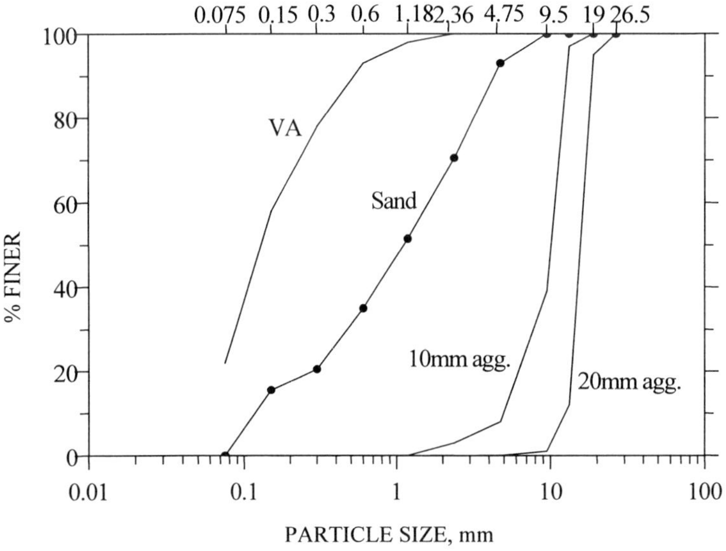

Figure 1 Particle size distribution

Effect of Volcanic Ash on the Properties of Cement

A series of tests had been carried out to study the effect of VA on the normal consistency and setting times of cement. Total ten mixes having different % of VA ranging from 0% to 50% by weight were used. Three tests for each item in each mix were carried out. Tests were performed according to Australian standard AS 2350.3, 1980. The variation of normal consistency and setting time with percentage of VA are presented in Figure 2. The normal consistency is decreased by 13.8% while initial setting time is increased by 90% and final setting time is increased by 58% when VA content is increased from 0% to 50%. This is due to the reduction of cementitious material as VA contains maximum 31% cementitious compound compared to about 75% in cement. On the other hand, specific gravity of VA is less than cement which resulted in the larger volume of VA compared to the same weight of cement replaced. As a result the overall volume was increased needing more water to form a paste of same consistency. The increase in both setting times with the increase of VA content is due to the reduction of cement in the paste and also due to the decrease of surface area of cementitious material as VA is much coarser than replaced cement. As a result hydration process slows down causing setting time to increase. The slow hydration means low rate of heat development. This is of great importance in mass concrete construction and it is there that Portland Volcanic ash cement or a partial replacement of Portland cement by VA can be mostly used.

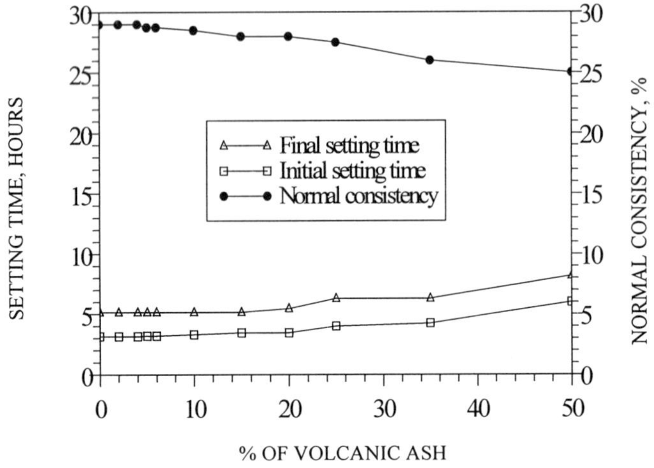

Figure 2 Effect of VA on normal consistency and setting time

Same mixes were used to study the effect of VA on the compressive strength of mortars. Mortar cubes were manufactured with standard Leighton Buzzard sand (Type1: strength cement) and locally available river sand (Type 2: local mortar manufacture). Standard 70 mm cubes (Cement + VA : sand = 1:3 by weight) were cast strictly following AS 2350.6, 1980. The water used in each mix was that derived from normal consistency of that mix.

The cubes were removed from the mould after 24 hours and then cured under water until they were tested immediately after removal from water while still wet. The variation in the compressive strength of two types of mortars tested is shown in Figure 3. The compressive strength is found to be decreased with the increase of VA content. This is reasonable due to the reduction of cement content in the mix with the increase of volcanic ash content. The finely divided silica (60%) in VA can combine with calcium hydroxide (liberated by the hydrating Portland cement) in the presence of water to form stable compounds like calcium silicates which have cementitious properties. Although VA has its own (31%) and can form cementitious compound, the reduction of cement due to replacement by VA is not compensated. The type1 mortar with standard sand shows higher strength than that of type 2 with local river sand. The results suggested [6] that the manufacture of Portland Volcanic Ash Cement (PVAC) similar to Portland fly ash cement of type FC (AS1317, 1982) is possible with maximum replacement of up to 20% . Type 2 mortar reflects the practical circumstances and provides a guideline to guess how much mortar strength can be possible to obtain when locally available sand is used. It is possible to obtain a mortar having a strength of 21MPa (7 day) and 24 MPa (28 day) using locally available sand and replacing 35% of cement by VA.

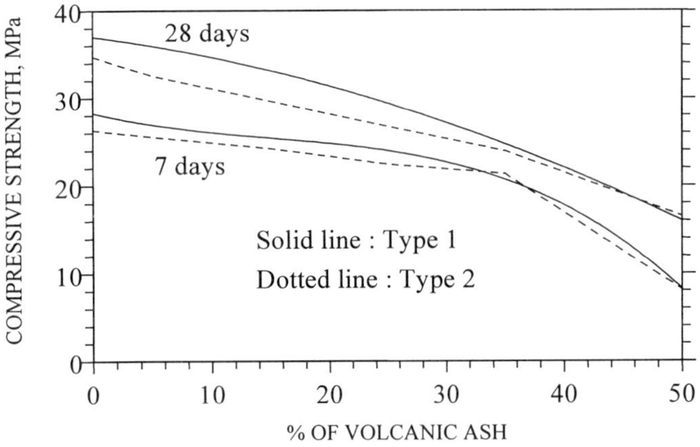

Figure 3 Effect of Volcanic ash on the compressive strength of mortar

Effect of Volcanic Ash on The Properties of Volcanic Ash Concrete (VAC)

This series of tests were performed to investigate the effect different percentages of volcanic ash on the strength of VAC. Initially the tests ware performed based on constant workability using water-cement ratio that produce a constant slump value of 75 mm as shown in Table 2.

The concrete mixes are classified into six mixes according to the % of cement replaced by volcanic ash. All concrete mixes had the same over all ratio of 1:2:4 (cement including VA : sand: aggregate) on the volume basis. After calculation of the volume required for the cement including volcanic ash, it was transformed into weight on the basis of cement.

Then from the equivalent weight of cement, the amount of volcanic ash was calculated based on % cement replacement by weight. The coarse aggregate consisted of 70% of 20 mm with 30% of 10 mm maximum size crushed gravel and river sand was used as fine aggregate.

Table 2 Volcanic ash concrete mix details

MIX DESIGNATION	CEMENT kg/m^3	VA kg/m^3	WATER* $Litre/m^3$	WET DENSITY kg/m^3	W/C+VA
Aggregate-total cementitious (C+VA) ratio =4.95; Constant Slump = 75 mm					
0-4.95	438	0	155	2463	0.354
5-5.21	416	22	161	2436	0.368
15-5.83	372	66	161	2388	0.368
25-6.61	328	110	167	2384	0.381
35-7.61	285	153	179	2367	0.409
50-9.90	219	219	202	2352	0.461

* excludes water absorbed by the aggregate , W=Water, C=Cement

The mix parameters and some characteristics of the fresh concrete are presented in Table 2. The first numeric in the mix designations represents % of volcanic ash and the second numeric represents aggregate cement ratio by weight.

Effect of VA on Fresh VAC

The results show that more water is needed to get similar workable mix as the % of VA is increased from 0 to 50% and at the same time wet densities are found to decrease as shown in Table 2. The more water requirement can be linked to the factors affecting the normal consistency of cement described earlier.

Effect of VA on Hardened VAC

The test specimens were cast using the optimum water-cement ratio for each mix. The specimens were 100 x100x100 mm cubes for compressive strength, 150 x 300mm cylinder for indirect tensile strengths. Total 15 cubes and 3 cylinders were cast from each mix. The samples were compacted by external vibration using a table vibrator. The specimens were demoulded 24 hours after casting and cured under water at a temperature of 23 ±2°C until testing at 7, 28 and 91 days. Five cubes for compressive strength and three cylinders for tensile strength were tested at a given age.

The variation in the compressive strength of VAC with different percentages of VA (having same workability) is shown in figure 4. As expected, strength decreased as the VA content increased. It is possible to obtain a concrete having 28 day strength of 25 MPa with VA content of 35%.

The density of VAC decreased with the increase of VA due to the replacement of comparatively heavier cement by lighter VA. The fresh density is found to be higher than those at 7 and 28 days due to subsequent removal of water from the fresh specimen with age.

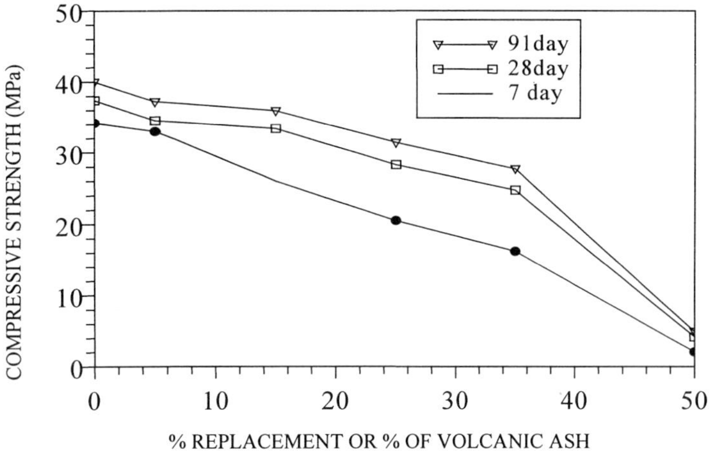

Figure 4 Effect of VA on the compressive strength of VAC

The density is found to be decreased with ages. The decrease in strength may also be linked with higher water cement ratio needed to have constant workability in all the mixes. Therefore, the decrease in strength may not be due to increase in VA alone but also may be due to the effect of increased water content. The research is now ongoing where the effect of volcanic ash on the strength of VAC is investigated by keeping water-cement ratio constant. Preliminary investigation suggested that both slump and workability of the mixes as well as the strength of the concrete increases with the increase of VA up to about 10%. Beyond 10%, both workability and strength of VAC decreases with the increase of VA. Tests are planned to improve the workability of the mixes by using admixtures and in the process, it may be possible to increase the strength of VAC.

A typical design chart for VAC for 28 day compressive strength suggested for the ordinary construction in the context of Papua New Guinea is presented in Figure 5. The chart can be used to relate all the parameters involved in the mix design such as: water-cement ratio, aggregate cement ratio, amount of cement and amount of volcanic ash to manufacture a VAC of particular strength. The numeric superscripts in the y-axis label identify the designated curve number shown in the body of the graph. The figure illustrates through arrow diagrams, the determination of mix design parameters for a 30 MPa VAC. For 30 MPa concrete, mix design parameters are: amount of cement (curve 3) =336kg/m^3, amount of volcanic ash (curve 4)=102 kg/m^3, % of volcanic ash=23.6, aggregate cement ratio(curve 2)=6.3 and W /(C+VA) ratio (curve 5) =0.38 (38%). This chart is valid only for the particular type of aggregate having similar gradation as used in this study. Investigations are in progress to generate design charts for VAC which can be applicable for general mix design purposes.

FEEDBACK FROM THE MANUFACTURING AND CONSTRUCTION INDUSTRY

The only cement manufacturing industry in PNG called "Halla Cement Factory" shows their interest and are providing assistance in this project. The recommendation of potential manufacture of blended cement using up to 20% VA is well received and under careful consideration by the authority.

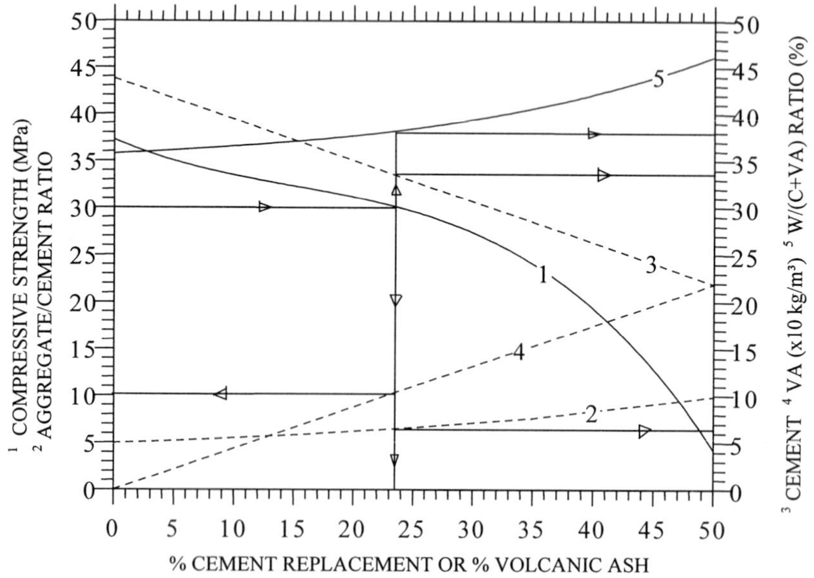

Figure 5 Typical mix design chart for VAC

The PNG Ready Mixed Company Ltd already using 5 to 8% VA as an admixture in concrete. It is believed that VA like fly ash and silica fume can be a useful material in the manufacture of concrete.

CONCLUSIONS

Manufacture of Portland Volcanic Ash Cement (PVAC) similar to Portland fly ash cement of type FC (AS1317, 1982) is possible with maximum replacement of up to 20%. The tests on fresh VAC, showed that more water is needed to get same workable mix as VA is increased from 0 to 50%.

Strength of both mortar and VAC decreased with the increase of VA. Results showed that by using 35% VA, it is possible to obtain a VAC of 25 MPa (28 day) and a mortar having a strength of 24 MPa (28day) using locally available sand. Use of 50% VA can produce a VAC of only 5% lighter than the normal concrete.

More investigations are needed to formulate general mix design guidelines for VAC.

ACKNOWLEDGEMENT

The author is grateful to the Papua New Guinea University of Technology for providing financial assistance in this project. The author is also grateful to the Technical staffs of the materials laboratory of the Department of Civil Engineering. Sincere thanks to Mr. Tom Moring of The PNG Halla Cement Factory, Mr. Christopher Sariman of PNG Harbours Board, and The PNG Ready Mixed Concrete Co. Ltd. for assistance.

REFERENCES

1. AL-ANI, M, AND HUGHES, B. Pulverized-fuel ash & it's uses in concrete. Magazine of Concrete Research, Vol. 41, No. 147, June, 1989.

2. MEHTA, P K. Properties of Blended Cements Made from Rice Husk ash. Journal of ACI, Vol. 74, 1979, pp. 440-442.

3. SWAMY, R N. Cement Replacement Materials. Concrete Technology and Design, Vol.3, Surrey University Press, Great Britain,1986.

4. SWAMY, R N. New Concrete Materials. Concrete Technology and Design, Vol.2, Surrey University Press, Great Britain, 1983.

5. BERRY, E E, AND MALHOTRA, V M. Fly ash for use in concrete – A critical review. Journal of ACI, 77, No. 8,1980, pp.59-73.

6. HOSSAIN, K M A. Volcanic Ash and Pumice based blended cement. Proce. of 23rd OWICS Conference, CI-Premier Conference Organisation, Singapore, 24-26 August, 1998, pp. 297-302.

THE INFLUENCE OF GROUND BRICK ON THE PHYSICAL PROPERTIES OF MORTAR AND CONCRETE

J Golaszewski

S Suchon

J Szwabowski

Silesian Technical University

Poland

M O'Farrell

B B Sabir

S Wild

University of Glamorgan

United Kingdom

ABSTRACT. This paper reports the results of investigations conducted on mortars and concretes in which the cement was partially replaced by ground brick procured from four different sources, one from each of the four participating countries. The influence of the degree of fineness of the ground brick on the workability of mortar was examined for mixtures with different water/binder ratios and for several cement replacement levels. The influence of ground brick on the porosity and pore size distribution in the mortar was determined by mercury intrusion. Amongst the other variables considered in the present study were the level of cement replacement by ground brick and the water/binder ratio. The paper reports the findings of these investigations and compares the compressive strength development in mortars and concretes containing ground brick procured from the four sources. The results show that, with judicious use, ground brick may be utilised as a pozzolanic replacement material for cement without detrimental effects on the mechanical properties.

Keywords: Ground brick, Mortar, Concrete, Workability, Yield, Viscosity, Pore size, Compressive strength.

Dr J Golaszewski is Lecturer in Building Materials at Silesian Technical University, Poland.

Mr M O'Farrell is Research Assistant at the University of Glamorgan.

Dr B B Sabir is Reader in Engineering Mechanics in the School of the Built Environment at the University of Glamorgan.

Dr S Suchon is Lecturer in Construction at the Silesian Technical University, Poland.

Professor J Szwabowski is professor of Civil Engineering at the Silesian Technical University, Poland.

Professor S Wild is Head of Building Materials Research Unit in the School of the Built Environment at the University of Glamorgan.

INTRODUCTION

The use of pozzolans as partial cement replacement materials in mortar and concrete has been a subject of intensive research for several decades. The interest in the utilisation of waste materials such as fly ash and silica fume was initially a consequence of environmental and economic considerations. In addition to the enhancement in strength, these pozzolans have been found to result in much improved durability properties. Because of these improvements the search for other pozzolans, especially waste materials, continues and recently the potential of the utilisation of waste clay bricks has been the subject of a detailed study by the authors. The use of crushed brick as a pozzolan has been practised in India for many years. When used with lime, in the production of mortar and concrete, crushed bricks result in a reduction in permeability, improved resistance to sulphate attack and alkali-aggregate reaction [1]. These improvements are attributed to a pozzolanic activity, the level of which depends on the chemical composition and stability of the clay structure, both of which depend on the firing temperature. At the high firing temperatures (about 1000 °C), clay building bricks form a liquid phase which on cooling solidifies to an amorphous glass phase which can exhibit high pozzolanicity. The present study presents a comparison of the performance, as pozzolanic material, of clay bricks produced from four different row materials, fired at approximately the same temperatures, using different manufacturing processes. The bricks were ground to particle sizes approximately the same as that of cement and the resulting materials were used in mortars and concretes as partial replacements for cement. The paper reports the results of workability measurements, strength development and pore size distribution of mortars and concretes in which the cement in the control mixtures was replaced by up to 30% ground brick (GB). The study also considers the influence of water/binder ratio on the performance in the fresh and hardened states.

EXPERIMENTAL DETAILS

Materials and Mix Proportions

Waste clay bricks from four sources, one from each country participating in the research programme (Britain - B, Denmark - D, Lithuania - L and Poland – P) were ground and the resulting materials were used as partial replacements for cement in mortar and concrete. The GB was obtained by plate grinding 20-50 mm pieces of brick and then subjecting the resulting material to ball milling to the required particle size. The GBs had specific surfaces (measured using the standard constant flow meter air pearmeability method) in the range 320-350 m^2/kg [2-3], that are approximately the same as that of OPC. Chemical and X-ray diffractometry analyses were carried out on the various GBs [3]. The chemical compositions (x-ray fluorescence spectrometry) and mineralogical (x-ray powder diffractometry) differences between the four GBs are shown in Tables 1 and 2, respectively.

The workability measurements were made on mortars containing GB type B (Hanson Brick, UK.), which was ground to three different particle sizes i.e. coarse (C), medium (M) and fine (F). The grinding was performed in a Mixermill 2000 mill for approximate times of 1, 5 and 20 min for the three gradings, C, M and F, with specific surfaces (measured by nitrogen adsorption using the BET multipoint technique) of 2.56, 3.55 and 4.91 m^2/g respectively. This method tends to give values up to one order of magnitude greater than the air permeability method. The particle size distributions for the three gradings of GB are given in

Table 1 Chemical compositions of the ground bricks

OXIDE (%)	GB - B	GB - D	GB - L	GB - P
SiO_2	54.83	69.99	68.79	72.75
TiO_2	0.97	0.55	0.85	0.84
Al_2O_3	19.05	10.62	15.23	15.89
Fe_2O_3	6.00	4.02	6.28	4.97
MnO	0.06	0.08	0.07	0.02
MgO	1.77	1.39	2.02	1.20
CaO	9.39	8.86	1.79	0.87
Na_2O	0.50	1.02	0.26	0.27
K_2O	3.15	2.61	3.71	2.17
BaO	0.04	0.05	0.04	0.05
P_2O_5	0.20	0.11	0.07	0.10
Cr_2O_3	0.03	0.01	0.02	0.02
SrO	0.05	0.03	0.01	0.01
SO_3	2.90	0.04	0.13	0.07
L.O.I.	1.48	0.25	0.19	0.36

Table 2 Crystalline mineral compositions of ground bricks

MINERAL (%)	GROUND BRICK B	GROUND BRICK D	GROUND BRICK L	GROUND BRICK P
Quartz	49	60	43	88
Feldspar	5	23	24	3
Haematite	10	6	20	5
Cristobalite	12	12	13	4
Spinel	+	+	+	-
Gypsium	17	-	-	-
Anhydrite	6	+	+	-

\+ denotes a trace   - denotes absence

Table 3. The cement was portland cement (PC) grade 42.5 complying with DIN 1164 and PN-88/B-04300. Natural sand complying with PN-88/B-11000 was used. Water/binder (w/b) ratios 0.45, 0.5 and 0.55 with cement replacement levels by GB of 10, 20 and 30% were employed in the workability tests.

In the next series of tests GBs from the four different sources were used to investigate the compressive strength development in mortar. The mortars were made using sea-dredged sand complying with DIN EN 196-1 and PC complying with BS3892: Part 1: 1982. The w/b ratio was 0.5 and the blend:sand ratio was 1:3. In these mixtures the cement in the control mortar was partially replaced by 10, 20 and 30% of each of the four GBs B, D, L and P.

Table 3 Grain size distribution of ground brick B

FRACTION (μm)	COARSE (C)	MEDIUM (M)	FINE (F)
>250	36.8	3.2	3.8
250-45	4.7	4.0	0.4
45-20	20.9	49.8	32.4
20-10	15.9	29.0	50.7
10-5	7.9	4.8	3.3
5-2.5	6.4	2.9	2.4
<2.5	5.0	5.1	6.5

In the last test series concrete mixtures were prepared with cement replacements by D and P ground bricks to study the compressive strength development and the static modulus of elasticity. The grain size distributions of the two GBs are given in Table 4. Three partial cement replacement levels of 10, 20 and 30% were again employed with two w/b ratios, i.e. 0.4 and 0.55. The coarse aggregate was crushed granite of 10 mm maximum size. The mass proportions of binder:sand:aggregate was 1:1.3:2.8. The cement content in the control concrete was 430 kg/m^3.

Table 4 Grain size distributions of ground bricks D and P

FRACTION (μm)	GROUND BRICK D	GROUND BRICK P
>250	0.2	0.2
250-45	27.0	29.5
45-20	25.5	25.5
20-10	14.0	14.0
10-5	10.5	11.5
5-2.5	8.0	6.0
<2.5	15.0	13.0

Sampling and Testing

Approximately 2.2 kg of mortar was prepared for each of the workability tests. The mortars were mixed in a Hobart mixer according to a procedure specified in DIN EN 196-1, briefly described in Table 5. The tests were conducted in a Viscomat PC rheometer developed by Shleibinger Gerate. The equipment is a variable speed viscometer, which consists of a rotating cylindrical container and a centrally positioned stationary blade paddle and a supporting frame. As the container rotates, the material flowing between the blades of the paddle produces a torque, due to the flow resistance, which is registered by a dynamometer. The equipment is controlled by a PC and the data, which includes test duration, speed, torque and temperature, are collected by the internal processor. The test and the data interpretation is based on the Bingham model for visco-plastic materials.

This requires the determination of two parameters; yield value (***g***) and plastic viscosity (***h***), and is thus generally referred to as the two-point workability test. The results of a single test are the parameters of the best-fit straight line for readings of torque taken on a 'down-curve' according to the equation:

$$\mathbf{T} = g + h\,\mathbf{N}$$

In the above equation **T** is the torque registered at rotation speed **N** of the impeller. The intercept on the **T**-axis gives the yield value ***g*** (Nmm) and the gradient of the straight line is a measure of the plastic viscosity ***h*** (Nmm min). The rheological tests were executed at the intervals of 5, 12 and 20 min after first contact between the water and the dry incredients. All tests were conducted at a temperature of 20±3°C.

Table 5 Mixing procedure for the rheology tests

OPERATION	PADDLE ROTARY SPEED (RPM)	DURATION (S)
Cement and water mixing	140±5	30±2
Adding sand		
Mortar mixing	140±5	30±2
Scraping of material adhered to sides of container	285±5	30±2
Further mixing	-	90±2
	285±5	60±2

Mortar cubes, of size 100 mm, were prepared in steel moulds from each of the thirteen mixtures prepared for the compressive strength tests. Three specimens were prepared for testing at each of the ages of 7, 28, 90 and 365 d. After casting, the moulds were covered by cling film to prevent evaporation. The specimens were stripped after 24 h and then cured in water at 20±2°C until testing.

Samples of approximately 2g of hardened mortar were extracted from the broken mortar cubes used for the compressive strength. These were taken from the same regions of the cubes, i.e. the narrowest part of the typical 'hourglass' failure common in mortar and concrete cubes. The samples, which were used for the porosimetry work, were dried at 40 °C under silica gel until a constant weight was achieved. The total drying time was approximately 20 d with approximately 80% of the weight loss occurring in the first 24 h. Porosity and pore size distribution measurements were performed on the dry samples using mercury intrusion. Two Fisons (Italy) instruments were used for this work; the Macropore Unit 120 and the Porosimeter 2000WS. Details of the testing procedure can be found in [4] and [5].

The compressive strengths of the concretes were obtained by testing 100 mm cube specimens prepared and cured in the same manner as the mortar samples. The modulus of elasticity of the concretes was determined by testing cylindrical specimens 300 mm long and 150 mm diameter.

Three cylinders, which were cured under the same conditions as those used for the compressive strength specimens, were tested for each mixture. The test was conducted according to ITB Instruction 194 "Guidelines for testing mechanical concrete properties on samples made in moulds". The testing is controlled by a PC and is based on two loading conditions, which employ stresses that are 1/10 and 1/3 of the compressive strength of the cylinder under test.

RESULTS AND DISCUSSION

Workability of Mortar

It was found that there were no significant differences in the values of ***g*** and ***h*** obtained at the three intervals of 5, 12 and 20 min and the values reported are the averages of the three measurements. The 'yield' and 'viscosity' parameters were obtained from the slopes of the 'best-fit' straight lines of the recorded measurements. High values of correlation coefficients were obtained for the vast majority of the mortars tested. The only exceptions were for the mortars with the higher cement replacement levels by GB (20 and 30%) at the lower w/b ratio of 0.4. It was noticed during the tests, that these two mixtures had a tendency to 'plug flow' rather than 'fluid flow'. The average value of the correlation coefficients was 0.97. Analysis of variance carried out on the basis of the w/b, grading and replacement level of GB showed that both ***g*** and ***h*** were predominantly influenced by the w/b ratio. To a lesser extent ***g*** was influenced by the grading and level of cement replacement by GB. The two- and three-parameter analysis showed that the interactions of w/b ratio and replacement level of GB have the greatest influence on ***g***. The analysis also showed that ***h*** was not influenced by the factors considered other than the w/b ratio.

Figures 1 and 2 give the variation in ***g*** and ***h*** of the mortars against GB content at the three w/b ratios. This shows that ***g*** increases gradually with increasing level of GB in all cases. The increase is greatest at the lower w/b ratios. In general ***g*** was not found to be sensitive to the fineness of the GB except for the mixtures with w/b=0.55 where significantly higher values are obtained for the mortars containing coarse GB. Figure 2 shows that decreasing values of ***h*** are obtained with increasing w/b ratio. The tests did not reveal consistent relationships for ***h*** with either the fineness of GB or the replacement level of cement by GB. This is believed to be due to the complex mechanisms associated with the degrees to which the GB adsorbs or absorbs the mix water, which varies in a non-systematic manner from one mixture to another. There were insignificant changes in ***g*** and ***h*** during the first 20 minutes after mixing. The principal finding which can be deduced from these observations is that unlike silica fume or metakaolin, ground brick has only a minor influence on the rheology of fresh mortar.

Strength Development in Mortar

Table 6 gives the average compressive strengths obtained from testing three cubes. Figure 3 gives the relative strength versus curing time for mortars containing GB of types B, D, L and P. Relative strength is defined as the ratio of the strength of the mortar at a particular age to the strength of the control mortar at the same age. The positive gradients of the curves up to 90 days are indicative of pozzolanic activity associated with the presence of GB and confirm earlier work on pozzolanicity conducted by the authors [3]. With 10 % replacement of cement by GB, the compressive strength at 90 days is close to or greater than that of the control mortar. It is apparent that with 20 and 30% replacements, ground brick D results in considerably reduced strengths compared to those of the mortars containing ground bricks B, L and P. Mortars containing 30% ground brick D develop a only 75% strength of the control.

Strength Development in Concrete

The results for the compressive strength of the GB concretes are given in Table 7. Figure 4 gives the variation in the relative strength of the concretes with curing time for the two types of GB and at the two w/b ratios. It can be seen that for w/b=0.4, both D and P GBs can be incorporated in the concrete by up to 20% without detrimental effects in the long

Table 6 Compressive strength of GB mortars

MIXTURE	7 DAYS	28 DAYS	90 DAYS	365 DAYS
Control	52.2	65.3	67.0	78.8
B10	43.8	61.2	74.0	87.3
B20	39.4	55.4	66.2	82.7
B30	31.2	45.2	55.7	69.2
D10	40.3	55.0	65.2	78.6
D20	30.1	45.0	55.0	66.0
D30	28.9	35.6	49.1	59.8
L10	44.7	59.4	69.1	81.3
L20	33.5	48.5	61.5	72.5
L30	27.8	40.5	55.1	65.2
P10	39.8	55.6	65.5	78.9
P20	34.5	50.1	62.7	74.2
P30	28.3	39.8	57.5	67.8

term. At the higher w/b ratio (=0.55), however, the use of GB results in more than 20% reductions in strength. The results also show that the reductions in the strengths of the GB mortars observed at the early stages of curing (Figure 3) are not manifested in GB concrete. It is suggested that this is associated with the influence of the GB on the interfacial zone packing and also on the inherent strength of the coarse aggregates in the concrete which act compositely with the hardened cement paste.

Porosity and Pore Size Distribution

Porosimetry tests were conducted on samples of mortar extracted from the broken cubes used for the compressive strength tests.

For brevity only the results conducted on the B-type GB mortars, used for the workability tests, are given here. Figures 5 and 6 give the changes in the total intruded pore volume and the % of pore sizes <0.05 μm radius respectively, over a curing period of one year. It can be seen that at the initial stages (7 d)

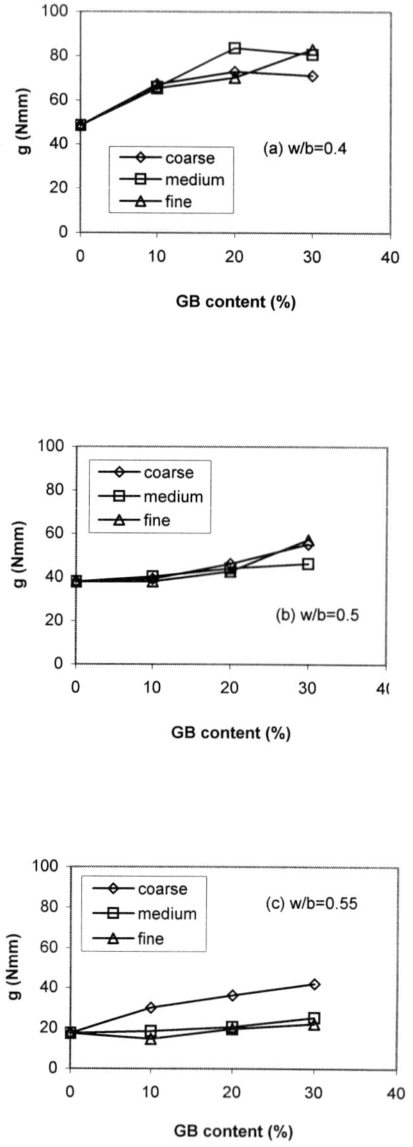

Figure 1 Effect of GB content on the yield value, g

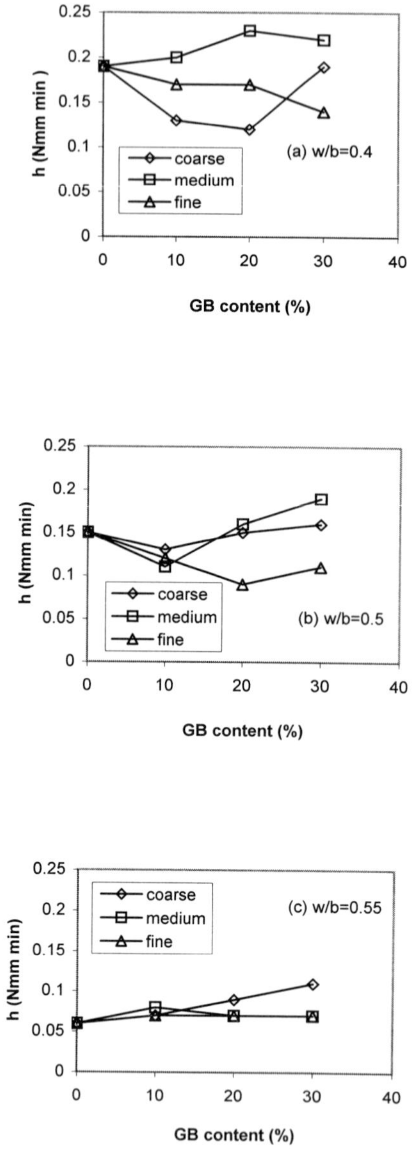

Figure 2 Effect of GB content on the yield value, h

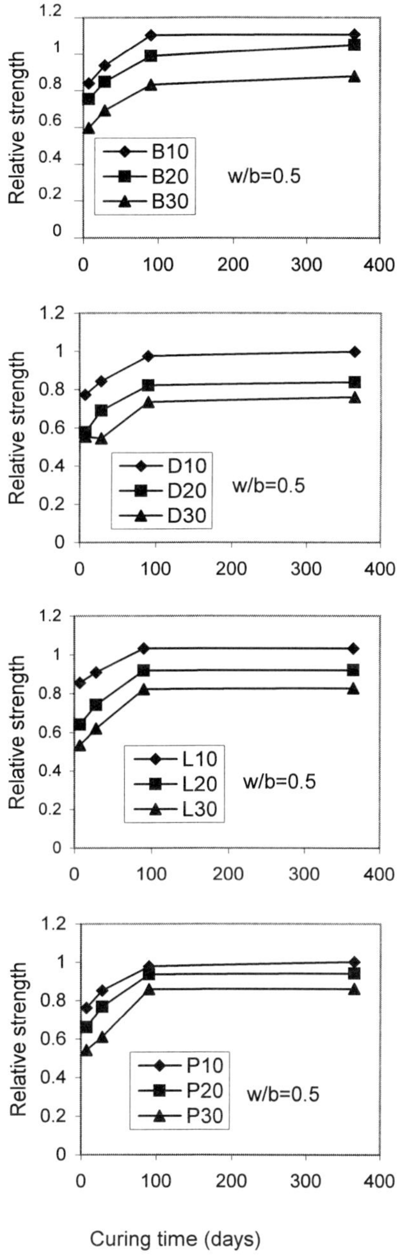

Figure 3 Relative strength of GB mortars

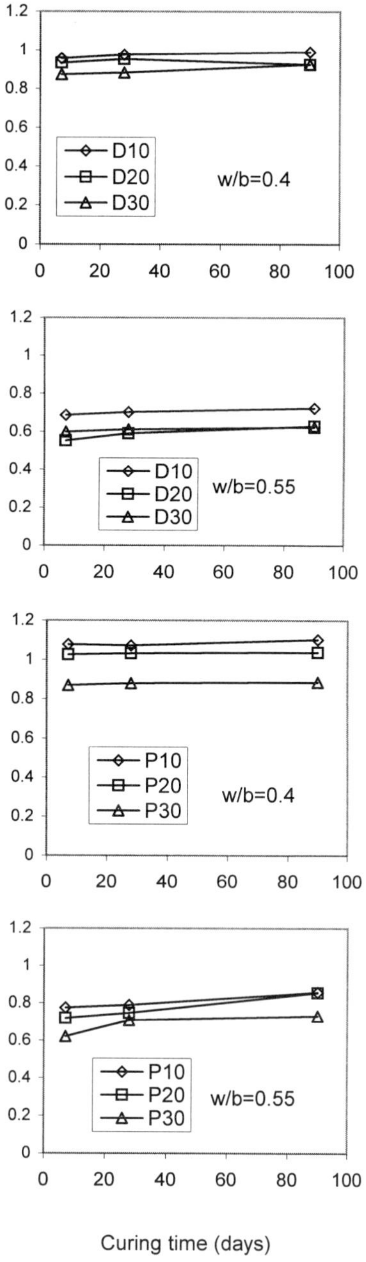

Figure 4 Relative strength of GB concrete

Table 7 Compressive strength of GB concretes

MIXTURE	COMPRESSIVE STRENGTH (MPa)		
	7 Days	28 Days	90 Days
CON40	48.1	53.1	58.3
CON55	43.3	51.5	53.6
D1040	46.0	51.8	57.7
D2040	44.9	50.6	53.9
D3040	42.0	46.8	53.9
D1055	29.7	36.1	38.7
D2055	23.9	30.4	33.7
D3055	25.9	31.5	33.4
P1040	51.8	56.9	64.2
P2040	49.3	54.8	60.4
P3040	41.8	46.7	51.6
P1055	33.5	40.6	45.9
P2055	31.1	38.4	45.8
P3055	26.9	36.4	39.1

the addition of GB results in an increase in the total pore volume. The pore volume increases with increasing level of GB and decreases with curing time. At 365 days, however, the total pore volumes of mortars with up to 20% GB are approximately the same as that for the control mortar. The mortar with 30% GB results in a significant increase in the pore volume over that of the control. The pore refinement of the mortars was assessed on the basis of the percentage of pore volume comprising pores <0.05 μm. It is seen, from Figure 6 that initially (at 7 d) the addition of GB results in greater proportions of the coarser pores than that in the control mortar. With time, however, the mortars with 30% GB give values, which are similar to those of the control. This is in contrast to the results of the total pore volume which are significantly increased when using this replacement level, particularly at 365 d. Replacement of the cement by up to 20% GB, however, does not result in a detrimental effect in the long term (at 365 d).

Modulus of Elasticity

The variations in the elastic modulus of concrete, measured at 28 d, with the level of replacement of cement by GB are shown in Figure 7. It is seen that GB leads to reductions in the values of the modulus with larger reductions taking place with increasing GB. There seems to be no significant difference between the performances of the two GBs used. As with the compressive strength (Figure 4), the elasticity modulus shows marked reductions when the w/b ratio is increased from 0.4 to 0.55.

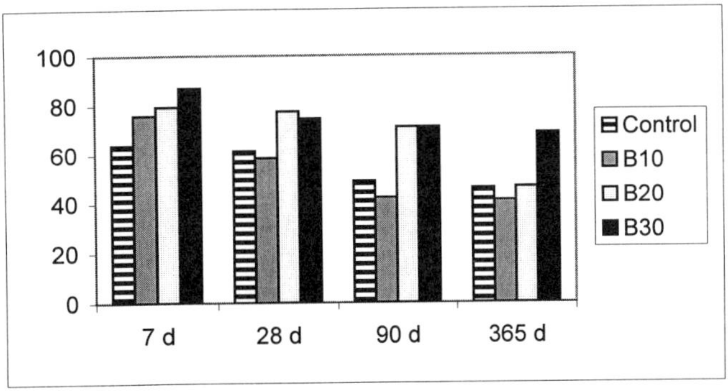

Figure 5 Total intruded pore volume of GB B mortar (mm³/g)

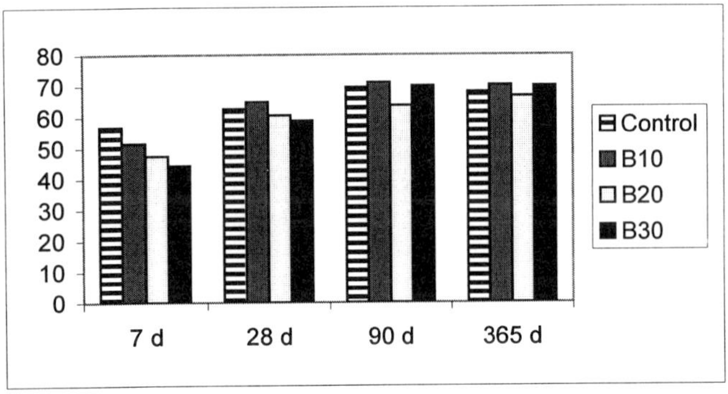

Figure 6 Percentage of pores <0.05mm in GB B mortar

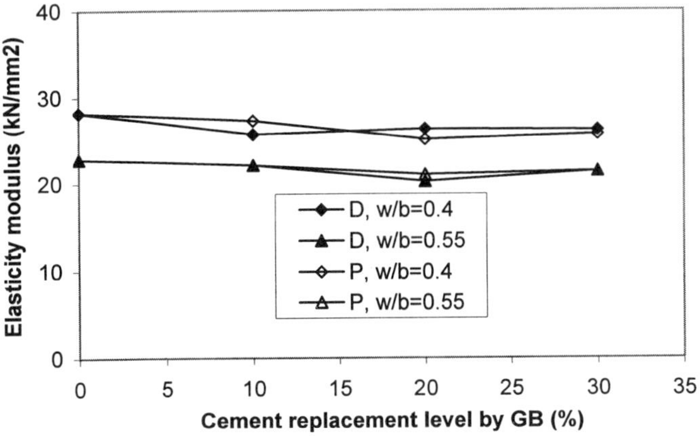

Figure 7 Modulus of elasticity of D and P GB concrete

CONCLUSIONS

The following conclusions may be drawn from the results presented:

The flow characteristics of mortars modified by ground brick follow the Bingham model. The yield and viscosity values are predominantly affected by the water/binder ratio and are not strongly influenced by ground brick content. The viscosity did not show a consistent behaviour with regard to ground brick content and fineness.

The cement in the control mortar may be replaced by 10% ground brick without a detrimental effect on the compressive strength. This is attributed to the pozzolanicity of the ground brick, which gradually builds up during the initial stages of curing. Some ground brick mortars produce strengths at 90 d which are greater than those of the control.

Although ground brick increases the total porosity of the mortar at the initial stages of curing, at extended ages 20% replacement can be used without adverse effects on pore structure.

Ground bricks can be used as partial cement replacement materials with no detrimental effect on the compressive strength of mortar and concrete at low water/binder ratios (0.4-0.5).

ACKNOWLEDGEMENTS

The authors gratefully acknowledge the following. Scientific and Medical Products Ltd for supplying BET data on ground brick specific surface, the European Commission for providing funding for Professor Albinas Gailius for grinding facilities and Helge Hansen for the chemical and mineralogical analysis.

REFERENCES

1. GUPTA, J. S. Technology Option for Manufacture of Calcined Clay Pozzolan (Surkhi). In Lime and other Alternative Cements. Intermediate Technology Publications Ltd., London. Ed. N. Hill, S. Holmes and D. Mather. 1992, pp 191-198.

2. COPERNICUS RESEARCH PROJECT "Recycling of Waste Clay Brick and Tile Material for the Partial Replacement of Cement in Concrete", Contract No. CIPA-CT94-0211, First Annual Report, European Commission, Brusels, 1996.

3. WILD, S., GAILUS, A., HANSEN, H., PEDERSON, L. and SZWABOWSKI, Y. Pozzolanic Properties of a Variety of European Clay Bricks, Building Research Information, Vol. 25, No. 3, 1997, 170-175.

4. KHATIB, J. M. and WILD, S. Pore Size Distribution of Metakaolin Paste, Cement and Concrete Research, Vol. 26, No. 10, 1996, pp 1545-1553.

5. O'FARRELL, M., KHATIB, J. M. and WILD, S. Porosity and Pore Size Distribution of Mortar Containing Ground Calcined Brick Clay, Proceedings of the 5th International Conference on Modern Building Materials, Structures and Techniques, Vilnius, Lithuania, May 1997, pp 33-38.

AN INNOVATIVE METHOD IN PRODUCING HIGH EARLY STRENGTH PFA CONCRETE

C S Poon S C Kou L Lam

Hong Kong Polytechnic University

Hong Kong

Z S Lin

Wuhan University of Technology

China

ABSTRACT. The paper introduces a new method (patent pending) in producing high early strength PFA concrete by accelerating the hydration of the PFA particles. The method involves the incorporation of a small amount of plaster stone (a kind of anhydrite gypsum), a natural mineral widely available in some parts of China. The formulated concrete mixes were allowed to cure in an accelerated environment for a short period of time before normal water curing. Significant (up to 100%) strength improvement compared to the control mixes was observed as early as after 3 days' curing. The results showed that the addition of 10% anhydrite gypsum, by the total weight of cement and fly ash, increased the 3-day compressive strength by up to 100% for the mix with high fly ash contents. It also increased the strength at later ages for these concretes. The results also indicated that the activating effect of anhydrite gypsum was insignificant to the concrete with low fly ash contents.

Keywords: Fly ash, Concrete, Curing, Steam, Gypsum, Anhydrite, Compressive Strength

Dr C S Poon is an Associate Professor at the Department of Civil and Structural Engineering at The Hong Kong Polytechnic University.

Mr S C Kou was a Research Assistant at the Department of of Civil and Structural Engineering, The Hong Kong Polytechnic University. He is currently a Lecturer at the Wuhan University of Technology.

Mr L Lam is a PhD candidate at the Department of Civil and Structural Engineering, The Hong Kong Polytechnic University.

Professor Z S Lin is Professor in cement materials at the Wuhan University of Technology.

INTRODUCTION

Fly ashes produced by coal-fired power plants are generally suitable for use as a partial replacement of cement in mortar and concrete. However, the different nature of coal and combustion processes produce fly ashes with different chemical and physical characteristics. Studies on the properties of these materials and their influence on mortar and concrete properties are necessary for their effective use. Recently, fly ash is being increasingly used in concrete because it lowers the cost and improves the properties of concrete. However, it is generally regarded that the replacement of Portland cement with fly ash, especially in high volume, decreases the early strength of concrete.

In general, fly ash improves the properties of concrete in three ways: (1)by reduction of water requirement for a given slump; (2) by increase of volume of paste thereby improving the workability and (3) by a pozzolanic reaction between fly ash and $Ca(OH)_2$ [1]. As pozzolanic reactions occur at later ages, fly ash contributes little to the early strength of concrete.

Different methods have been used to accelerate the pozzolanic reaction between fly ash and $Ca(OH)_2$ to increase the early strength of the concrete containing fly ash. Increasing the fineness of fly ash apparently increases the pozzolanic reactivity of fly ash [2-4]. Elevating the curing temperature is also beneficial to the early strength development of fly ash concrete [5-6]. Some studies indicated that the addition of chemical activators can effectively accelerate or improve the pozzolanic activity of natural pozzolans [7-10]. It was reported that addition of 3% to 6% gypsum to a low alkali low C_3A cement blended with varying amount (30%-60%) of class F fly ash [11] resulted in a distinct increase in strength, when compared with the concrete blends without the addition gypsum. Although a few studies were carried out on the chemical activation of the reactivity of fly ash [12-14], there is little information available on using anhydrite gypsum as an activating agent.

EXPERIMENTAL DETAILS

Material Properties

The materials used to prepare the concrete mixtures were ASTM Type I Portland cement, ASTM Class F fly ash (low calcium), a crushed granite aggregate, and natural river sand. Natural anhydrite gypsum from Jiao Cheng, Shanxi, China was used as an activating agent. The gypsum was ground in a mortar to achieve the desired fineness before use. The chemical and physical properties of the cement, fly ash and anhydrite gypsum are given in Tables 1 and 2.

Table 1 Chemical composition of cement, fly ash and anhydrite gypsum

MATERIALS	LOSS ON IGNITION (%)	SiO_2 (%)	Fe_2O_3 (%)	Al_2O_3 (%)	CaO (%)	MgO (%)	SO_3 (%)
Cement	2.97	19.61	3.32	7.33	63.15	2.54	2.13
Fly Ash	3.90	56.79	5.31	28.21	<3	5.21	0.68
Anhydrite Gypsum	7.53	1.88	0.02	0.04	37.94	1.57	51.02

Table 2 Physical properties of cement, fly ash and anhydrite gypsum

PROPERTIES	CEMENT	ANHYDRITE GYPSUM	FLY ASH
Density	3.16	2.91	2.31
Specific Surface (cm^2/g)	3519.5	12503.2	3960

Concrete Mix Design

Three series of concrete mixtures were prepared. Series 1 was prepared with 15 to 55% fly ash (based on the total weight of the cement and fly ash in the concrete), without the addition of anhydrite gypsum. Series 2 was with 15 to 55% fly ash, and an addition of 10% anhydrite gypsum. Series 3 was the concrete mixtures containing a fixed fly ash content (35%), and varied (6% to 14%) anhydrite gypsum contents. All the above percentages were based on the total weight of the cementitious materials (cement + fly ash) in the concrete mixture. All the concrete mixtures were designed to have the same cementitious materials content of 500 kg/m^3, a water content of 150 kg/m^3, and a sand content of 725 kg/m^3. The water-to-cementitious materials ratio was 0.3. Superplasticizer was used to achieve an adequate workability of the concretes. The details of the concrete mix design are given in Tables 3.

Specimen Preparation, Curing and Testing

The concrete mixtures were prepared in the laboratory using a pan mixer. 100 x 100 x 100 mm cubes were cast in steel moulds and compacted by a vibrating table. The cubes were removed from the moulds at 24 hours after casting. The cubes were subjected to a period of steam curing. The conditions of the steam curing step had been optimized in a preliminary test in which a 6-hour steam curing at 65 ^{o}C produced the best results for the early and later compressive strength for the concrete mix with 35% fly ash with 10% anhydrite gypsum. After curing in a steam bath for 6 hours, the cubes was cooled to room temperature, and were then transferred to a water curing tank at 27 ^{o}C, in which the cubes were allowed to continue to cure until the testing age.

Compression strength test was performed on the cubes using a Denison compression machine, at the ages of 3, 7, 28 and 90 days.

RESULTS AND DISCUSSION

In this section, the average values of the test results are presented. Each value was obtained from three measurements.

Concrete Mixes without Addition of Anhydrite Gypsum

The results of the compression strength test of the concrete mixes without the addition of anhydrite gypsum are shown in Table 4. In a previous study by the authors [15], the strength development of the fly ash concretes prepared with the same materials, the same water/binder ratio, and the same fly ash replacement levals, as the concrete mixes in the present experiment, but without the initial steam curing was studied.

The results are given in Table 5 for reference. When comparing Table 4 and Table 5, it can be noticed that at the age of 3 days, the initial steam curing step resulted in about 15% increase in compressive strength, but at the later ages, the beneficial effect of the initial steam curing step was not so significant.

Table 3 Mix proportion of concrete

SERIES	MIX	WATER (kg/m^3)	CEMENTITIOUS MATERIALS* (kg/m^3)	FLY ASH CONTENT (%)	ADDITION OF ANHYDRITE GYPSUM (%)	CRUSHED GRANITE (kg/m^3)	SAND (kg/m^3)
	PC1	150	500	15	0	1020	725
	PC2	150	500	25	0	970	725
1	PC3	150	500	35	0	922	725
	PC4	150	500	45	0	873	725
	PC5	150	500	55	0	824	725
	C1	150	500	15	10	977	725
	C2	150	500	25	10	920	725
2	C3	150	500	35	10	880	725
	C4	150	500	45	10	831	725
	C5	150	500	55	10	782	725
	Y1	150	500	35	6	870	725
	Y2	150	500	35	8	888	725
3	Y3	150	500	35	10	880	725
	Y4	150	500	35	12	871	725
	Y5	150	500	35	14	863	725

*Cementitious materials = cement + fly ash

Table 4 Compressive strength of the fly ash concrete mixes without addition of anhydrite gypsum (cured in steam at 65 °C for 6 hours before water curing)

MIX	FLY ASH (%)	ANHYDRITE GYPSUM (%)	COMPRESSIVE STRENGTH f_C (MPa)			
			3 days	7 days	28 days	90 days
PC1	15	0	59.08	59.08	88.47	107.60
PC2	25	0	54.24	54.24	86.66	110.30
PC3	35	0	46.43	46.43	78.93	101.47
PC4	45	0	39.42	39.42	70.79	94.47
PC5	55	0	25.75	25.75	62.28	85.20

Concrete Mixes with 10% Anhydrite Gypsum

The results of the compression strength test of the concrete mixes with the addition of 10% anhydrite gypsum are shown in Table 6. Comparing with Table 4, it can be noticed that the addition of 10% anhydrite gypsum resulted in significant compressive strength increase. The extent of the increase is dependent on the percentages of fly ash replacement and curing ages.

At 3 days, the addition of 10% anhydrite gypsum resulted in 115% increase in compressive strength for the mix with 55% fly ash.

However, the increase was only 23% for the mix with 15% fly ash. As the curing age increased, the strength enhanchment effect of anhydrite gypsum became smaller. At 90 days, the addition of anhydrite gypsum resulted in 14% increase in compressive strength for the mix with 55% fly ash, but only a few percents for the mixes with lower levels of fly ash replacement. The percentages of the strength increase are plotted against the curing age in Figure 1.

Table 5 Compressive strength of the fly ash concrete mixes (cured in water at 27 °C) [15]

WATER/BINDER RATIO	FLY ASH (%)	COMPRESSIVE (MPa) STRENGTH f_C			
		3 days	7 days	28 days	90 days
0.3	0	64.9	75.5	86.8	95.7
0.3	15	52.1	66.4	86.0	107.6
0.3	25	48.0	65.7	85.4	110.3
0.3	45	34.0	49.2	71.8	94.5
0.3	55	22.3	36.4	54.7	85.2

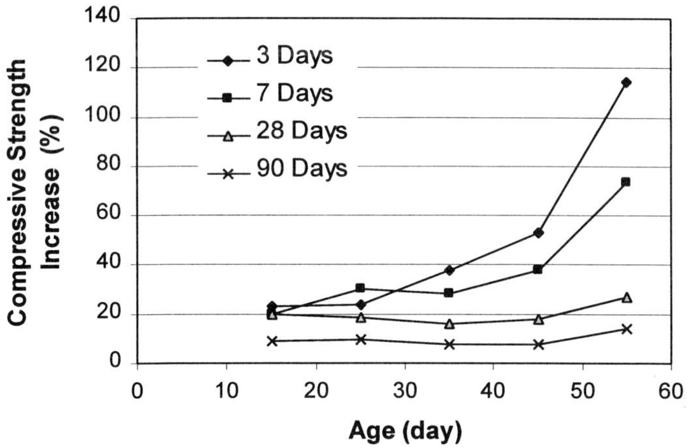

Figure 1 Compressive strength increase due to the addition of 10% anhydrite gypsum

The above results have shown that anhydrite gypsum has an activating effect on the reaction of the fly ash in concrete. With an addition of 10% anhydrite gypsum and an initial curing at 65 °C, the compressive strength of the mix with 55% fly ash reached 55 MPa at 3 days and 97 MPa at 90 days which was almost the same as that of the plain cement concrete mixes [15] in which fly ash, anhydrite gypsum and initial steam curing were not used. These results indicate a new approach for improving the strength properties of high volume fly ash concrete.

It should be noted that the anhydrite gypsum used in this study had a very small particle size (12503 cm^2/g in specific surface area, see Table 2). It may also serve as a binder material with micro filler effects. As the anhydrite gypsum was used as an addition to the cementitious materials rather than a replacement, it lowered the actual water/binder ratio, which may also result in higher compressive strength.

Optimization of Anhydrite Gypsum Content and Steam Curing Conditions

The effects of anhydrite gypsum content and steam curing conditions (curing temperature and curing time) on compressive strength were investigated with the mix with 35% fly ash replacement. Table 7 shows that 10% anhydrite gypsum resulted in higher compressive strength than lower or higher levels of addition of anhydrite gypsum. Table 8 shows that the 3-day compressive strength of the specimens with the initial curing at 65 °C was about 50% higher than that of 45 °C. Table 9 shows that at 65 °C, a period of 6 hours initial curing period is sufficient. From Tables 8 and 9, it can also be noticed that an initial curing at the temperature higher than room temperature is necessary and essential when anhydrite gypsum is used.

Table 6 Compressive strength of the fly ash concrete mixes with addition of 10 anhydrite gypsum (cured in steam at 65 °C for 6 hours before water curing)

MIX	FLY ASH (%)	ANHYDRITE GYPSUM (%)	COMPRESSIVE STRENGTH f_c (MPa)			
			3 days	7 days	28 days	90 days
C1	15	10	72.62	83.49	106.13	117.33
C2	25	10	67.07	80.70	102.97	121.23
C3	35	10	63.87	75.07	91.40	109.20
C4	45	10	60.42	68.70	83.30	101.63
C5	55	10	55.29	63.39	78.86	97.37

SUMMARY

An attempt has been made to increase the early compressive strength of fly ash concrete. The experimental results showed that anhydrite gypsum has an activating effect on the pozzolanic reaction of the fly ash in concrete. An addition of 10% anhydrite gypsum increased the 3-day compressive strength by about 100% for the concrete with high fly ash contents. It also increased the strength at the later ages for these concretes.

Table 7 Effect of anhydrite gypsum content on compressive strength (cured in steam at 65 °C for 6 hours before water curing)

MIX	FLY ASH (%)	ANHYDRITE GYPSUM (%)	COMPRESSIVE STRENGTH (MPa)			
			3 days	7 days	28 days	90 days
Y1	35	6	57.93	68.72	79.56	99.69
Y2	35	8	59.92	70.07	85.28	101.06
Y3	35	10	63.87	75.07	91.40	109.90
Y4	35	12	55.68	67.84	86.45	109.45
Y5	35	14	54.35	64.34	82.39	103.65

Table 8 Effect of initial steam curing temperature on compressive strength (cured in steam for 6 hours before water curing)

FLY ASH (%)	ANHYDRITE GYPSUM (%)	STEAM CURING TEMPERATURE (°C)	COMPRESSIVE STRENGTH (MPa)			
			3 days	7 days	28 days	90 days
35	10	45	42.14	54.80	72.46	97.56
35	10	55	56.31	61.59	85.50	105.75
35	10	65	63.87	75.07	91.40	109.90
35	10	75	67.46	78.80	91.92	107.69
35	10	85	67.40	79.74	93.43	108.22

Table 9 Effect of steam curing time on compressive strength (cured in steam at 65 °C before water curing)

FLY ASH (%)	ANHYDRITE GYPSUM (%)	STEAM CURING TIME (HOUR)	COMPRESSIVE STRENGTH (MPa)			
			3 days	7 days	28 days	90 days
35	10	4	54.66	70.38	84.32	102.46
35	10	6	63.87	75.07	91.59	109.90
35	10	8	64.79	76.05	91.89	108.74
35	10	10	65.59	76.32	91.40	108.54
35	10	12	66.79	78.42	93.18	110.25

In these cases, the initial curing at the temperature higher than room temperature is necessary and essential. However, the activating effect of anhydrite gypsum is insignificant to the concrete with low fly ash content.

The results of present study indicate a new approach for improving the compressive strength of high volume fly ash concrete. More microstructure work is being carried out to devise the mechanism for strength enhancement. Also as other properties such as durability of the concrete containing anhydrite gypsum are not clear, further investigation is needed.

REFERENCES

1. S. POPOVICS, Proceeding of 2nd International Conference on Fly ash, Silica Fume, Slag and Natural Pozzolans in Concrete, SP-9, PP. 313-33, American Concrete Institute, Chicago, 1986.

2. R. K. DHIR, F.H. HUBBARD, J. G. L. MUNDY AND M. R. JONES, Proceeding of 2nd International Conference on Fly ash, Silica Fume, Slag and Natural Pozzolans in Concrete, SP-91, PP. 693-721, American Concrete Institute, Chicago, 1986.

3. Z.GIERGICZNY AND WERYNSKA, Proceeding of 3rd International Conference on Fly ash, Silica Fume, Slag and National Pozzolans in Concrete, SP-114, PP. 97-115. American Concrete Institute, 1989.

4. J. D. WATT AND D. J. J. THORNE. Journal of Applied Chemistry, Vol, 16, No.1, PP. 33-39, 1966.

5. V. R. RIVERA, Proceeding of 2nd International Conference on Fly ash, Silica Fume, Slag and Natural Pozzolans in Concrete, SP-91, Vol..1, PP. 219-230, American Concrete Institute, Chicago, 1986.

6. J. D. WATT AND D. J. J. THORNE, Journal of Applied Chemistry, Vol, 15, PP.595-604, 1966.

7. C. SHI AND R. L. DAY, Cem. Concr. Res., Vol.23., No.4, PP.824-832, 1993.

8. C. SHI AND R. L. DAY, Cem. Concr. Res., Vol.23., No.6, PP.389-396, 1995.

9. C. SHI AND R. L. DAY AND A. HUIZER, A., Journal of British Masonry Society. Vol. 6, PP. 1-3, March 1994.

10. C. SHI AND R. L. DAY, Roceeding of 10th International Bricks/ Blocks Masonry Conference, Vol.3, PP. 1153-1162, Calgary, Canada, July 5-7, 1994.

11. X. AIMIN, S. L. SARKAR, Cem, Concr. Res, Vol. 26, No 6, PP1137-1138, 1991.

12. F. M. LEA, The Chemistry of Cement and Concrete, 3rd Edition, Edward Arnold, 1974.

13. N. G. DAVE, Proceeding of Australian Conference on Engineering Materials, Cook. D. J. ed., Sydney, PP.623-638, 1981.

14. C. SHI AND R. L. DAY, Cem. Concr. Res., Vol.25., No.1, PP.15-21, 1995.

15. L. LAM, Y. L. WONG, AND C. S. POON, Effect of Fly Ash and Silica Fume on Compressive and Fracture Behaviours of Concrete. Cem. Concr. Res., Vol. 28, No. 2, PP. 271-283, 1998.

EFFECT OF BLENDING WITH KAOLIN OR DIATOMITE ON CHARACTERISTICS OF PORTLAND CEMENT PASTE AND MORTAR

P Stroeven

Delft University of Technology

Netherlands

V D Dau

Hanoi University

Vietnam

ABSTRACT. Partial replacement of Portland cement by calcined kaolin can lead to savings in countries such as Vietnam where clays with high alumina content are indigenous. Experiments were therefore performed to assess the potentiallities of using kaolin in combination with cementitious pastes, mortars and concretes. Optimum replacement percentages could be established leading to significant increases of compressive strength at lower Portland cement ratio's. This paper reports on data pertaining exclusively to the pastes and mortars. Incinerated diatomite was added to the research program for as a reference mineral admixture. It was however experimentally shown to be able competing with the kaolin.

Keywords: Bending, Cement blending, Compression, Consistency, Diatomite, Kaolin clay, Mineral admixture, Portland cement.

Professor Piet Stroeven is Head of the Group on Materials Engineering of the Division of Mechanics of Materials and Structures, Faculty of Civil Engineering and Geosciences, Delft University of Technology, The Netherlands. He has published widely on subjects in concrete technology relevant for industrialized as well as developing countries, such as micromechanics, mechanisms of damage evolution, bond, fibre reinforcement, cement blending. He serves as a member in RILEM committee TC162-TDF. He is a member of the Advisory Board of two International Journals.

Vu Dinh Dau MSc is assistant professor at the Faculty of Building Materials, Hanoi University of Civil Engineering in Vietnam. His teaching responsibilities and research activitics arc in thc ficld of concrctc tcchnology, with spccial cmphasis on thc optimum use of indigenous resources in Vietnam, like mineral admixtures, and natural and artificial pozzolanas

INTRODUCTION

Indigenous resources are an asset of developing countries that require optimum exploitation. In a long-ranging cooperation program with Hanoi University of Technology, some of Northern Vietnam's natural resources which may provide economical solutions to the cementitious building materials market were investigated. Particularly, the use of rice husk ash was emphasized. Also some attention was paid to available fly ashes. Another potentially interesting mineral admixture is kaolin clay. Various clays with high alumina content are found in the region and could be exploited. It is well-known that by proper combustion, an artificial pozzolana can be obtained. The raw material could even by mixed with the rice husks during incineration, so that no additional fuel would be required, enhancing the economy of the production process.

The blending of Portland cement by mineral admixtures has also a durability aspect. In the presence of moisture, the free lime in cementitious materials makes the material vulnerable to sulphate attack or leaching. Blending the Portland cement with a mineral admixture capable of combining with the free lime reduces therefore such durability risks. The objective was to appeal to indigenous resources for this purpose. The artificial pozzolona obtained by incinerating rice husks fulfilled this requirement. It was shown to lead to economic and technologically interesting solutions [1]. Alumina-rich clays are abundant in Northern Vietnam. By incinerating such kaolin clays, the kaolinite is transformed into the artificial pozzolana metakaolin. Since properties of such pozzolanic materials fluctuate widely, depending on their composition and structure [2-4], it was considered relevant to assess the potentialities of representative samples of kaolin by an experimental approach.

An experimental program was set up for that purpose at Delft University of Technology, using in Northern Vietnam available mineral admixtures, such as the ones presented in Tables 1 and 2. The applied kaolin and diatomite were representative as to the alumina content in the average Vietnamese material (respectively, Al_2O_3-contents of 35.7 and 11.9%).

Table 1 Composition of local mineral resources: Clay (C)/ Kaolin (K)

COMPOSITION % BY WEIGHT	REGION					
	Tuyen Quang (C)	Vinh Phu (C)	Hai Hung (C/K)	Quang Ninh (K)	Thanh Hoa (K)	Son Tay (K)
SiO_2	50-60	44-46	55-60/45-63	43-70	48	54-55
Al_2O_3	36-42	34-37	36-40/24-35	61-63	31	27-29
Fe_2O_3	2.5-3.5	0.9-0.3	4.0-5.0/0.8-1.3	0.1-0.7	1.8	0.8-0.9
CaO	0.5	0.6-0.7	0.4/0.3	0.5-1.2	1.3	1.1-1.5
MgO	1.5-1.8	0.5	0.3/0.4-1.2	0.2	0.7	0.3-0.6
K_2O	0.6-0.8		1.7-2.4/1.3-3.5	0.2	3.1	0.8-5.9
TiO_2	0.8-3.0		1.0-2.0/ -	1.0		
LOI	9-11	14	7-11/7	14	16	

Table 2 Composition of local mineral resources: Diatomite (D)

COMPOSITION % BY WEIGHT	REGION			
	Contum	Tuy Duong	Tuy Hoa	Quang Ninh
SiO_2	48-55	53-67	66	54-59
Al_2O_3	29-30	15-24	15	11-15
Fe_2O_3	3.4-3.9	2.0-9.2	3.2	6.5-9.7
CaO	0.1	0.9-1.1	0.6	0.3-0.6
MgO	0.5-0.7		0.6	0.4-0.7
TiO_2	1.2		0.6	1.1-1.5
LOI	9-14	8-12	9	7-8

The present study more specifically focussed on [5,6]:

- pozzolanic activity of calcined kaolin;
- effect of blending by kaolin and diatomite on consistency and setting time;
- efect of blending by kaolin and diatomite on strength of pastes and mortars;
- effect of combined use of kaolin and superplasticizer on strength of mortars;
- effect of the amount of mixing water or water to binder ratio on blended mortars.

MATERIALS

Aggregate in Northern Vietnam consists of broken rock (granite) and very fine river sand (with a fineness modulus around 1), leading to gap-graded mixes [7,8]. Further, only a low quality PC (similar to ASTM type I) is presently available. Since the experiments were performed in The Netherlands, these indigenous products were replaced by similar 'Dutch' alternatives. This holds for the sand (S) and the Portland cement (PC). Tables 3 to 5 specify the relevant characteristics of the used material components.

- Sand was composed of two fractions, *i.e.* 0.1-0.25 mm and 0.2-0.5 mm, in the proportions 3:7 by weight. It was employed for making mortars. The grading of the sand was changed to fulfil the requirements of ASTM-C778/80 for assessing pozzolanic activities.
- A Portland cement CEM I 32R was used (equivalent to ASTM type I).
- The kaolin (K) was of Czech origin.
- A superplasticizer type TM OFT-3 was employed.

The clay was dried, whereupon the powder-like kaolin was incinerated for two hours at about 800 °C. It was allowed to cool down in the oven. The diatomite was crushed and thereupon subjected to the same firing and cooling cycle. In this way, a proper quality of artificial pozzolan was obtained from the clay, i.e. the pozzolanic activity index according to ASTM C595-81 amounted 93.5 % (minimum requirement is 75 %). Incinerating at 750 °C and at 850 °C yielded slightly lower values, *i.e.* 93.0 and 86.3 %.

The reactivity of the incinerated kaolin primarily depends on metakaolin content. This will also govern the improvements in paste, mortar and concrete strength [9-11]. The diatomite was ground for 8 hours in a laboratory ball mill before being used.

Table 3 Chemical and mineral composition of Portland cement

COMPOSITION	CONTENT (%)
CaO	64.0
SiO_2	21.0
Al_2O_3	5.0
Fe_2O_3	3.0
MgO	2.0
Na_2O	0.6
C_3S	63.0
C_2S	13.0
C_3A	8.8
C_4AF	9.0
SO_3	3.0

EXPERIMENTAL DETAILS

The Portland cement was blended by 10, 15, 20, 30 and 40 % calcined kaolin, or by 10, 20, 30 and 40 % diatomite, both by weight. The setting time of the blended cements was determined according to ASTM C191-79, and normal consistency in agreement with ASTM-C187. Results are presented in Table 6. For the mechanical tests, maximum blending percentage was restricted to 30. The tests were conducted according to ASTM C438-80. 40x40x160 mm prisms (for the mortars) and 40 mm cubes (for the pastes) were made. Pastes were tested in compression only. Mortars were tested in bending and compression (on the pieces of the broken prisms).

Center-point loading was used for determination of tensile strength. Strength data are the average of 3 specimens, except for the compression test on 6 pieces of mortar. Pastes and mortars were produced in a mixer, mixing time being 150 seconds in case no superplasticizer (SP) was employed and 180 seconds when SP was used.After casting, the specimens were stored in vertical position for 22 to 24 hours under controlled conditions of temperature (20 ^{o}C) and humidity (99 %RH). Thereupon, they were demoulded and re-stored under the same conditions until the time of tensile and compression testing at 1, 3, 7, 28, 60 and 90 days, respectively. Due to space restrains, only part of the data can be discussed.

RESULTS

The compressive strength data pertaining to the koalin-blended Portland cement pastes, defined in Table 6, are presented in Table 7.

Table 4 Chemical composition of calcined kaolin and diatomite

COMPOSITION	CONTENT K (%)	CONTENT D (%)
Free water at 105 °C	1.10	9.70
LOI at 800 °C	11.90	7.00
Na_2O	0.07	0.12
K_2O	1.65	0.47
SiO_2	46.60	57.90
Al_2O_3	35.70	11.90
Fe_2O_3	0.85	8.90
MgO	0.35	0.42
CaO	0.35	0.30
TiO_2	0.26	1.17
P_2O_5	0.11	0.27

Table 5 Particle size distribution of sand, and calcined materials

	ACCUMULATED PERCENTAGE BY WEIGHT RETAINED		
SIEVE (mm)	Sand	Kaolin	Diatomite
0.6	4.2		
0.425	40.2		
0.3	65.1		
0.15	82.7		
0.08		5.68	6.69
0.06		8.28	6.99
0.045		11.78	7.99
0.02		28.74	15.92
0.01		40.48	21.65
0.005		57.18	28.65

Compressive strength data of blended mortars are collected in Table 8 (mixes with superplasticizer) and in Table 9 (mixes without superplasticizer). For comparison reasons, the 28-days bending tensile strength is added. The effects revealed by both loadingmodes were found similar. Note that the code indicates the amount of superplasticizer (as % by wt). Figure 1 shows the development of compressive strength with increasing blending percentage of mortars with and without superplasticizer. At older age, the optimum in the curves tends to shift to higher blending percentages [12].

Table 6 Normal consistency and setting times of kaolin-blended Portland cement

CODE	PC (%)	K (%)	W/(PC+K)	SETTING TIME, MINUTES Initial	Final
P	100		0.25	215	275
PK1	90	10	0.30	210	270
PK2	85	15	0.33	215	280
PK3	80	20	0.36	215	280
PK4	75	25	0.38	220	285
PK5	70	30	0.41	225	290
PK6	60	40	0.45	245	305

Table 7 Compressive strength of K-blended PC pastes

CODE	BLENDING % BY WT	COMPRESSIVE STRENGTH, MPa 1 Day	3 Days	7 Days	28 Days	60 Days	90 Days
P	0	38.5	64.5	77.0	83.7	92.4	99.4
PK1	10	33.3	58.9	69.7	78.6	89.5	91.8
PK2	15	26.4	47.5	62.1	68.9	73.9	80.2
PK3	20	22.3	40.3	59.1	66.1	67.3	74.6
PK4	25	14.1	31.8	46.2	63.1	66.6	68.8
PK5	30	11.1	27.4	41.1	58.2	61.4	62.3

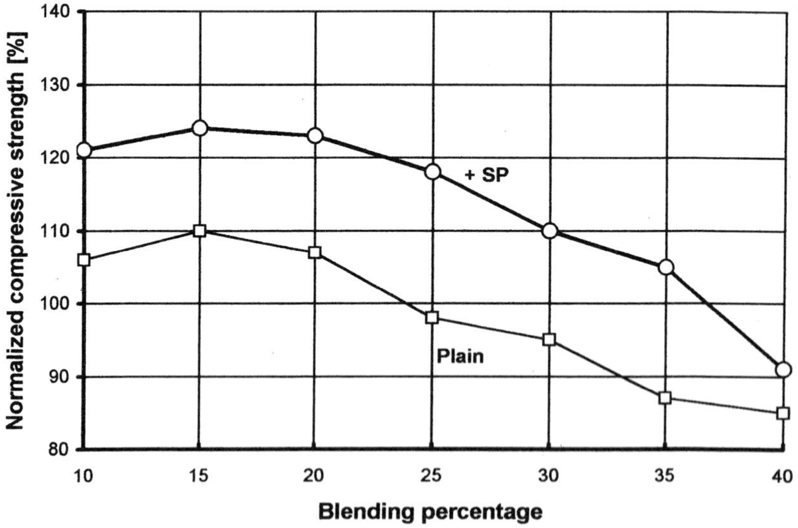

Figure1 Normalized 28-days compressive strength of kaolin-blended PC mortars with ('+ SP') and without ('plain') superplasticizer

Table 8 Compressive and 28-days bending strength of K-blended-PC mortars + SP

CODE	BLEND % BY WT	W/B-RATIO	COMPRESSIVE AND 28 DAY BENDING STRENGTH, MPa					
			1 Day	3 Days	7 Days	28 Days	60 Days	90 Days
PK1-1	10	0.43	4.2	12.6	21.0	28.3/5.0	31.9	33.7
PK2-1	15	0.44	4.3	11.9	22.6	28.3/4.9	33.0	34.5
PK3-1	20	0.45	3.0	11.5	23.0	28.7/5.0	33.0	34.6
PK4-1	25	0.46	2.9	11.4	20.8	27.4/4.9	32.0	33.9
PK5-1	30	0.47	2.4	11.2	16.4	26.4/4.8	30.8	32.3
PK1-1.5	10	0.41	1.7	11.3	20.3	28.6/5.3	32.0	33.1
PK2-1.5	15	0.42	2.0	11.3	21.4	29.3/5.1	32.4	34.4
PK3-1.5	20	0.43	1.8	11.5	18.9	29.2/5.4	32.1	34.5
PK4-1.5	25	0.44	1.6	11.7	18.2	28.0/5.0	31.9	34.6
PK5-1.5	30	0.45	1.5	10.2	14.7	26.0/4.9	31.4	32.8
PK1-2	10	0.40	2.9	14.2	21.3	28.6/5.0	32.3	34.2
PK2-2	15	0.41	1.6	12.3	19.8	28.7/5.0	32.8	34.7
PK3-2	20	0.42	1.5	10.8	18.3	28.7/5.3	32.3	34.4
PK4-2	25	0.42	1.4	7.3	16.4	27.1/4.8	32.6	33.1
PK5-2	30	0.43	2.1	6.8	15.6	24.7/4.6	31.5	32.8

Table 9 Compressive and 28-days bending strength of blended-PC mortars, no SP

CODE	BLEND % BY WT	W/B-RATIO	COMPRESSIVE AND 28-DAYS BENDING STRENGTH, MPA					
			1 Day	3 Days	7 Days	28 Days	60 Days	90 Days
PD1	10	0.48	4.5	12.1	18.1	24.5/5.0	28.3	31.9
PD2	20	0.50	4.1	10.2	15.2	24.0/5.0	28.5	32.4
PD3	30	0.52	4.0	9.6	13.6	22.3/4.8	27.6	31.7
P	0	0.48	4.0	12.9	16.6	23.7/4.8	28.2	29.7
PK1	10	0.48	3.5	13.2	19.0	25.2/5.0	29.2	31.2
PK2	15	0.50	3.4	12.2	18.2	26.2/5.0	30.9	32.1
PK3	20	0.51	3.3	10.9	18.1	25.4/4.9	30.0	32.2
PK4	25	0.52	2.4	9.6	16.3	23.2/4.5	28.4	32.3
PK5	30	0.53	2.2	9.5	15.4	22.5/4.4	28.0	30.4

Separately, the effect of a variation in the water to binder ratio on compressive and tensile strength has been investigated in the ranges 0.47 to 0.53, and 0.40 to 0.44 for mortars without and with superplasticizer, respectively. They were found in agreement with Taylor's law.

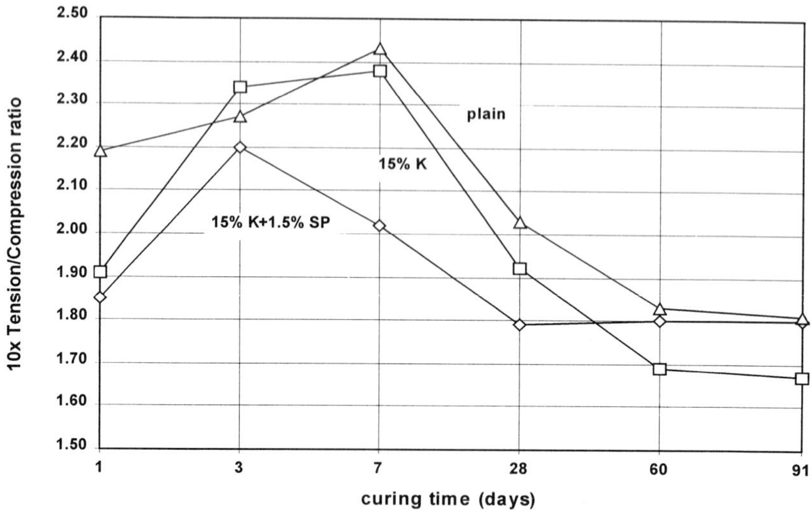

Figure 2 Development of tensile-compressive strength ratio of plain and 15 % kaolin blended PC mixes, the same mixes, but with 1.5 % SP, are represented by the third curve (W/B=0.50 for the first two series, and W/B=0.42 for the last series)

The superplasticizer allowed a reduction in W/B-ratio (maintaining the consistency in the range of 105 to 115 mm), leading to higher strength data [5,13]. Blending by a pozzolan like kaolin delays the strength development of young mortars. The effect on the tensile strength exceeds that of the compressive strength.

This is visualized in Figure 2, where the strength ratio is given as a function of curing time. The effect of the SP is additionally indicated. The same phenomena can be observed directly in Figure 3, where the bending and compressive strength vs curing time curves are given.

CONCLUSIONS

1. Calcined kaolin is a highly reactive pozzolan, which can be economically used to replace part of the Portland cement. The present tests - and additional ones on blended PC concretes [13] - demonstrate the optimum blending percentage to be higher at larger maximum grain size of the mix (*i.e.* from pastes, via mortars to concretes).

2. Blending by kaolin and diatomiteenhances the normal consistency and the setting times of Portland cement pastes, the latter however only at higher replacement percentages (*i.e.* 25, or more). Water demand of PC with kaolin is higher than when blended by the same percentage of diatomite (fly ash, or silica fume).

3. Blending PC by 10 % of kaolin or diatomite will only insignificantly influence the compressive strength of pastes. At higher replacement percentages, the compressive strength markedly declines. Hence, 10 % is the optimum blending percentage for PC pastes.

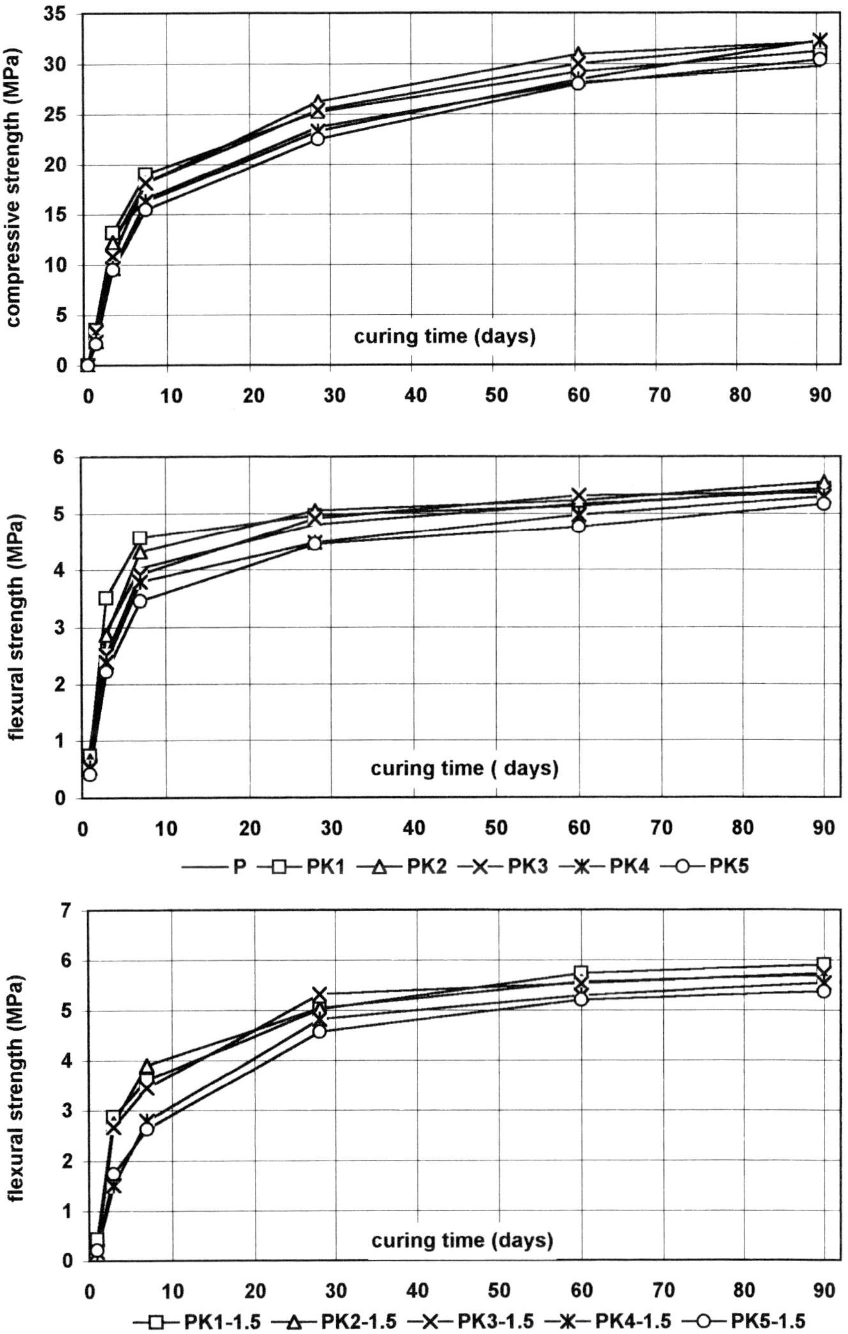

Figure 3 Strength development with curing time. From top to bottom: compressive strength, followed by bending strength of K-blended mixes, and bending strength of K-blended mixes + 1.5 % SP

4. Optimum blending percentage of mortars increased at growing maturity. 10 % at 28 days increased to 25 % at 90 days. Mortars of 28 days or older yielded a higher resistance in bending and compression when based on diatomite-blending, then on kaolin-blending. At a replacement percentage of 30, the strength in both loading modes still exceeded that of the non-blended mixture.

5. Strength was reduced at increasing water to binder ratio. Addition of a superplasticizer allowed a reduction in this ratio at all replacement percentages. As a consequence, strength was increased in all cases by addition of a sufficient amount of superplasticizer.

REFERENCES

1. BUI, D D, AND STROEVEN, P. Experimental assessment of development potentialities of cement-based building materials in Vietnam using indigenous resources. Proc.XXIV-th} IAHS World Housing Congress, Eds. O Ural,D Altinbek and T Birk"{o}n"{u}l, METU, Ankara, 1996, pp 1094-1101.

2. AMBROISE, I, MARTIN-CALLE, S AND PERA, J. Pozzolanic behaviour of thermally activated kaolin. Proc. Fourth Int. Conf., Istanbul, SP 132-40, 1, 1992, pp 731-739.

3. KOSTUCH, J A, WALTHERS, G V AND JONES, T R. High performance concrete incorporating metakaolin- a review: Concrete 2000, Economical and Durable Construction through Excellence, Eds. R K Dhir and M R Jones, 2,Chapman and Hall, London, 1993, pp 1799-1811.

4. ASBRIDGE, A AND JONES, T R. High performance metakaolin concrete: results of large scale trials in aggressive environments. Concrete in the Service of Mankind - Radical Concrete Technology, Eds. R K Dhir AND P C Hewlett, E & FN Spon, London, 1996, pp 13-24.

5. VU, D D. The effect of kaolin on of raw materials on the properties of artificial pozzolanic blended cement. Blended Cements in Construction, Ed. R N Swamy, Elsevier Appl. Sc., London, 1991, pp 45-55.

6. PERA, J, AMBROISE, J AND JONES, T R. Investigation of pozzolanic bindercontaining calcined laterites. Materials Research Society, Pittsburgh, 245, 1992, pp 25-35.

7. ZHANG, M H AND MALHOTRA, V M. Characteristics of characteristics of blended portland cement paste and mortar. Report 03.21.1.32.30, Faculty of Civil Engineering, Delft University of Technology, Delft, 1996.

8. VU, D D, BUI, D D AND STROEVEN, P. Experimental study of the use of calcined clay, rice husk ash and gap-graded aggregate in PC-based mortars and concretes. Proc. 2nd Int. Conf. on Non-Conventional Construction Materials (NOCMAT-97), Natural Resources Development Foundation. Bhubaneswar, India, 17-20 June, 1997, pp 96-110.

9. STROEVEN, P, QUY, N T AND DONG, A V. High strength gap-graded concrete using very fine sand in Vietnam. Concrete in the Service of Mankind, Concrete for Environment Enhancement and Protection, Eds. R K Dhir and T D Dyer, E & FN Spon, London, 1996, pp 559-568.

10. DALHUISEN, D H, STROEVEN, P, BUI, D D AND QUY, N T. Replacement ofcondensed silica fume by rice husk ash for the production of high strength concrete in developing countries. Utilization of High Strength/High Performance Concrete, Presse Pont et Chauss'{e}es, Paris, 1996, pp 235-244.

11. FATHY, H M AND MABROUK, M R. Effect of firing temperatures and durations thermally activated alumino-silicate pozzolanic material and its use in concrete. Cem. Concr. Res., 8, 25, 1995, pp 1713-1725.

12. STROEVEN, P AND VU, D D. Strength of diatomite- and kaolin-blended Portlandcement pastes and concretes. Proc.Materials Science and Concrete Properties, LMDC INSA-UPS, Toulouse, 1998, pp 11-18.

13. STROEVEN, P AND VU, D D. Effect of kaolin in concrete, Report 03.21.1.32.05, Faculty

MECHANICAL PROPERTIES AND DURABILITY OF MAE MOH-FLY ASH MORTAR

B Chatveera

S Thasanakosol

Thammasat University

Thailand

ABSTRACT. The objective of this paper was to investigate the depth of carbonation, autogenous shrinkage, drying shrinkage and compressive strength of mortar containing lignite fly ash from Mae Moh. The water-cementitious materials ratio was kept constant. The main parameter was the percentage replacement of fly ash, 30% and 60%. For carbonation, autogenous shrinkage and drying shrinkage tests, the specimens were made with the replacement of only one type of fly ash whereas for compressive strength test, 3 types of fly ash were used. The test results were indicated that when pertentage of fly ash replacement increased all types of shrinkage decreased. However, fly ash did not improve the carbonation resistance. For the compressive strength result, the strength of fly ash mortar can be predicted when the CaO-Equivalent in mortar was known.

Keywords: Autogenous shrinkage, Carbonation, Compressive strength, Drying shrinkage.

Dr Burachat Chatveera is an assistant professor in the Department of Civil Engineering at Thammasat University, Rangsit Campus, Pathum Thani 12121, Thailand. He received his D.Eng. from Asian Institute of Technology, Thailand.

Mr Sayan Thasanakosol is a graduate student in the Department of Civil Engineering at Thammasat University, Rangsit Campus, Pathum Thani 12121, Thailand. He received his B.Eng. from Chiang Mai University, Thailand.

INTRODUCTION

It is known that fly ash is one of pozzolanic materials, which has great potential to be used to improve concrete quality in construction. In recent years, energy consumption in Thailand rises appreciably due to the industry expansion and economic growth, causing the increase of coal consumed to produce electricity, therefore increasing the amount of fly ash which is waste product. Mae Moh electricity generating plants use 3 million tons of lignite coal each year to produce the electricity. It generates a lot of amount of coal ash but a little amount is used. The remained amount may harmful to the environment in term of dispersing in the air if keeping condition is unsuitable. If the using of fly ash increases, environmental problem will be solved.

This paper presents some advantages of using fly ash as a constituent in mix in term of mechanical properties and durability.

EXPERIMENTAL DETAILS

Materials

Portland cement: Ordinary Portland cement type 1 was used. The properties and composition are shown in Table 1.

Fly ash: Three type of fly ash from Mae Moh electricity generating plant were used. The properties and composition are shown in Table 1.

Fine aggregate: Natural river sand passing ASTM sieve number 4 was used. It was prepared in accordance with ASTM C128.
Mixing water: Ordinary tap water was used.

Table 1 Composition and properties of cement and fly ash

COMPOSITION/PROPERTIES	OPC	FH	FM	FL
SiO_2, %	21.45	26.23	30.39	42.86
Al_2O_3, %	5.35	13.72	15.9	23.05
Fe_2O_3, %	3.01	9.99	10.53	8.75
CaO, %	67.33	29.97	23.12	10.13
MgO, %	1.52	3.16	3.26	2.93
K_2O, %	0.33	1.7	2	2.55
Na_2O, %	0.11	0.21	0.19	0.2
SO_3, %	2.31	6.01	4.17	1.17
MnO, %	-	0.139	0.132	0.087
Specific Gravity	3.15	2.65	2.23	1.94
Bulk Density, gm/L	1.02	1.13	1.00	1.00
Fineness Passing 45μm sieve, %	8.6	30.22	35.68	42.68
Moisture content, %	0.11	0.08	0.14	0.07

FH, high SO_3 fly ash FM, medium SO_3 fly ash FL, low SO_3 fly ash

Mix Proportions

The mix proportions are summarized in Tables 2 and 3

Table 2 Mix proportion for shrinkage and carbonation tests

% FLY ASH	W/(C+F)	F/(F+C)	(F+C)/S	TYPE OF FLY ASH
0(OPC)	0.5	0	0.36	-
30	0.5	0.3	0.36	FL
60	0.5	0.6	0.36	FL

Table 3 Mix proportion for strength test

MORTAR	W/(C+F)	F/(F+C)	(F+C)/S	TYPE OF FLY ASH	CURING CONDITION
OPC(W)	0.5	0	0.36	-	W
OPC(W-D)	0.5	0	0.36	-	W-D
FH-30(W)	0.5	0.3	0.36	FH	W
FH-30(W-D)	0.5	0.3	0.36	FH	W-D
FH-60(W)	0.5	0.6	0.36	FH	W
FH-60(W-D)	0.5	0.6	0.36	FH	W-D
FM-30(W)	0.5	0.3	0.36	FM	W
FM-30(W-D)	0.5	0.3	0.36	FM	W-D
FM-60(W)	0.5	0.6	0.36	FM	W
FM-60(W-D)	0.5	0.6	0.36	FM	W-D
FL-30(W)	0.5	0.3	0.36	FL	W
FL-30(W-D)	0.5	0.3	0.36	FL	W-D
FL-60(W)	0.5	0.6	0.36	FL	W
FL-60(W-D)	0.5	0.6	0.36	FL	W-D

Curing Environments

For strength test, two curing conditions were used. These were : (W) water curing, (W-D) cyclic wetting and drying condition. One cycle consists of 1-week wetting and 1-week drying period. For wetting, thc spccimens were submerged in water. For drying, the specimens were put into the specimen curing unit having temperature of 25 ± 2°c and 60±5% relative humidity.

For drying shrinkage test, water curing for 28-day after demolding. Then, the specimens were put into the specimen curing unit having temperature of 25 ± 2°c and 60±5% relative humidity.

For autogenous shrinkage test, after demolding, the specimens were wrapped with plastic sheet and put into the specimen curing unit having temperature of 25±2°c and 60±5% relative humidity.

For carbonation test, water curing for 28-day after demolding. Then, the specimens were placed in the location containing CO_2.

RESULTS AND DISCUSSION

Strength of Mortar

It was obviously observed from Figures 1 and 2 that strength of mortar having 30 and 60 percent replacement of fly ash to ordinary Portland cement are lower than the control in early age. In long term, mortar having 30 percent replacement of fly ash to ordinary Portland cement have nearly value of strength due to the pozzolanic reaction which occurred in long-term. Furthermore, the value of strength varied by the amount of SO_3 in material compositions. It was found that high SO_3 fly ash has a tendency that strength will higher because fly ash in this study has the amount of CaO regarding to the amount of SO_3. The higher amount of CaO in material composition implied that the strength of mortar is higher [1].

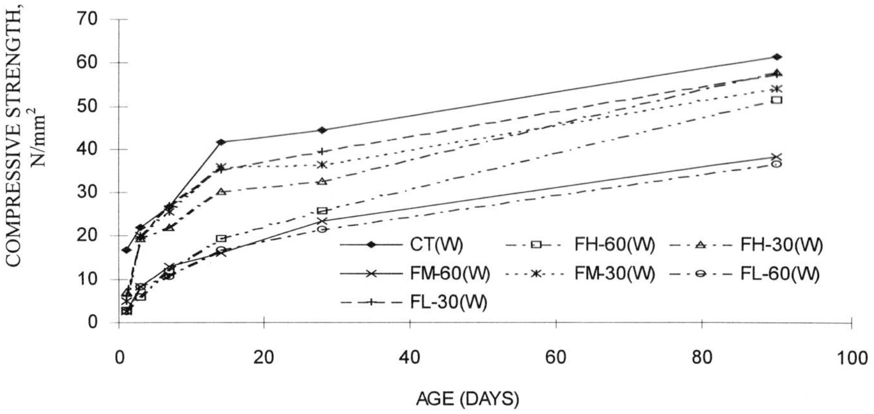

Figure 1 Compressive strength of mortar having varied amount of SO_3 and percent replacement of fly ash to cement in water curing condition

Drying Shrinkage of Mortar

Figure 3 showed that the mortar with fly ash display smaller drying shrinkage than the control especially when the percentage of replacement is larger because the water requirement of mortar with fly ash is smaller. So, it can be noticed that the water to cementitious materials ratio of the mixture with fly ash can be reduced in practice, therefore, drying shrinkage is reduced [2].

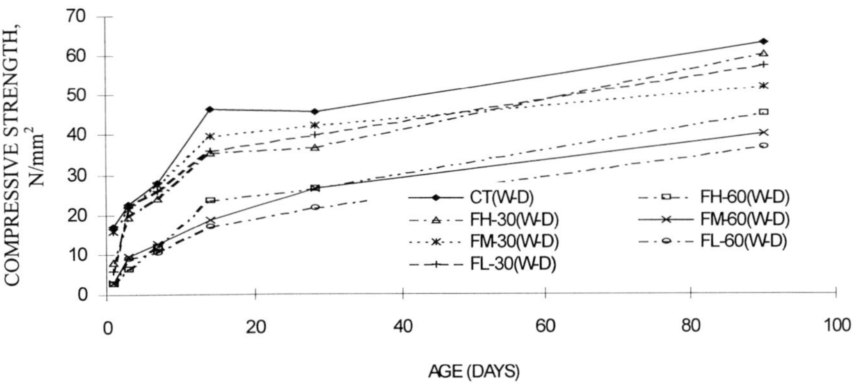

Figure 2 Compressive strength of mortar having varied amount of SO_3 and percent replacement of fly ash to cement in cyclic wetting and drying curing condition

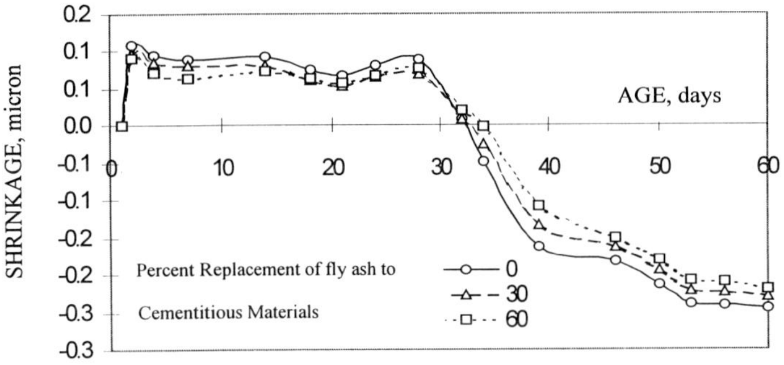

Figure 3 Drying shrinkage of mortar containing fly ash

Autogenous shrinkage of mortar

Figure 4 showed the comparison of length change of specimens containing fly ash 30 and 60 percent to the control. It was shown that all specimens with fly ash exhibit smaller autogenous shrinkage than the control, especially when the replacement percentage is higher. It can be seen that the higher content of fly ash is the more reduction of autogenous shrinkage. It can be explained that it is the result of spherical shape of fly ash resulting in less retained water than the cement particles [2]. This result is larger free water content in the mixture with fly ash than the mixture without fly ash when prepared with the same water to binders ratio. As the autogenous shrinkage is a result of water consumption in hydration process.

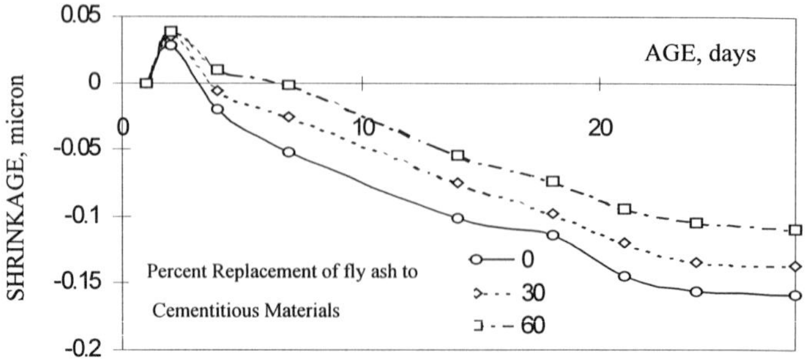

Figure 4 Autogenous shrinkage of mortar containing fly ash

Carbonation of mortar

Figure 5 illustrated that carbonation of mortar containing fly ash is larger than the control and higher when the amount of fly ash is increased [1].

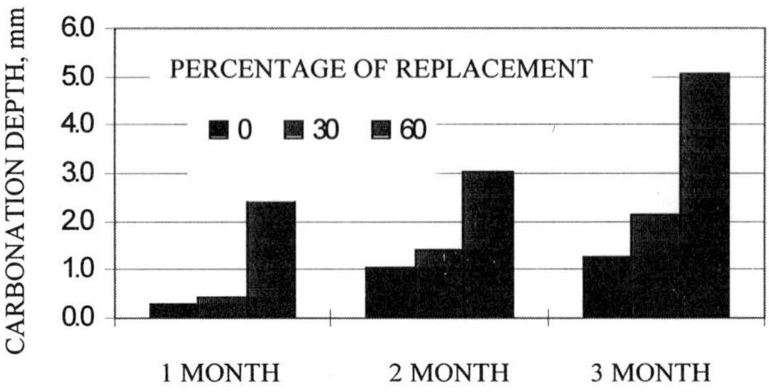

Figure 5 Carbonation depth of mortar

A model for strength prediction of mortar containing Mae Moh-fly ash

The parameters affecting the strength of mortar are type of fly ash and percent replacement of fly ash to cement. When the affecting parameter are changed, strength of mortar will be changed. Generally, the crucial portion of chemical composition affecting to the strength of mix is CaO and SiO_2 in cementitious materials contents, as known in the hydration reaction. Refer to Table 1, the control compound affecting the strength of mix is CaO. The CaO affecting the strength of mix is called as $CaO_{Equivalent}$ comprising CaO from cement powder and fly ash.

In this study, the strength of mortar having the water to cementitious materials ratio equal to 0.5 can be predicted by linear regression analysis from strength data, percent replacement of fly ash to cementitious materials and $CaO_{Equivalent}$ in fly ash.

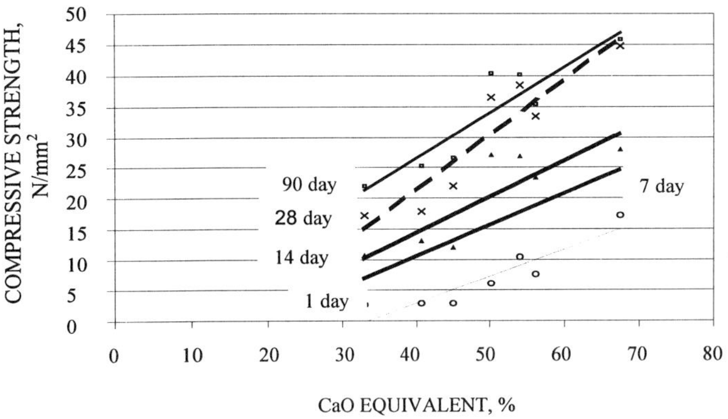

Figure 6 Relationship between compressive strength of mortar and $CaO_{Equivalent}$

From Figure 6, strength of mortar can be predicted by the following equations.

$$ST_1 = 0.4399(CaO_{Equivalent})-14.658$$
$$ST_7 = 0.5261(CaO_{Equivalent})-10.442$$
$$ST_{14} = 0.5982(CaO_{Equivalent})-9.4456$$
$$ST_{28} = 0.9094(CaO_{Equivalent})-14.989$$
$$ST_{90} = 0.7466(CaO_{Equivalent})-3.241$$

Where $CaO_{Equivalent} = (1-r)(CaO_c)+(r)(CaO_f)$
CaO_c is percent by weight of CaO in cement powder
CaO_f is percent by weight of CaO in fly ash
r is the weight ratio of fly ash by cementitious materials [F / (F+C)]

CONCLUSIONS

1. By replacing the fly ash to some parts of cement at 30 percent, early age strength is lower whereas long-term strength is the same as the control. However, at the percentage of fly ash replacement equal to 60 percent, strength is reduced.

2. At the same percentage replacement of fly ash, the compressive strength of mortar is regarding to the content of CaO in fly ash. That is the more content of CaO in fly ash, the more strength in mortar.

3. Under saturated and cyclic wetting-drying curing conditions of mortar containing fly ash at 30 or 60 percent by weight, strength is slightly different.

4. The autogenous shrinkage and drying shrinkage can be reduced by increasing the content of fly ash in mixture.

5. The carbonation of mortar containing fly ash is larger than the control and higher when the amount of fly ash in mixture is increased.

6. The strength of mortar can be predicted when the $CaO_{Equivalent}$ was known.

REFERENCES

1. DECHNANTARAT, T, TANTIWAT, S AND YOTHMAT, S, Mechanical Properties and Durability of Mae Moh-Fly Ash Mortar. B.Eng. Civil Engineering Project, Thammasat University, Bangkok, Thailand, 1995.

2. TANGTERMSIRIKUL, S, Durability of Concrete using Pozzolans. Seminar of Advance in Concrete Technology, ACI (Thailand Chapter) and Engineering Institute of Thailand, Bangkok, 1996.

UPGRADE OF HIGH CALCIUM HELLENIC FLY ASHES FOR THEIR USE IN CONCRETE

S Tsimas

A Moutsatsou

National Technical University of Athens

Greece

ABSTRACT. Even though the technical and economical benefits from the use of high calcium Hellenic fly ashes in concrete, have been verified, their use presents some difficulties which inhibit their addition. The problems in use are mainly focused to variations in chemical and mineralogical composition, necessity for supplementary grinding, high proportion of free CaO (CaO_f) and periodically high proportion of SO_3 content. The final result is that Greece has not yet established specifications concerning the addition of fly ashes in concrete.

To overcome these problems, the quality of fly ash was cheaply upgraded by grinding, with simultaneously hydration for the reduction of CaO_f. The so treated fly ash was successfully used as the main material for the construction of a dam, which is already completed and has excellent behavior referring to strength.

In this paper details for the milling plant and the industrial production are given and the physicochemical characteristics of the product are compared with raw material.

Keywords: High calcium fly ash, Portland cement (PC), Roller compacted concrete (RCC), Free CaO, SO_3, Dam durability.

Associate Professor Stamatis Tsimas is former Director of the laboratory of Inorganic and Analytical Chemistry, NTU, Athens. His major interests are in chemistry and technology of aluminosilicates, specialising in cement technology and upgrade techniques of industrial minerals and by- products. Prof. Tsimas has published over 25 papers. He is member of CEN WG11/TC51.

Assistant Professor Angeliki Moutsatsou is a Chemical Engineer with great experience in chemical analysis. The research field where she specialises refers to Chemistry and Technology of metals, to recycling of materials as well as to methods for the treatment of solid and liquid wastes. She has published over 20 papers.

INTRODUCTION

The use of fly ashes in concrete began forty years ago while the evaluation of fly ashes with elevated CaO content begun later (since middle of '70) [1,2,3,7]. The technical and economical benefits from the use of high calcium fly ashes in concrete have been verified by researche dated from early sixties. In Greece their addition in large quantities as the fourth material in concrete still presents some difficulties, mainly pertaining to chemical composition (CaO_f, SO_3, as discussed below) as well as to their inhomogeneity. These difficulties delay the establishment of relative standards and thereby inhibit their use.

Hellenic ashes and especially those of Ptolemais origin are classified in type C, e.g. into class with high proportion of CaO as compounds like 3 $CaO{\cdot}Al_2O_3$, $CaO{\cdot}SiO_2$, $4CaO{\cdot}Al_2O_3{\cdot}SiO_2$ or sulfur-calcium-aluminates are predominant. They appear not only pozzolanic but also hydraulic behavior. In all cases the sum of Al_2O_3+SiO_2 +Fe_2O_3 is greater than 50% but less than 70% as have ashes classified in type F. The high level of CaO_f inhibit their incorporation in EN 450 [2,4,5,6].

Eventhough the well known advantages of the partially substitution of cement with high calcium fly ashes, Hellenic fly ashes must be upgraded by specific treatment in order to overcome their negative characteristics. Untreated ash characteristics are mainly focused to i) the variations in the chemical and mineralogical composition, as they are by-products of a generation of energy process with no respect to the quality of ash, ii) the necessity for supplementary grinding for better reveal of their pozzolanic and hydraulic properties, iii) the elevated proportion of their CaO_f as its hydration cause sound-ness problems as well as significant temperature increase and iv) the periodically elevated proportions of SO_3 content with negative consequences similar to cement.

The treatment of raw fly ash is carried out through a specific designed and constructed ball mill grinding plant with simultaneous water spraying system inside.

This paper mainly comprises details concerning the milling plant, but examines also the different behavior of untreated (raw) and treated high calcium fly ash concerning strength development as well as other physicomechanical properties.

THE MILLING PLANT

Although the above mentioned negative effects, in 1992 Power Public Corporation (PPC) decided to go forward to the construction of Platanovrissi dam with the RCC technique and the use of 150000t of Ptolemais fly ash as basic cementitious material. For the realization of this decision and for overcoming the mentioned particularities of high calcium fly ashes of Ptolemais origin, PPC had imposed strict requirements. Their covering has as basic assumption the construction of a milling plant with simultaneous partial hydrolization of fly ash, near Ptolemais Power Station. By this process the problems of high CaO_f and the necessity for supplementary grinding are successfully confronted [7, 8].

The untreated (raw) fly ash was controlled every two hours at the point of production. The percentage of SO_3 content was measured and the parties with values higher than 7% have been rejected. This max value of SO_3 was established after a wide research program in mortars and concretes.

Additionally for the inhomogeneity problem, after a series of preliminary tests, two units of Ptolemais Power Station were choosen. One of them (The 4th unit) was the basic and had offered the 85% of raw ash while the other was supplementary. For minimizing the deviations of lignite quantities, these units were continiously fed with the same rate from two lignite quarries.

The study of the project and the construction of milling plant have been carried out according to PPC's specifications by AEGEK SA which was the main contractor for the construction of the dam. The grinding unit consists of a closed cirquit ball mill with the relevant supplementary equipment. The three compartment ball mill has been constructed for through-put limits between 16.7 and 23 t/h.Its dimensions are 2.6 X 12 m and has a total filling degree of 30% covered by balls of 30mm and 40mm and the material which has a bulk density of 660 kg/m^3 when aerated [9].

The flyash temperature was controlled in the mill by a combination of the heat of exothermic hydration reaction, the frictional heat, the radiation and convection from the shell and the heat removal by turning excess water into steam

Due to continious ash variations, the suitable proportion of water to be added is automatically controlled by feedback from flyash temperature. Mill temperature is thereby controlled by balancing heat removal, in converting excess water to steam with heat generated from reacting a variable lime content.

The importance of residence time is obvious as it demonstrates the necessary time for hydrolization. For that reason the injection must performed in the begining of the first compatment of the mill.

DETAILS FROM THE INDUSTRIAL PRODUCTION

Fly ash feed has a retained on 45μm sieve of 40-45% (with S_B =2000-2400 cm^2 /g) and a composition with a CaO_f variable in the range of 5-13%. The feed temperature is 110-130°C as it passes first from the preheater. Finally its max moisture content before preheating is 0.3%. Water is sprayed into the mill, so that the most of the free lime is hydrolized to $Ca(OH)_2$ under hot moist conditions. It is essential that the mill must operate at a relatively high temperature (~100°C) to speed the reaction and to boil off surplus water. The water spraying system was designed for 5.0-12.5% water addition i.e. up to 48L/min .

Basic demands from mill were the reduction of size so that the retained on 45 μm sieve was 20-30%, the partial (up 75%) hydration of CaO_f content so that the finally CaO_f did not exceed 3% and the percentage of free water which did not exceed 1%.

The production of fly ash was complete into 19 months (October 95-April 97). In this period 135000t of upgraded fly ash were produced and transported with 5500 silo vehicles for a distance of 380 Km in the place of dam in Platanovryssi. There, it was mixed with type I cement at a rate of C/F = 20/80 and this mixture was applied for the construction of the main body of dam with RCC method. Until now (after 3 years), excellent results concerning strengths and other relevant properties have been obtained.

During the construction period, very strict quality control program was followed aiming to control four different levels of the production.
At level 1, the acceptability of flyash was tested every two hours in continious sampling and 18000 measurements of SO_3 and CaO_f have been carried out. A percentage of 20% of the samples led to the rejection of relevant quantities of fly ash, as the requirement of SO_3 (<7%) was not covered. At level 2, the ability of the mill for the reduction of CaO_f was controlled every two hours and in the same time the appropriate corrections of the water injection were performed. In this level about 6000 measurements of CaO_f and fineness from the exit of the mill, have been carried out. At level 3, the quality control of the product was checked. Each silo vehicle was controlled (CaO_f, SO_3, fineness and humidity) during its loading, before leaving the plant and If any vehicle stands beyond the limits, its load was rejected. Measurements of physical and chemical properties every 250t or daily were performed and the results of the total production (after 471 batches) are presented in Figures 1, 2, 3. Finally at level 4, the main parameters were checked after the homogenization of treated ash in six silos of 3000 each, in Platanovryssi.

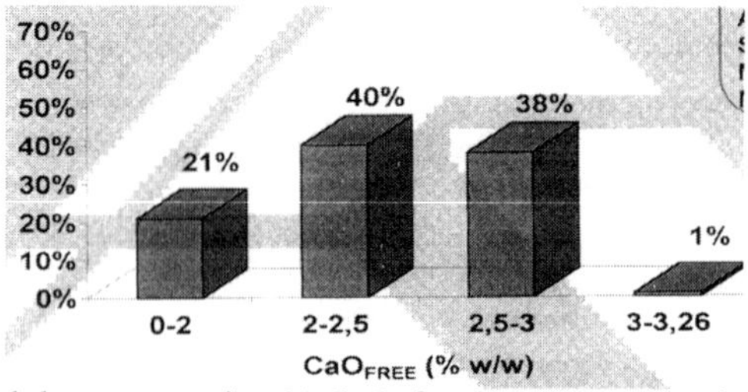

Figure 1 Statistical data concerning % CaO_f for the total production of milling plant

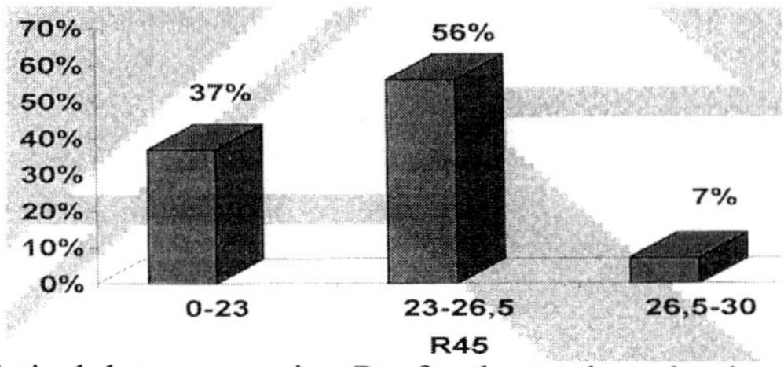

Figure 2 Statistical data concerning R_{45} for the total production of milling plan

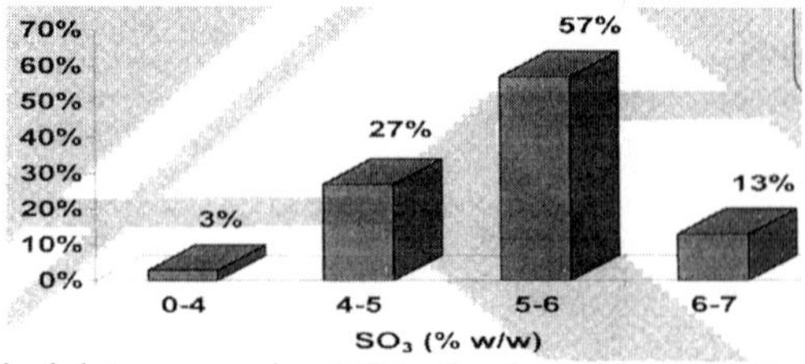

Figure 3 Statistical data concerning %SO_3 for the total production of milling plant

EXPERIMENTAL DETAILS

In order to examine the influence of fineness and CaO_f of industrial produced fly ashes on strength development in different proportions with OPC, six different fly ashes were produced. Their characteristics are summarized in table 1. It is obvious that F1 refers to untreated fly ash. In figures 5 and 6 are compared the untreated fly ash (F1) with an ash after treatment as is F6.

Table 1 Characteristics of industrial produced fly ashes with different treatment

FLY ASH CODE	CaO_f (+/- 0.2%)	R45μm (+/-1.5%)
F1	6.8	41
F2	6.8	22
F3	3.5	22
F4	3.5	11
F5	1.8	22
F6	1.8	11

These six fly ashes were mixed with OPC (C) in proportion C:F= 80:20, while F3 –F6 were additionally mixed with OPC at percentages varying from 40 to 80%. Compressive strength (3, 7, 28, 90 days) as well as setting time and soundness were measured to all mixes according EN 196.

RESULTS AND DISCUSSION

In Table 2 are summarized the results from the compressive strength tests. The results obtained from the compressive strengths test showed, as it was expected, the positive effect that fineness has in strength development (CF2>CF1, CF4>CF3, CF6>CF5). From the same samples derives also that intermediate CaO_f (CF3, CF4) values show the best results (CF3>CF5, CF4>CF6), indicating the hydraulic behavior of high calcium fly ash. On the other hand samples of the higher CaO_f (as are CF1, CF2), showed less compressive strengths compared to CF3,CF4 due to the formation of $Ca(OH)_2$ in large quantities. The hydration also of such samples showed much heat generated which is extremely prohibited in mass constructions as in the case of dams.

Same observations show the examination of the other mixture proportions. All mixtures show the following relations referring to strength : CF4>CF3, CF6>CF5 as well as CF3>CF5, CF4>CF6.

In all cases, except CF5 mixtures, strength development showed better results comparing to untreated fly ash (CF1). Small values of CF5 show that there is not need for further reduction (>75%) in CaOf and that values near 3% are the optimum. In case that CaO_f is less than 2% the mixture must be additionally ground. No unexpected observations derive from the examination of setting time and expansion test.

Table 2 Compressive strength results (N/mm^2)

MIX	PROP.	COMPRESSIVE STRENGTHS				MIX	PROP.	COMPRESSIVE STRENGTHS			
		3D	7D	28D	90D			3D	7D	28D	90D
C:F1	80:20	16.1	22.4	31.2	33.2	C:F4	60:40	20.1	33.1	43.8	49.8
C:F2	80:20	19.2	24.4	33.0	44.3	C:F4	40:60	19.4	32.9	42.8	52.7
C:F3	80:20	20.2	26.3	36.5	46.5	C:F4	20:80	14.9	29.4	42.7	44.8
C:F4	80:20	26.0	27.8	42.6	47.2	C:F5	60:40	12.9	23.2	26.1	35.4
C:F5	80:20	18.3	28.3	34.2	37.0	C:F5	40:60	10.6	15.2	22.4	31.8
C:F6	80:20	20.7	28.2	36.5	38.5	C:F5	20:80	7.5	12.2	18.7	26.2
C:F3	60:40	17.0	29.7	40.7	54.0	C:F6	60:40	21.2	32.5	43.4	47.8
C:F3	40:60	16.6	27.1	39.0	49.9	C:F6	40:60	18.1	30.5	41.3	45.4
C:F3	20:80	10.5	23.0	37.5	39.9	C:F6	20:80	13.1	26.5	36.1	39.6

XRD Patterns of the same fly ash in its untreated and treated form, indicate that the industrial treatment was successfully completed as in the pattern of treated form the peaks for CaO are much less intense comparing with the relevant of the untreated. In XRD of the treated fly ash appear also clear the peaks for $Ca(OH)_2$ derived from the hydration into the mill of CaO_f. These peaks are not present in the untreated form.

CONCLUSIONS

1. It is possible to upgrade high calcium fly ashes by a cheap method consisting to simultaneous to their grinding, hydration with a automatically controlled water spraying system.

2. By suitable adjustment of mill operation parameters, it is possible to obtain several qualities of treated fly ashes. This differentiation is concerned to their fineness as well as to their CaO_f.

3. The use of 135000t of industrial produced treated fly ash as the main material for the construction of a dam by RCC method, showed excellent behavior referring to strengths and other properties.

4. Industrial experiments showed that best results concerning strength development, in mixtures of OPC and high calcium fly ashes, are obtained when CaO_f of fly ashes is near 3%.

REFERENCES

1. BERRY,E.E AND MALHOTRA, V.M. Fly ash for use in concrete–A critical review ACI Journal, Proceedings Vol. 77, No 2, 1980, pp 59-73.

2. PAPAYIANNI, J. Strength and bond data for Greek high-lime flyash. Proceedings of the 2nd International Conference, CANMET/ACI, Madrid, 1986, pp 367-386.

3. ACI COMMITTEE, Use of fly ash in concrete, ACI Material Journal, Vol. 84, No 5 pp 389-409.

4. TSIMAS, S. ET AL, Study on the evaluation possibilities of Hellenic fly ashes, Technical Chamber of Greece, Athens, 1985, pp 238.

5. PAPAYIANNI, J. An investigation of the pozzolanicity and hydraulic reactivity of a high lime fly ash, Magazine of Concrete Research, Vol. 39, No 138, London, 1987.

6. PAPADAKIS, V.G ET AL, Hydration and carbonation of pozzolanic cements, ACI Materials Journal, Vol. 89, No 2, 1992.

7. DUNSTAN, M.R.H. Fly ash as the fourth ingredient in concrete mixtures. Proceedings of the 2nd International Conference, CANMET/ACI, Madrid, 1986.

8. DUNSTAN M.R.H. Future trends in Roller Compacted Concrete Dam Construction, ACI SP144-16, Detroit, 1994, pp 307-325.

9. KRAVARITIS, A., TSIMAS, S., MOUTSATSOU,A., AND TSIKNAKOU Y. Utilization of fly ash for construction of dams by RCC method, Proceedings Power - Gen'96 Europe, Boudapest, 1996, Vol. 1, pp 833-858.

EVALUATION OF PROPERTIES OF LOCALLY PRODUCED SILICA FUME FOR USE IN CONCRETE

A R Khaloo

M R Houseinian

Sharif University of Technology Tehran

Iran

ABSTRACT. In this study, the chemical and physical properties of silica fumes produced by Ferrosilice and Ferroalloys factories in Iran are investigated. The test results show that the properties of the two silica fumes correspond with that according to ASTM C-1240 standard. An extensive experimental program was also conducted to determine the compressive strength of concrete containing the locally produced silica fumes. The test variable included: percentage of silica fume (0% to 15%); water to cement ratio (0.3 to 0.57); weight of cement (300 to 520 kg/m^3), and curing period (1 to 91 days). In total, over 50 concrete mixes were designed and more than 600 cubical specimens were tested for compressive strength evaluation. Test results indicate that 5 to 10 percent replacement of silica fume for cement provides the highest strength for short and long terms. Generally, the mixes containing low cement resulted in higher percentage of strength gain. The range of compressive strength gain for silica fume concrete including the above-mentioned variables varies between 20 to 67% at 28 days, and 20 to 55% at 91 days with respect to control concrete strength.

Keywords: Silica fume concrete, Compressive strength, Curing period, Water-cement-ratio, Cement content, SF content, Superplasticizer, Tests.

Professor Ali R Khaloo is Director of Materials and Structural laboratory of Civil Engineering at Sharif University of Technology, Tehran. His research and consulting activities include various topics related to reinforced concrete structures and concrete composite materials. Dr. Khaloo is Vice President of the Iran chapter of ACI. He is an active member on several national technical committees.

Mohammad R Houseinian has an MS degree in chemical engineering. His present work is related to cement, pozzolanic materials and admixtures.

INTRODUCTION

Silica fume (SF) is by-product of the production of silicon metal or some ferrosilicon alloys. It is a super pozzolan, since its small replacement with Portland cement enhances most of the concrete properties [1,2,5,6]. It was about 5 years ago, when SF was produced in Iran for the first time. The research study aims at evaluating the chemical and physical properties of locally produced silica fumes, and also investigates its influence on enhancing concrete strength, by its partial replacement with Portland cement. Table 1 provides some information on factories producing Ferrosilice and silica fume.

Table 1 Factories producing ferrosilice (75%) and silica fume

FACTORY	LOCATION	BUILT BY	OPERATING DATE		YEARLY PRODUCTION (TON)	
			FERROSILICE	SILICA FUME	FERROSILICE	SILICA FUME
Ferro-Alloys of Iran Industry	Azna	ELKEM (Norway)	Feb 1993	Sep 1996	25000	9000
Ferro-Silice of Iran	Semnan	DEMAG (Germany)	May 1994	May 1994	25000	9000
Navid Alloys	Esfahan	Steel Melting of Esfahan	1996	1996	2000	700

EXPERIMENTAL DETAILS

The experimental program consisted of testing chemical and physical properties of silica fume and its use as some percentages for cement replacement in concrete mix designs.

Materials, Mix Proportions, Variables and Curing

Portland cement type II, complying with ASTM C-150 and locally produced silica fumes tested according to ASTM C-1240 were used. The coarse and fine aggregates consisted of crushed stone and natural sand with maximum size of 25 mm and 5 mm, respectively. Sand had a fineness modules of 2.86, and specific gravity and water absorption of 2.60 and 5.50% for sand, and 2.60, and 1.50% for coarse aggregate, respectively. The concrete mixes were designed according to recommendations of ACI-211 Committee and ASTM C-33 Specifications. Various dosages of superplasticizer were used in the concrete mixes to reach a target slump of 4 to 7 cm.

Test variables for making concrete specimens included the silica fumes from two different factories, silica fume as a percentage of total cementitious materials, weight of cement and water to cement ratio. The cement content was selected according to the usual cement consumption in the field. Also, the water to cement ratio is based on minimizing the use of superplasticizer for reaching a desired slump (3,4,6). Moreover, importance of use of superplasticizer was demonstrated, by testing samples made without this chemical admixture. All specimens were tested at ages of 1, 7, 28 and 91 days after casting. For each variable, at least three 15x15x15 cm cubical specimens were cast. Over 50 concrete mixes were designed, and in total more than 600 specimens were tested to evaluate the silica fumes based on the compressive strength of silica fume concretes. All the specimens were kept in the cubical molds for 24 hrs, and then were cured in water until one day before testing. One day specimens were tested after demoulding and drying the surface with air pressure.

RESULTS AND DISCUSSION

The results obtained from the chemical and physical tests on the three kinds of silica fumes are given in Tables 2 and 3, and compared with ASTM standards (5). The properties of the silica fumes correspond with that according to ASTM C-1240 standards. Typical compressive strength results for concrete containing SF-1 are presented in Table 4. Influence of different mix design parameters on compressive strength of silica fume concrete are discussed separately in the following paragraphs.

Silica Fume and Cement

Test results indicate that 5 to 10 percent replacement of silica fume (i.e., SF-1 and SF-2) for cement provides the highest strength for short and long terms. Generally, the mixes containing low cement resulted in higher percentage of strength gain, compared to the mixes containing higher cement. For example, percentage increase for mixes with 300 kg/m^3 and 520 kg/m^3 cement, W/C= 0.57 and 0.3, and 10% and 7% SF replacement for cement at 28 days is 67 and 42 percent , respectively, with respect to control mixes.

Table 2 Chemical ingredients of SF samples and standard

TYPE	SiO_2	Al_2O_3	Fe_2O_3	CAO	MGO	Na_2O	K_2O	SO_3	CI	P_2O_5	LOSS
SF -1	88.09	0.71	1.85	1.29	1.37	0.65	2.36	0.33	0.01	0.21	2.35
SF -2	89.00	2.60	1.70	1.00	2.40	0.65	2.10	0.34	0.036	0.051	1.44
SF -3	96.40	0.75	0.30	0.23	0.74	0.20	1.10	0.11	0.02	0.087	1.35
ASTM C-1240	Min 85	---	---	---	---	Max 1.5 for Na_2o	---	---	----	----	Max 6.00

Water to Cement Ratio and Superplasticizer

An average of 0.5 to 1.0 percent superplasticizer was consumed for addition of each 5% SF in concrete mixes designed with 300 and 450 kg/m^3 of cement. In all the mixes, compressive strength of SF concrete increased with reduction in water to cement ratio, however, higher dosage of superplasticizer was consumed. For an optimized mix proportion on the basis of the strength and economy, the relatively expensive item of superplasticizer may be reduced and compensated by increase of the water to cement ratio and silica fume content.

Table 3 Physical properties of SF samples compared to cement and ASTM standard

SAMPLE PHYSICAL PROPERTY	SF-1	SF-2	SF-3 C-1240	ASTM	CEMENT
Unit Wt. (gr/cm^3)	2.25	2.40	2.15	-----	3.15
Bulk den (kg/m^3)	250-300	200-250	350-400	-----	1.300
Specific Surface (m^2/gr)	----	17.1 - 17.9	----	Min. 15	0.3
Retained on Sieve 45 μ m (%)	5.5	5	7.5	Max 10	----
Average size of grains (μ m)	0.85	0.73	0.89	-----	----

Table 4 Compressive strength of SF-1 concrete

CEMENT (kg)	W/C+ SF	% SF	% SP*	SLUMP (mm)	COMPRESSIVE STRENGTH (N/mm²)			
					1 Day	7 Days	28 Days	91 Days
300	0.57	0	0	60	3.4	18.0	26.4	29.7
		5	0.5	55	7.8	26.8	43.4	46.9
		10	1.5	65	7.8	25.0	44.0	482
		15	2.5	65	5.9	23.6	39.7	43.8
300	0.50	0	1.25	55	2.8	20.4	28.4	32.0
		5	2.25	55	7.8	30.3	46.2	52.1
		10	3.50	65	7.8	28.8	44.5	49.6
		15	4.25	55	2.9	26.2	40.7	45.6
450	0.35	0	1.33	60	10.8	41.1	48.2	52.2
		5	2.33	60	13.3	46.9	67.0	75.1
		10	3.50	60	4.3	45.2	66.1	72.9
		15	4.25	55	2.8	43.6	63.9	67.5
520	0.30	0	0.95	40	19.4	40.7	53.1	61.9
		5	1.10	45	22.7	52.6	74.0	83.8
		7	1.18	45	21.8	52.8	75.9	84.9
		10	1.21	40	20.1	49.6	67.5	79.4
		15	1.30	45	17.2	49.2	63.2	73.9

* Superplasticizer

Concrete Age and Presence of Superplasticizer

Normalized range of strength gain at various ages on the basis of 28 day strength is shown in Figure 1. The slope of strength gain from 7 to 28 days for SF concrete is higher than the control concrete, however, the slope form 28 to 91 days is almost the same for the two cases. The highest percentage of strength gain is 67% (17 MPa), which belongs to a mix with 300 kg/m^3 cement content, W/C=0.57 and 5% SF replacement for cement at 28 days. A mix was used without superplasticizer having a slup of 40 to 45 mm. The slump of SF concrete was maintained by adjusting water to cement ratio. Test results indicate that due to increase in water requirement for addition of SF, the highest strength gain compared to control specimens is about 5%. Compressive strength of mixes without superplasticizer are 25 to 40 lower than mixes containing superplasticizer at all ages.

TECHNICAL AND ECONOMICAL CONSIDERATIONS

Technical evaluation of the test results reveal that SF enhances concretes strength by a relatively high percentage with respect to ordinary concrete. However, high cost of SF compared to cement and specially the need to use the expensive item of superplasticizer may raise serious questions regarding SF concrete. A comparative study on the two concrete mixes with and without SF having close compressive strength is made (Table 5). Table 6 shows that, for 42 MPa the final expenses are very close for the two types of concrete. It should be also mentioned that SF reduces permeability and consequently greatly enhances durability of concrete.

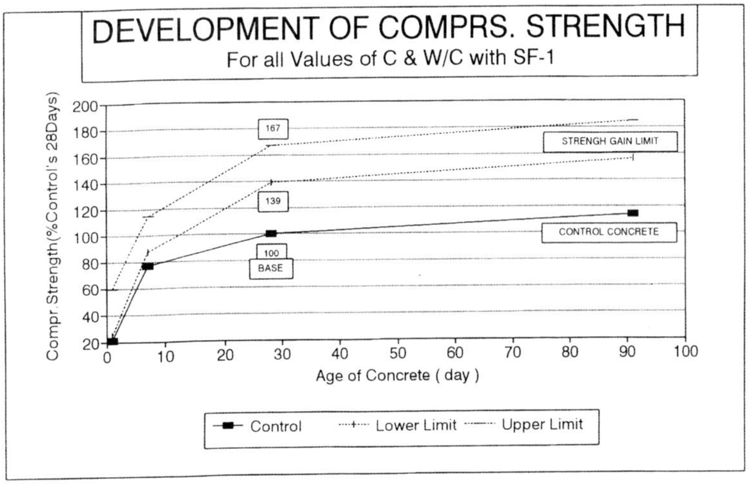

Figure 1 Range of strength gain for specimens containing SF compared to control specimens

STRENGTH DEVELOPMENT AND WATER PERMEABILITY OF HIGH STRENGTH TIA CONCRETE

Y L Lee

Institut Teknologi Tun Hussein Onn

K O Khoo **S Chong**

Petro Pipe~Daido Concrete (KL) Sdn

M W bin Haji Hussin

Universiti Teknologi Malaysia

Malaysia

ABSTRACT. This paper reports some tests results of strength development and water permeability of high strength Timber Industrial Ash (TIA) concrete subjected to four different curing conditions. Grade 80 TIA concrete were prepared under controlled conditions in a quality assured factory environment. A total of 96 concrete test cubes were subjected to air-curing, water-curing, steam-curing and autoclave-curing. Steam-cured TIA concrete outperformed the rest with water permeability of around 1×10^{-12} m/s in 91 days. Surprisingly, air-cured TIA concrete containing 5% TIA as a partial cement replacement material showed better water impermeability compared to the water-cured TIA concrete. The air-cured autoclaved TIA concrete also produced marginally higher value of water permeability at 28, 56 and 91 days although its compressive strength was higher than others. An average compressive strength of around 100 MPa was achieved in 7 days for the air-cured autoclaved TIA concrete, followed by a decline in compressive strength of around 10 % until 91 days. An extensive collaborative research is being planned to study the long term engineering properties of high strength TIA concrete prototypes.

Keywords: TIA, Strength development, Water permeability, Curing

Mr Y L Lee, Head of Concrete Laboratory, ITTHO, Parit Raja, Batu Pahat, Johor, Malaysia. He is a past Vice President of the Concrete Society of Malaysia.

Ir K O Khoo, General Manager, Petro-Pipe~Daido Concrete (KL) Sdn. Bhd., Nilai, N. Sembilan, Malaysia.

Mr Stuart Chong, Technical Manager, Petro-Pipe~Daido Concrete (KL) Sdn. Bhd., Nilai, N. Sembilan, Malaysia.

Professor Dr Mohd Warid bin Haji Hussin, Deputy Dean, Civil Engineering Faculty, Universiti Teknologi Malaysia, Skudai, Johor, Malaysia.

Addition of too much fines into the raw material mix creates micro-cracking as a result. Cracks are a consequence of shrinkage of the CSH-gel structure. The more fines are added, the higher is the resulting CSH-gel concentration. Cracks appear either along quartz grains or inside of the CSH-matrix as the following micrographs show in detail. Here it can be seen easily that macro- and microdefects are in direct correlation to each other. Magnification of Figures 10 and 11 are different, as the bars at the lower left of the photographs indicate.

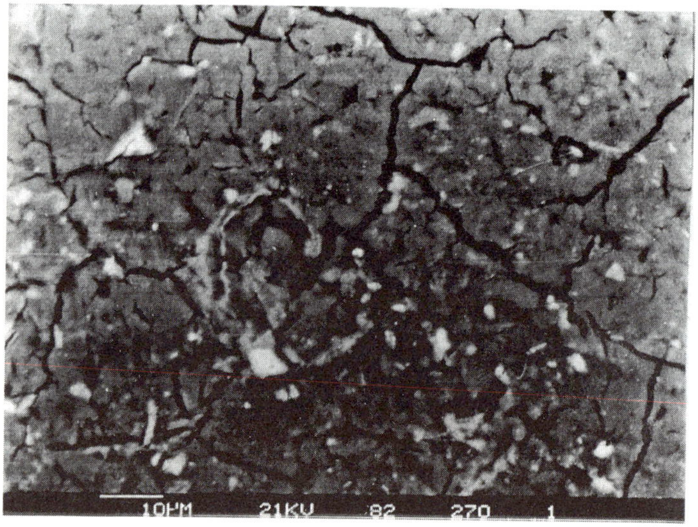

Figure 10 Microstructure of a 1 a old blended cement incorporating 70 mass% grade U

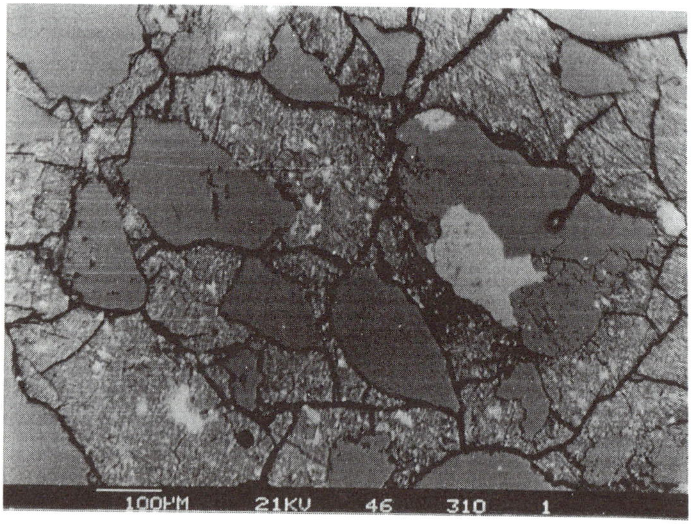

Figure 11 Microstructure of a 1a old blended cement incorporating 70 mass% grade U

DURABILITY

Durability was characterised by observing the corrosive reactions of CO_2-gas. Carbonation progress is directly dependent on density of the mortar. CO_2 has to penetrate the microstructure to be able to react with gel-CSH-phases and portlandite in a humid environment to form calcite. If microstructure is absolutely dense, no carbonation can occur. Therefore, progress of carbonation is an indicator for density.

Figures 12 to 14 show the degree of carobonation after 8 weeks for mortar blocks containing various cement types.

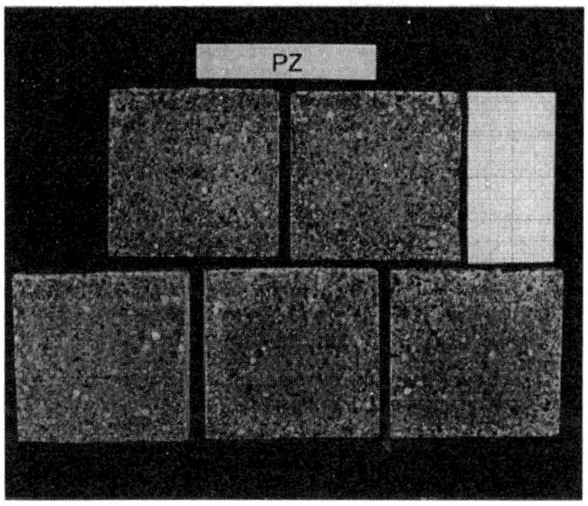

Figure 12 Mortar containing ordinary portland cement (OPC) carbonated 8 weeks

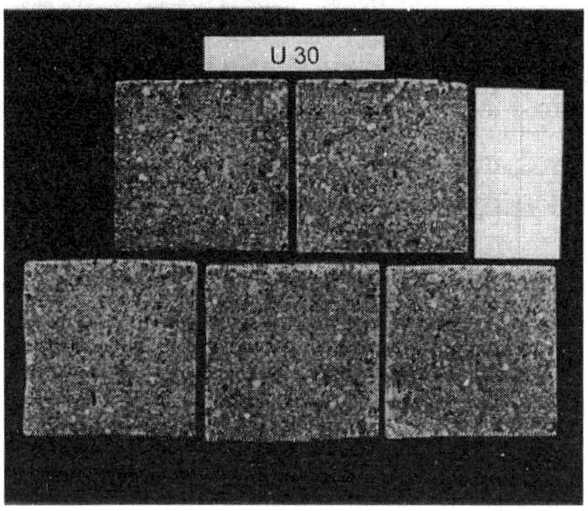

Figure 13 Mortar containing 30 mass% grade U, carbonated 8 weeks

Except for 50% PFA and 30% RHA mixtures, the other ones located in the high performance group are PFA ternary mixtures. One may consider that PFA alone, was the only pozzolan that presented, concomitantly, inferior performance regarding fineness and regular pozzolanic reactivity index and, at the same time, superior performance related to the ternary mixtures durability This apparent paradox may be explained by the synergic effect of PFA when its content increased in the mixtures. The physical action (deflocculation, dispersion) that the fly ash exert, concurrently with the pozzolanic action of RHA and SF, resulted in higher nucleation sites, microstructure with less porosity (tortuosity, pore blockage) and more homogeneous texture (grain and pore refinement). More details about synergic effect of PFA ternary mixtures are found in references [8, 10].

In the medium performance group the classification by pozzolan quantity showed that the amount of addition exerted influence not only on the pozzolanic reactivity but also on the physical effect. For the low performance group it was evident that the absence of PFA and its respective synergic effect decreased the concrete performance related to the properties studied, except for the PFA+SF mixture. It was also observed that the high content of SF seemed to provide less protection to the reinforcing steel by increasing the possibility of depassivation.

CONCLUSIONS

The following conclusions are drawn from this study:

1. The best durability of HPC mixtures from the point of view of steel corrosion was shown by the PFA+RHA (20+30)% ternary mixture;

2. The RHA in the 30% binary mixture and with PFA in ternary mixtures presented high performance;

3. The SF concrete showed good results only with the PFA+SF (15+10)% ternary mixture;

4. The SF 10% and RHA 20% binary mixtures, the PFA+SF (30+20)% ternary mixture presented lower performance, in an inferior degree of the reference concrete;

5. The high contents of pozzolan were beneficial in the RHA binary and ternary mixtures, except for the RHA 20% mixture

6. The high contents of SF, in binary and ternary mixture, seems not to be appropriate to rebar protection against corrosion;

7. The high content of PFA 50% binary mixture presented the second best performance among the 13 mixtures tested;

8. In general, the performance regarding the properties studied, were better with increasing pozzolanic content, except for the SF;

9. The better performance of ternary mixtures was due to the synergetic plasticizing effect presented by PFA.

REFERENCES

1. MEHTA, P K. Pozzolanic and cementitious by-products in concrete -Another look. Proc. Int. Conf. on PFA, SF, slag and natural pozzolans in concrete, 3rd, Trondheim, 1989. ACI, v. 1, p. 1-44 (SP-114).

2. JONES, M R, DHIR, R K, MAGEE, B J. Concrete containing ternary blended binders: resistance to chloride ingress and carbonation. Cement and Concrete Research, v. 27, n. 6, 1997, p. 825-831.

3. BUTLER, W B. Durable concrete containing three or four cementitious materials. Proc. Int. Conference CANMET/ACI, 4th, Sydney, 1997. ACI, v. 1, p. 309-330 (SP-170).

4. SWAMY, R N. Design for durability and strength through the use of fly ash and slag concrete. Proc. Int. Symp. CANMET/ACI, 3rd, Auckland, 1997. ACI, p. 1-72 (SP-171).

5. MEHTA, P K, FOLLIARD, K J. Rice husk ash – A unique supplementary cementing material: durability aspects. Proc. Int. Symposium CANMET/ACI, 2nd, Las Vegas, 1995. ACI, p. 531-542 (SP-154).

6. CEB-FIP Model Code 1990. Final Draft. Lausanne, 1991 (Bulletin n. 205).

7. DHIR, R K. Additional materials and allowable contents in cement. Proc. Nat. Seminar, Dundee, 1994. Eds R. K. Dhir and M. R. Jones. E & FN SPON, 1994, p. 57-68.

8. ISAIA, G C. Synergic action of PFA in ternary mixtures with SF and RHA. Proc. Int. Congress on the Chemistry of Cement, 10th, Gothenburg, 1997, v. 3, p. 3ii110.

9. DAN, W. et al. Comportement hydrique d'un composite ciment-polymère-verre (CCPV) et influence de l'humidité sur ses proprieté mechanique. Cahiers du CSTB, 1988, p. 1-16.

10. ISAIA, G C. Synergic action of PFA in pozzolanic ternary mixtures in HPC: a new durability approach. Paper accepted for presentation at II International Conference on High-Performance Concrete, Gramado, RS, Brazil, June 1999 (ACI Proceedings).

LIMESTONE, FLY ASH, SLAG AND NATURAL POZZOLANA: A COMPARATIVE STUDY OF THEIR EFFECT ON THE CEMENT PROPERTIES

N Voglis S Tsvilis G Kakali

National Technical University of Athens

E Chaniotakis

C Meletiou

TITAN Cement Company S.A.

Greece

ABSTRACT. In this paper the effect of limestone, fly ash, slag and natural pozzolana on the cement properties is studied. Four composite cements containing limestone, natural pozzolana from the Milos Island, slag and fly ash have been produced by intergrinding clinker (85%), the above main constituent (15%) and gypsum. The grinding process was designed in order to produce cements of the same 28d compressive strength. The mechanical and physical properties of the cements were measured and hydrated pastes were studied by means of TG and XRD. It is concluded that composite cements, containing the above materials, present significant differences as far as the clinker fineness, the hydration rate, the development of the early strength and the water demand are concerned.

Keywords: Portland-composite cements, Limestone, Pozzolana, Slag, Fly ash, Mechanical properties, Physical properties, Hydration, XRD, TGA.

N Voglis is Chemical Engineer and post-graduate student in Chemical Engineering Department, National Technical University of Athens, Greece.

Dr S Tsivilis is an Assistant Professor in Chemical Engineering Department, National Technical University of Athens, Greece.

E Chaniotakis is Director of the Research and Development Department, TITAN Cement Company S.A., Greece.

Dr G Kakali is a Lecturer in Chemical Engineering Department, National Technical University of Athens, Greece.

Dr C Meletiou is Head of Concrete Technology Laboratory, Department of Research and Technology, TITAN Cement Company S.A., Greece.

The strength development of the tested cements is illustrated in Figure 1. As all the cements have the same 28d compressive strength, the interesting point is the strength development rate. It is seen that PLC and PSC have the higher early strength while PFC the lower one. This phenomenon is correlated with the clinker fineness in each composite cement, as well as the behavior of the added material (pozzolanic, hydraulic or inert).

Table 3 Characteristic properties of the cements

CEMENT	GRINDING TIME, min	FINENESS		28D COMPR. STRENGTH (N/mm^2)
		Specific surface (cm^2/g)	Residue at 32μm (%)	
PC	41	3030	33.5	40.3
PLC	60	5710	23.5	40.5
PPC	52	4180	24.5	41.2
PSC	85	3590	22.3	42.5
PFC	40	3880	30.4	41.0

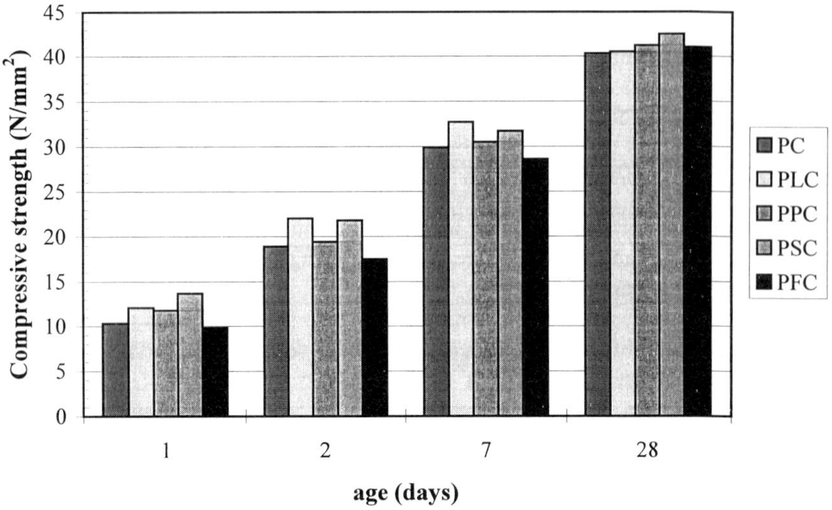

Figure 1 Strength development of the cements

The physical properties of the cements are given in Table 4. The PLC (despite its higher specific surface), PSC and PC demand less water than the PPC and PFC. The wider particle size distribution of PLC and PSC is a significant reason for the lower water demand [7, 12]. The initial and final setting of all cements do not indicate any important differences. The soundness of the cements (expansion according to Le Chatelier process) is satisfactory excluding the PFC that reaches the limit value prescribed by the prEN 197-1.

Table 4 Physical properties of the cements

SAMPLE	PASTE WATER DEMAND (%)	SETTING TIME (min) Initial	SETTING TIME (min) Final	EXPANSION (mm)
PC	24.3	125	170	1
PLC	24.5	90	170	2
PPC	26.2	95	140	3
PSC	24.0	80	150	3
PFC	26.4	100	160	10

The results of the shrinkage/expansion tests are presented in Table 5. The linear shrinkage of the tested cement mortars is satisfactorily low. In addition, the linear expansion is low in all samples, with PFC indicating the higher value.

Table 5 Linear shrinkage/expansion of the samples

SAMPLE	SHRINKAGE (10^{-6})				EXPANSION (10^{-6})			
	3d	7d	14d	28d	3d	7d	14d	28d
PC	50	80	170	200	10	20	30	40
PLC	20	100	150	210	35	45	45	50
PPC	30	100	130	230	15	20	25	40
PSC	60	120	180	180	10	15	20	25
PFC	10	80	140	190	35	40	45	75

The hydrated phases formed in pastes of blended cements are, in general, the same as those found in Portland cement pastes. Figure 2 presents, indicatively, the XRD pattern of PFC paste, hydrated for 28 days. Unhydrated calcium silicate compounds of clinker and hydration products such as CH, C-S-H and ettringite are clearly observed. Quartz, the main constituent of fly ash, is also found. In samples of limestone cement, monocarboaluminate ($3CaO \cdot Al_2O_3 \cdot CaCO_3 \cdot 11H_2O$) has been already formed in the first 24 hours and is still present after 28 days, as it is shown in Figure 3. Although there are not significant differentiation among samples hydrated for the same period of time, some modifications of calcium aluminate hydrates are indicated by the XRD patterns. Further investigation is needed in order to identify the exact nature of hydrates, in the presence of the added materials.

Table 6 presents the content of $Ca(OH)_2$, measured by means of thermogravimetry, in pastes hydrated for 28 days. The higher content of calcium hydroxide is found in PSC paste while the lower one is found in PFC paste.

8. TSIVILIS, S, KAKALI, G, CHANIOTAKIS, E, SOUVARIDOU, A. A study on the hydration of Portland limestone cement by means of TGA. Journal of Thermal Analysis, Vol. 52, 1998, pp 863-870.
9. BARKER, A, CORY, H. The early hydration of limestone-filled cements. Proc. Int. Conf. Blended Cements in Construction, Ed. N. Swamy, Elsevier Science, London, 1991, pp 107-124.

10. KLEMM, W, ADAMS, L. An investigation on the formation of carboaluminates. Proc. Conf. Carbonate Additions to cement, Eds. P. Kligger & D. Hooton, ASTM STP 1064, Philadelphia, 1990, pp 60-72.

11. TEZUKA, Y, GOMES, JR, MARTINS, J M, DJANIKIAN, J G. Durability aspects of cements with high limestone filler content. Proc. 9th Int. Congress on the Chemistry of Cement, New Delhi, Vol. V, 1992, pp 53-59.

12. SCHILLER, B, ELLERBROCK, H G. The grinding and properties of cements with several main constituents. Zement-Kalk-Gips, Vol. 45, 1992, pp E223-E231.

HIGH PERFORMANCE CEMENT: DISCOVERING NEW HORIZONS

K G Sobolev

KOCA Group

S V Soboleva

Summa Group

Turkey

ABSTRACT: The production of special and custom- made cements is becoming one of the powerful trends of a modern market. New types of high- tech cements based on high performance cement technology can be concerned as one of the promising materials for future construction projects. Developed approach to control the cement properties allows significantly improvement of high performance cement based systems strength and durability.

High strength and durability of high performance cement are provided by the application of silica fume based complex admixture. This phenomenon allows to use the required amount (up to 50%) of a granulated blast furnace slag as mineral admixture in the cement composition that results in the increasing of chemical and thermal resistance. The analysis of a wide scale investigation demonstrates the opportunity to produce high performance cement and mortars with compressive strength comprised between 95 and 145 MPa, respectively.

According to the results available, a 28-day compressive strength of 90 - 135 MPa was obtained for high performance cement based mortars. High performance cement based mortars possesses low permeability, high resistance to chemical attack and thermal resistance.

Keywords: High performance cement, Silica fume, Complex admixture, Granulated blast furnace slag, Compressive strength, Resistance to chemical attack, Thermal resistance.

K G Sobolev PhD works as a Research Adviser for KOCA Group, Turkey. He received his Doctoral degree from the Moscow Scientific and Research Institute of Concrete and Reinforced Concrete, Russia. He has been a head of more than 20 research projects in the field of high-performance cement and concrete.

S V Soboleva MSc graduated from the Moscow Civil Engineering University, Russia and took part in 17 projects as a researcher and public relations officer. Her scientific interests are in the field of high-performance cement. She currently works for Summa Group Turkey.

INTRODUCTION

There are numerous of examples for successful development and application of high strength and high performance cement based materials, when basic principles of portland cement - silica fume - superplasticizer systems were formulated by H. Bache [1]. Extensive work of many research teams worldwide allowed to transfer the laboratory findings in this area to practice [2, 3].

New types of high- tech cements based on high performance (HP) cement technology can be concerned as one of the promising materials for future construction projects [4-7]. In HP cement and mortars high strength (up to 145 MPa) and durability are provided by the application of silica fume based complex admixture. This phenomenon allows to utilize a large quantity (up to 70%) of indigenous mineral admixtures in a high volume mineral admixtures (HVMA) cement composition. The natural pozzolanic materials, sand, limestone, granulated blast furnace slag, fly ash, broken glass and ceramic can be used as indigenous mineral admixtures in the cements. This approach allows manufacturing of a wide range of special and custom- made cements with unique performance [4-7].

The main idea of the HP cement technology can be presented as physical-and-chemical activation of the cement grinding process with complex admixture. The increasing of dispersion and reaction ability of the cement components, as well as modifying of the cement surface with complex admixture are achieved due to complex admixture formulation. The following processes can be expected:

- pulverisation of clinker and mineral admixtures due to development of micro-distractions
- creation of high active amorphous structures and pre-hydrates
- physical-and-chemical modification of the minerals by complex admixture

The HP cement is manufactured through blending and grinding of a certain amounts of clinker, gypsum, complex admixture and mineral admixture of industrial or natural origin. There are some alternatives in HP cement production modes:

- basic HP cement (Type A) production mode
- blended HP cement (Type B) production mode

Basic Type A production mode is to manufacture HP cement by grinding of clinker, gypsum and complex admixture combination. Blended production mode covers the wide range of HP cements with mineral admixtures.

The mineral admixtures content in blended HP cement depends on required level of properties and type of the admixture used. Research results demonstrated the opportunity to produce the HP cements with mineral admixture content closed to standard limitations (within the range of 25-50%), as well as to extend frames of mineral admixture application for HVMA cement manufacture. Some new types of mineral admixtures such as sand, broken glass and ceramic can be utilized as a component of the blended HP cement [4-5]. The usage of granulated blast furnace slag (GBFS) in blended HP cement composition provides very high resistance to chemical attack.

Sulphate resistance of these systems is better than existing requirements for sulphate resistance cement due to super low permeability and more resistant chemical structure.

Wide scale investigations of strength behaviour of the system is demonstrating the opportunity to produce blended HP cement with compressive strength more than 80 MPa for GBFS content up to 50%.

The present report is intended to observe some properties detected for HP cement and mortars.

CHEMICAL COMPOSITION

Chemical composition of HP cement depends on the type and composition of components used. The chemical composition of HP cements (HPC) in comparison with normal portland cement (NPC) and GBFS used are presented in Table 1.

Table 1 Chemical composition of HP cements

	SiO_2	Al_2O_3	Fe_2O_3	CaO	MgO	K_2O	Na_2O	SO_3	L.O.I
NPC	19.44	4.78	3.57	63.74	1.90	0.77	0.21	2.70	2.41
GBFS	40.09	11.48	1.56	31.85	9.49	1.00	0.00	2.10	0.17
HPC-A	28.47	4.18	3.36	55.01	1.89	1.02	0.38	2.61	2.83
HPC-B	34.28	7.83	2.46	43.43	5.69	1.01	0.19	2.35	1.50

HP CEMENT PHYSICAL PROPERTIES

The test results of different types of HP cement (HPC) and normal cement (NPC) in accordance with corresponding ASTM procedure are summarized in Table 2. The particle size distributions of investigated cements are presented in Figure 1.

Table 2 Physical properties of HP cement

	FINENESS		SETTING TIME min		NORMAL CONSISTENCY %	COMPRESSIVE STRENGTH, MPa @ AGE, DAYS					
	BLAINE m^2/kg	45 µ %	INITIAL	FINAL		1	2	3	7	28	90
NPC	310	8.50	165	205	27.1	26.2	36.4	42.4	48.5	57.1	64.7
HPC-A	570	5.40	100	145	18.5	44.3	55.9	62.2	74.1	94.4	96.2
HPC-B	580	5.20	175	225	17.5	35.2	44.8	54.2	65.6	92.7	105.5

It was found that in spite of the higher fineness the HP cement demonstrated lower normal consistency. There is no much difference in setting time of Type B HP cement and normal cement. Type A HP cement demonstrated some reduction of setting time duc to rapid hardening.

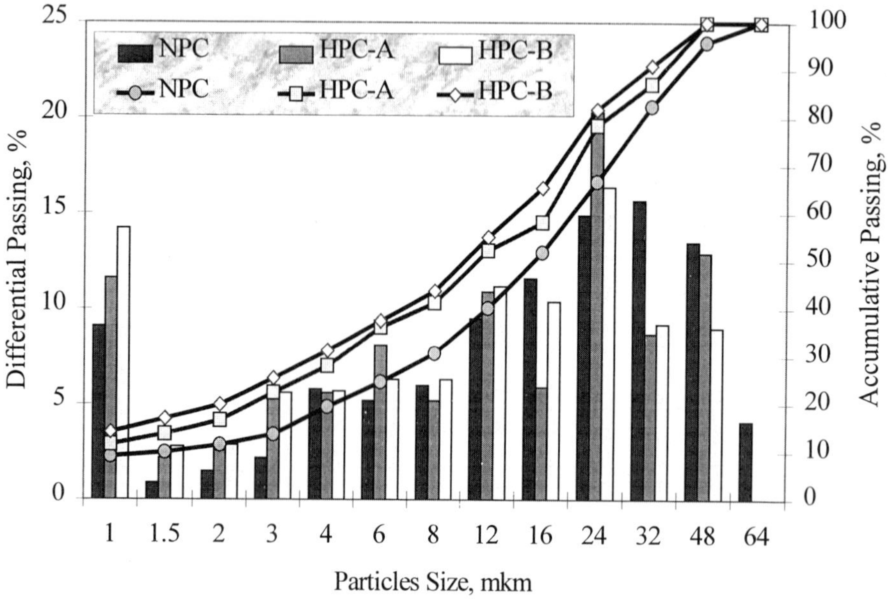

Figure 1 Particles size distribution of investigated cements

HP cement possesses a 28- day compressive strength, which is at least 65% higher than the strength of normal strength. Early age strength development of the HP cement allows to classify this cement as a super rapid hardening cement. HP cements demonstrated an ability to develop a long-term strength, especially, in case of Type B based on 50% of GBFS (Figure 2).

STRENGTH OF HP CEMENT BASED MORTARS

High strength phenomenon allows to apply the HP cement for production of the wide range of mortars. The 28- day compressive strength of HP cement mortars was found in the range of 40 to 145 MPa depending on sand to cement ratio (S/C).

The test results of strength behaviour of Type B HP cement based mortars in comparison with normal portland cement mortars in accordance with ASTM C389 "Standard Specification for Packaged, Dry, Combined Materials for Mortar and Concrete" are summarized in Table 3.

According to the obtained results, the strength of HP cement mortars at S/C=7 can be compared with the strength of NPC mortars at S/C=5 (Figure 3).

Table 3 Strength of HP cement based mortars

	28-DAY FLEXURAL STRENGTH, MPa @ S/C				28-DAY COMPRESSIVE STRENGTH, MPa @ S/C			
	1	3	5	7	1	3	5	7
NPC	12.2	8.9	5.3	3.7	70.2	43.8	22.3	12.9
HPC-B	24.5	12.8	8.3	5.9	148.1	96.4	60.8	39.8

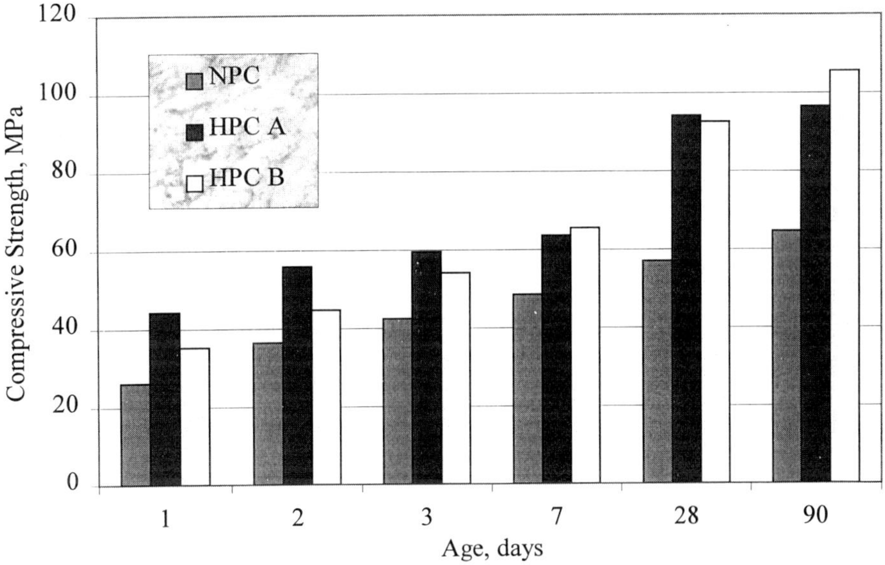

Figure 2 HP cement strength development

DURABILITY ASPECTS OF HP CEMENT BASED SYSTEMS

Water Absorption and Impermeability

Permeability of hardened cement stone is one of the basic properties determining the durability of concrete. Due to low W/C of HP cement based system, the capillarity is very low and water absorption value was found in the range of 0.3-0.8%.

In case of usage of GBFS for Type B HP cement, an additional reduction of pore size takes place. These factors provide impermeability of HP cement system at low diffusion coefficient of $1.2\text{-}0.8*10^{-10}$ cm/s.

Resistance to Chemical Attack

Very low permeability of HP cement systems provides high resistance to chemical attack. Investigation of acid resistance of mortars based on GBFS Type B HP cement demonstrated the opportunity to apply such systems in HCl solutions for 1½ month without risk of failure, while normal cement samples were completely destroyed after two weeks.

Freezing and Thawing Resistance

HP cement based mortars possess excellent freezing and thawing resistance. There was no visible destruction of HP cement samples after 140 cycles of freezing and thawing at $-50\ ^0C$. Bar expansion was 0.02%, than normal portland and GBFS cements length change was indicated at 0.1% after 45 and 35 cycles, respectively.

Thermal Resistance

HP cement mortars demonstrated high resistance to elevated temperatures. Only 10- 20% reduction of strength was found for the GBFS Type B HP cement mortars after step by step temperature rising (each step was 100^0C with 24- hour exposure).

Under the same conditions mortars based on normal cement have lost more than 50% of strength 500^0C.

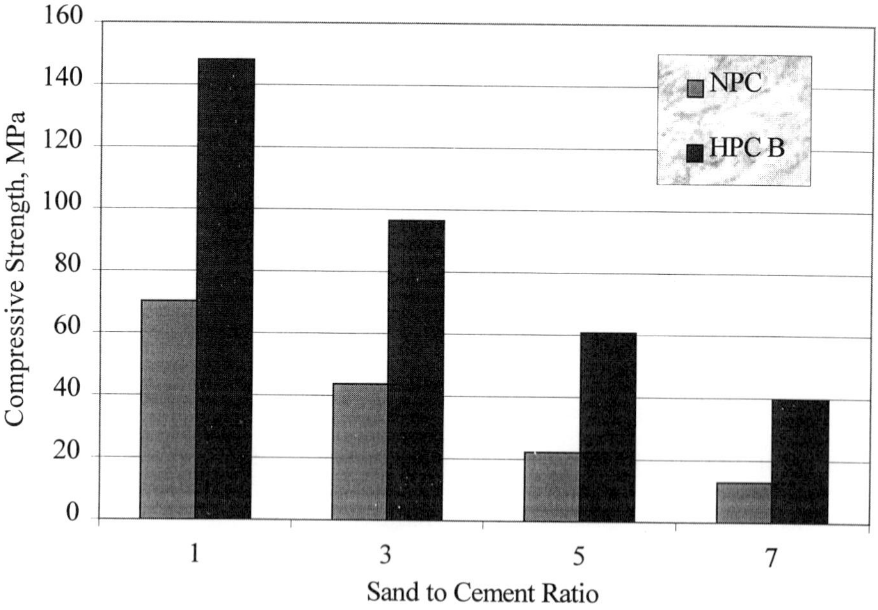

Figure 3 Strength of HP Cement Based Mortars

CONCLUSIONS

1. HP cements developed in accordance with presented approach have demonstrated better performance than existing normal cement in such aspects as strength at all age of hardening, permeability, resistance to chemical attack, freezing and thawing resistance, resistance to elevated temperature.

2. HP cement technology allows to utilize high volumes of mineral admixtures like GBFS for manufacturing of the cement based products with high strength and significantly improved durability. composition.

3. High level of HP cement properties allows to recommend this relatively new material for application in the unique projects such as high-rise buildings, airport runway structures, bridges, marine and offshore structures, tunnels, parking desks, shotcrete and repairing of structures, underwater concrete, special floors.

REFERENCES

1. BACHE, H H. "Densified Cement/Ultrafine Particle Based Materials". The Second International Conference on Superplasticizers in Concrete, Ottawa, 1981.

2. SHAH S P. "Recent Trends in the Science and Technology of Concrete". Concrete Technology: New Trends, Industrial Applications, Proceedings of the International RILEM Workshop, E&FN Spon, London, 1993, pp 1-18.

3. FIDJESTOL P AND KNUDSEN K. "High Performance Concrete for Durability, Using Microsilica". ERMCO-95, Proceedings of the XIth European Ready Mixed Concrete Congress, Istanbul, 1995, pp 379-389.

4. SOBOLEV K G AND SOBOLEVA S V. "High Performance Indigenous Cement". XXIVth World Housing Congress, Ankara, 1996, pp 189-202.

5. SOBOLEV K G AND SOBOLEVA S V. "Complex Admixture and Method of Cement Based Materials Production". TR Patent Application No 97/00426, Ankara, 1997.

6. SOBOLEV K G AND SOBOLEVA S V. "High Performance Cement: Form Idea to Industrial Trials". 7th International Conference on Management of Technology, Orlando, USA, 1998.

7. SOBOLEV K G AND SOBOLEVA S V. "Environmental Aspects of High Performance Cement Production". 1st International Symposium on Cement Industry, Assiut, Egypt, 1997.

EARLY HYDRATION OF ACTIVE MINERAL ADMIXTURE CONTAINING CEMENT SUSPENSIONS

V M Lilkov

S Gaidardzhiev

University of Mining and Geology

E D Krusteva

Bulgarian Academy of Sciences

Bulgaria

ABSTRACT. The study forms part of a project supported by the "Scientific Investigations" Fund of the Bulgarian Ministry of Education and Science aimed at the utilisation of various waste materials in cement-containing or cement-less slurries. The structure formation in the compositions is investigated by means of laser-granulometric and microscopic analyses for the dilute suspensions and by rheological properties analysis for the concentrated ones. The fractal geometry approach is used to describe the aggregate formation in the systems containing active mineral additives of the 'Pozzolit" type - classified fly ash (FA) and silica fume (SF) in mass ratio 1:1. Conclusions are drawn about the structure formation during the first hours of hydration.

Keywords: Active mineral additives, Pozzolit (Pz), Classified fly ash (FA), Silica fume (SF), Microscopic, laser-granulometric and rheological analysis, Structure formation, Fractal dimensions.

Associate Professor Vili M Lilkov is a lecturer at the Department of Physics, University of Mining and Geology, Sofia, Bulgaria. He specialises in the use of active mineral additives in concrete and in nuclear physics.

Research Associate Ekaterina Dimitrova Krusteva is a member of the Rheology Department of the Central Laboratory of Physico-Chemical Mechanics, Bulgarian Academy of Sciences. Her main research interests are in the field of rheology of inorganic materials and of waste product utilisation for ecological purposes.

Dr Stoyan Gaidardzhiev is a Chief Assistant at the University of Mining and Geology, Sofia, Bulgaria. He specialises in laser-granulometric and microscopic methods.

The particle size distribution obtained as a result of disintegrating processes has been successfully interpreted using the concepts of fractal geometry. The three-dimensional cubic model proposed by Moscovits [10] for the case of grinding of ordinary Portland cement clinker proves to be adequate to describe the fractal nature of the particle distributions of self-similar objects, moreover that the α ratio for the greater part of the cement, SF, FA and Pz particles is smaller than 1.16, i.e. their shape could be assumed as spherical, respectively cubic one. However, when trying to assess the efficiency of chemical admixtures intensifying the grinding process [11], the particle size distribution is no longer fractal, which means that with one crushing event a multitude of almost equal particles is obtained and self-similarity is no longer observed.

The applied model which divides a centre into P^3 parts yields the following expression for the cumulative weight distribution:

$$G(S) = \int_0^{\infty} W(S)ds = C.S^{(4-d_F)} \tag{1}$$

where, $W(S)$ = weight fraction distribution
S = particle size
d_F = fractal dimension

The smallest theoretical value of d_F is 2 for the lowest reasonable fragmentation value in two parts, and the highest - for an infinite number of fragmentations, is $d_F = 3$.

An attempt is made here to apply this approach to the opposite process - the aggregation of particles. It follows form Equation (1) that a log $(G(S))$ versus log (S) plot should produce a straight line over a certain range with a slope of 4 - d_F. The obtained d_F values for the curves in Figure1 are as follows:

Table 1 Values of d_F for the active mineral additives (in unreactive medium)

ADDITIVE	SF	FA	PZ
Laser-granulometric analysis	3.47*	3.245*	2.888
Microscopic analysis	3.483*	3.252*	2.94

* not a fractal dimension

As seen from the data in Figure1 and Table 1 the silica fume and the fly ash themselves do not exhibit self-similarity in their particle size distribution, i.e. their granulometry is characterised by a high content of fine particles of equal size. When mixed to obtain the Pozzolit additive they form self-similar aggregates with a fractal dimension typical for the ordinary Portland cements. The large FA grains are observed separately in the SF-FA mixture, while the smaller ones (size of up to 10 μm) are covered by fine SF particles. There is a strong tendency for the SF and FA particles to adhere to each other, thus forming stable complexes of small particles. A part of the SF particles stuck to the coarser FA grains are set apart in the suspension. The d_F value of the cement used in this investigation is very close to 3 (d_F= 2.96) which is an indication for a rather finely ground initial clinker.

The granulometric analysis of the hydrating suspensions (Figure 2a,b,c) shows the development of the aggregates in the course of time. The values of the fractal dimensions at the age of 0 h, 1 h and 4 h are presented in Table 2. The specific surface of the newly formed hydration products after the first contact of cement with water is increased more than 500 times [12]. The presence of the abundant fine particles of almost the same size during the initial stage of the hydration of the plain cement suspension is reflected in the d_F value, which is greater than 3 and does not correspond to a fractal nature. The granulometric distribution has therefore an entirely different character compared to that in unreactive medium. The SF containing cement suspension resembles the behaviour of the plain cement one at the very beginning (0 h) when the role of the numerous fine silica fume particles is predominating. Their high reactivity leads to a sharp change as early as during the first hour and to the formation of different in size self similar aggregates, which is seen from the d_F value for the first hour - 2.8 (Table 2). On the contrary, the effect of the fly ash is not felt substantially and the fractal dimensions are very close to those of the plain cement suspensions.

The Pozzolit containing suspensions exhibit a stable fractal structure during the whole investigated time interval. At the 4th hour of hydration the particle size distributions of all the suspensions are very similar (Figure 2c) and they all display a fractal character. However, these results should be interpreted more carefully since at this age the shape of the particles becomes more irregular, the share of the aggregates having an α ratio less than 1.16 [7] is diminished and the assumption of spherical (respectively cubic) form of the particles is no longer valid.

Table 2 Values of d_F for the cement suspensions with and without active mineral additives (in reactive medium - water)

TIME	C	C + SF	C + FA	C + PZ
0 h	3.178*	3.391*	3.295*	2.968
1 h	3.36*	2.8	3.04*	2.858
4 h	2.848	2.836	3.029*	2.975

* not a fractal dimension

Rheological analysis

The apparent viscosity dependence on the water/solid ratio for the compositions with 10 % of additive content is shown in Figure 3 for a fixed shear rate $\gamma = 100\ s^{-1}$. The component exerting the greatest effect on viscosity is the silica fume [8], its specific surface being higher by an order than that of cement and fly ash. The SF content contributes to the increased viscosity and improved stability of the suspensions. At the same time the typical for the amorphous silica high thixotropy under strong agitation [13] renders good workability to the SF containing compositions. It is clearly seen that the Pozzolit additive represents a compromise with regard to viscosity, hence workability of the suspensions.

REFERENCES

1. MASSAZA, F. Chemistry of pozzolanic addition and mixed cements. Proceedings of the 6th International Congress on Chemistry of Cement, Moscow, 1974, pp III-209-221 (in Russian).

2. MALHOTRA, V M (Ed). Supplementing Cementing Materials for Concrete. 1997.

3. ROY, D M. Fly ash and silica fume chemistry and hydration. Proceedings of the 3rd International Conference on Fly Ash, Silica Fume, Slag and Natural Pozzolans in Concrete, Trondheim, 1989.

4. STOICHKOV, V, ABADJIEV, P, LILKOV, V, VASILEVA, V. Effect of the "Pozzolit" active mineral admixture on the properties of cement mortars and concretes. Part I: Physical and mechanical properties. Cem. Concr. Res., Vol. 26, No. 7, 1996, pp 1065-1072.

5. LILKOV, V M, STOICHKOV, V. Effect of the "Pozzolit" active mineral admixture on the properties of cement mortars and concretes. Part 2: Pozzolanic activity. Cem. Concr. Res., Vol. 26, No. 7, 1996, pp 1073-1081.

6. LILKOV, V, DIMITROVA, E, PETROV, O. Hydration process of cement containing fly ash and silica fume: the first 24 hours. Cem. Concr. Res., Vol. 27, No. 4, 1997, pp 577-588.

7. LILKOV, V, DIMITROVA, E, GAIDARDZHIEV, S. Microscopic and laser granulometric analyses of hydrating cement suspensions. Cem. Concr. Res. (in press).

8. DIMITROVA, E, LILKOV, V. Effect of active mineral admixtures on the rheological properties of cement pastes. Proceedings of the 12th International Congress on Rheology, (Eds A. Aït-Kadi, J M Dealy, D F James and M C Williams), Quebec, 1996, 473.

9. NEGRO, A, MONTANARO, L, BACHIORRINI, A. On the reactivity of the cement clinkers components by means of laser granulometer. Cem. Concr. Res., Vol. 15, No. 2, 1985, pp 315-319.

10. MOSCOVITS, M. The fractal nature of particle size distributions of ground clinker. Cem. Concr. Res., Vol. 20, No 4, 1990, pp 499-505.

11. KRUSTEVA, E D. Description of the grinding processes of cement clinker using the fractal geometry concepts. Proceedings of the 6th International Conference on "Theory and Practice of Disintegration Processes", August, 1998, Odessa, Ukraine (submitted).

12. POWERS, T C. The properties of fresh concrete, John Wiley & Sons, Inc., N.Y. 1968, 664 p.

13. DIMITROVA, E, SERBEZOVA, R. Rheological properties of thixotropic castables. Progress and trends in rheology IV (Ed C. Gallegos), Steinkopff, 1994, 639-641.

PERFORMANCE OF CONCRETE CONTAINING TERNARY BINDERS IN CHLORIDE-LADEN ENVIRONMENTS

B J Magee

Purdue University

United States of America

M R Jones

R K Dhir

University of Dundee

United Kingdom

ABSTRACT. Based on the premise that if one additional binder material in concrete provides improved performance, then further benefits are likely when using two, this study was carried out to examine selected properties of ternary binder concrete (TBC). Various fresh concrete properties were considered as well as strength development and chloride resistance of TBC. In addition, the nature of the near surface pore structure of the concrete has been inferred from its initial surface absorption. Performance in all instances was compared to PC and binary PC/PFA concrete of equivalent 28-day cube strengths of 20, 40 and 60 N/mm2. In general, no practical dissimilarity existed between TBC, PC and PC/PFA mixtures in terms of fresh concrete properties. When designed for equivalent 28-day strength, TBC exhibited lower strengths than the controls. Equations enabling estimates of strength reductions are proposed. Improved long-term strength was achieved by TBC, with PC replacement levels of around 60% by mass identified as an optimum. Using accelerated electrochemical and penetration test methods, it has been shown that the chloride resistance of all the ternary binder concrete (TBC) is significantly higher than corresponding PC and PC/PFA mixes. This improved performance is attributed mainly to the improved microstructure of TBC mixes.

Keywords: Ternary binder concrete, PFA, GBS, SF, Mix handling/placing/finishing characteristics, Strength development, Chloride ingress.

Dr Bryan J Magee is currently a post-doctoral research associate at the School of Civil Engineering, Purdue University, Indiana, USA. Prior to this appointment, he held a similar position at the Department of Civil Engineering, University of Cape Town, South Africa. He received his BSc(Eng) and PhD degrees from the University of Dundee, Scotland. His research interests include cement and concrete technology, with emphasis on the use of supplementary cementitious materials and the issue of long-term durability

Dr Rod Jones is a chartered civil engineer and senior lecturer in the Concrete Technology Unit in the Department of Civil Engineering at the University of Dundee. His research focuses mainly on binder technology, concrete durability and repair and maintenance.

Professor Ravindra Dhir is the Director of the Concrete Technology Unit and Professor of Concrete Technology at the University of Dundee. He is a member of numerous national and international technical committees and has published extensively on many aspects of concrete technology, binder science, durability and construction methods.

INTRODUCTION

Premature corrosion damage of embedded reinforcement in concrete continues to present one of the most serious durability problems facing engineers. Although it is established [1-3] that improved resistance to corrosion damage is achievable through the use of materials such as pulverised-fuel ash (PFA), granulated blastfurnace slag (GBS) and silica fume (SF) in binary blends, case studies indicate that many existing structures continue to suffer [4].

Encouragingly, recent research [5,6] indicates that ternary blended binders in concrete allow high levels of durability to be achieved, particularly with respect to chloride ingress. These findings assume particular significance with the adoption of the cement standard ENV 197-1 (1996), which presents European engineers with the opportunity to tailor concrete for high levels of durability via binder optimisation. Indeed, ENV 197-1 allows cocktails of three or more binder materials to be prescribed for any given concrete mix. The value of using ternary binder concrete (TBC) has already been realised in many countries, with its advantages exploited for some major infrastructure projects. For example, the Great Belt Link in Denmark was constructed using a PC/PFA/SF blend [7]. In this case, TBC was used as an integral component of a multistage protection strategy designed to provide a service life of 100 years.

this study was undertaken to examine both the practicality and performance of TBC. Although the main focus of the study was the chloride resistance of TBC, selected fresh properties and strength development were additionally considered. Despite ENV 197-1 identifying a total of 25 types of cement (of which Types II-N and V allow ternary blended binders) based around 8 different additional materials, only those of immediate relevance to the UK were considered. Combinations of Portland cement (PC) to BS 12 (1991), PFA to BS EN 450 (1995), GBS to BS 6699 (1991) and SF were, therefore, considered within the scope of this work, used in combinations based on the requirements of ENV 197-1.

EXPERIMENTAL MIX PROPORTIONS

On a practical basis and in order to limit large numbers of possible binder combinations, additional materials were proportioned based on replacement levels similar to those used in typical binary mixes. Such binder proportions were also considered suitable due to their potential to provide practical mixes with high levels of chloride resistance [1-3]. Using a mix design method developed at the University of Dundee [8], TBC mix proportions were selected to give standard 28-day cube strengths of 20, 40 and 60 N/mm^2. The control mixes were PC and PC/30% PFA concrete of equal cube strengths. The mix proportions used are given in Table 1, which additionally includes a binder proportion summary where each binder material is represented as a percentage of the total binder content by mass.

Performance of each binder combination chosen was highlighted by maintaining fixed free water and coarse aggregate contents (185 l/m^3 and 1200 kg/m^3 respectively). Unit volumes of the mixes were subsequently achieved by varying the fine aggregate contents. It should be noted that with the free water content selected, 28-day cube strengths of 60 N/mm^2 were not achievable with the PC/PFA/GBS blends. All mixes obtained a nominal slump of 75 mm without the need for a plasticising chemical admixture except for those using SF, where the very cohesive nature of the concrete required the use of a minimal dosage of superplasticiser (see Table 2 for details). All concrete specimens were standard water cured for 28-days and all tests were initiated at that time.

Table 1 Summary of concrete mix proportions

MIX CODE	28-DAY CUBE STRENGTH, N/mm^2	CONSTITUENT MATERIALS, kg/m^3					BINDER PROPORTION, % BY MASS
		Binder				Fine Aggregate	
		PC	PFA	GBS	SF		
PC control							
C 1	20	225	-	-	-	820	100/0/0
C 2	40	325	-	-	-	745	100/0/0
C 3	60	430	-	-	-	630	100/0/0
PC/PFA control							
C 4	20	220	95	-	-	600	70/30/0
C 5	40	270	115	-	-	540	70/30/0
C 6	60	405	175	-	-	360	70/30/0
PC/PFA/GBS							
TBC 1	20	140	50	140	-	595	42.5/15/42.5
TBC 2	20	100	120	100	-	565	32.5/35/32.5
TBC 3	20	195	70	195	-	475	25/15/60
TBC 4	20	85	150	195	-	475	20/35/45
TBC 5	40	230	55	100	-	545	60/15/25
TBC 6	40	155	175	155	-	425	32.5/35/32.5
TBC 7	40	155	90	370	-	330	25/15/60
PC/PFA/SF							
TBC 8	20	180	20	-	20	725	80/10/10
TBC 9	20	140	140	-	30	580	45/45/10
TBC 10	40	250	30	-	30	615	80/10/10
TBC 11	40	220	220	-	45	410	45/45/10
TBC 12	60	265	265	-	65	355	45/45/10
PC/GBS/SF							
TBC 13	40	195	-	85	25	625	65/25/10
TBC 14	40	155	-	155	35	580	45/45/10
TBC 15	40	105	-	245	35	545	25/65/10
TBC 16	60	160	-	370	50	370	25/65/10

FRESH PROPERTIES

Mix Observations

Owing to the unfamiliarity of TBC, qualitative observations for ease of compaction, cohesiveness, colour and finishing characteristics were initially documented, see Table 2. Clearly, TBC mixes exhibited no visual dissimilarity to good quality PC concrete. In fact, with regards to fresh properties, TBC was generally of higher quality than the PC and PC/PFA controls. Of practical significance, TBC was easily compacted, exhibited no visible bleeding and produced an excellent surface finish.

Table 2 Fresh properties and visual observations of TBC

MIX CODE	SLUMP, mm	SP DOSAGE[1]	AIR CONTENT, %	OBSERVATIONS OF MIX HANDLING CHARACTERISTICS
PC control				
C 1	85	-	1.2	Normal mix appearance. Good workability and compaction.
C 2	75	-	1.1	Normal mix appearance. Slight bleeding.
C 3	60	-	1.1	Normal mix appearance. Good workability and compaction.
PC/PFA control				
C 4	90	-	1.0	Normal mix appearance. Very workable, easy compaction.
C 5	75	-	1.2	Normal mix appearance. Good cohesion and compaction.
C 6	65	-	1.1	Normal mix appearance. Handling and compaction good.
PC/PFA/GBS				
TBC 1	75	-	1.1	Normal mix appearance, with good workability, compaction and finish
TBC 2	65	-	1.3	Normal mix appearance. Good workability, slight bleeding.
TBC 3	80	-	1.2	Normal mix appearance, no bleeding, good cohesion and surface finish.
TBC 4	60	-	1.1	Good workability, compaction and cohesion. Appeared normal.
TBC 5	60	-	1.1	Normal mix appearance, quite stiff, no bleeding, good cohesion.
TBC 6	50	-	1.0	Normal mix appearance.
TBC 7	50	-	1.3	Normal mix, good workability, compaction, finish and cohesion.
PC/PFA/SF				
TBC 8	65	0.7 (7.0)	0.9	Average workability but mix easily compacted. Dark in appearance.
TBC 9	60	0.9 (9.0)	0.8	Fairly harsh mix, no bleeding. Easy concrete compaction. Dark grey.
TBC 10	60	0.9 (9.0)	0.6	Mix appeared normal. Good workability, easy handling and workability.
TBC 11	55	1.1 (17.0)	0.9	Mix stiff, with low slump. Compaction achieved. Very dark colour.
TBC 12	50	1.5 (13.0)	0.9	Mix stiff, with low slump. Compaction achieved. Very dark colour.
PC/GBS/SF				
TBC 13	75	1.0 (12.0)	0.8	Discolouration, average workability, mix compacted satisfactorily.
TBC 14	70	1.0 (10.0)	0.9	Harsh, stiff mix but good compaction. No bleeding, discolouration.
TBC 15	70	1.1 (12.0)	0.7	Mix appeared normal. Good workability, handling and compaction.
TBC 16	65	1.4 (16.0)	0.9	Stiff, harsh mix. No bleeding, high cohesion. Slight discolouration.

[1] Superplasticiser dosage as % of total binder content (and SF content) by mass

However, there was some variation in the colour of TBC when assessed at 28 days. Mixes rich in GBS and PFA tended to range from lighter shades of grey to typical shades. Alternatively, and due to the nature of the slurry used, mixes rich in SF ranged from darker shades of grey to almost black.

Entrapped Air Contents

As the potential for TBC to entrap air was uncertain, all concrete was tested in compliance with BS 1881: Part 106 (1983) after undergoing full compaction. As shown in Table 2, results indicate minor differences existing between the controls and PC/PFA/GBS mixes. In contrast, entrapped air contents obtained for all TBC mixes containing SF were on average 30% lower than the controls. Why the observed reduction in entrapped air occurs is unclear, although some hypotheses have been described for binary mixes. As superplasticisers are dispersants they tend to repel air that has become, in part, stabilised by orientation of the surfactant. For this reason, high workability concrete containing superplasticisers are often characterised by foam, caused by a rapid release of entrapped air [9]. As all TBC mixes inclusive of SF contained superplasticiser, this theory may, in part, explain the reduced air contents recorded.

STRENGTH DEVELOPMENT

Specimens (100 mm cubes) were tested at intervals up to 180-days in accordance with BS 1881: Part 116 (1983). TBC mixes generally exhibited lower early strength development than PC and PC/PFA control concrete. Indeed, at 1 and 3-days, respective TBC strengths were on average 50% and 30% lower than PC concrete, and 25% and 15% lower than PC/PFA concrete. These strength differences decreased with time, however, and by 90 days the TBC strengths were on average 25% and 20% higher than the PC and PC/PFA concrete respectively. This level of strength improvement was maintained up to 180 days.

Cube strengths obtained for TBC mixes at 1, 3, 90 and 180 days are given in Figure 1, expressed as a percentage of PC control results and plotted with respect to the percentage of PC replacement used for each mix.

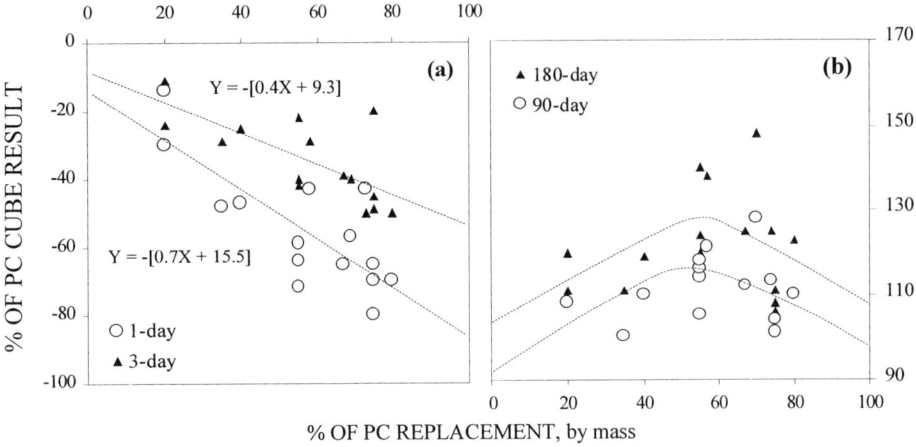

Figure 1 Relationship between TBC strength and percentage of PC replacement used

Figure 1(a) indicates that as PC replacement in the TBC mixes increased, strength development at 1 and 3 days decreased with respect to the PC control. By calculating the equations of the best-fit lines shown, it is possible to estimate early strength development of TBC.

For example, consider a TBC mix with a PC replacement of X%. Strength at 1 and 3 days would be approximately [0.7 X + 15.5]% ± 9.0% (at a 90% confidence) and [0.4 X + 9.3]% ± 8.0% (at a 90% confidence) lower than a typical PC mix respectively.

In contrast, the relationship between improved TBC strength and PC replacement at 90 and 180-days was not consistent. Indeed, Figure 1(b) indicates that an optimum PC replacement exists, in terms of ultimate strength development, of around 60% by mass. Beyond this level, it is likely that $Ca(OH)_2$ available for pozzolanic reactions may be limited. It should be recognised, however, that as all concrete specimens were standard water cured until the required test age, continued pozzolanic activity was not prevented.

ACCELERATED CHLORIDE RESISTANCE TESTS

Electrochemical Chloride Ingress

Predicting a potentially high chloride ingress resistance of the TBC mixes, an accelerated 14-day electrochemical chloride transmission test [10] using 12V DC potential difference (PD) was initially adopted. PD indices were subsequently calculated using Fick's First Law and these are compared for all mixes in Figure 2.

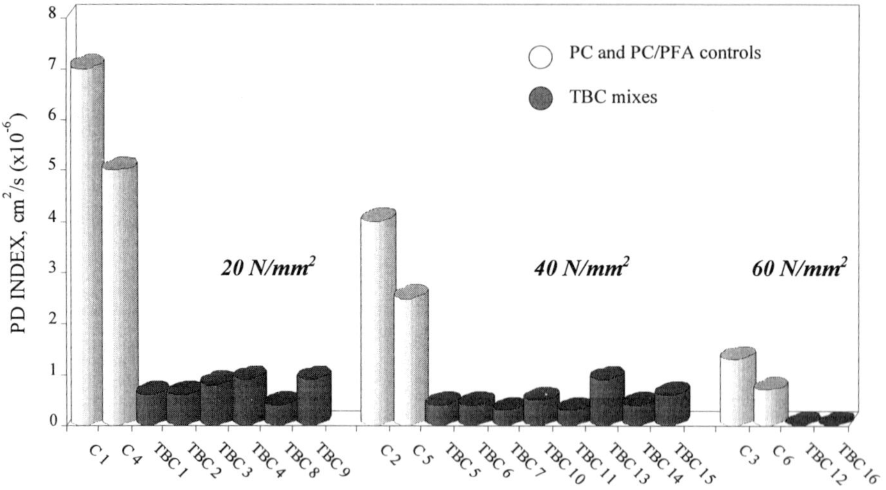

Figure 2 Comparison of PD indices obtained for control and TBC mixes

In comparison to both the PC and PC/PFA control mixes, the use of ternary blended binders resulted in a dramatic reduction in PD index. Considering the 20 N/mm^2 mixes for example, PD indices obtained for TBC were on average 90% and 85% lower than those obtained for the control PC and PC/PFA concrete respectively.

The degree of improved performance decreased slightly with an increase of concrete strength to 40 N/mm^2 and corresponding reductions for these TBC mixes were on average 89% and 82%. At the conclusion of the 14-day test period no chloride transmission was measured for either of the 60 N/mm^2 TBC mixes considered.

Comparing the PD indices obtained it is evident that only minor differences existed (results ranged from between 0.30 to 0.90), irrespective of the concrete strength or the binder combination used. These variations are, in fact, within the accuracy of the test method and may, therefore, not be significant [10]. This trend most likely reflects equally very high levels of chloride resistance for all TBC. It is interesting that as the level of PC replacement ranged from between 20 to 80% by mass, the results from this test method suggest that no optimum PC replacement exists in order to achieve maximum chloride resistance.

Chloride Penetration

Owing to the rapid nature of the electrochemical chloride transmission test and the resulting likelihood of limited chloride binding [11], a restricted series (i.e. mixes TBC 13-16 omitted) of chloride immersion testing was additionally carried out. All test specimens (75 x 75 x 300 mm prisms) were immersed in 5 M. NaCl solution and sealed on 5 faces to give uniaxial chloride penetration. After an immersion period of 28, 56 and 91 days, powder samples were obtained by incremental drilling at depths of 0-5 mm, 6-15 mm and 16-30 mm from the as-cast surface. X-ray fluorescence was subsequently used to analyse these samples for their total chloride content.

Using results obtained after 91 days of immersion, chloride profiles were constructed as shown in Figure 3.

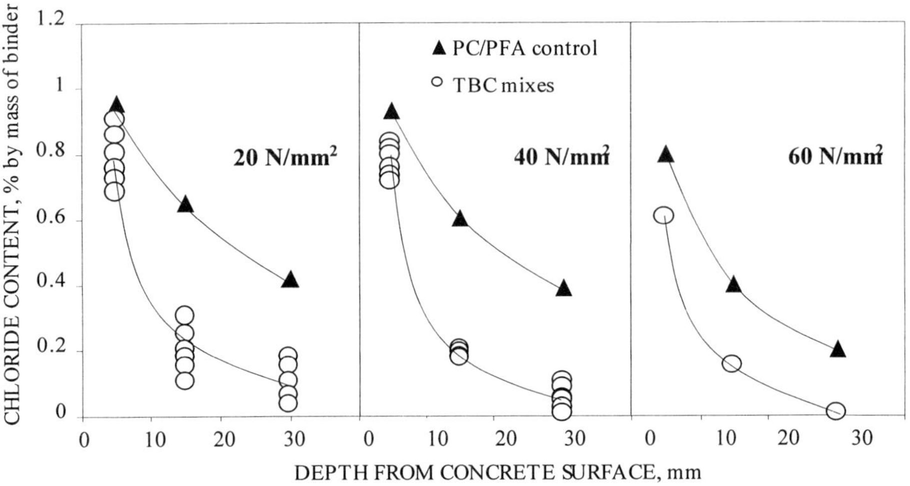

Figure 3 Chloride profiles constructed from 91 day accelerated chloride penetration results

The significant advantage of using TBC mixes is, once again, clearly apparent. For example, considering the 16-30 mm results collectively over the range of 28 day strengths, the TBC mixes had on average 80% lower chloride concentration than the PC/PFA controls. As would be expected due to shorter exposure periods, increasingly lower chloride levels, in comparison to Figure 3, were obtained for all mixes at 56 and 28 days. However, profiles constructed from results at each of these ages produced very similar trends to those shown in Figure 3, with TBC outperforming the controls in all instances.

CHLORIDE DIFFUSION COEFFICIENTS

Owing to the accelerated nature of the chloride tests considered, the results obtained (PD indices and percentages of total chlorides by mass of concrete for the electrochemical and penetration tests respectively) are not comparable and cannot easily be used to evaluate concrete's resistance to chloride ingress. For this reason, coefficients of chloride diffusion were estimated or calculated from the results obtained from each test.

Curves have been reported in the literature [10], correlating values of PD index and coefficients of chloride diffusion for PC and PC/30% PFA concrete. As no established relationship exist between coefficients of diffusion and PD indices for TBC, the curves given for the PC/30% PFA concrete were adopted for both the PFA control and TBC mixes. Using these curves, coefficients of chloride diffusion have been estimated as shown in Table 3. It is recognised that further research is merited in this area, to establish relationships between coefficients of diffusion and PD indices for various TBC mixture combinations. Alternatively, coefficients of chloride diffusion were calculated from the chloride profiles shown in Figure 3, using Fick's Second Law. Clearly from Table 3, and most likely reflecting the assumptions made when estimating values from the electrochemical test data, variations existed in coefficients of chloride diffusion obtained from the two accelerated test methods.

The results again illustrate, however, the greatly increased performance of TBC in comparison to the controls. Nomograms enabling estimates of service life for a given set of conditions have been established [12] and indicate that poor, good, very good and excellent performance is likely for coefficients of chloride diffusion in the ranges >10.0, 5.0-10.0, 1.0-5.0 and <1.0 x 10^{-9} cm^2/sec respectively. From this study, TBC mixes indicate the likelihood of very good and excellent performance in all cases, whereas only 60 N/mm^2 PC/PFA concrete achieves very good performance. The PC control concrete results indicate the probability of poor performance at strengths up to 40 N/mm^2. The 60 N/mm^2 PC concrete achieves only good performance status.

DISCUSSION OF TEST RESULTS

Fresh Concrete

From the qualitative analysis carried out, the use of ternary blended binders in concrete presents no immediate technical difficulties. Indeed, the compaction, cohesiveness, and finishing characteristics of TBC were all favourable. However, the colour variability noted has aesthetic implications, for example, in structures where selected elements use TBC with conventional or binary mixes in close proximity. TBC that is almost black in colour is not likely to be practicable in many cases and probable application for such mixes is limited to underground works. For structures where aesthetics are critical, suitable materials and material combinations should be chosen and trial mixed.

Table 3 Coefficients of chloride diffusion calculated or estimated from accelerated tests

MIX CODE	COEFFICIENT OF CHLORIDE DIFFUSION, $cm^2 \times 10^{-9}$	
	Calculated from accelerated chloride penetration test [1]	Estimated from accelerated electrochemical chloride test [2]
PC control		
C 1	65.0	95.0
C 2	25.0	40.0
C 3	9.5	11.0
PC/PFA control		
C 4	15.0	35.0
C 5	8.5	17.0
C 6	5.5	11.0
PC/PFA/GBS		
TBC 1	2.0	4.0
TBC 2	1.0	4.0
TBC 3	1.0	5.0
TBC 4	1.7	5.5
TBC 5	0.5	3.5
TBC 6	1.0	3.5
TBC 7	1.1	2.0
PC/PFA/SF		
TBC 8	2.1	3.5
TBC 9	3.0	5.5
TBC 10	0.5	4.0
TBC 11	0.4	2.0
TBC 12	0.4	0

[1] Calculated using Fick's Second Law.
[2] Estimated using established relationships between PD-index and coefficient of chloride diffusion, with relationship for PC/30%PFA assumed to be valid for TBC mixes.

Strength Development

The inherent low strength development of TBC mixes at early ages may have to be accounted for in practice, particularly if designing for applied pre-stressing forces, formwork pressures and stripping times etc. In addition to estimating early strength reductions using the equations proposed, it is likely that this drawback might be minimised by using accelerating admixtures, raised curing temperatures or by reducing the level of PC replacement. In adopting the latter course of action, however, the producer may sacrifice longer-term concrete performance, particularly with respect to durability.

Clearly, the results indicate an optimum PC replacement level with respect to the longer-term strength development of TBC. It is likely that optimum performance in terms of durability may also be achieved using this PC replacement level (around 60% by mass).

In many cases, however, achieving maximum strength development may not be the main design criterion and factors such as cost, heat evolution etc. may be the controlling parameter. This may result in selecting a TBC mix for a minimum total binder content or a minimum PC content.

Chloride Resistance

In terms of chloride resistance and due to either physical and/or chemical influences [5], the improved TBC performance is striking, with 20 N/mm^2 TBC mixes outperforming 60 N/mm^2 controls in many cases. Physically, pozzolanic reactions associated with additional materials are likely to refine the microstructure of concrete [12]. The results recorded after 10 minutes (ISAT-10 results) indicated a large reduction of TBC surface absorption in comparison to the controls. ISAT-10 values of 75.0, 55.0 and 40.0 $ml/m^2/sec \times 10^{-2}$ were obtained for the 20, 40 and 60 N/mm^2 PC concrete respectively, with reductions of 14%, 18% and 50% noted for the corresponding PC/PFA mixes. In contrast, average reductions in ISAT-10 values for the TBC mixes were 63%, 67% and 80% at each strength grade respectively. This trend essentially mirrors that obtained for chloride resistance, as given in Figure 2.

It has been reported [11] that some aluminate phases in concrete may play a role in the binding of chloride ions. Although the bulk Al_2O_3 content may be a relatively poor estimate of the actual amount of chloride binding sites available, it is at least generally indicative. Despite variations in total binder content existing between mixes, calculations indicated that the quantity of Al_2O_3 (per kg/m^3) in the majority of TBC mixes was not significantly different than that in the PC/30% PFA controls and indeed in many cases was lower, in particular for the SF mixes. This suggests that overall it is probably the improved microstructure of TBC that has the dominant effect on increased chloride resistance.

CONCLUSIONS

1. In terms of the fresh properties examined, no practical difference existed between TBC, PC and PC/PFA mixes. TBC was, however, very cohesive but easily compacted and produced a high quality surface finish. The use of superplasticiser was found necessary in all TBC mixes containing SF. TBC containing PC/PFA/GBS entrapped similar quantities of air to PC and PC/PFA control mixes. All TBC containing SF entrapped on average 30% less air than control mixes. TBC had a wide range of colours, ranging from light grey to almost black, depending on the combinations of binder materials used.

2. When designed for equivalent 28-day strength, TBC exhibited on average 50% and 30% lower strength than PC control concrete and 1 and 3 days. These differences reduced with time, however, and the TBC strength results were on average 25% higher than the PC control at 90 and 180 days. For optimum long-term strength development, PC replacement levels of around 60% by mass were identified for TBC.

3. The results obtained from both the electrochemical ingress and penetration tests indicated that the chloride resistance of TBC was significantly better than both the PC and PC/PFA control mixes. The initial surface absorption of TBC was reduced in comparison to the controls. Since there were only minor differences in bulk Al_2O_3 content, this suggests that the chloride resistance of TBC is mainly the result of microstructural refinement. No optimum replacement levels for PC or combinations of binder materials for chloride resistance were found and it is recommended, therefore, that mix proportions are chosen to maximise strength development (i.e. by selecting a PC replacement of around 60%).

REFERENCES

1. DHIR, R K, and MATTHEWS, J D. Durability of PFA Concrete. The Use of PFA in Construction, Ed. RK Dhir and MR Jones, pub E&FN Spon, London, 1992.

2. BIJEN, J. Blast Furnace Slag Cement for Durable Structures. Stichting BetonPrisma, 1996, 62pp.

3. ZHANG, M H, and GJØRV, O E. Cement and Concrete Research. Vol. 23, no. 411, 1985.

4. WALLBANK, E J. The performance of concrete in Bridges: A survey of 200 highway bridges. Department of Transport, April 1989, 96pp.

5. JONES, M R, DHIR, R K, and MAGEE, B J. Concrete Containing Ternary Blended Binders: Carbonation & Chloride Resistance. Cemt & Concrete Research, Vol.27, No.6,pp.825-831, 1997.

6. MAGEE, B J, and ALEXANDER, M G. Durability performance of concrete containing condensed silica fume. Submitted for publication.

7. VINCENSTEN, L J, and HENRIKSON, K R. The Great Belt Link – Built to last. Concrete International, Vol.14, No.25, 1992, pp.30-33.

8. JONES, M R, MAGEE, B J, and DHIR, R K. Concrete Containing Ternary Blended Binders: Fresh Properties, Strength Development and Mix Design. In preparation

9. HEWLITT P C. Private communication with Dundee University, 1998.

10. DHIR, R K, JONES, M R, AHMED, H E H. and SENEVIRATNE, A M G. Rapid estimation of chloride diffusion in concrete. Mag of Concrete Research. Vol.42, No.177, 1990, pp.177-185.

11. ALONSO, C, and ANDRADE, C. Corrosion of steel reinforcement in carbonated mortar containing chlorides. Advances in Concrete Research, Vol.1, No.3, pp.155-163, 1988.

12. DHIR, R K, JONES, M R, and AHMED, H E H. Concrete durability: estimation of chloride concentration during design life. Mag. of Concrete Research, Vol. 43, No.154, 1991,pp.579-590.

13. ROY, D M. Hydration of blended cements containing slag, fly ash or silica fume. Proceedings Institution of Concrete Technology. Coventry, UK, 1987.

ADDITIVES FOR INCREASED CAESIUM TRAPPING IN CEMENT

S Bagosi

L J Csetenyi

University of Veszprém

Hungary

ABSTRACT. Cementitious materials can be used to immobilise low and medium level radioactive wastes. In practice, the hard to immobilise elements are usually cemented after being sorbed on Ion Exchange Resin or other media. For Cs, however, binding the resinated waste is not a successful procedure as the high ion concentration of the pore fluid soon removes Cs from the active sites of the resin at great extent resulting in a pore fluid high in caesium. This pore fluid together with the low concentration ground water in case of an ingression would represent a large driving force leading to an easier release of the isotope into the environment. A suitably tailored matrix composition may solve this problem, for which present paper intends to give guidelines. Effects of zeolite, PFA and BFS admixtures on caesium trapping is investigated via measuring the pH and chemistry of pore fluid.

Keywords: Caesium, Immobilisation, Pore fluid, Zeolite, Ordinary portland cement (OPC), Blast furnace slag (BFS), Pulverised fly ash (PFA).

Sandor Bagosi is a newly elected research assistant at the University of Veszprém, Hungary. He has finished his PhD studies recently under supervision of the co-author Dr Laszlo J Csetenyi in the frame of a three year COST-PECO research project with Aberdeen University and he currently prepares for his degree examination.

Dr Laszlo J Csetenyi is an assistant professor at Veszprem University. He specialises in hazardous waste immobilisation in cement. He obtained his PhD degree in Aberdeen, Scotland under supervision of Prof. Fredrik P Glasser on immobilisation of boron wastes in cement. During a half year study trip to Dounreay, Scotland he investigated special waste disposal methods and was project manager of a recently finished three year COST-PECO project.

INTRODUCTION

Cementitious materials can be used to immobilise low and medium level radioactive wastes, notwithstanding almost every waste stream requires individual investigation partly due to the multicomponent nature of the solidifying matrix making clarification of the binding mechanisms even more complicated [1, 2]. In practice, the hard to immobilise elements are usually cemented after being sorbed on Ion Exchange Resin or other media [3, 4, 5]. For Cs, however, binding the resinated waste is not a successful procedure as the high ion concentration of the pore fluid soon removes Cs from the active sites of the resin resulting in a pore fluid high in caesium, sometimes 70-75% of the previously sorbed amount is released [6]. A suitably tailored matrix composition, however, may solve this problem, for which present paper intends to give guidelines. Our interest is turned to one of the most difficult to confine isotope, Cs and the effects of zeolite, PFA and BFS admixtures were investigated on its confinement by measuring the pH and chemistry of pore fluids extracted regularly.

EXPERIMENTAL DETAILS

In order to investigate behaviour of caesium isotope in cementitious solidifying matrix, three different binding material composition series were prepared. Namely, pure OPC (CEM I 42.5, originating from DDCM, Vác, Hungary) [7], 50-50% OPC/PFA as well as 50-50% OPC/BFS blends, designated as 0C, 0P and 0S, respectively. Using these binder material compositions, matrices containing 40 and 60% zeolite were also prepared. Moreover, two particle size fractions of zeolite were used, the one with particle diameter under 0.125 μm (zeolite A) and the other with particle diameter between 0.125-0.25 μm (zeolite B). Chemical composition of materials used and the mineralogical composition of the mainly clinoptilolite containing zeolite originating from Rátka region, Hungary are given in Table 1.

Table 1 Chemical analysis of Portland Cement (OPC), Pulverised Fuel Ash (PFA), Blast Furnace Slag (BFS) and Zeolite, in % (by weight)

CHEMICAL COMPOSITION					MINERALOGICAL COMPOSITION	
Component	PC [%]	PFA [%]	BFS [%]	Zeolite [%]	Zeolite [%]	Zeolite [%]
SiO_2	20.82	55.12	37.8	72.69	Clinoptilolite	50
Al_2O_3	7.4	24.26	4.68	12.67	K-feldspars	20
K_2O	0.93	1.78	0.8	3.32	Quartz	10
CaO	60.25	3.28	42.09	2.23	Christobalite	10
Fe_2O_3	4.41	7.68	1.06	1.28	Volcanic glass	10
MgO	1.2	1.17	8.67	0.89		
Na_2O	0.2	2.12	0.4	0.28		
TiO_2	0.1	0.12	0.29	0.07		
SO_3	3.16	0.58	3.7	-		
Ign, loss	1.4	3.7	-	6.14		
Total	99.87	99.81	99.49	99.57		

Thus 15 different matrix compositions were obtained belonging to three groups. The first group (C-series) comprises the following five matrices: pure OPC (0C), OPC substituted by fine particle size zeolite in 40 and 60%, designated as A40C and A60C, respectively and mixtures in which OPC is substituted by coarser particle size zeolite in 40 and 60% designated as B40C and B60C, respectively. The second group (P-series) is built up similarly, but instead of OPC a binder material of 50-50% by mass OPC/PFA was used resulting mixtures designated 0P, A40P, A60P, B40P and B60P and the third group (S-series) likewise replaces pure OPC with a 50-50% by mass, but now OPC/BFS mixture giving matrices 0S, A40S, A60S, B40S, B60S. The compositions of the samples are shown in Figure 1. The required quantities of powdered raw materials were weighed, mixed dry, and combined with 10^{-2} M CsCl (about 1300 mg/l caesium) solution to form pastes at 0.5 solution/solid ratio applying a mixing procedure according to the EN196-1 standard [8]. The pastes were placed in plastic bag lined ∅70 × 150 mm PVC cylinders and capped. This double sealing avoided evaporation of bleed water, if any, and also eased demoulding.

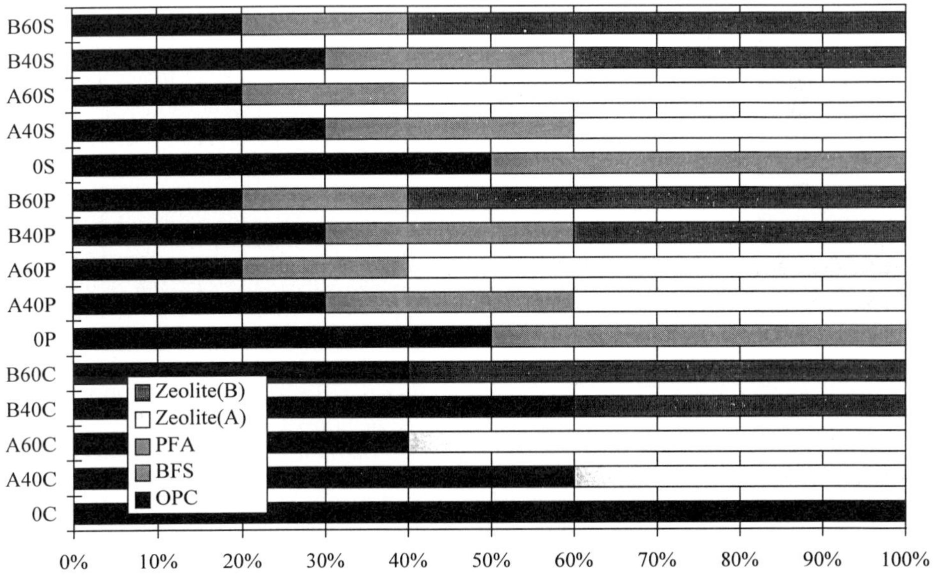

Figure 1 Compositions of the samples

RESULTS AND DISCUSSION

The samples were stored in a humidity cabinet at room temperature and their pore fluid extracted after 2, 10, 36 and 95 weeks. Immediately after the extraction of usually 5-10 ml pore fluids, their pH were measured by titrimetry using 0.01 M HNO_3 and m-cresol purple as indicator (Figure 2). The ∅70 × 150 mm samples, depending on their composition, required significantly different pressures for pore fluid extraction ranging between 1500 to 2300 kN. Cs concentration of the obtained pore fluids were determined by flame emission photometry and shown in Figure 3 as a function of age of sample.

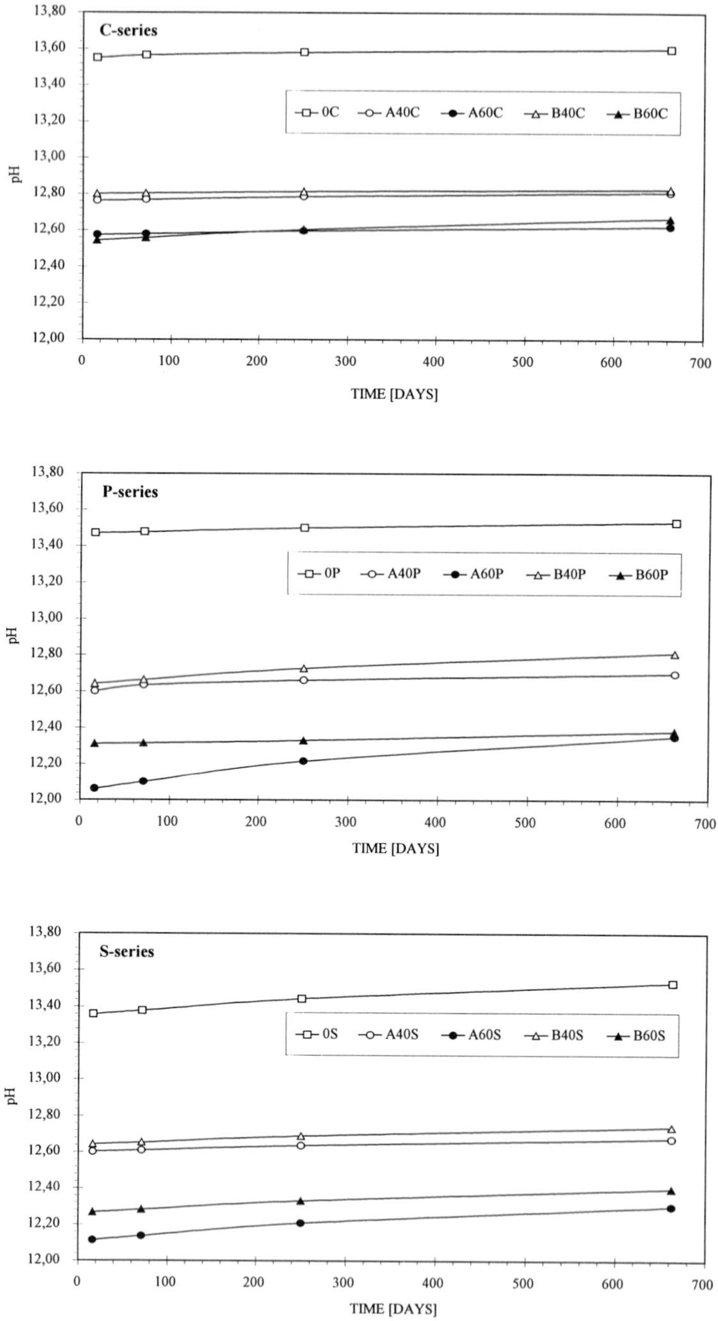

Figure 2 pH of pore fluids as a function of time

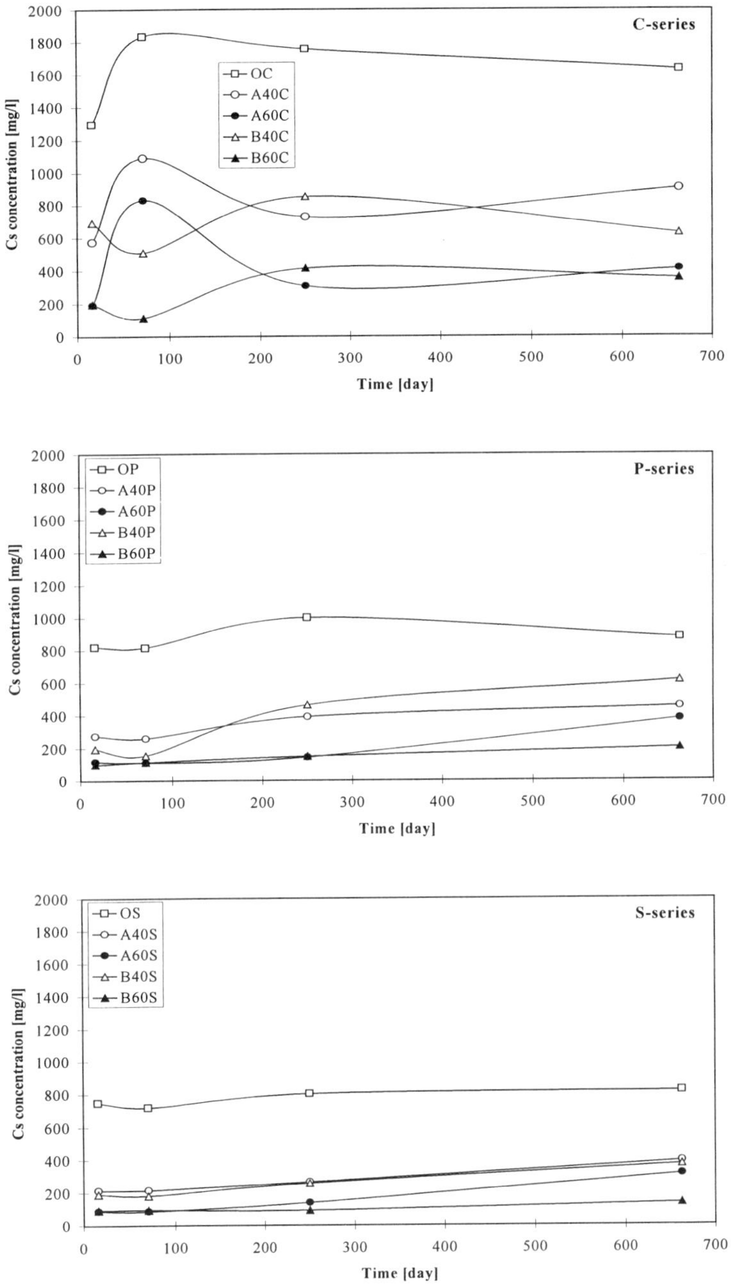

Figure 3 Caesium concentration of pore fluids as a function of time

Beyond their Cs concentration, pore fluids were also analysed for their K, Na, Si, Al and Ca content using flame emission photometry and ICP. Small parts were chipped off from the samples immediately before pore fluid extraction and used for quasi-dynamic leaching experiments whose results are currently under evaluation.

CONCLUSIONS

Basis of Pore Fluid pH and Matrix Composition

The pH of the pore fluid is consistently lower if BFS or PFA are also present in the matrix composition, especially for the slag used in our experiments. As hydration proceeds, pH of the pore fluid for each sample increases as expected. The ration of this increase is less pronounced for the cement/zeolite matrices containing no BFS or PFA (Figure 2).

It is obvious that pH of the pore fluid becomes lower if the zeolite content of the matrix is increased at the expense of other constituents. It is noteworthy, however, that for samples made using a zeolite of fine particle sizes (zeolite A) the pH of the pore fluid is somewhat, but consistently lower than for samples prepared using coarser zeolite (zeolite B). The larger active surface of the fine zeolite must have played a role in this. The effect of particle size of the zeolite on the value of pH is more pronounced for blends also containing BFS or PFA (S series and P series, respectively).

Basis of Pore Fluid Cs Concentration and Matrix Composition

As far as Cs concentration of the pore fluid is concerned, one has to remember that broadly speaking the pore fluid is the excess mixing water remaining in the matrix after hydration takes place. The water consumption of the hydration process basically determines how concentrated the pore fluid becomes. Without any immobilisation one should expect an increase in the concentration of the admixed waste material to be disposed of, which increase will be adversely proportional to the rate of reduction of the mixing water, provided the solubility limit is not reached. If moderate degree of immobilisation takes place, the concentration of the waste in the pore fluid may still be higher than it was originally in the mixing water, which is the case for the 0C sample in our experiments (Figure 3) clearly demonstrating the low Cs binding capacity of OPC. We have taken this matrix (coded 0C) as reference and all observations made on the average Cs concentration of pore fluids for a given sample were evaluated on the basis of the percentage changes in relation to the average Cs concentrations measured for 0C.

One can conclude that for the C-series the presence of 40% and 60% zeolite (without any slag or fly ash) results in 50% and 80% decrease of Cs concentration in the pore fluid, respectively. The effect of zeolite particle size is relatively easy to observe showing about 10% lower concentrations when the coarser zeolite fraction was used (zeolite B) instead of the fine one (zeolite A).

For the P-series the 50% replacement of OPC by PFA reduced the Cs concentration within the pore fluid by about 40%, whereas the 40% and 60% replacement of the OPC/PFA blend by zeolite resulted in a further 40% and 50% concentration drop, respectively.

For the slag containing samples (S-series) concentration reductions are the most significant. The presence of slag itself results in a reduction of 50%, the substitution of the binder material by zeolite in 40 and 60% gives a further 30 and 40% drop, respectively.

In the case of the latter two series (P and S-series) the alteration of the zeolite particle size does not result in a significant decrease in the Cs concentration of the pore fluid (0-3%), but even for these small changes one may note that lower Cs concentrations were measured for the samples made using coarser zeolite (zeolite B).

By considering the tendencies observed altogether, one may state that in the case of hazardous isotopes with similar behaviour to caesium, simple solidification in cement causes a greater problem than disposing of the waste sorbed e.g. in resin, because hydration processes turn mixing water into a concentrated pore fluid very rich in Cs and this high concentration together with the low concentration ground water in case of an ingression would represent a large driving force leading to easier release of the isotope into the environment.

Therefore care should be taken that when solidifying Cs containing wastes the matrix material be admixed with suitable constituents to decrease the Cs enrichment in the pore fluid during hydration at a rate as high as possible. Zeolites are particularly suitable for this purpose not only by their remarkable selectivity for Cs, but also by their excellent binding capacity maintained even in high ion concentration conditions experienced in pore fluids [9].

ACKNOWLEDGEMENTS

The authors wish to acknowledge the financial support received from the European Union through the ERBCIPDCT 930447 COST-PECO programme and thank Prof. F P Glasser (University of Aberdeen) and Prof. J Tritthart (Technical University of Graz) for their gratefully appreciated help throughout the above programme and pore fluid expressions, respectively.

REFERENCES

1. F, P, GLASSER. Progress in the Immobilization of Radioactive Waste in Cement; Cem. Conc. Res., Vol. 22, 1992, pp 201-216.

2. S, L, HOYLE, AND M, W, GRUTZECK. Incorporation of Caesium by Hydrating Calcium Aluminosilicates; J Am. Ceram. Soc., Vol. 72, [10], 1989, pp 1938-1947.

3. H, AALTO, AND A, IPATTI. Leach Test of Spent Ion Exchange Resins Solidified in Concrete. Cem. Conc. Res., Vol. 22, 1992, pp 375-380.

4. M, FÖLDESOVÁ, P, LUKÁČ, J, MAJLING, V, AND, TOMKOVÁ. Deposition of Caesium and Cobalt Sorbed on Zeolite in Matrices of Blast Furnace Slag. J. Radioanal. Nucl. Chem., Letters 212, (4), 1996, pp 293-302.

5. S, BAGOSI. Leaching Study of Cs Ions from Zeolite Admixed Cement, MSc Thesis, University of Veszprem, 1992.

6. S, BAGOSI, AND L, J, CSETENYI. Immobilisation of Cs-Loaded Ion Exchange Resins in Zeolite-Cement Blends. Proc. 10th Int. Conf. Chem. Cem. 4, 1997, 4iv045.

7. STANDARD EN 197-1. Cement - Composition, Specifications and Conformity Criteria. Part 1, Common Cements.

8. STANDARD EN 196-1. Methods of Testing Cement - Part 1. Determination of Strength.

9. S, BAGOSI. Commission of the European Communities, Annual Project Report ERBCIPDCT 930447, 1997.

THEME THREE:

NON-PORTLAND CEMENT-BASED BINDERS

Keynote Paper

APPLICATIONS OF CALCIUM ALUMINATE AND OTHER SPECIAL CEMENTS

K L Scrivener

S Butler

Lafarge

France

ABSTRACT. As the name implies, special cements are not replacements for Portland cement, but cements which provide particular properties for specialist applications. This presentation discusses some of the areas in which special cements are used, with particular reference to the applications of calcium aluminate cement. Calcium aluminate cements have a very wide range of applications, which include:

- use in conventional concrete form for severe and aggressive environments where it can provide resistance to acids, abrasion and heat;
- use in blends to provide shrinkage compensation;
- use in conventional concrete form or in blends to provide rapid hardening compounds.

Other special cements have a more limited range of applications which include rapid setting and hardening and shrinkage compensation.

Keywords: Calcium aluminate cement, Special cements, Acid resistance, Abrasion resistance, Heat resistance, Shrinkage compensation, Rapid hardening.

Dr Karen Scrivener is Head of the Aluminate Department in the Central Research Laboratory of Lafarge, responsible for a wide range of research on the properties, applications and development of calcium aluminate cements.

Mr Steve Butler is a Research and Applications Engineer for Lafarge Aluminates.

INTRODUCTION - WHY SPECIAL CEMENTS?

Portland cement is by far and away the most widely used cementitious material. Indeed, world wide there is more Portland cement used than any other material, around 2 billion tonnes per year. Despite the dominance of Portland cements a number of special cements have been developed over the last century for use in niche applications where they have advantages over Portland cement which justify their higher cost. In many cases such niche applications fall outside of the traditional domain of reinforced concrete for structures.

Amongst these niche applications, three important areas of demand can be identified where the performance of Portland cement concrete has some limitation:

1. Need for durability in extreme or aggressive environments
2. Need to reduce the effects of shrinkage, which occur in Portland cement
3. Need for high early strengths due to low temperatures or rapid in service requirements.

In each case special cements can, under particular circumstances be competitive even when the performance Portland based cements is modified by various admixtures (liquid or mineral). However, as with high performance (Portland) concretes, specialist applications demand a high degree of quality control and impose high demands on cement performance and regularity.

This presentation discusses the most important special cements, defined here as those cements having chemistries and mineralogies outside of the normal range for Portland cement and its derivatives such as "sulfate resisting", "low heat" and "rapid hardening" cements. In addition to discussion of their applications, some issues related to their production are also considered. Despite the obvious bias of the author the emphasis on the uses of calcium aluminate cements reflects the considerable diversity of applications of this 'special' cement.

TYPES OF SPECIAL CEMENTS

The most important special cements, in terms of their technical and geographical range of applications are calcium aluminate cements. The range of chemistries of these cements is extensive, as shown in Figure 1. The principal reason for this range of chemistries is the use of these cements in monolithic refractories. Calcium aluminate cements can be, and are, used in conventional concrete, in which the paste component consists of just cement and water with minor amounts of admixtures. In this case, the formation of stable hydrates may be preceded by metastable hydrates and the inevitable process of conversion must be taken into account. Nowadays, however, a more important use of calcium aluminate cements is as part of blended binders also containing ingredients such as calcium sulfates (gypsum, anhydrite or hemi-hydrate), Portland cement, or lime as well as several admixtures. Such blends find applications in specialised building products such as self levelling floor screeds and tile cements. In these blends the hydration products are ettringite or AFm phases which are also products of Portland cement hydration. Whether in conventional concrete form or as part of a complex blend, calcium aluminate cements have applications in all of the areas identified above where the performance of Portland cement may be limited.

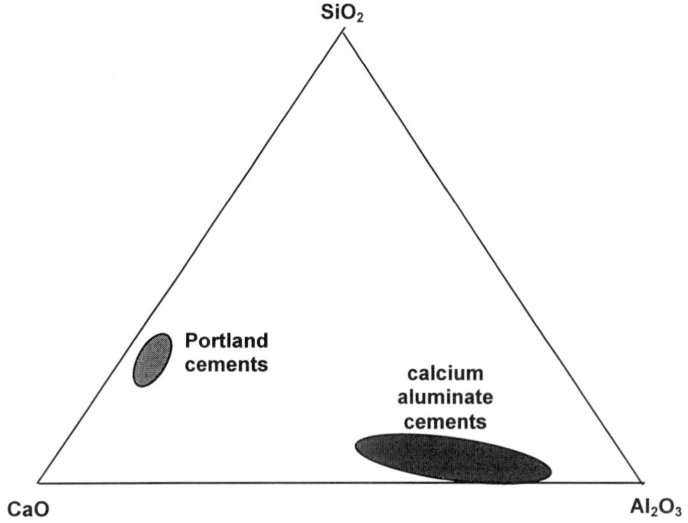

Figure 1 Schematic representation of relative chemistries of calcium aluminate and Portland cements

Other than calcium aluminate cements, the most significant group of special cements are those in which the aluminate phase of Portland cements (C_3A) is partially or wholly replaced, by other anhydrous phases containing alumina. The alternative alumina containing phase may be $C_4A_3\overline{S}$ (Klein's compound) or hybrids of $C_{12}A_7$ such as $C_{11}A_7CaX_2$ (where X is Cl or F). Despite their categorisation as a single group, there are many different cements of this type available with a wide range of different chemistries and $C_4A_3\overline{S}$ contents ranging from 20 – 70 % . However, for all these cements, ettringite is an important hydration product and the control of the morphology and reaction kinetics of this phase plays an important role in determining performance. In this respect there are certain applications of these cements which overlap those of blends containing calcium aluminate cement.

Generally this category of cements are used in applications which demand rapid hardening or reduction of shrinkage. $C_{11}A_7CaX_2$ phases are usually found in rapid setting cements, while $C_4A_3\overline{S}$ may be a component of either shrinkage compensated (type K) or rapid setting cements according to the amount and type of calcium sulfate addition.

These two groups cover most special cements which rely on hydraulic bonding for their performance. That is to say that the anhydrous cement minerals formed in a kiln react with water to form hydrates which cause the concrete or mortar to be transformed from a fluid suspension to a rigid solid. Outside these groups there are many other special cements which rely on other forms of chemical hardening, such as magnesium, zinc or aluminium oxychloride cements, magnesium sulfate or phosphate cement and dental cements. Such cements are produced in much smaller quantities, and have very specialised applications, generally with very quick setting. It is not possible to discuss all these different types in this presentation, which will concentrate on the first two groups of special cement.

PRODUCTION OF SPECIAL CEMENTS

Before looking at the applications of special cements from the perspective of performance, it is worth considering briefly their production. Over the past fifty years, tremendous advances have been made to optimise the production of Portland cement in a rotary kiln. Also, advances have been made to reduce costs through economies of scale, improved energy efficiency and computer aided operation. A modern efficient rotary kiln for Portland cement may have a capacity of over a million tonnes a year, similar to the worldwide production of calcium aluminate cements.

Calcium aluminate cements are produced in reverberatory furnaces and electric furnaces as well as in rotary kilns, depending on the production and chemical requirements. The use of the reverberatory furnace for the production of CACs was developed by Jules Bied in the Lafarge laboratories and patented in 1908.

Modern versions of these furnaces have advantages in less raw material preparation and great mineral consistency because the clinker is melted to a liquid. However, the energy consumption per ton is somewhat higher than for a rotary kiln which, together with the expense of the bauxite raw material, adds to the cost of these cements.

Most of the volume of calcium aluminate cement production world wide is produced this way. These are dedicated furnaces which produce only calcium aluminate cement enabling good consistency in chemistry and mineralogy to be achieved.

Calcium aluminate cements with very high alumina content (>60%) are also produced in dedicated rotary kilns, but these products, made from high purity raw materials, find their application almost exclusively in monolithic refractory concretes.

Cements containing $C_{11}A_7CaX_2$ or $C_4A_3\overline{S}$ are produced in rotary kilns and more details of the production of the various types can be found in [1]. In a few cases due to a local supply of suitable raw materials such special cements are produced in dedicated small rotary kilns. However, in general, they are produced in short runs in normal Portland cement kilns by adapting the raw meal — with addition of CaF_2 to produce $C_{11}A_7CaF_2$ and of calcium sulfate to produce $C_4A_3\overline{S}$.

When production is switched between clinker types it may take several days to stabilise the production process, therefore it can be difficult to produce a regular final product. This problem is compounded by the greater demand for regularity in special cements necessitated by their specialised properties.

The presence of sulfate in the raw meal poses particular challenges. In these cases a viscous liquid phase may form, which can cause ring formation and kiln blockages. Sulfates are also volatile, this poses a challenge for control of the chemistry of the final products as well as the potential problem of emissions, which could contribute to acid rain..

For these reasons many companies have developed proprietary blends of Portland, calcium aluminate, and calcium sulphate which can be tailored to give similar performance to many special cements and to give properties not achievable with a single clinker.

APPLICATIONS IN SEVERE AND AGGRESSIVE ENVIRONMENTS

Some of the earliest problems with Portland cement, at the end of the last century, were lack of durability when exposed to sea-water or sulphate bearing ground waters. In fact it was this problem, and the offer of a prize for its resolution, which led to the development of calcium aluminate cements. One of the earliest uses for these cements was in the construction of a tunnel through a solid mass of anhydrite. Illustrating the excellent resistance of these cements to sulfate attack.

Nowadays the principal reason for poor durability – high porosity – is well understood (although sometimes ignored). Well made Portland based concrete, with a low water cement ratio, well cured, and probably containing fly ash, slag or silica fume, will generally be durable in most commonly encountered environments. On the other hand even sulphate resisting Portland cements, will have poor durability in sulfate containing environments if made at high water to cement ratios.

In General, the group of special cements containing $C_4A_3\overline{S}$ or $C_{11}A_7CaX_2$ would be expected to show durabilitites in the same range observed for Portland cements. Type K cements have been shown to have good resistance to sulfate attack as the alumina content is already in the form of ettringite. However these observations cannot be generalised to other cements of this type. The overall alumina to sulfate ratio will determine whether the alumina is present as ettringite or monosulfate, which in turn would be expected to affect the resistance to sulfate attack.

On the other hand calcium aluminate cements, with their very different chemistry, can have significant advantages over Portland cements in certain areas where the durability of the latter is limited, for example:

- acidic environments;
- environments subject to severe mechanical abrasion;
- environments subject to extremes of temperature.

Applications in Acidic Environments

In Portland based concrete, calcium hydroxide is dissolved at even weakly acidic pH and the C-S-H loses calcium to leave a structureless gel. This process leads to rapid degradation and loss of aggregates.

Calcium aluminate cements have good resistance to acidic environments due to the nature of the hydrates formed. The calcium aluminate hydrates (CAH_{10}, C_2AH_8 and C_3AH_6) are dissolved by acids and the calcium component is lost. However, the alumina component, together with AH_3 hydration products are unaffected by acids down to pH 3 - 4. This hydrous alumina gel protects the concrete from further attack. Calcium aluminate cement concretes can also perform well at pH below pH 3-4, where the dissolution of alumina neutralises a high amount of acid. This effect can be further exploited by the use of a concrete containing synthetic aggregate of similar composition to the CAC (Alag®).

The good performance of calcium aluminate cements in acidic environments finds application particularly in sewage networks subject to biogenic corrosion caused by acid producing bacteria (Figure 2). This application is described in more detail in another paper in these proceedings [2]. It is worth noting that performance of CAC concrete in such applications is unaffected by conversion and most pipe linings are heat cured before entering service (Figure 3).

Figure 2 Application of calcium aluminate cements for lining sewage networks subject to biogenic corrosion

Figure 3 Pipes lined with calcium aluminate cement mortar for sewage and waste water applications.

Applications Involving Abrasion Resistance

The factors determining the abrasion resistance of concrete are not fully understood. There is a strong relationship to compressive strength, but even at equivalent compressive strength calcium aluminate cement concretes show better resistance to abrasion than Portland cement concretes. The full benefits of this resistance to abrasion are realised in concrete made with synthetic calcium aluminate aggregate (Alag®). In this case, there is a slight reaction of the surface of the aggregate resulting in a dense interfacial transition zone (ITZ).

The excellent resistance of CAC concretes, to hydraulic corrosion in particular, has led to their application in parts of dams subject to high wear (Figure 4).

In such applications the ITZ of conventional and even silica fume concretes is vulnerable to erosion and the aggregate particles can be torn away, leading to more rapid wear. The similarity between paste and aggregate in Alag concrete leads to even wear and longer service life. In such applications CAC concretes are laid in small checkerboard sections to counteract shrinkage effects and to avoid surface cracks.

This somewhat more time consuming installation process is justified by the longer service life, which is competitive with that of materials such as granite blocks or steel plates, with much higher installation costs. These applications are discussed in more detail elsewhere in these proceedings [3].

Figure 4 Use of calcium aluminate cement concrete in a dam spillway where it provides excellent abrasion resistance

Figure 5 Fire training building, built with calcium aluminate cement concrete

Applications Involving Extremes of Temperature

Conventional Portland cement concretes cannot withstand repeated exposure to heat, due mainly to the dehydration of calcium hydroxide to free lime. On cooling the free lime will rehydrate from moisture in the air, which results in swelling and cracking. Repeated firings cause this dehydration and repeated expansion to destroy the surrounding matrix. Calcium aluminate cements retain their form on repeated heating and cooling.

The CAC with the highest alumina contents, which also have negligible silica and iron contents, have extremely high softening temperatures enabling them to be used in refractory concretes up to 1600°C and more. Outside specialised monolithic refractories used for lining kilns and reaction vessels, CAC concretes also find uses in associated applications for foundry floors and the like. A more unusual application which makes use of the temperature tolerance of CAC concretes is their use in buildings used for training fire fighters, which can tolerate numerous fires (Figure 5).

EXPANSION AND SHRINKAGE COMPENSATION

Expansive Mechanisms

Hydraulic cements all undergo or must compensate for shrinkage, partly due to the reduction in total volume involved in the hydration reaction:

$$V_{\text{anhydrous}} + V_{\text{water}} > V_{\text{hydrates}}$$

But mainly due to the evaporation of water from the pore structure.

When wet curing takes place some hydration reactions can provide sufficient expansion to compensate for subsequent drying shrinkage. The most common of these are the formation of calcium hydroxide from free lime:

$$CaO + H_2O \rightarrow Ca(OH)_2$$

and of ettringite from an anhydrous calcium aluminate and sulphate source, e.g.:

$$CA + 3C\bar{S} + 2C + 32H \rightarrow C_3A(C\bar{S})_3H_{32} \qquad \text{(Type M)}$$

$$C_4A_3\bar{S} + 9C\bar{S} + 5C + 96H \rightarrow 3\,C_3A(C\bar{S})_3H_{32} \qquad \text{(Type K)}$$

$$C_3A + 3C\bar{S} + 32H \rightarrow C_3A(C\bar{S})_3H_{32} \qquad \text{(Type S)}$$

It is most important to note than none of these reactions are *intrinsically* expansive, in so much as the total volume of the reactant, including the water, is always more than the volume of products. Their capacity to generate an expansive force depends firstly on the uptake of water and secondly on the presence of anhydrous phases with significantly greater free energy and much higher solubility than the products. During reaction these translate into a super-saturation with respect to the reaction products, which can provide a crystallisation pressure.

Although the broad aspects of the expansion mechanism are known, the precise details of the mechanisms are not. The rate and amount of expansion are affected by parameters such as the relative proportions of the reactive ingredients and by their fineness. Control of the expansive process is generally regulated by empirical experiments to develop a formulation for a binder which works in practice.

In general it has been found the reaction of C_3A occurs too fast and becomes blocked by ettringite precipitation close to the surface of the grains, so it is not suitable for most practical purposes. Free lime is so reactive it is an unreliable agent to be used alone because of the difficulty of controlling the speed of the reaction and its susceptibility to ageing due to reaction with carbon dioxide or water vapour in air.

There are many cementitious systems which use $C_4A_3\bar{S}$ or CA or combinations of these with free lime to give early expansion for shrinkage compensation. It is important that at least one of the reactive ingredients is totally consumed within the initial expansive phase, so as not to cause uncontrolled expansion later.

Applications

In reinforced concrete, drying shrinkage can cause the formation of micro-cracks. The impact of such micro-cracks is not clear, but in aggressive environments they might be expected to facilitate the penetration of aggressive agents into the concrete. In unreinforced slabs, joints must be cut at regular intervals to avoid macro cracking, curling and sub-grade deterioration. In shrinkage compensating concrete the intention is for the expansion generated during curing to counterbalance the subsequent drying shrinkage so that there is no net change in dimension at the end of curing and drying (Figure 6).

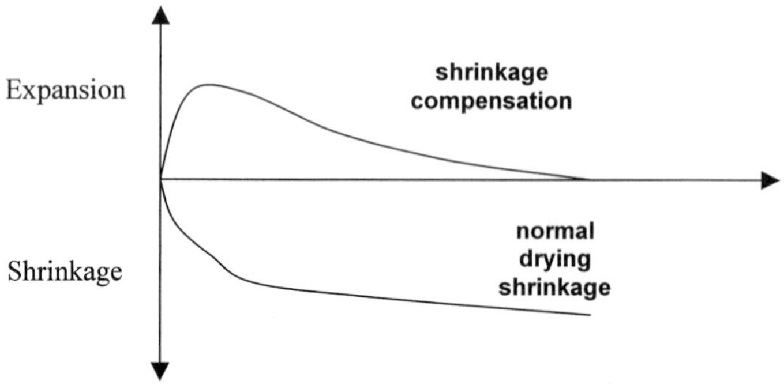

Figure 6 Schematic dimensional evolution of concrete under normal drying shrinkage and in shrinkage compensated systems

The problem is that the mechanisms of the two processes are completely different and so will be affected differently by factors such as temperature, mix design and ageing and other variations in the cement. This means that the production and curing conditions of shrinkage compensating concretes must be very closely controlled if they are to be successful.

In North America, type K cements have been produced and used in bridge decks to reduce the number of joints. There are many examples of successful applications over the past twenty years. However, the need for a carefully controlled curing process, with 7 days wet curing, has been a major impediment to the widespread utilisation these materials. In Japan and some other countries, shrinkage compensating cements are used in pre-casting where the production process can be carefully controlled, to pre-stress reinforcement. The increased strength resulting from pre-stressing is important for earth quake resistance. Around the world, shrinkage compensating systems are more generally found in proprietary repair systems. In these limited size repairs it is important to avoid shrinkage of the material in order to maintain a good bond between the substrate and the repair.

An important application for calcium aluminate cements is in blends with calcium sulfate and small amounts of Portland cement and other additives to produce proprietary floor levelling compounds to provide a smooth flat crack-free base for carpet or other flooring (Figure 7). In such blends the formation of ettringite is controlled to combine several effects:

- dimensional stability;
- rapid hardening, the surface may be walked on in a matter of hours;
- rapid decrease in humidity, through the chemical combination of water in hydration products in order that tiles or carpet can be placed on top.

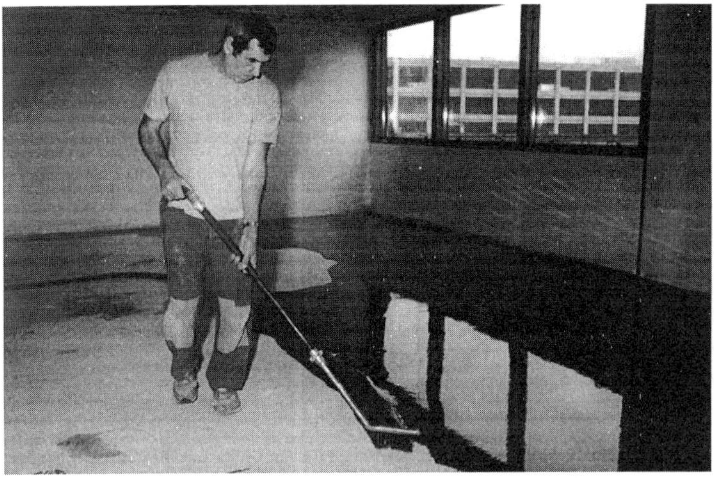

Figure 7 Application of self levelling floor compoundcontaining calcium aluminate cement as part of a blend

In such systems the regularity of the mixed product, and therefore of the cements within it, is of the utmost important. This requirement for regularity makes calcium aluminate cements the materials of choice for such applications.

RAPID HARDENING

Various modifications can accelerate the hydration reactions and hardening of Portland cement, including:

- temperature;
- cement fineness;
- chemical additives;
- contents of C_3S and C_3A.

The use of temperature is only practical in pre-cast plants, where adequate control is possible. Increased fineness, particularly for cements with higher than average C_3S and C_3A content, is the basis of rapid hardening or type 3 cements which have strength about 90% higher than type 1 cements at one day. Chemical additives can also provide significant increases in strengths in the first few days. However, in order to achieve very early strength development, in say 6-8 hours, special cements with different hydration reactions have a role to play.

The main impediment to early strength development in Portland cements is the pattern of deposition of the hydration products. The main hydration product, C-S-H, forms as a layer around the cement grains. This has two effects. Firstly the spaces furthest from the cement grains are not efficiently filled with hydration products and secondly the C-S-H layer quickly slows the rate of reaction of the cement grains, (Figure 8).

As a consequence the degree of hydration is typically less than 30% at 12 hours and only around 50% at 24 hours. Special cements giving early strength work through the formation of different hydration products which have a different distribution in space.

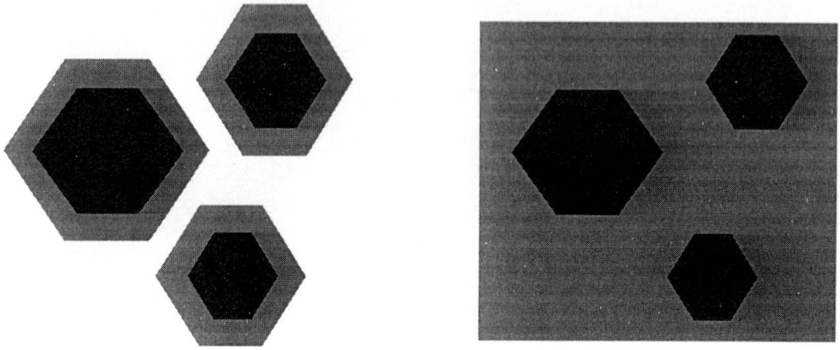

Figure 8 Schematic representation of the distribution of hydration products in Portland cement (left) and calcium aluminate cements (right)

Hydration and Strength Development in Calcium Aluminate Cements

Calcium aluminate cements are unique in exhibiting rapid hardening, but with set times comparable to Portland cements. This is due to the different hydration reactions which occur. In calcium aluminate cements both calcium and aluminium ions pass into solution rapidly. However, there is a significant nucleation period before the precipitation of the hydrates, during which the concrete is workable. At the end of the induction period, nucleation of the hydrates occurs throughout the water filled space so the rate of reaction tends to be limited by the availability of water and space, rather than by the formation of a diffusion barrier around the cement grains as with Portland cement (Figure 8). During early reaction, strengths of 20-30 MPa can be achieved in as little as 6-8 hours. If hydration occurs at relatively low temperatures, in small sections, very high strengths up to 80-100 MPa can be achieved in the first few days. But these are not utilisable due to the effects of conversion.

Conversion is the thermodynamically driven and inevitable process of reaction of the metastable hydrates, CAH_{10} and C_2AH_8 to the stable phase assemblage of C_3AH_6 and AH_3:

Low temperatures: $CA + 10H \rightarrow CAH_{10}$

or $2CA + 11H \rightarrow C_2AH_8 + AH_3$

Conversion: $6CAH_{10} \rightarrow 3C_2AH_8 + 3AH_3 + 27H \rightarrow 2C_3AH_6 + 4AH_3 + 36H$

High Temperatures: $3CA + 12H \rightarrow C_3AH_6 + 2AH_3$

As the density of C_3AH_6 is higher than that of the metastable hydrates the porosity will be higher for a given degree of hydration, resulting in lower strength. This effect can be partially mitigated by the reaction of water released by the conversion reaction with any remaining unhydrated material. However the converted strengths of calcium aluminate cements are slightly lower than those of typical Portland cements at the same water/cement ratio. A well made CAC concrete with a water/cement ratio of 0.4 will have a minimum long term strength of around 30-35 MPa (compared to a typical value of around 40 MPa for Portland cement concretes). In sections of any significant size, the self heating effects of the hydration reaction are significant and lead to temperatures in the concrete of around 70°C within the first 24hours. In such cases, conversion occurs almost immediately and the strength remains stable or increases thereafter.

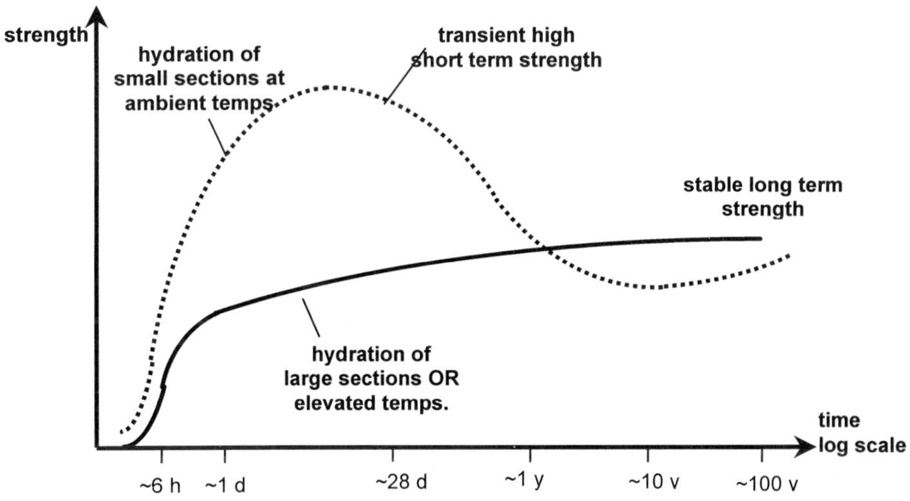

Figure 9 Schematic strength development in calcium aluminate cement concrete under different curing conditions

Figure 9 shows strength development curves for concretes subjected to 20°C and a thermal cure up to 70°C. Due to conversion it is particularly important to control the water to cement ratio of CAC concretes and this should be less than 0.4. Such water to cement ratios have also been shown to have good durability including in some of the specialist applications discussed above.

After conversion, CAC concretes are not of exceptionally high strength and this together with their high cost means that they are not the materials of choice for general structural purposes. However, their ability to develop their ultimate strengths very quickly makes them of interest for rapid repairs of slabs on grade in places such as airports, where it is important to avoid long closure times. Some applications, such as on industrial flooring, exploit both the superior resistance of CACs to chemical and mechanical degradation and the rapid hardening properties to minimise down times.

A further application for the special hardening properties of calcium aluminate cements are concretes which harden at low temperatures. When at 0°C, strengths of some 30 MPa can be developed in 16 hours. This has led to their use for example in arctic Canada for grouting radio masts.

Rapid Hardening Based on Ettingite Formation

As described above, the formation of ettringite can be used to provide shrinkage compensation. In this case, the reaction is controlled to occur after the paste has set due to the normal calcium silicate reactions. Similar systems can be designed in which the ettringite formation occurs much earlier leading to rapid setting and early strength development. The detailed mechanisms and control of these reactions is complex.

The result of this complexity is a multitude of proprietary products often produced by blending Portland cement with around 10-30% of a special cement containing $C_4A_3\overline{S}$, CA or both and a source of calcium sulfate such as anhydrite. After the early formation of ettringite the strength development of these cements is controlled by the formation of the normal calcium silicate hydrates, C-S-H and CH. Depending on the ratio of aluminate to sulfate phases, some or all of the ettringite formed may react to form monosulfo-aluminate hydrates.

SUMMARY

The aim of this presentation has been to highlight some areas of application of special cements, which extend the range of performance attainable with Portland cements. These specialist applications can justify the higher cost of special cements, but often demand a higher degree of regularity, which may be difficult to achieve for cements made in small quantities.

Calcium aluminate cements have very good repeatability of mineralogy and chemistry as they are made in dedicated furnaces with a high degree of quality control. They also have unique flexibility in terms of applications as they can be used in both conventional concrete and in blended form. In conventional concrete form, conversion must be taken into account. But even when converted they show excellent reistance to acids, resistance to abrasion and high temperature performance. In blended form, the hydrates formed are the same as those found in Portland cements and conversion is not an issue. In blends, the performance can be adapted to the requirements of the application and may include, shrinkage compensation, rapid hardening and drying through combination of the added water.

Other types of special cement find applications mainly in the fields of shrinkage compensation and of rapid hardening.

REFERENCES

1. KURDOWSKI, W AND SORRENTINO, F. Special Cements. pp 471-554 in Structure and Performance of Cements, ed. P Barnes, Applied Science, London, 1983.

2. LETOURNEUX, R AND SCRIVENER, K L. These Proceedings.

3. CABIRON, J-L. These Proceedings.

NEW CONCRETES BASED ON CALCIUM SULFOALUMINATE CEMENT

L Zhang

F P Glasser

University of Aberdeen

United Kingdom

ABSTRACT. The development and application of new types of hydraulic binders based on calcium sulfoaluminate is described. Outside China these binders are used mainly as additives to Portland cement, to reduce autogenous shrinkage, and in special formulations, e.g. in screeds and repair materials. Inside China they mainly are used to replace Portland cement in formulating mortars and concretes, especially where rapid strength gain is important. This presentation describes the technical basis of their formulation and use and gives examples of their durability in real constructions.

Keywords: Calcium sulfoaluminate, Novel cements, Rapid-hardening cements, Low energy cements, Durability.

Liang Zhang is a Senior Engineer at the China Building Materials Academy, Beijing, Peoples Republic of China, where he leads a group working on calcium sulfoaluminate cements and concretes. He is presently on leave to enable him to undertake PhD studies at Aberdeen.

F P Glasser is Professor of Chemistry at The University of Aberdeen. He has published a number of research papers on materials science, many of which relate to cement and concrete science.

INTRODUCTION

The development of new inorganic cementitious binders for the construction industry has been inhibited by the low cost of Portland cement coupled with its other advantages: good performance in a wide range of service environments coupled with reasonable compatibility and protection for a range of other engineering materials, e.g., steel.

Alternative binder systems to Portland cement do exist. High alumina cement (HAC) concretes, for example, exhibit more rapid strength gain than Portland concretes. However problems of "conversion", a reaction conditioned by mineralogical change, leads to diminution of specific volume of the solids. This conversion, coupled with loss of strength and increased permeability, has limited their structural applications. HAC cement is also more expensive to produce than Portland cement.

The flexibility and versatility of both Portland and HAC systems have been greatly extended in recent decades as engineers have increasingly accepted blended cements, containing supplementary cementing materials such as silica fume, slag and fly ash, for a wide range of applications: the cement paste can be specifically tailored to achieve the desired properties. Organic admixtures have enabled large improvements in ultimate compressive strength to be achieved, mainly by lowering the water: cementitious materials ratio of engineering mixes. Thus while the performance envelope of traditional cements has been much extended, scope still exists for the development of new cementitious binders.

An alternative to Portland cement, which forms the subject of this presentation, is a cement based on calcium sulfoaluminate. The modern development of these cements began in China in the 1970's and, although production has now reached ~ 1 m tonnes/yr, structural uses of these cements outside China are relatively uncommon; applications are largely confined to special cements. Nevertheless, their properties commend their potential use to engineers, particularly as (i) their production makes environmentally less demands than Portland cement and (ii) they are likely to become more widely available in the next decade.

The published literature on calcium sulfoaluminates is quite limited, except in Chinese. We draw upon unpublished information but also on conference proceedings [1,2] and on a special issue of a Journal containing conference papers [3]. Another conference, to be held late in 1998, is expected to contain important research papers [4]. One of us (LZ) is co-editing a book on the production and engineering properties of sulfoaluminate cements [5]. This book will appear initially in Chinese, but a parallel version in English will follow shortly.

CLINKERING

Sulfoaluminate cements are, like Portland cement, made in rotary kilns. Typical raw materials for their production include calcium carbonate, perhaps as limestone, bauxite and $CaSO_4$. Bauxite is an alumina-rich earthy material which occurs widely as surficial deposits. Bauxites high in alumina are reserved for aluminium production but lower purity grades, often rejected for Al production, can be utilised for sulfoaluminate cement. $CaSO_4$ is abundant in nature, occurring as gypsum and anhydrite, but may also be available as a by-product from other industrial processes, e.g., flue gas desulfurisation.

Typical bulk compositions for the two grades of calcium sulfoaluminate cement made in China are shown in Table 1

Table 1 Clinker chemistry of $C\bar{S}A^{+}$

CLINKER TYPE*	WEIGHT %				
	CaO	Al_2O_3	SiO_2	Fe_2O_3	SO_3
$C\bar{S}A$	36-45	30-40	3-13	1-3	8-15
$C\bar{S}FA$	43-50	25-35	5-13	3-13	7-12

+$C\bar{S}A$, calcium sulfoaluminate, will be used as a collective term
*$C\bar{S}A$, Calcium sulfoaluminate clinker
*$C\bar{S}FA$, Calcium sulfo ferroaluminate clinker

The energy input for calcium sulfoaluminate cements (for which the generic term $C\bar{S}A$ will be used) is less than for Portland cements and the specific CO_2 output much less because (i) lower firing temperatures are required for clinkering, typically 1325°C for $C\bar{S}A$ cements but ~ 1450°C for Portland cements (ii) less energy is required for grinding $C\bar{S}A$ clinker, which is friable (iii) less energy is required for decarbonation; the $CaCO_3$ required per unit mass is less than for Portland cement and finally, (iv) prior to use, $C\bar{S}A$ clinkers are interground with much gypsum, which has a low energy requirement for production. The lower $CaCO_3$ requirement for $C\bar{S}A$ and $C\bar{S}FA$ clinkers relative to Portland cement thus make a much greater reduction in emissions than would be expected from the saving in thermal energy.

The mineralogy of the fired clinkers is given in Table 2. Typically clinkers are rather finer-grained than Portland cements; whereas alite crystals in Portland cement may exceed 10 μm, maximum crystallite sizes in $C\bar{S}A$ clinkers are typically only 1-2 μm. The silica component of $C\bar{S}A$ clinkers is belite, not alite. Like Portland cement, they contain an iron-bearing phase, the so-called ferrite phase, which concentrates in interstitial spaces between belite and $C_4A_3\bar{S}$ crystallites. However, the most distinctive feature of $C\bar{S}A$ cements is their content of $4CaO.3Al_2O_3.SO_3$; this phase is not a normal constituent of either Portland or HAC clinkers

Table 2 Clinker Mineralogy

CLINKER TYPE†	PHASE COMPOSITION %			
	$C_4A_3\bar{S}$	C_2S	C_4AF*	C_6AF_2*
$C\bar{S}A$	55-75	8-37	3-10	
$C\bar{S}FA$	33-63	14-37		15-35

*The composition of the iron-bearing phase is approximate. Abbreviations: †$C\bar{S}A$: calcium sulfoaluminate cement: $C\bar{S}FA$ calcium sulfo ferro aluminate cement.

Figure 1 shows a typical clinker microstructure by backscattered electron imaging. The frequent pores are responsible for the ease of grinding. The equant shaped crystals are of two types: belite (grey) and $C_4A_3\overline{S}$ (brighter). The groundmass is mainly ferrite, together with minor unreactive phases such as perovskite.

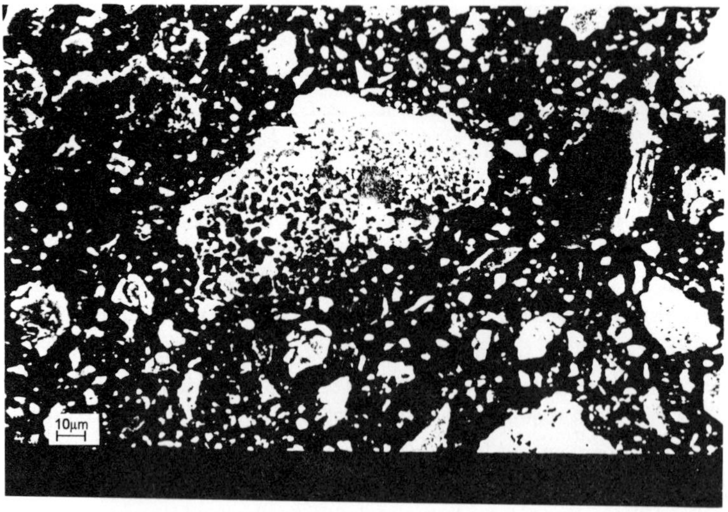

Figure 1 Backscattered electron image of a $C\overline{S}A$ clinker embedded in resin (black). The microstructure of clinker grains is particularly evident in coarser grains such as that near the centre of the micrograph

CLINKER FINISHING

Sulfoaluminate clinkers are ground to high specific surface area, typically > 3500 cm^2/g, prior to delivery. Thus particle size and size distributions are very similar to those of Portland cement. Another point of similarity is the introduction of other materials during grinding. Whereas Portland cements are normally interground with 2-5% calcium sulfate, calcium sulfate is added to sulfoaluminate clinkers at much higher levels, typically 10-35% by weight. These large additions affect the hydration process and mineralogy of the paste fraction, as will be shown. Incorporation of so much gypsum also helps reduce the environmental impact of manufacture and dilutes the unit cost of clinker. For special purposes limestone and slag may, if required, be interground at this stage.

HYDRATION REACTIONS

Calcium sulfoaluminate cements are mixed with aggregate and water immediately prior to use in the same way as Portland cement concretes. The same batching, mixing and transport systems can be used for both cement types as the rheology of fresh mixes is comparable.

As water is added an initial burst of heat occurs, similar to the initial heat evolution encountered in Portland cement systems, the so-called heat-of-wetting. Thereafter both types of system exhibit a period of reduced heat evolution.

During this interval fluid mixes have similar properties: both Portland and $C\bar{S}A$ mixes behave as non-Newtonian fluids. The rheology of $C\bar{S}A$ mixes is much influenced by additives; this seems to be a characteristic of aluminate - rich systems and in this respect, $C\bar{S}A$ mixes resemble HAC. Thus calcium sulfoaluminate cement systems can be accelerated or retarded, depending on application, rather more readily than equivalent OPC systems. For normal, low-to moderate strength concretes, Chinese experience is generally satisfactory of using $C\bar{S}A$ mixes without admixtures. Rapid hardening formulations generally incorporate a proprietary water-reducing additive and retarder, to enable use of a low w:c ratio and for high strengths to be achieved. Where fluid concretes require to be stored or pumped, it may be essential to use a set retarder in conjunction with a plasticiser. The specific chemical formulations required to control these properties are easily achieved but, because of the proprietary nature of the chemicals involved, some experimentation with locally - available materials may be required to adjust dosage and achieve a satisfactory balance of properties.

Detailed accounts of the mineralogical changes occurring during hydration have been presented elsewhere [6]. Briefly, during the early states of reaction, calcium sulfoaluminate, gypsum and clinker anhydrite react with water to form ettringite. The period between initial and final set is marked by development of a skeletal network of ettringite, diminution of free water and substantial heat evolution. As strength gain occurs, belite and ferrite phases begin to hydrate, albeit slowly. Eventually a dense paste consisting of an intimate mixture of ettringite and C-S-H, develops; the Ca:Si ratio of the C-S-H is low, *ca* 1.2, reflecting the absence of portlandite, $Ca(OH)_2$, amongst the hydration products and the high calcium content of ettingite.

The hydration processes of $C\bar{S}A$ cements thus differ in several significant ways from those operative in Portland pastes. Firstly, major mineralogical differences occur: $Ca(OH)_2$, abundant in Portland cement, is absent from sulfoaluminate pastes. Ettringite is not a major phase in Portland cement but is often microscopically conspicuous, forming needles and prisms in cavities and pores, whereas in sulfoaluminate pastes, ettringite is abundant although it is not generally possible to resolve individual ettringite crystals within the dense and relatively featureless matrix which forms after the first few days of hydration. Another important difference concerns water requirement for hydration. It will be recalled that the water demand of a Portland cement, defined as the water required for chemical hydration, expressed as a water:cement weight ratio, lies in the range 0.3-0.4 with a median value of 0.36. Typical constructional Portland concretes will be formulated to w/c ratios exceeding 0.4, so will contain an excess of water, termed pore water from its characteristic distribution. In sulfoaluminate cements, calculation of the water demand for chemical hydration has been made [7]. It is much influenced by the potential to form ettringite which, in turn, is affected by the amount of gypsum interground with clinker. As a consequence of the high specific water content of ettringite, the chemical water demands will exceed the amount of mix water, with the result that hydration reactions cease before clinker hydration can be completed.

Figure 2 shows a specimen calculation in which the chemical water demand (CWD) is depicted as a function of the amount of interground gypsum.

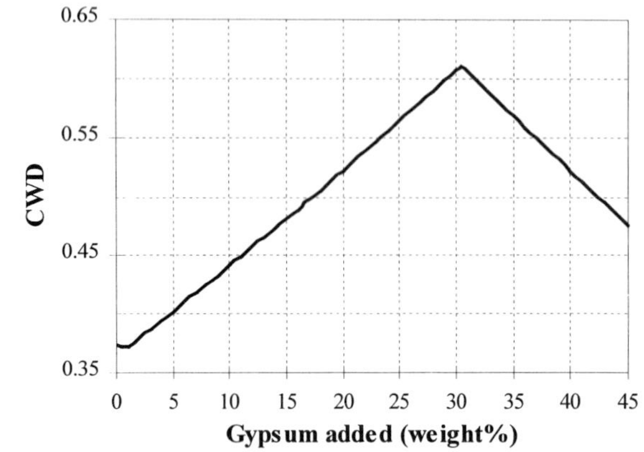

Figure 2 Chemical water demand, CWD, expressed as a water:cement ratio, as a function of gypsum addition (wt.%), from [7]

Chemical water demand tends to peak at the gypsum content required to give a maximum yield of ettringite. This arises from consideration of the water content necessary to form ettringite. If for example, ettringite were formed from hydrating a stoichiometric mixture of $C_4A_3\overline{S}$ and anhydrite, the water:solid ratio required for hydration would be ~0.6.

Given the high chemical water demand of sulfoaluminate cements relative to the water required for plasticity, it is easy to prepare mixes which upon reaction, develop a "dry" internal environment, i.e., liquid water is totally consumed during hydration. Under these circumstances the hydrate mineralogy is unaffected, although the total amount of hydrates is decreased because much unconsumed clinker remains as a microaggregate. This abundance of unhydrated clinker is another characteristic feature of hardened sulfoaluminate pastes. The writers have examined $C\overline{S}A$ cement concretes continuously immersed in water for 12-16 years and despite the abundance of external water, clinker hydration is only complete in the near-surface layers, extending to depths of < 1.0 mm.

The strength gain of $C\overline{S}A$ concretes, like those of Portland concretes depends on many factors: cement fineness and content, w:c ratio and cure. Several representative datasets are shown in Table 3 for concretes. A notable feature is the high early strengths which are achieved, but strength gain typically continues to longer ages. Fig 3 shows the cube compressive strengths of a concrete made using cement:coarse aggregate:sand:water mix in the ratio 1:1.33:2.81:0.42, i.e., 440 kg cement/m^3 and a w:c ratio 0.42, subject to moist cure at ~ 20°C for 6 years. It is apparent that the strength develops rapidly and continues to improve with age.

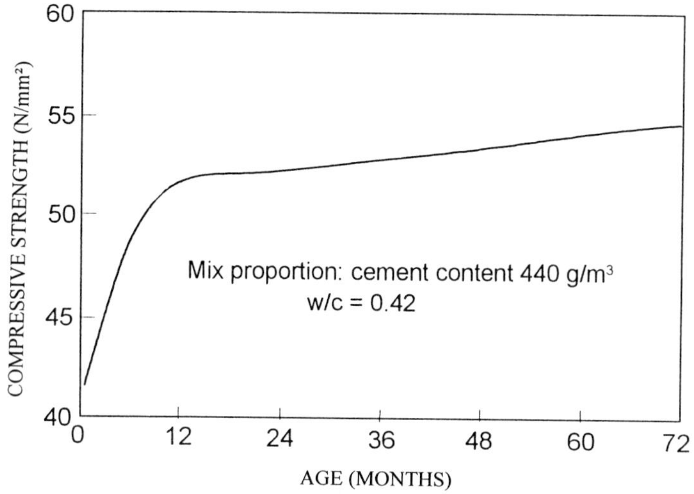

Figure 3 Long term strength gain of a concrete made with commercial C$\bar{S}$A cement

Table 3 Strength development of rapid hardening C$\bar{S}$A concretes at early ages

CEMENT†	CEMENT CONTENT*	W/C	SPECIAL ADMIXTURE	COMPRESSIVE STRENGTH (MPA)					CURING CONDITION
				12h	1d	3d	7d	28d	
C$\bar{S}$A	300	0.58	No admix	14.2	29.7	36.4	36.9	41.3	20°C, RH
C$\bar{S}$A	460	0.35	Yes		50.6	57.8	64.3	76.7	>90%
C$\bar{S}$FA	500	0.32	Yes	42.0	58.0	64.5	70.2	81.3	

†Cured at 20°C and relative humidity >90%. *kg/m3

C$\bar{S}$A = rapid hardening calcium sulfoaluminate cement

C$\bar{S}$FA = rapid hardening ferro-aluminate cement

"Special admixture" refers to proprietary admixture reducing water content and retarding setting time

During the early stages of hydration, C$\bar{S}$A cements evolve rather more heat than equivalent Portland cement concretes. If we compare a rapid hardening Portland with a C$\bar{S}$A mix, typical 3 day heat evolutions may be on the order of 300J/g and 400-450 J/g, respectively. The hydration heat can be both an asset and liability; it is an asset in cold-climate concreting, but a potential liability in large pours, which may require careful sequencing to prevent excessive thermal excursions.

Since sulfoaluminates are widely used in expansive and shrinkage-compensating additives, it may surprise readers that the normal commercial grades of calcium sulfoaluminate cement shrink when used in mortars and concrete subject to isothermal cure. The shrinkage is, however, characteristically somewhat less than from an equivalent Portland mixture but by altering the formulation, non-expansive sulfoaluminate cements can be made expansive. Perhaps the simplest method is to increase the gypsum content during clinker grinding, or to add CaO, or to proportion and fire the clinker so as to develop free CaO and/or calcium aluminates. While the topic of expansive and self-stressing formulations does not generally lie within the scope of this paper it should be noted that the rules for proportioning clinkers so as to achieve controlled shrinkage are well understood, at least empirically. Thus expansion is easy to achieve, but precise control of the expansivity is more difficult to control.

The internal alkalinity of $C\overline{S}A$ cements is characteristically about 1pH unit lower than in Portland cement. The reasons for this pH decrease are complex and arise from a combination of circumstances of which the most important are the generally low alkali contents of the raw $C\overline{S}A$ clinkers, the low Ca:Si ratio of the C-S-H and the absence of $Ca(OH)_2$ from amongst the hydration products. The internal pore fluid pH is probably in the range 11-12 but as much of the mix water is chemically combined, it becomes difficult to express pore fluid from mixes made to normal engineering w:c ratios [8]. Indeed, many formulations made to low w/c ratios probably become dry within a few days or weeks of mixing. Once $C\overline{S}A$ cement matrices achieve a dry internal environment. pH concepts are of course inapplicable. Hence pore fluid can only be obtained at (i) early ages, or (ii) at very high w:c ratios or (iii) some combination of the first two conditions. The pH of hydrated cement can be measured indirectly, by crushing a mature paste and stirring the powder with CO_2 - free water, followed by pH measurement but the method is unreliable. Nevertheless, observation of real structures, coupled with laboratory, electrochemical corrosion studies, disclose that the matrix pH provides passivation for embedded steel, yet is low enough to inhibit alkali-aggregate reaction (AAR). While the corrosion of mild steel in Portland cement concrete has been much studied in matrix conditions for steel passivation in $C\overline{S}A$ still require to be defined. We know many factors which lead to depassivation in OPC; ingress of chloride, oxygen access to steel, severe loss of pH arising from carbonation, etc., although we are unable to quantify their interactions. The consequences of achieving and maintaining a dry internal environment must affect transport processes in the matrix and, especially, in the vicinity of the steel-concrete interface. Not surprisingly, therefore, much of what we know about the performance of newer binders such as sulfoaluminate cement is derived from practical experience.

IN-SERVICE PERFORMANCE

Concretes are used in a wide range of service environments and undergo a variety of degradation mechanisms. Several are considered here.

Carbonation does not much affect the physical performance of Portland cement concrete but does lower internal pH with the result that the passivation afforded to embedded steel is reduced. Sulfoaluminate cements, like OPC, undergo carbonation. The iron-rich FAC types are particularly resistant to carbonation.

Table 4 shows data obtained from test prisms, 10x10x10cm. These were removed from moulds as soon as practicable after casting and thereafter stored and cured at 70% relative humidity, 20°C, for 28d and thereafter in a test chamber containing 20% CO_2, 80% air at 90% rh and 20°C. After 28 days of accelerated carbonation the depth of carbonation was measured using the phenolphthalein test. The control specimens were cured in CO_2 - free air at 20°C and 90% rh. Despite the accelerated nature of the test and immature cure state of the paste at the commencement of the tests the depth of carbonation was small. Not surprisingly, lower w/c ratios decreased the carbonated depths.

Table 4 Carbonation of iron-rich calcium sulfoaluminate concrete

CEMENT	ADMIXTURE	W/C	SLUMP (cm)	CARBONATED DEPTH (mm)*	CONTROL STRENGTH (N/mm²)	CARBONATED STRENGTH (N/mm²)
$C\bar{S}$ FA from Tangshan	No	0.52	7.0	1.29	51.4	49.3
$C\bar{S}$ FA from Tangshan	Yes	0.40	18.0	0.98	69.4	64.5
$C\bar{S}$ FA from Lengshuitan	Yes	0.30	5.0	0.26	73.6	71.8

* see text for description of cure and accelerated test.

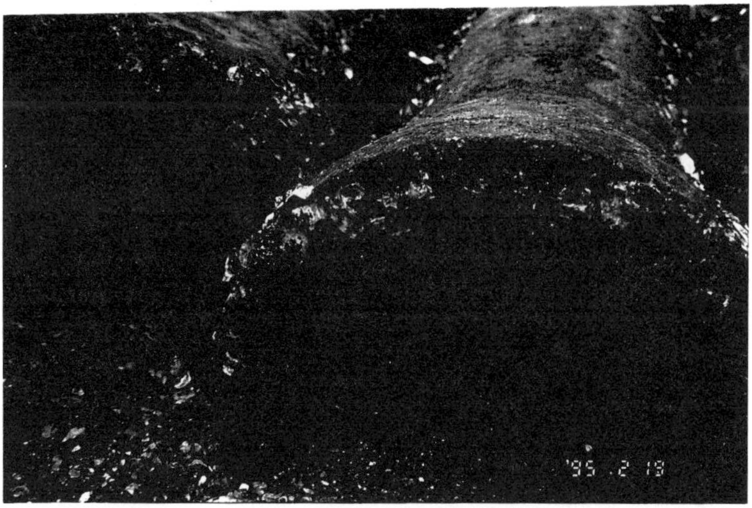

10cm

Figure 4 General appearance of ~ 0.5 m diameter centrifugally - cast concrete pipes made with calcium sulfoferrialuminate cement after removal from sea water

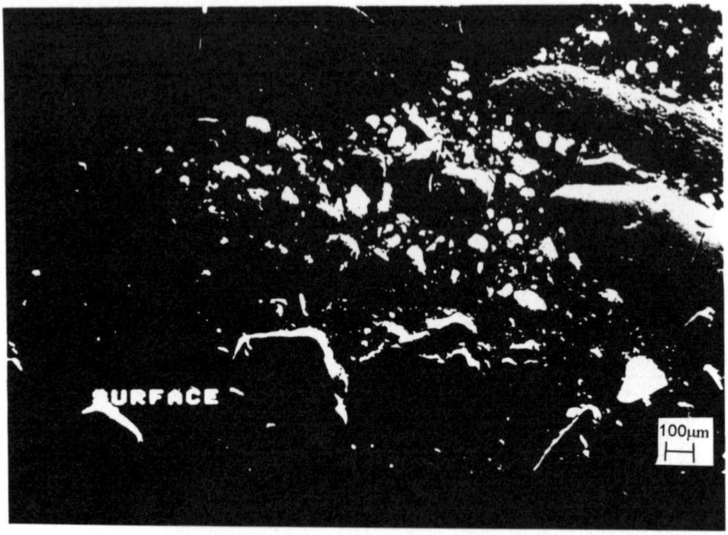

Figure 5 Backscattered electron image of a concrete pipe exposed to sea water for ~ 16 years, see text for an explanation of its features

Figure 6 Backscattered electron image of a concrete pipe exposed to sea water for ~ 16 years, see text for an explanation of its features

Figure 7 View of a concrete pipe exposed to sea water for ~ 16 years, the concrete has been mechanically broken away from the steel reinforcements which are uncorroded

Resistance to aggressive salt solution has also been determined. Figure 4 shows the physical appearance of a centrifugal cast concrete pipe removed from service after 16 years immersion in sea water: Figures 5 and 6 show micrographs of the near-surface layers of this pipe. The concrete formulation was not known precisely except that it was a $C\overline{S}AF$ self-stressing cement. In any event, the centrifugal casting process leads to some segregation so the effective w:c ratio is low, probably 0.3 to 0.4. Figures 5 and 6 show backscattered electron images from sections cut from the pipe normal to its exposed surface. The concrete - sea water interface is located at the bottom of the photographs. The magnification scale of Figure 6 is approximately double that of Figure 5 and , moreover, the exterior surface of the pipe, Figure 6, is partly occupied by a mollusc shell which looks smooth and bright in the backscattered electron image. Within the cement matrix both photographs show coarse aggregate particles and finer sand grains. However attention focuses on the altered matrix which has developed a weakly zoned microstructure. The first 0.6 to 0.8 mm depth has been slightly leached and is porous and free from unhydrated clinker. However, at greater depths, the matrix is uniformly dense, and contains much unhydrated clinker which shows as optically bright grains. Figure 6, at slightly higher magnifications, shows a leached zone, perhaps partly protected by a shell, which extends only ~ 0.3 to 0.4 mm into the pipe. At this magnification pores are more readily apparent in the leached zone as is the persistence of unhydrated clinker grains at greater depths.

As leaching occurs, hitherto unhydrated clinker grains react with permeating water and the hydrates thus formed are believed to contribute to pore blocking and maintaining the integrity of the outer layers. The hydration of residual clinker is, however, confined to the extreme outer zones which after 16 years immersion is not observed to exceed 0.8 mm.

Some of the concrete was mechanically broken away from the steel reinforcements. Figure 6 shows their condition; they are smooth and uncorroded, despite the often poor cover.

However, what is remarkable is the resistance to leaching of the matrix and the relative lack of reaction with the magnesium and sulfate components of sea water, as typically occurs even with sulfate resisting - Portland cement. The authors accept that proper trials involve comparison of samples containing different cement types, made and exposed under equivalent conditions. Lacking these control samples, we can only call attention to the apparently excellent resistance of the $C\bar{S}A$ and especially of $C\bar{S}FA$ pastes to degradation by sea water. The paste constituents do not react with sulfate because they are fully sulfated and hence passivated at the outset of exposure.

Owing to the lower effective alkalinity in $C\bar{S}A$ cements relative to Portland cement, they have been widely used in two applications involving potential materials incompatibility. The first is in situations where alkali-aggregate reaction may occur. These is lack of consensus about rapid testing of aggregates for potential alkali susceptibility and these concerns particularly affect inclusion of new cement types within the envelope of test conditions. Therefore only preliminary statements can be made about the potential for alkali aggregate reaction. Nevertheless some 20 years of experience, including those of concretes made with aggregates known to be susceptible to AAR in Portland-based concretes, have failed to disclose the occurrence of expansion when the same aggregate is used in $C\bar{S}A$ formulations. This is attributed to the lower internal pH and dry internal environment of the latter. The lower intrinsic alkalinity of sulfoaluminate pastes is also related to the improved retention of strength of glass-fibre reinforced shapes. It will recalled that glass fibre is an ideal material for reinforcement of cement, at least from the mechanical point of view. Unfortunately even special glass fibre degrades in the wet, high pH environment of Portland cement [9]. However, development of low-alkali sulfoaluminates-belite cements affords new opportunities for specifying durable composites. Not only is it easier to make internally-dry compositions but, even when wet, the matrix alkalinity of $C\bar{S}A$ based formulations is lower than Portland cement. Moreover, $Ca(OH)_2$ crystals, known to damage fibres in Portland-based composites, are absent from $C\bar{S}A$ matrices.

SUMMARY

The technology of production and use of new cement types based on calcium sulfoaluminate, ferrite and belite phases are described. They have five qualities that commend then to engineers:

1. they can be handled and used like Portland cement, with which they are compatible
2. they have good strength at both ultimate and early ages
3. they are resistant to carbonation and to the action of sulfate and sea water

4. they have better compatibility to glass fibre and alkali - susceptible aggregates than Portland cement, while at the same time having sufficient alkalinity to protect embedded steel
5. they are susceptible to tailored property alteration using admixtures: accelerators, retarders, plasticisers, anti-bleed and anti-freeze agents, etc.

These advantages, combined with low cost and low specific CO_2 emissions in the course of production, suggest that use of these materials will increase markedly, especially as several decades of experience of their structural use in China has been generally favourable.

Are there are negative factors which suggest that $C\bar{S}A$ cements should not be used in certain applications? At present, we have only limited data about their durability but these data derive from experiences including a wide range of commonly - encountered environmental exposures. After a century of experience of Portland cement concretes, we still have much yet to learn about its behaviour but we know better what questions to ask and how to resolve them. Work on $C\bar{S}A$ continues at the China Building Materials Academy, at Aberdeen and in the Czech republic to establish the long-term performance of $C\bar{S}A$ materials. However, we welcome more general interest in these materials and regard other research activities as complimentary rather than competitive with our own.

ACKNOWLEDGEMENTS

One of us (L.Z) thanks The China Building Materials Academy for leave of absence to pursue PhD studies at The University of Aberdeen. We gratefully acknowledge financial support from Frosroc, a member of the Burmah-Castrol Group and The China Building Materials Academy for concrete specimens removed from service trials.

REFERENCES

1. ZHAOQI, W, JIAFEN, JIANG, SHIYUAN, H, et.al (Editors) Proceedings of the 3rd Beijing International Symposium on Cement and Concrete, 3 vol. (1993). China Building Materials Academy, Beijing, Peoples Republic of China.

2. JUSTNESS, H. (Editors) Proceedings of the 10th International Congress on the Chemistry of Cement and Concrete, Gothenborg, Sweden (1997). 5 Vol. Amarkai AB and Congrex Goteborg AB (1997). ISBN 91-630-5495-7.

3. GLASSER, F P. (Editor). Advances in Cement Research, Jan. 1999 (Special issue, in press).

4. Proceedings of the 4th Beijing International Symposium on Cement and Concrete (in press) Oct. 1998.

5. WANG, U M, SU, M Z, and ZHANG, L. "Calcium Sulfoaluminate Cement" (in Chinese) in press, to be published in 1999.

6. ANDAC, O, and GLASSER, F P. "Microstructure and Microchemistry of Calcium Sulfoaluminate Cement" Mater. Res. Soc. Symp. Vol 370 (Ed., S. Diamond, et.al) Mater. Res. Soc., Pittsburgh PA (1995) pp 135-142.

7. ZHANG, L, and GLASSER, F P. Calculation of the Water Demand for Hydration of Calcium Sulfoaluminate Cement. To appear in ref. 4, 1998.

8. ANDAC, M, and GLASSER, F P. "Pore solution composition of Calcium Sulfoaluminate Cement". Advances in Cement Res. 10, 1-4 (1998).

9. YILMAZ, V T, LACHOWSKI, E E. and GLASSER, F P. "Chemical and Microstructual Changes at Alkali - Resistant, Glass Fibre-Cement Interfaces". Jour. Amer. Ceram. Soc. 74 3054-3060 (1991).

THE RESISTANCE OF CALCIUM ALUMINATE CEMENTS TO ACID CORROSION IN WASTEWATER APPLICATIONS

R Letourneux

K Scrivener

Lafarge

France

ABSTRACT. Bacteriogenic corrosion of concrete is a well-known phenomenon in wastewater networks. Studies carried out by the American Environmental Protection Agency have shown that it occurs world-wide, having been identified in more than 20 countries and in more than 70% of American cities. Proliferation of bacteria produces sulphuric acid which causes deterioration of many of the materials used in wastewater networks. This deterioration has serious consequences in terms of the high cost of repair. Calcium Aluminate Cements (CAC) and mortars offer a cost-effective method of prolonging the life of affected networks. Theoretical understanding, laboratory studies and experience in the field confirm the advantages of CACs in this area.

Keywords : Calcium aluminate cement, Acid attack, Neutralisation capacity, Bacteriogenic corrosion, Wastewater networks.

Mr Roger Letourneux is the Market Manager for Pipes and Environment with Lafarge Aluminates

Dr Karen Scrivener is head of the Aluminate Department in the Central Research Laboratory of Lafarge, responsible for a wide range of research on the properties, applications and development of calcium aluminate cements.

This capacity to neutralise acid is another important factor in the resistance of CAC to acid attack. The neutralisation capacity is defined as the amount of acid (in mM) which is necessary to destroy an amount of material (in g). The higher the neutralisation capacity, the more acid is needed for destruction of the material or, for the same amount of acid, the material takes longer to destroy.

Different types of cement have different neutralisation capacities as do different aggregates. When aiming for acid corrosion resistance, the obvious approach is to choose the material with the highest neutralisation capacity. Silica sand aggregates, for example, are inert to acid attack but thereby contribute nothing to the neutralisation capacity of the mortar as a whole. Although the aggregate may remain unattacked if the cement around it is destroyed it will fall out of the concrete and not provide protection from further attack. Synthetic calcium aluminate aggregate, when combined with calcium aluminate cement, contributes greatly to the overall neutralisation capacity of the mortar and results both from experimental work and from the field have illustrated its effects.

Neutralisation capacities may be determined by calculation from the chemical composition of the cementitious material. Thus each 2 moles of H+ ions are needed to neutralise each mole of calcium present and 3 moles for each mole of alumina, whilst the silica content does not contribute to the neutralisation capacity as it is unattacked. The calculated neutralisation capacities for Portland cement, calcium aluminate cement, and calcium aluminate cement mortars with synthetic calcium aluminate mortars are shown in Figure 3.

As indicated in Equations (3) and (4), the full neutralisation capacity of CAC is developed in two stages, Figure 4:

1. the first at a pH above 3.5, where C_3AH_6 is attacked to give precipitated $Al(OH)_3$
2. the second at a pH below 3.5, where the precipitated $Al(OH)_3$ together with that formed on initial hydration is attacked, further contributing to the overall neutralisation capacity of the cement

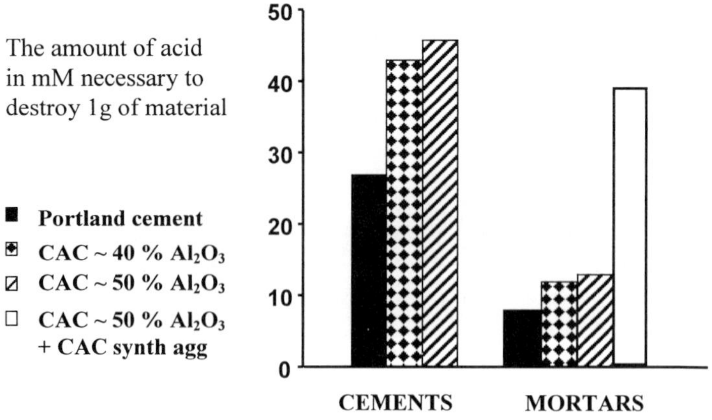

Figure 3 Chemical corrosion - Neutralisation capacity in mM/g

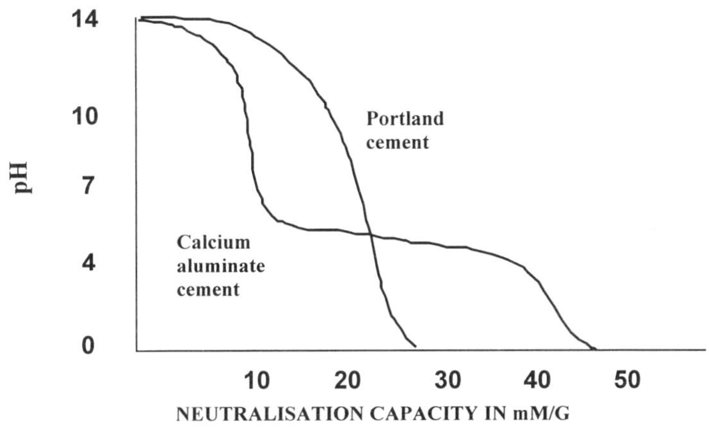

Figure 4 Development of neutralisation capacity

Experimental Determination of Neutralisation Capacity

Neutralisation capacities are found in the form of 'Alkalinity' in equations commonly used by wastewater network designers to determine the suitability of a material for a particular application in the design. Alkalinity is simply the neutralisation capacity of a material divided by that of calcium carbonate, 20 mM/g.

The most common method to be found in the literature for determination of the neutralisation capacity of Portland cement involves the addition of excess hydrochloric acid followed by back titration to neutral pH with sodium hydroxide to determine the amount of added acid which was neutralised by the cement. This method effectively only determines the neutralisation capacity of the calcium component of the cement and so gives fairly accurate results for Portland cements. However the method gives false results when the cementitious material contains significant amounts of alumina, as with blends containing slag or with calcium aluminate cements. (See annex for calculation method).

When the alumina component is initially attacked by acid, it forms aluminium ions (Al^{3+}). However, during back titration to return the pH to neutral these ions combine with hydroxide and precipitate out as $Al(OH)_3$ above pH 4 and the neutralisation capacity due to the alumina which is developed at lower pHs is not measured.

During the back titration the method assumes that the sodium hydroxide reacts with the excess hydrochloric acid, not neutralised by the cementitious material:

$$3H^+ + 3NaOH \text{ --------> } 3Na^+ + 3H_2O$$

But when alumina was present in the cementitious material, part of the sodium hydroxide is used to reprecipitate the aluminium ions:

$$Al^{3+} + 3NaOH \text{ ---------> } Al(OH)_3 + 3Na^+$$

The neutralisation capacity determined by this method gives a figure of 26 mM/g for OPC compared to 28 mM/g by calculation, but only 13 mM/g instead of 46 mM/g for CAC .

A number of solutions may be proposed to modify the experimental method in the case of materials containing significant quantities of alumina:

a) sodium fluoride (NaF) may be added to the solution used for back titration. This combines with the aluminium ion released and stops the formation of $Al(OH)_3$. The calculation of the neutralisation capacity resulting from this experiment then becomes accurate;

b) measurement of Al^{3+} release by Induction Coupled Plasma Atomic Emission Spectroscopy (ICP), but this method calls for much more elaborate and expensive equipment; or

c) titration as above but using an insufficient quantity of acid (as opposed to an excess of acid) and adding a further quantity of acid to maintain the pH at a given level, avoiding the use of sodium hydroxide and a subsequent false result. This, however, takes much longer than titration with sodium hydroxide.

LABORATORY TESTS OF BACTERIOGENIC CORROSION

For many years, the only evaluation tests available were chemical tests which subjected the material to attack by sulphuric acid at various pH levels, but bacteriogenic corrosion as found in wastewater networks is different from chemical corrosion. Whilst the end result of bacterial action is sulphuric acid, chemical corrosion tests do not take into account the activity of the bacteria which produce the acid, which affects the rate of acid available for corrosion.

Corrosion of wastewater networks has mainly been found in countries with hot climates, but countries with temperate or Nordic climates, such as that found in Hamburg, are not spared this phenomenon. In order to simulate such environments, the Microbiological Department of the University of Hamburg has developed a simulation chamber to enable comparisons of the performance different materials in controlled accelerated conditions. Comparison of the rate of deterioration in the simulation chamber with samples exposed in sewers, indicates that this chamber accelerates deterioration by a factor of more than 20. Thus 250 days in the simulation chamber is equivalent to about 16 years in service.

Figure 5 shows a diagram of the apparatus. Tests samples are inoculated with bacteria and placed in the chamber (volume - 1 m3). Temperature - 30°C, H2S - 10 ppm, Relative humidity - 98 %. Six nozzles deliver a solution containing nitrogen and phosphorus in order to optimise conditions. At the base there is a bath of water 10 cm deep. The atmosphere in the chamber is circulated continuously by a powerful fan. Each experiment lasts up to 12 months.

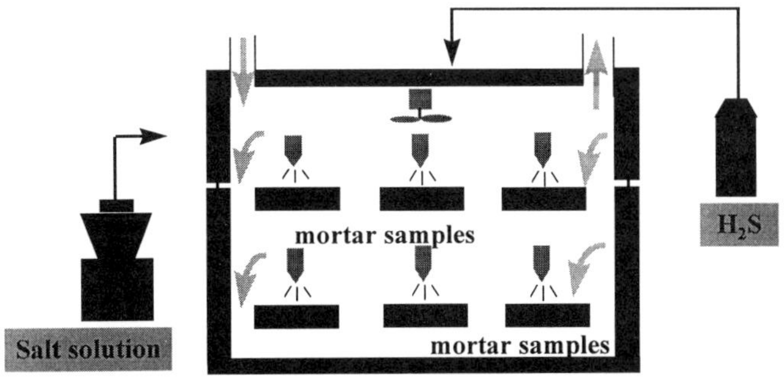

Figure 5 Bacterial corrosion simulation chamber

Studies carried out with this equipment show that bacteria on a CAC substrate produced measurably less acid than bacteria on a Portland substrate and that the use of synthetic CAC aggregate (Alag®) in the mortar reduced acid production further.

As shown in the results, Figures 6 and 7, the combination of high neutralisation capacity and reduced production of sulphuric acid translates to a much higher surface pH and consequently leads to much lower weight loss.

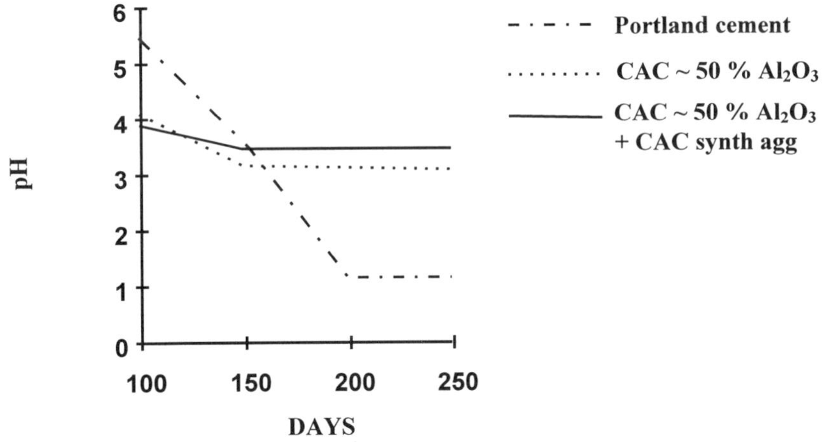

Figure 6 H_2S bacterial corrosion : influence on the substrate on pH

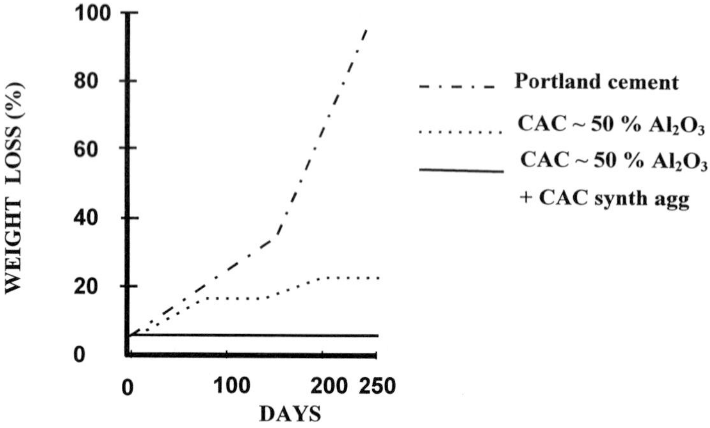

Figure 7 H_2S bacterial corrosion : influence on the substrate weight loss

FIELD TESTS

Tests in the field have shown the same trend. A study by TNO in Rotterdam, Holland, where samples were hung in manholes (Figure 8) in conditions known to be aggressive showed that CAC samples gave better results than other hydraulic mortars. The samples were tested over two years, during which the H2S level s were monitored.

The depth of neutralised concrete, measured by a pH indicator, is added to the thickness loss to give the total degradation depth. The results are shown in Table 1.

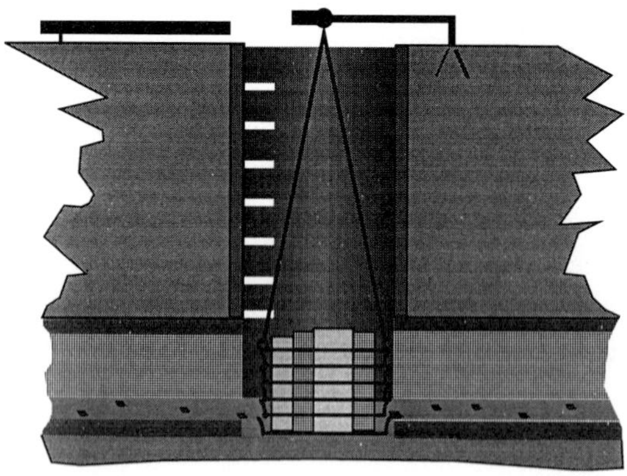

Figure 8 H2O bacterial corrosion : field test

Table 1 Results of field testing

	CAC + CAC AGGREGATE	CAC + SILICEOUS SAND	SLAG CEMENT + SILICEOUS SAND	PORTLAND + SILICEOUS SAND
Weight loss (%)	0	0.1	1.3	1.2
Total degradation depth (mm)	0	0.2	1.6	1.0

Long-term monitoring of a connection chamber in Hampton Roads in the USA, and preliminary work carried out at the Virginia testing station for evaluation of pipe materials in South Africa have given similar results. A new series of tests is about to start at Virginia, led by the University of Cape Town.

Backing up these results is the fact that many thousands of kilometres of ductile iron pipe have been protected over the last 15 years with CAC mortar linings and references of linings on centrifuged concrete wastewater pipes go back to the 1950's. The countries involved include South Africa, Brazil, Malaysia, Taiwan, France, Germany, UK, USA, etc., covering a wide range of climatic conditions. Calcium aluminates have always been useful materials in wastewater applications and the work mentioned above now explains fully the reasons for this.

CONCLUSIONS

The combination of three main factors :

1. stable hydrates containing no Portlandite
2. higher neutralisation capacity
3. lower bacteriogenic acid production

lead to much greater corrosion resistance for CAC than for OPC of all types. CAC concretes and mortars demand no special machinery or equipment for installation apart from what is commonly found in handling Portland cement and thus provide a simple and cost-effective solution to the problems of bacteriogenic corrosion which tend to occur in wastewater networks.

Similar applications in other fields where the same mechanisms apply, such as the cooling towers of electric power stations, could be envisaged.

STUDY OF THE EFFECT OF ALKALIS ON THE SLAG-CEMENT SYSTEMS

B Mezghiche

A Guettala

R Chebili

L Zeghichi

University of Biskra

Algeria

ABSTRACT. The replacement of natural materials by industrial waste has given a variety of new construction materials that are economical and in the same time harmless for the environment, thus ecological. High furnace slag is one of these industrial wastes that has its objectives and evident reasons of use; such as its important properties or characteristics and rapid growth of production in Algeria. This research deals with the effect of alkaline activation of slag-cement with clinker. Tests carried out, using some agents that activate the process, show a considerable increase in compressive strength at 28 days that reaches 50% with respect to compressive strength obtained from slag-cement at different percentage of slag (50 to 80%).

Keywords: Slag- cement, Binding medium, Activation agent, Alkalis, Compressive strength

Dr B Mezghiche is a Lecturer in Construction Materials, Civil Engineering Institute, and Director of the Concrete Technology Institute, University of Biskra (Algeria). He specialises in the use of binders and durability of concrete, published researches on Binders.

Mr A Guettala is Director of the Civil Engineering Institute, Lecturer, University of Biskra, (Algeria). He specialises in the permeation properties of concrete, and has performed researches on the durability of construction materials.

Dr R Chebili is The President of the Scientific Committee of the Architecture and Civil Engineering Institute, Lecturer, University of Biskra (Algeria). He specialises in materials and structural mechanics.

L Zeghichi is a Research Assistant in Concrete Technology, University of Biskra (Algeria).

INTRODUCTION

The slag of high furnace is obtained, by a physicochemical process, usually as waste in heavy industry. Mainly, it can be obtained when producing cast- iron, and in a great part from ores, ferrous minerals, coal impurities and additives.

Granulated slag is widely used as an additive to the Portland cement, due to the similarity of its chemical constituents with this later. It can also be used separately as a binding medium in some other cases. It can influence directly the mechanical strength and time setting of the binding medium. This is because of a weak pH that can influence solubility of slag constituents. A pH grater then or equal to 12 can easily alters the constituents of the slag by permitting the formation of stable hydrates and the structure paste. Such pH is due to the presence of active base, [1, 2, and 3].

Alkaline activation is the most efficient method to activate the slag, and is principally performed using:

- Alkalis (R-OH)

- Salts, non-silicates obtained from weak acids (R_2CO_3 - R_2SO_3 - R_2S). Radical R represents the Alkaline metal's ions (Na^+, K^+, and L^+), [4].

EXPERIMENTATION

Materials

Portland cement

Is obtained from grinding of clinker using about 5% of gypsum. In the present experiment, the following chemical composition of clinker has been used;

Table 1 Chemical Composition of Clinker

CONTENT %				
SiO_2	Al_2O_3	Fe_2O_3	CaO	MgO
21.14	5.46	3.57	64.2	1.30

Granulated slag of high furnace

The present slag used in the experiment is provided by the metallurgic unit of (El Hadjar) Annaba, Algeria. It is obtained by rapid immersing in cold water basin, presenting a base modulus (M_b = 1.074), ground to a specific surface of 3000 g/cm^2 and has a chemical composition presented in table 2.

Table 2 Chemical Analysis of the Slag

CONTENT %						
SiO_2	Al_2O_3	CaO	FeO	MgO	MnO	SO_3
40.8	5.20	43.01	0.53	6.4	3.02	0.8

Activation agents (Alkalis constituents)

Several types of solutions having different densities have been prepared, see table 3 below, by adding quantities of water to obtain a wanted densities;

1st group of activation agents:(Caustic alkalis) NaOH - KOH.
2nd group " " : Na_2CO_3 - K_2CO_3 - $Na_2 SO_3$.
3rd group " " : NaCI - K_2O.
4th group " " : K_2SO_4 - Na_2SO_4.

Table 3 Densities of Used Solutions

SOLUTION	1ST GROUP			2ND GROUP			3RD GROUP		4TH GROUP	
	water	NaOH	KOH	Na_2CO_3	K_2CO_3	Na_2SO_3	K_2O	NaCI	K_2SO_4	Na_2SO_4
Indices	A	B	C	D	E	F	G	H	I	J
Density g/cm^3	-	1.18	1.18	1.18	1.19	1.17	1.19	1.09	1.19	1.18

Sand

The used sand is provided from Biskra (the Southeast region of Algeria) and has an apparent density of 1530 kg/m^3 and absolute density of 2600 kg/m^3.

Crushed-sand

SAMPLES PREPARATION

Tests carried out are, strength compressive tests on cement paste and mortar (cement/sand ratio 1/3).

Compression tests are carried out on 2*2*2* cm cubes and prismatic of 4*4*16 cm for different percentage of slag in the binding medium (50% - 60% - 70% - 80%) with different activation agents. The first category of tests is composed of six samples and the second is composed of three samples. These samples are all stored in water under a specified temperature 20 °C for the period of tests. Other tests also have been retained to evaluate the activated slag-cement quality, among which we state; setting time of slag-cement and expansion.

RESULTS AND DISCUSSION

Effect of Activation Agents on the Mechanical Strength

The slag percentage is fixed to 80% and CPA cement (Ordinary Portland Cement) is fixed to 20%, solution/cement ratio is 0.3.
Tests carried out on slag-cement cubes show the effect of the activation agents on the evolution of the compressive strength. Results are presented, for natural hardening during 7 and 28 days, in the diagram below, Figure 1.

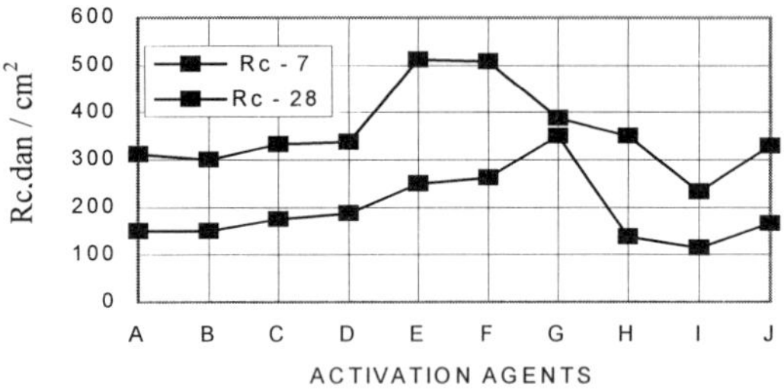

Figure 1 Compressive strength with different activation-agent

It is noticed that the activation agent's effect remains in the same level either in 7 days or in 28 days tests. This means that the activation agents that cause increase or a decrease in strength during 7 days have the same effect during 28 days.

Slag addition of 80% in the binding medium without activation agents, affect the strength because of the weak percentage of lime that is contained in the 20% of CPA cement, and cannot perform real reactions. With the activation agents of the first category, we notice a slight increase in the strength compared to non-activated binding medium. Moreover, the activation agents of second category i.e. (R-CO_3) increase neatly the strength, whereas (R-SO_3) diminishes the strength.

The evolution in strength is related to the increase in pH, which permits the solubility of the binding medium constituents and leads to the formation of the structure.

Effect of Slag Content on the Mechanical Strength.

In this part of the present work, efficient activation agents have been used, i.e. (NaOH, Na_2SO_4, Na_2CO_3) with a density of 1.15 g/cm^3.
Compressive strength test's results after 1 day, 7 days, 28 days, 120 days and 365 days of hardening are represented in figures 2, 3, 4 and 5.

Figures (2 and 3), show that with increasing the percentage of slag, the mechanical strength diminishes with or without activation agent NaOH, and this for different sample ages.

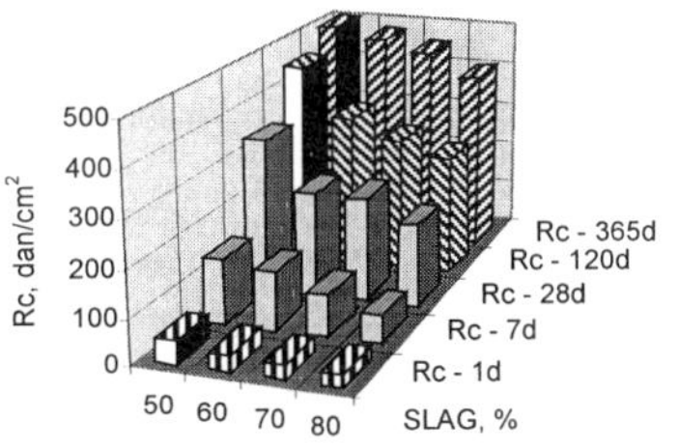

Figure 2 Natural hardening of binding medium non activated function of different percentage of slag

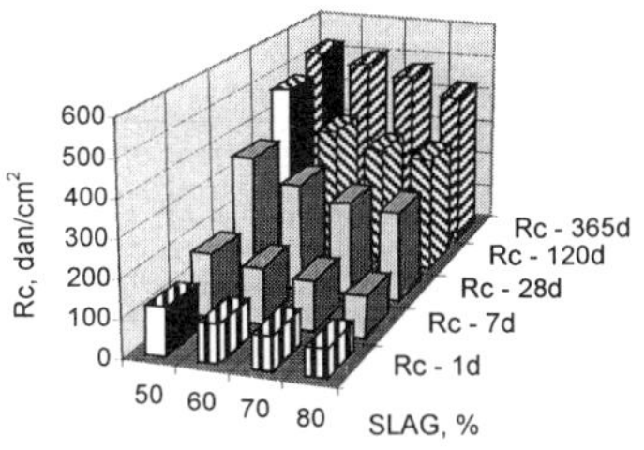

Figure 3 Strength evolution of a binding medium activated by NaOH of different percentages of slag

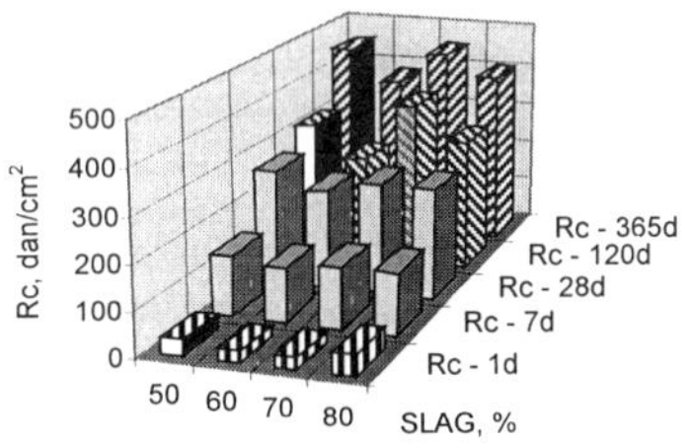

Figure 4 Natural hardening of binding medium activated by Na_2CO_3 function of different % of slag

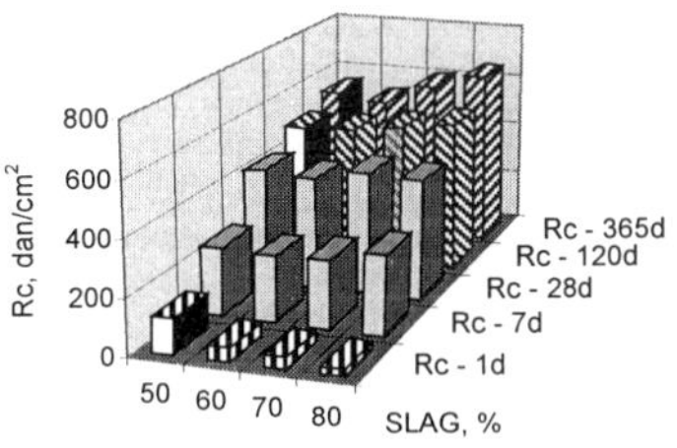

Figure 5 Natural; hardening of a binding medium activated by Na_2SO_4 function of different % of slag

When adding Na_2CO_3 and Na_2SO_4 with greater percentage of slag (70 and 80%), the strength increases, as shown in figure 4 and 5. This can be explained by the positive effect of the used activation agents, i.e. they permit reactions with the slag.

For an equal percentage of C.P.A. cement and slag i.e. the binding medium has 50 % of cement and 50% of slag, the lime that is freed from the C.P.A. cement during the hydration process is sufficient to permit reactions and gives an important mechanical strength. Beyond 70 % of slag an activation agent is necessary for reactions.

When adding the activation agent NaOH, the strengths are very significant during the first days with respect to non-activated binder, and activated binder by Na_2SO_4 and Na_2CO_3. This can help when performing prefabrication of elements.

Effect of the activation agent's density on the strength of the binding medium

For this experiment, three different compositions of binding medium have been chosen; 50%, 70% and 80% of slag, using only one activation agent (NaOH) with densities $d_1 = 1.20$ g/cm^3 and $d_2 = 1.25$ g/cm^3. Tests are carried out on a mortar (1/3).

The evolution of compressive strength, flexion and tension by flexion is represented on the Figures 6, 7 and 8.

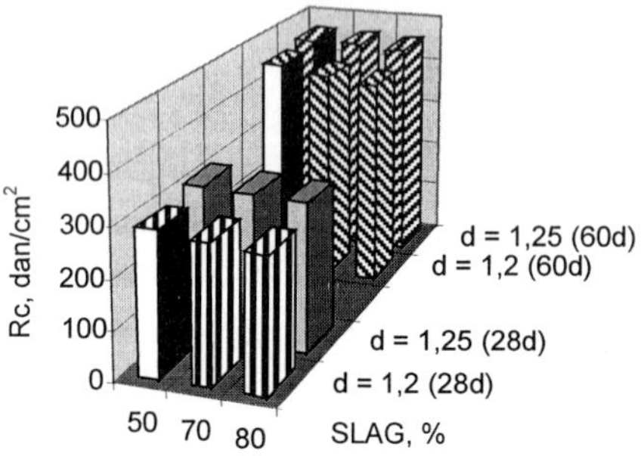

Figure 6 Compressive strength evolution function of slag percentage activation agent density (NaOH) . (Natural hardening)

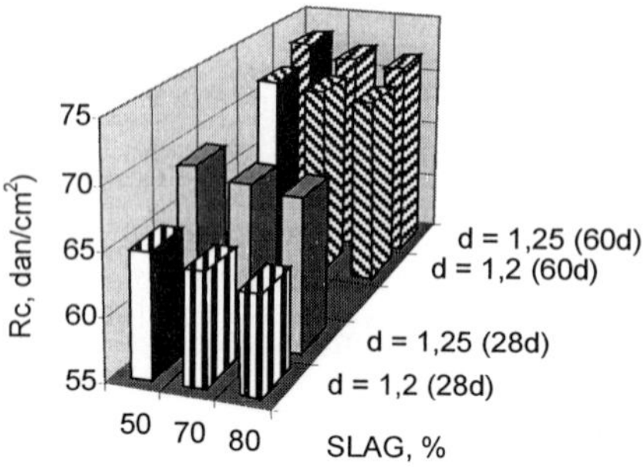

Figure 7 Flexural strength evolution function of slag percentage and activation agent density (NaOH) . (Natural hardening)

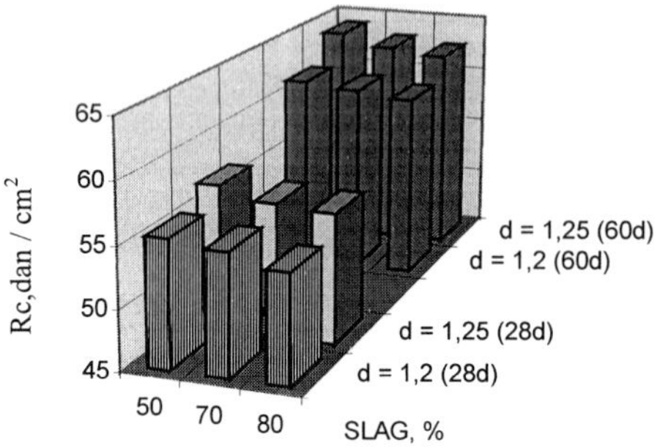

Figure 8 Tensile strength by flexion evolution function of slag percentage and activation agent (NaOH)

Figures 6, 7 and 8 show an increase in the strengths (compressive strength, flexural strength and tensile strength) when increasing the activation agent density, either during 28 days or 60 days of natural hardening.

The NaOH density increasing, causes the concentration of the solution, which is favorable for the rapid hydration process of the slag and hence the strength evolution.

The different strength values of activated mortar are important for the different percentages of slag.

Study of Physical Characteristics of the Activated Binding Medium

For this analysis, only one sample has been studied; a binding medium constituted of 20% of cement and 80% of slag with only one activation agent (NaOH).

Setting of activated binding medium

The setting of slag is usually slow; addition of activation agent and the fine grinding of cement can better improve the time setting. Table 5 shows that the use of activation agent NaOH with density equal to 1.18 g/cm^3 can improve neatly the setting time.

Table 5 Setting Time of Activated Binding Medium

	BIND. MEDIU.	T . C	WAT. QUAN. ML	INITIAL SET	FINAL SET
non activated	20% C.P.A 80% slag	25	30.5	2H.50min.	5H.45 min
activated	20% C.P.A 80% slag	25	29	0H.20 min.	1H.50 min

The study of Figure 9, shows that the increasing of the percentage of slag can negatively affect the time setting process and this is due to the slow hydration of non-activated slag cement.

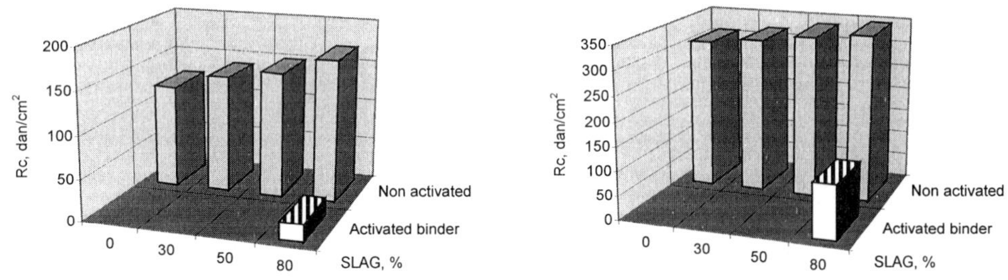

Initial set Final set

Figure 9 Setting time function of the slag percentage and the effect of the activation agent

Expansion of the Activated Binding Medium

The expansion is measured when the sample is hot to obtain significant results. It is noticed that the expansion of slag cement diminishes when rising the percentage of the slag, which permits to conclude that the cement presents volume stability. The activation agent does not influence the expansion.

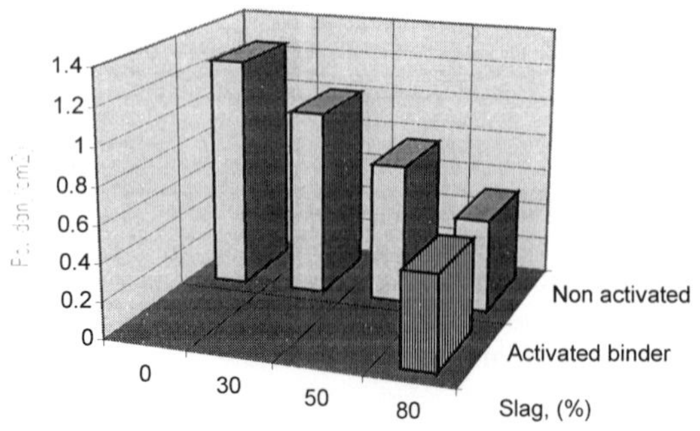

Figure 10 Expansion function of slag percentage

Shrinkage Test

When cement starts hardening, it is usually accompanied with shortening phenomenon. Cement hardening, when immersed in water diminishes the shrinkage phenomenon. Certain type of cement is more sensible than other to desiccation, mainly slag-based cement.

Two different compositions of cement samples have been tested;
Shrinkage tests have been conducted on prismatic samples 4*4*16 cm, using caustic soda of density equal to 1.25 for activated samples.
- 50% of CPA and 50% of non-activated slag, and activated slag (50 and 50b).
- 20% of CPA and 80% of non-activated slag, and activated slag (20 and 80b).
The obtained results are shown in Figure 11.

CONCLUSIONS

Following the results obtained, it can be said that the activation of slag improves the physico-mechanical characteristics of slag- cement.

In order to perform economy when using OPC (C.P.A.), it is possible to replace the cement par 50 % of slag without activation. If the quantity of slag is increased to 80%, then the activation agent is most important for an acceptable strength.

Slag activation by (NaOH) improves the time setting of slag-cement, and gives notable strength earlier, which can help when producing prefabrication elements.

When using the activation agents, the mechanical strength of slag- cement increases with 10% to 60% of the strength of non-activated slag-cement.

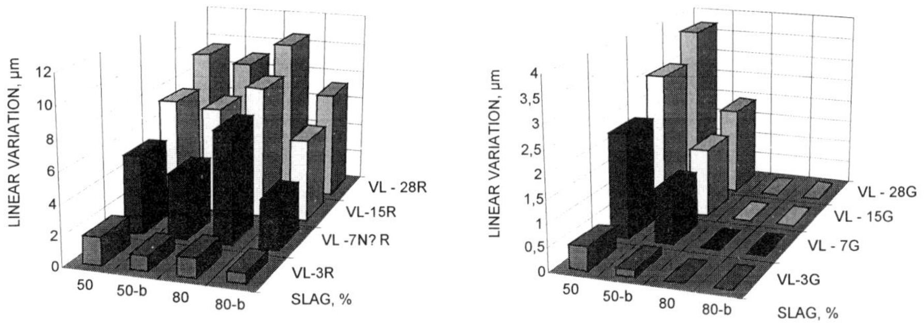

Non-Immersed   Immersed

Figure 11 Linear variation of cement function of time and slag percentage

REFERENCES

1. MICHEL VENANT. La pratique des ciments mortiers et bétons. Editions du Moniteur, Paris, 2ème édition, T1, 1989, 282 pp.

2. MEZGHICHE BOUZIDI. Technologie des bétons a laitier basique pour les conditions de la Republique Algerienne. Kiev, 1989, 181 pp.

3. YUAN RUNZHANG, GAO QIONGYINE AND WYANG SHYKI. Study on structure and latent hydraulic activity of slag and its activation mechanisms. Silicates industrials, 1988, N° 3, pp 55 – 59.

4. RONNY ANDERSSON AND HANS ERIK GRAN. Proprieties of alkali activated slag-concrete. Nordic Concrete Research, 1987 pp 7 – 18.

5. JAN BIJEN. Blast furnace slag cement for durable marine structures. Association of the Netherlands Cement Industry, The Netherlands 1996, 62 pp.

6. MEZGHICHE BOUZIDI. Résistance et deformabilite des bétons basiques, Premier séminaire de génie civil Biskra Aligeria décembre 1996, pp 86 – 93.

THE LIME-ASH BINDER STRUCTURE FORMATION

M S Garkavi

Magnitogorsk State Technical University

N B Glazatova

Magnitogorsk Cement-Refractory Factory

Russia

ABSTRACT. In the given work some features of the hardening of the lime-ash binder are considered. Research the hardening of the lime-ash binder, containing up to 80% of ashes has shown, that the process of structure formation in it's similar to the Portland cement and proceeds the one step mechanism. The lime-ash binder, containing 90% of ashes has two extremes on a curve modified from the amplitude of an electrical signal, indicating a multistage structure formation. Lime-ash binder with multistage structure formation gives a stone of great strength. The criteria of multistage structure formation development are established in mixed binding.

Keywords: Mixed binding, Structure formation, Instantaneous potency of structure formation, Energy of activation.

Professor M S Garkavi is the manager of the faculty of building materials of Magnitogorsk State Technical University and an expert in the field of the thermodynamic analysis of hardening of binding substances and management of processes of structure formation,.

Mrs N B Glazatova is engineer-contributor of Magnitogorsk Cement-Refractory Factory and an expert in technology of mixed cements and their hardening processes.

INTRODUCTION

Application of ashes in quality the component of the mixed cements has received recently broad distribution. The use of lime is well known for activation of a response of hydration of a glassy phase of ashes. However the hardening of lime-ash binders is investigated unsufficiently, that is the constraining factor of their broad application in building practice. The hardening of binding substances is stipulated the course of two interacting processes - hydration and structure formation. These both processes closely are connected one another, i.e. the shaping of an artificial stone is determined by a kinetics and quantitative performances of the indicated processes, their mutual influence against each other.

The present work is devoted to research of process structure formation during hardening of lime-ash binder.

THEORETICAL WORK

During hardening of polymineral binding substances the chemical responses their hydration conjugate with the same process of structure formation having a physical nature. If the binding substance represents a mechanical mixture of components (mixed cement), which have various hydraulic activity, the interaction of their responses hydration with process of structure formation can be divided in time. In this case in a hardening system the origin multistage structure formation is possible. Multistage structure formation means cyclicity in emerging of separate structural condition of a knitting system, i.e there can be interparticle contacts of the same nature periodically take place, for example, coagulation or coagulation-condensation. The number of stages in structure formation depends on a ratio hydraulic activities of components and their amount in mixed binding.

The contribution of components mixed binding in shaping of contacts of the certain type and in development of process of structure formation can be evaluated on magnitude of an instantaneous potency of structure formation W_η, which represents a product of a constant of a velocity of process k_η on magnitude of apparent energy of activation $E_{a\eta}$, i.e. $W_\eta = k_\eta * E_{a\eta}$ [1]. When multistage structure formation during hardening of the mixed cement the following correlation is executed

$$\mathbf{W_{\eta mc} = W\eta a + W\eta n} \qquad (1)$$

Where Wηmc, Wηa, Wηn - instantaneous potencies of structure formation of the mixed cement, active component and microfilling accordingly.

The correlation (1) means, that sequentially participate in shaping of a structure of artificial stone components of the mixed cement. The initial contribution in this process introduces component with the greater instantaneous potency of structure formation. The character of structure formation of the mixed cement predetermines it physical-mechanical and service properties, therefore, allows to define it's rational structure and area of application.

MATERIALS AND TECHNIQUE OF A RESEARCH

In the present work is investigated hardening of lime-ash binder with a specific surface 500 m^2/kg. It is obtained by grinding in a vibrational mill lime, ashes and plaster stone. The content of lime in composition mixed binding was changed from 10 to 50%. The fly-ash has the following chemical composition(%): CaO - 5.02; Al2O3 - 26.04; SiO2 - 53.12; Fe2O3 - 4.79; MgO - 1.14; SO3 - 0.27.

From indicated mixed binding was prepared paste with water-binding by the relation equal 0,5. The structure formation was studied with the electrophysical method [2].

EXPERIMENTAL OUTCOMES AND THEIR ANALYSIS

The curves modification of amplitude of an electrical signal originating during hardening of lime-ash binders containing 50 and 90% of fly-ashes are represented in Figure 1. From represented datas follows, that on a curve of a modification of an electrical signal during hardening of lime-ash binding, containing 90 % of ashes are two extremums. It testifies to multigradualness structure formation in the given binding system. The account of an instantaneous potency of structure formation on a correlation (1) has shown, that the given parity is executed for lime-ash binding, containing 90 % of ashes (Figure 2).

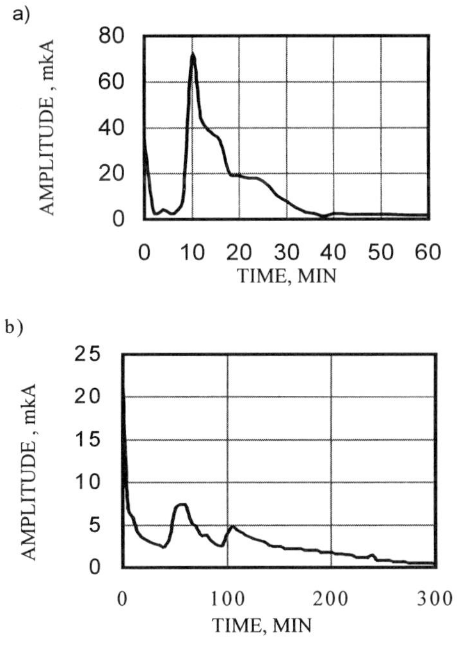

Figure 1 Modification of amplitude of an electrical signal during hardening of lime-ash binder at 20°C a) content of ashes 50 %; b) content of ashes 90%

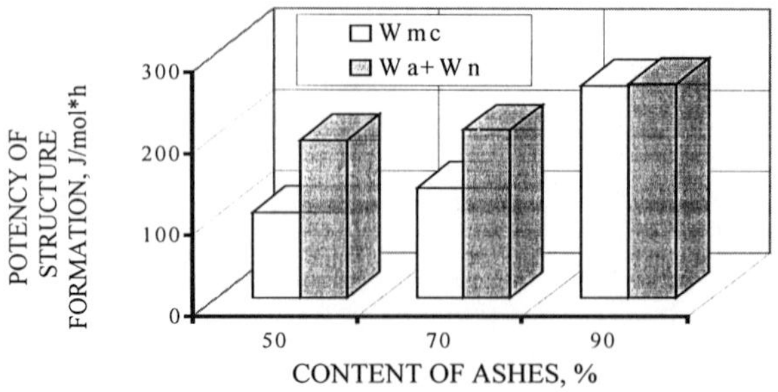

Figure 2 Instantaneous potency of structure formation during hardening of lime-ash binders

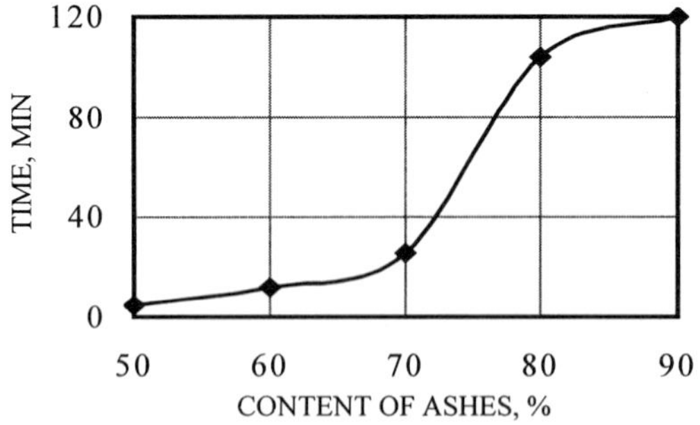

Figure 3 Duration of coagulation-condensation structure during hardening lime-ash binding

The emerging of the second extremum on a curve of a modification of amplitude of an electrical signal (Figure 1, b) is stipulated intensification of process hydration of lime-ash binding for the account puzzolana reaction of ashes. This increases a duration of period of existence in a hardening system a transitional structure with coagulation-condensation contacts (Figure 3). With a small content of ashes in lime-ash binding this response develops practically simultaneously with hydration of lime. In this case process structure formation develops on the single-stage mechanism. With of magnification of the content of ashes for accumulation necessary amount Ca $(OH)_2$ for development the puzzolana responses it's need the certain space of time, as the apparent energy of activation hydration of ashes in 1.46 of time exceeds energy of activation hydration of lime. It results in violation of synchronism of interaction of the indicated responses with process structure formation, as determines development of the latter on the multistage mechanism and it results in magnification of duration of existence of an intermediate coagulation-condensation structure. The magnification of period of existence this structure promotes more deep reorganization preceding of a coagulation structure, as around of particles of ash are present spherical pores, filled in a liquid phase and hindering to rapproachement of particles. The magnification of an amount of ashes, which particles play a role of centres of crystallization hydrating new forming, promotes filling indicated pores and shaping of a strong condensation structure of an artificial stone (Figure 4).

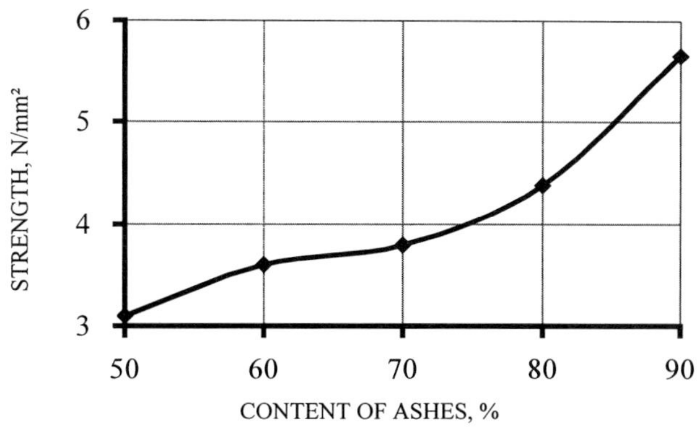

Figure 4 Strength of lime-ash binding at 28 dayS

The magnification of strength from lime-ash binding stone with multistage structure formation is stipulated by the stable condensation structure formation with adhesional contacts of high strength.

CONCLUDING REMARKS

In outcome of conducted researches is established, that a premise for multistage structure formation in mixed binding is the essential difference in apparent energy of activationof hydration of their components.

The criterion of development multistage structure formation in mixed binding is established - instantaneous potency of process structure formation.

During multistage structure formation lime-ash binding happens the deep reorganization of a coagulation structure. It promotes shaping of the thermodynamicaly stable condensation structure with strong adhesional contacts.

REFERENCES

1. GARKAVI M S, BELYCH V T. Multistage structure formation in binding systems. Modern problems of building materials science, Samara, 1995, c.77-79

2. GARKAVI M S, DOLDZENKOV A V, ZALHAROV A J. Electrophysical Control of Hardening Cement. Inetrnational Conference " CONCRETE - 2000 ", Dundee, 1993.

MATHEMATICAL MODEL FOR THE STRENGTH BEHAVIOUR OF ALKALI ACTIVATED BLAST FURNACE SLAG CEMENTS

F Puertas

A Fernández-Jiménez

Institute of Construction Materials "Eduardo Torroja"

J G Palomo

Escuela Universitaria de Arquitectura Técnica

Spain

ABSTRACT. The present work has been carried out for the purpose of determining a mathematical model of strength behaviour of alkali-activated blast furnace slags. The alkaline solution used was NaOH. The effect of several factors has been investigated: the curing temperature (25°C and 45°C), the fineness of slag (450 and 900 m^2/kg) and the concentration of alkali solution (4, 3, and 2,4% of Na_2O wt. with respect to the slag total weight). The results considered were flexural and compressive strengths to 3, 7, 28, 90 and 180 days.

The mathematical model enable to determine the optimum conditions of work and the activity coefficients of the different factors studied and their interactions. Also the correlation between flexural and compressive strength at the same time has been estimated.

Keywords: Alkali, Activation, Blastfurnace slag, Strength, Mathematical model, Analysis of variance.

Dr F Puertas is senior researcher of CSIC, in the Institute of Construction Materials "Eduardo Torroja", Madrid, Spain. She is specialised in clinker formation and hydration. She is experienced in microestructural studies of concrete. She is Director the scientific journal "Materiales de Construcción".

J G Palomo is a lecturer of mathematics in Escuela Universitaria de Arquitectura Técnica (UPM), Madrid. He has experience in experiments design and the application of these theories to the materials science.

A Fernández-Jiménez got a degree in chemistry. She has been working in the in the Institute of Construction Materials "Eduardo Torroja", CSIC, Madrid, Spain, for some years. She has experience in formation and hydration of alkaline cement.

INTRODUCTION

Alkaline cements based on alkaline activation of slags (AAS) initially were developed in the 50s by Glukhovsky [1] in Kiev, Ukraine. During the last decades their study and utilization has spread up all over Europe and the U.S.A. giving birth to a number of publications.

The manufacture of these cements based on AAS has an important impact both from an ecological point of view for opening a way out to industry by-products, and from an economical point of view since the cost of energy production is lower than the OPC. Moreover these AAS cements have also a series of technological advantages compared to OPC such as: development of faster and stronger mechanical strengths, less hydration heat, better resistance to chemical attacks, better behaviour regarding carbonation, higher resistance to aggregate-matrix interface, stronger resistance to ice-thaw cycles... [2-4]. However it is necessary to indicate that they also present some problems like excessively fast settings, high shrinkage and, consequently formation of microcracks, exposure to expansive processes due to aggregate-alkali reaction, and more formation of efflorescences.

The main factors affecting the development of mechanical strengths in AAS mortars and concretes are well known [5]. These factors are: nature of the alkaline activator, concentration of the activator, specific surface of the slag, curing temperature, alkaline activator/slag ratio... Until now the influence of these factors has been studied independently, reason why the objective of the present work has been to study the joint effect of some of these factors and interactions, establishing at the same time a mathematical model allowing to determine the rate of influence the factors considered.

MATERIALS AND METHODS

For this purpose, the alkaline activation of a blast furnace granulated slag with NaOH solution was carried out following the whole factorial design (3^1 x 2^2) for each age of the study. The follow factors were considered: specific surface of the slag (900 and 450 m^2/kg), curing temperature (25°C and 45°C) and activator concentration (4, 3 and 2.4% of Na_2O wt. with respect to the slag total weight).

Table 1 Chemical composition of blast furnace slag (% wt.)

NAME OF SLAG AREA SURFACE	SiO_2	CaO	Al_2O_3	MgO	Fe_2O_3	SO_3	S^{-2}	L.O.I.	I.R.	I.B.(F_3)
E-M 450 m^2/kg	35.5	41.45	12.15	8.34	1.01	2.47	0.92	-	0.21	1.75
E-F 900 m^2/kg	34.72	41.05	11.87	8.24	0.44	2.43	0.83	.10	0.29	1.76

The chemical composition of Spanish blast furnace slags used in this work are shown in Table 1. The mortars were prepared according to the Spanish standard UNE-80-101-88. The size of the prisms was 4x4x16 cm. The aggregate / slag ratio used was = 2 / 1 and the alkaline solution / slag ratio was equal to 0.51 for the 450 m^2/kg slag and 0.61 for the 900 m^2/kg slag.

The samples were cured at 25°C and 45°C for 20 hours, then they were kept at ambient temperature in a humid chamber (98%) until the testing day. The mechanical strengths (flexural and compressive) were measured at 3, 7, 28, 90 and 180 days.

The significance of the factors considered has been developed by applying the analysis of variance to the experimental results [6, 7]. All contrasts of the hypothesis were carried within at the 95% confidence level therefore, all factors whose significance level is > 0.05 are stadistically significant. Once the significance of the factors was known, the coefficient values of the models were calculated by minimal squares.

The mathematical model used is reckoned by Equation (1) [8]:

$$y_{ijk} = \mu + A_i + B_j + C_k + (AB)_{ij} + (AC)_{ik} + (BC)_{jk} + \varepsilon_{ijk} \qquad (1)$$

where:

y_{ijk} is the value of the response variable reckoned at levels i, j and k.
μ is the general average of the value.
A_i, B_j, C_k, represent the effects of the three factors considered.
$(AB)_{ij}$, $(AC)_{ik}$, $(BC)_{jk}$ represent the effect of binary interactions between the three factors.
ε_{ijk} represent the aleatoric errors made in measurements, submitted to the hypothesis that $\varepsilon_{ijk} \sim N(o, \sigma)$.
σ has been calculation using the freedom degrees disponible when considering the interactions the order higher two insignificant.

In Table 2 factors considered in this work and the levels connected with each factor are shown. Sign (+) has been used to indicate the higher level and (-) for the lower level. Factor C (activator concentration) defined at three levels, consider (0) as intermediate level.

Table 2 Factors considered in the process of alkaline activation

FACTOR	DEFINITION	VALUE		
A	Specific Surface of Slag (m^2/kg)	450 X_A=(-1)		900 X_A=(+1)
B	Curing Temperature (°C)	25 X_B=(-1)		45 X_B=(+1)
C	Activator Concentration (% Na_2O with respect to the weight of slag)	2.4 Z_C =(-1)	3 Z_C=(0)	4 Z_C=(+1)

Equation (1) now can be written more easily defining a variable indicating each factor, as follow (2).

$$R = R = \mu + A_1 X_A + B_1 X_B + Z_{NC} + \delta_{NAB} + \delta_{NAC} + \delta_{NBC} + \varepsilon \quad (2)$$

Variables X_A and X_B have values (-) and (+), depending on factors A and B being at level (-1) and (+1) respectively. In this design A_1 and B_1 represent the expected increase in responses related to the mean level when factors are established at level (+). By agreement in factorial designs (factors A and B, specific surface and temperature, respectively), there is only one parameter per factor; the effect of each factor is defined as the expected increase of the response when this factor move from (-) to (+). Factor C (defined at three levels) and the different interactions (AB, AC and BC) have different values for each level.

RESULTS AND DISCUSSION

In Figures 1 and 2 are presented, respectively, flexural and compressive strength values of the different mortars studied, and in Tables 3 and 4, equations of the model obtained in each age of the study.

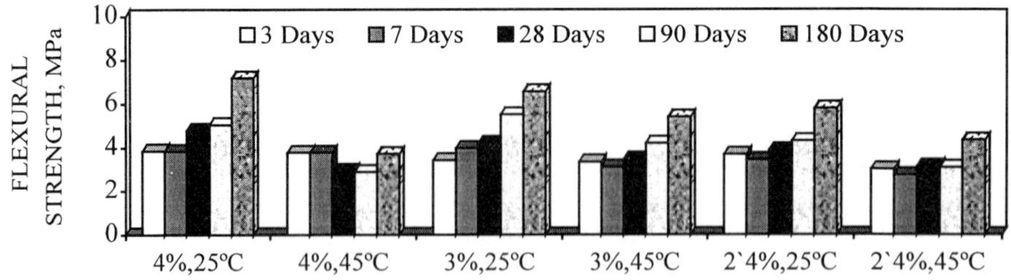

a) Slag of 450m2/kg

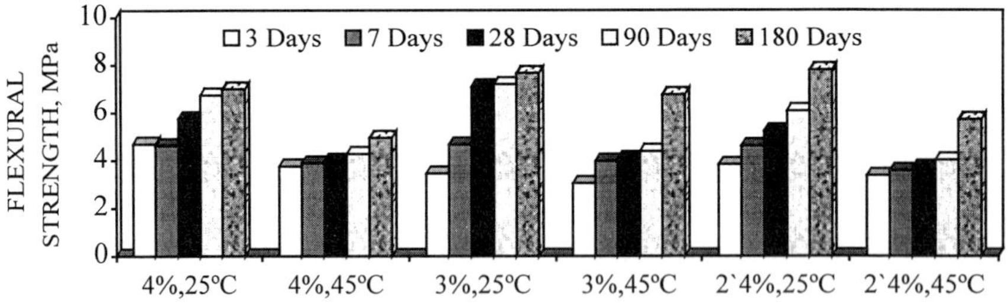

b) Slag of 900m2/kg

Figure 1 Flexural strengths, a) E-M, 450 m^2/kg, b) E-F, 900 m^2/kg

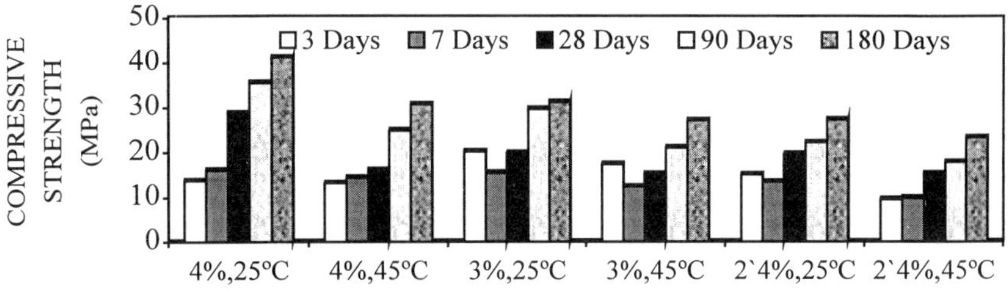

a) Slag 450 m2/kg

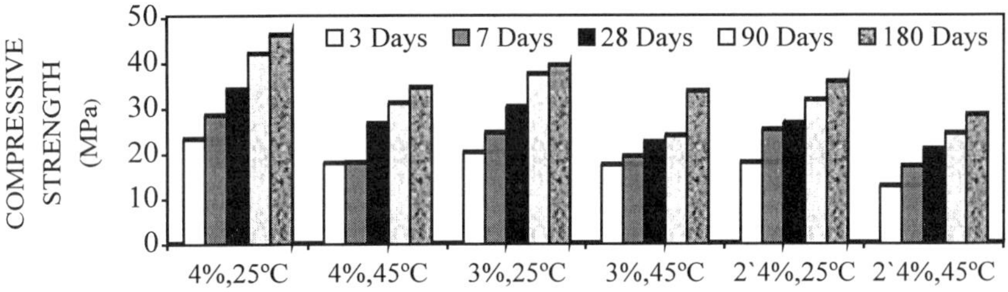

b) Slag 900 m2/kg

Figure 2 Compressive strengths, a) E-M, 450 m^2/kg, b) E-F, 900 m^2/kg

Table 3 Mathematical equations of flexural strengths

TIME, days	MATHEMATICAL EQUATIONS OF FLEXURAL STRENGTHS
3	$R_3^F = 3.61 + 0.08\ X_A + 0.21\ X_B + Z_{3C} + \varepsilon$ $(\sigma \cong 0.267)$
7	$R_7^F = 3.86 + 0.37\ X_A - 0.33\ X_B + Z_{7C} + \varepsilon$ $(\sigma \cong 0.21)$
28	$R_{28}^F = 4.36 + 0.62\ X_A - 0.80\ X_B + Z_{28C} + \varepsilon$ $(\sigma \cong 0.59)$
90	$R_{90}^F = 4.92 + 0.52\ X_A - 1.025\ X_B + Z_{90C} + \varepsilon$ $(\sigma \cong 0.53)$
180	$R_{180}^F = 6.07 + 0.57\ X_A - 0.92\ X_B + Z_{180C} + \varepsilon$ $(\sigma \cong 0.51)$

Table 4 Mathematical equations of compressive strengths

TIME, days	MATHEMATICAL EQUATIONS OF COMPRESSIVE STRENGTHS
3	$R_3^C = 16.50 + 1.65\ X_A - 1.85X_B + Z_{3C} + \varepsilon$ $(\sigma \cong 1.43)$
7	$R_7^C = 17.82 + 4.21\ X_A - 2.66\ X_B + Z_{7C} + \varepsilon$ $(\sigma \cong 1.68)$
28	$R_{28}^C = 23.21 + 3.53\ X_A - 3.27\ X_B + Z_{28C} + \varepsilon$ $(\sigma \cong 1.23)$
90	$R_{90}^C = 28.75 + 3.57\ X_A - 4.29\ X_B + Z_{90C} + \varepsilon$ $(\sigma \cong 2.28)$
180	$R_{180}^C = 33.18 + 3.03\ X_A - 3.60\ X_B + Z_{180C} + \delta_{180\ BC}^{C} + \varepsilon$ $(\sigma \cong 0.56)$

From the analysis of these figures and considering the equations of the model appearing in Tables 3 and 4, it can be inferred that:

a) Mean mechanical strengths increase as the reaction time increases. This effects appears when the independent term of the model equations augments as the reaction time augments.

b) The three main factors considered in this work do not have, statistically, the same level of relevance. The statistical analysis shows that the most important factor is the curing temperature, followed by the specific surface of the slag. Activator concentration does not appear relevant to flexural; to compressive it is only significant at ages over 28 days. However, as observed in Figures 1 and 2, strengths augment as activator concentration increase.

Curing temperature is, statistically, the most significant factor as for the development of mechanical strengths at all ages of the study (except at 3 days where flexural data are not significant). This makes the factor to generally show, in absolute value, the highest coefficient among the equations of the model. This coefficient at all stages of the study has a negative sign (except for 3 days to flexural). This indicates that mixes cured at higher temperature (T° = 45°C) have developed lower strengths than those cured at 25°C.

In earlier studies [9] it has been proved that an increase in the curing temperature accelerates the activation process of the slag. This effect is quite noteworthy at early reaction time (10), when an increase of strengths happens. However, as the reaction time goes by the temperature increase has a negative effect, causing a decrease of the final strength values. Similar results have been also found by other authors [5]. This phenomenon can be explained since there is a bigger amount of reaction products, at early ages, in those mixes cured at higher temperature, with the resulting consistency of the paste and modified structure. So when augmenting the reaction time, diffusion processes are more difficult to develop, and harder for the late reactions to speed up.

Considering the experimental conditions in which the work was carried out, where the first measures were taken at 3 days, the positive effect of the curing temperature is not any more significant. This justifies that in the practice, at ages over 3 days, mortars cured at 45°C develop lesser mechanical resistances than those cured at 25°C.

The second relevant factor is the specific surface of the slag. This factor is statistically relevant at ages over three days. When the specific surface augments from 450 to 900 m^2/kg a slight increase on flexural as much as on compressive strength happens. This fact is shown for the positive sign that precede to X_A in the equations of the model, Table 5.

Mixes with wider slag surface needed more mixing liquid content to reach plasticity conditions similar to those with smaller surface. This lead to guess that in equal conditions (same solution/slag ratio) differences would be bigger, obtaining higher strength with the wider specific surface slag. These results coincide with those expressed by other authors (5) since the increase of the slag specific surface of the reacting slag favours the increase of reactivity.

In spite of the increase of the mechanical strengths values resulting from increasing the specific surface of the slag from 450 to 900 m^2/kg, it seems it is not important enough as to justify the high energetic and economic cost it implies.

The concentration of the activator factor, as said, is not significant to flexotration, and to compression it is only significant at 28 days. However, since it is a main factor, it shall be considered in the model. The parameter has three different levels (see Table 2) which makes the value of the coefficients of the model to have three different values (see Table 5).

Table 5 Values of the activator coefficients, Zc, to flexural and compressive strengths

FLEXO	-1	0	+1	COMP	-1	0	+1
Z_{3C}	-0.12	-0.280	0.41	Z_{3C}	-2.74	2.260	0.48
Z_{7C}	-0.25	0.084	0.18	Z_{7C}	-1.44	0.075	1.36
Z_{28C}	-0.36	0.350	0.01	Z_{28C}	-2.64	-1.33	3.98
Z_{90C}	-0.44	0.400	0.04	Z_{90C}	-4.78	-0.79	5.57
Z_{180C}	-0.16	0.520	-0.360	Z_{180C}	-4.53	-0.37	4.90

Values of the activator coefficients (Table 5) are noticeably inferior than the coefficients of the other main factors. This indicates that the variation in activator concentration, within the limits studied, has generally an effect less significant than other factors in the development of mechanical strengths.

However, analysing the results of Figures. 1 and 2, an increase of strengths is observed as the activator concentration augments, despite this factor does not result statistically significant. These discrepancies can be due to an overlapping effect of the main factors. This fact necessary to carry out a different design experiment in order to value more exactly the role of the activator concentration in this type of mortars.

The statistical analysis of the experimental results has also revealed that of the three possible binary interactions (specific surface of slag-curing temperature, specific surface of slag-activator concentration, curing temperature-activator concentration) none of them are relevant, excepting in the results obtained from compressive strengths at 180 days. The significant interaction in compression at 180 days (see Table 6) is curing temperature-activator concentration. Experimental errors can be the cause for the interaction to be significant to compressive, but not to flexural strengths.

Table 6 Coefficient values of curing temperature-activator concentration interaction to compression at 180 days

INTER.	-1, -1	-1, 0	-1, 1	1,-1	1,0	1,1
$\delta_{180}{}^{C}{}_{BC}$	-0.77	-1.158	1.92	0.77	1.158	-1.92

Correlation Between Flexural and Compressive Strengths Values

The relationship of dependency between experimental flexural and compressive strengths values at the same age of the study has been calculated using a model of simple linear regression. The correlations statistically significant were reckoned first, with a degree of reliance of 95%, (P-value < 0.05), see Table 7. Among the variables studied the equation of the model of simple regression was later reckoned by square minima. Linear equations obtained are of the type [3]. Where F is the flexural strength value at N days and C the compressive strength value.

$$C_N = A + B\,F_N \qquad [3]$$

Results obtained, see Table 7, indicate that there is, statistically, a significant ratio at almost all ages of the study. This ratio is a more important one at 7, 28 and 90 days, where P-values <<0.05 are obtained and also good correlation coefficients, (0.93, 0.82 and 0.81 respectively). At 180 days correlation is noticeably smaller and ranges around 0.66. At three days, statistically, no significant ratios with 95 % reliance were obtained, but just with 90 % reliance.

The bad correlation obtained at 3 days is mainly related with temperature. Again, at three days flexotration is not a significant factor, but compression is.

Table 7 Linear equations of correlation between flexotraction and compression strengths and their correlation coefficient

TIME	LINEAL EQUATION	P-VALUE OF MODEL	COEF. CORRELATION
3	$C_{3d} = 0.91187 + 4.31873\ F_{3d}$	0.0960	0.502369
7	$C_{7d} = -16{,}0556 + 8{,}77769\ F_{7d}$	0.0000	0.930648
28	$C_{28d} = 5.12424 + 4.14350\ F_{28d}$	0.0009	0.825596
90	$C_{90d} = 5.86353 + 4.64768\ F_{90d}$	0.0013	0.812721
180	$C_{180d} = 12.7800 + 3.35953\ F_{180d}$	0.0182	0.665211

CONCLUSIONS

Supported by the statistical analysis of the results obtained, and applying the analysis method of variance with 95% reliance limit for hypothesis contrasts, it is inferred :

a) The models obtained describe with precision the mechanical strengths behaviour of the mortars of slag at the different ages of the study.

b) Mean value of the mechanical strengths as the reaction progresses with the reaction time.

c) Of the three main values studied, the priority range from a statistical standpoint is :

Curing temperature > specific surface of slag >>> activator concentration.

d) None of the possible interactions (specific surface of slag-curing temperature, specific surface of slag-activator concentration, curing temperature-activator concentration) is statistically significant, excepting results obtained at 180 days.

The increase of curing temperature does not favour the increase of strength. Initially, reaction processes are accelerated but at time > 3 days a decrease of mechanical strengths occur. In another hand the increase of the specific surface of the slag has a positive effect in the increase of mechanical strengths both to flexural and compressive.

Increasing the activator concentration does not result statistically significant, but when increasing it, an increase of resistances is observed. This suggests that a new experimentation focussing this effect and diminishing experimental errors shall be carried out.

Finally the significant linear correlation existing between flexural and compressive strengths values, has been determined, but not at 3 days. Such ratio allows reckon results of compressive strength, once those to flexural are known and vice versa.

ACKNOWLEDGEMENTS

The authors of the present work wish to thank the DGICYT for funding this research through the project PB94-0048. They thank also J. Lechuga for the blast furnace slag supply.

REFERENCES

1. GLUKHOVSKY, V.D. et all. "High strength slag-alkaline cements". Proc. 7th Int. Congr. On the Chemistry of Cement, Paris, 1980, 3,ppV164-V168.

2. WANG, S .D, PU, X C, SCRIVENER, K L, AND PRATT, P L. "Alkali-activated slag cement and concrete: a review of properties and problems". Adv. In Cem. Res. 1995, 7, No°27, pp93-102.

3. TAILING, B AND BRANDSTER, J. "Present sate and future of alkali-activated slag concretes",. Proc. 3rd Int. Conf. Fly Ash Silica Fume, Slag and Natural Pozzolans in Concrete, Trondheim. 1989, 2, SP114-74, pp1519-1546.

4. PUERTAS, F. "Cementos de escorias activadas alcalinamente: situación actual y prespectivas de futuro". Materiales de Construcción. 1995 , No239, Vol 45, pp53-64.

5. WANG, S D, SCRINENER, K L AND PRATT, P L. "Factors affecting the strength of alkali-activated slag". Cem. Conc. Res. 1994, Vol 245, No 6, pp1033-1043.

6. BOX, G E P, HUNTER, J S. "Statistics for experimenters". Ed. Wiley, New York, 1978.

7. JOHN P W M. "Statistical design and analysis of experiments". Ed. Macmillan, New York, 1971.

8. PEÑA SANCHEZ , D. "Estadística modelos y métodos". Ed. Alianza Editotial, S. A., Madrid, 1989.

9. TALLING, B,. "Effect of curing conditions on alkali-activated slags". 3rd Inter. Conf. Fly-ash, Silica Fume, Slag and Natural Pozzolans in Concrete, Norway,1989, Vol.2, SP114-74, pp1519-1546.

10. FERNANDEZ-JIMENEZ, A, PUERTAS, F. "Alkali-activated slag cements: kinetic studies". Cem. Conc. Res. 1997, Vol 27, No 3, pp359-368.

HIGH- AND LOW-CALCIUM FLY ASH BINDERS AND THEIR DURABILITY

C Freidin

Ben-Gurion University of the Negev

Israel

ABSTRACT. High- and low-calcium Fly Ash Binders (FAB) were prepared by mixing high-calcium oil shale fly ash (HCOSFA) and low-calcium coal fly ash (LCCFA). It was determined that the distinguishing feature of HCOSFA is the high amount of CaO_{free} and SO_3 in form of lime and anhydrite. This fly ash is characterized by high water requirement. A FAB including HCOSFA possesses high water requirement also. The influence of HCOSFA and LCCFA mixture ratio, quantity of mixing water and curing conditions on the compressive strength of FAB were examined. It was fixed that the higher the HCOSFA content as well as mixing water amount the higher the compressive strength. Moist air and water are most favorable conditions for curing and aging processes of FAB. In atmospheric air are three stages in strength change of cured FAB: increase of compressive strength for 1 month, sharp drop after 1 month up to the 3rd-6th month and stabilization or slow-down of loss in strength after 3-6 months of exposure. Duration of the second and third stages depends on HCOSFA content and doesn't depend on curing conditions. The microstructure and composition of the new formations of FAB after curing and long-term exposure in moist air and atmospheric air were studied. It was established the main binding matter of cured HCOSFA is ettringite, but ettringite is unstable in air. This can be a reason for the decrease in strength of FAB and a disintegration of the hardened system in atmospheric air. In the presence of LCCFA in FAB interaction of HCOSFA and LCCFA forms additional amount of stable calcium silicate hydrates (CSH). A positive effect of calcium silicate hydrates CSH on compressive strength and durability of FAB in atmospheric air is observed. It was concluded that high- and low-calcium fly ash binders have enough durability and could be used as cement substitute for building products.

Keywords: Fly ash, Binder, Curing, Durability, Moist air, Water, Atmospheric air, Decay, Ettringite.

Professor Constantin Freidin is Principal Investigator, the Jacob Blaustein Institute for Desert Research, Ben-Gurion University of the Negev, Sede-Boker Campus, Israel. His main research interests include the technology of the inorganic building materials as well as technology of concrete works and methods for control of their quality. He specializes in theoretical and experimental examinations of new cements, concretes and concrete-like products as well as in finding solutions to increase durability of concrete and other building materials, to utilize industrial and energy waste in construction.

INTRODUCTION

Fly ashes are waste of combustion process of pulverized lignite, bituminous subbituminous or anthracite coals and oil shales in power stations. Depending on their chemical and mineralogical composition, they can be divided on two great groups: low-calcium fly ashes and high-calcium fly ashes. While both types posses pozzolanic activity high-calcium fly ash such as oil shale fly ash from Jordan, Greek, Israel and other countries which contain free CaO and SO_3 in the form of lime and anhydrite exhibit hydraulic properties: can set and harden with water without any further additives [1-3]. Studies of high-calcium ashes, formed in fluidized bed combustion process of Israel's oil shales, showed that one of the products of their hydration is ettringite ($3CaO{\cdot}Al_2O_3{\cdot}3CaSO_4{\cdot}31H_20$) [4,5]. Ettringite is stable in an aqueous solution at temperatures up to 90°C and higher [6-8]. But it reacts differently in air. Dehydration and carbonization processes, which can take place in air reduce the amount of ettringite and gypsum, calcium carbonate and alumina gel form in its place [6,7]. In other words the ettringite-containing system can decompose and reduce its strength and consequently its durability.

In our opinion, these negative results of ettringite breakdown could be eliminated by the forming of an additional amount of more stable binder phases in the initial stage of high-calcium fly ash curing. This effect can be obtained by a combined hydration of high-calcium fly ash and low calcium fly ash. Their interaction to each other would give secondary high-strength and more stable calcium silicate hydrates (CSH), which should improve the durability of the hardened system in atmospheric air.

The present paper discussed the development of strength, the composition and microstructure of new formations of high- and low-calcium fly ash binders after curing in various conditions and long-term exposure to moist air and water as well as atmospheric air.

EXPERIMENTAL DETAILS

Materials

High-Calcium Oil Shale Fly Ash (HCOSFA) from the Pama plant, located in the Negev ("Developing Energy Resources LTD", Israel), served as high-calcium fly ash (Table 1).

Table 1 Chemical analysis of fly ashes, % (by weight) [ASTM 114]

	HCOSFA	LCCFA
SiO_2	16.3	45.6
Al_2O_3	5.6	33.5
Fe_2O_3	3.6	3.5
CaO_{total}	52.3	5.4
CaO_{free}	10.1	-
MgO	0.8	1.6
SO_3	10.7	0.4
P_2O_5	2.5	1.6
Na_2O	0.3	0.5
K_2O	0.4	0.3
Loss on Ignition 950°C	5.1	5.9

This plant burns local low-calorie oil shale containing 12-15% of organic matter in a matrix of calcite, quartz, fluorapatite [$Ca_{10}(PO_4)_6F_2$] and minor amounts of other crystalline and amorphous matters [12]. Prior to burning, oil shale is ground by special equipment. Ground oil shale consists of particles, equal and smaller than 4-6 mm in diameter. Most particles (55 – 60%) measure less than 0.8 mm and 20-25% measure less than160 $_{\mu}$k in diameter. Ground oil shale is combusted in a fluidized bed combustor at temperature in the range from 850 to 900∘C. Fuel particles remain in fluidized bed combustion process on average for a few minutes.

As was found by XRD analysis, anhydrite, calcite, lime, quartz, fluorapatite and clinker minerals (β-C_2S, C_3A and C_4AF) are the main crystalline phases of HCOSFA. X-Ray Diffraction patterns of High-Calcium Oil Shale Fly Ash (HCOSFA) and Low-Calcium Coal Fly Ash (LCCFA).These minerals were also discovered in the Israeli HCOSFA by other researchers [3, 4, 9, and 10]. It is important to point to the existence (and especially the co-existence) of three high-temperature clinker phases as well as calcite in such type of high-calcium fly ash. The formation of clinker minerals can be explained as follows: Although the nominal combustion temperature is around 850-900∘C, much hotter zones ("hot spots") with temperature well in excess of 1200∘C exist. This should also have produced melt and left glass in the combustion ash. In its liquid phase clinker, mineral compound reactions are possible. The presence of calcite in fly ash is conditioned to the parameters of oil shale combustion - fluidized bed combustion. Low combustion temperature (850-900∘C) and excess pressure don't allow for the complete calcination of calcite to lime and both minerals are present in HCOSFA. Properties of HCOSFA are showed in Table 2.

Low-Calcium Coal Fly Ash (LCCFA) was used as low-calcium fly ash. The LCCFA was obtained from the coal burning power station in Hadera (Israel). Its chemical composition is presented in Table 1. LCCFA consists of glassy spherical particles which incorporate mullite and quartz crystals recognized by X-Ray Diffraction (XRD) pattern. Results of LCCFA standard tests as defined by American Society for Testing and Materials are summarized in Table 2.

Table 2 Physical properties of fly ashes

	ASTM	HCOSFA	LCCFA
Specific Gravity, kg/m³	C 188	2300	2500
Bulk Specific Gravity, kg/m³	C 29	690	810
Fineness:			
passing the 45$_{\mu}$k (N°.325) sieve, %	C 430	89.1	75.3
specific surface area (on Blaine), cm²/g	C 204	15700	3400
Water requirement, %	C 187	81.7	51.3
Soundness: Autoclave Expansion, %	C 151	0.1	0.03

High- and low-calcium Fly Ash Binders (FAB) were prepared by mixing HCOSFA and LCCFA in a blade paddle mixer for 3-5 minutes. Standard tests by ASTM C187 indicated that FAB, containing from 20 to 80 % HCOSFA have a high water requiremet (from 57.4 to 75.5 %) compared with Ordinary Portland Cement (22-27%). It is to be expected that such high water requirement would give a low strength at testing FAB using normal consistency paste specimens. Similar results were discovered by A. Bentur and others which determined the strength of HCOSFA based binder in cast paste specimens with large quantity mixing water [4]. However their experiments, performed with compacted HCOSFA mortar specimens, showed that obtained strength values of binders are high enough to recommend their application as a cementing material for production of compacted (pressed) building units such as blocks, bricks and tiles [11]. Current research takes this recommendation into account. The durability tests of HCOSFA and LCCFA binders were performed using pressed specimens to simulate their possible utilization as cement substitute for pressed building products.

Natural quartz sand used as a part of the mortar specimens has a specific gravity of 2.61 g/cm^3, a bulk specific gravity of 1470 kg/m^3 and the fineness modulus of 2.83.

Experimental Procedure

Pressed mortar specimens-cylinders of 5.70 cm in height and diameter were used to determine the effect of HCOSFA content and amount of mixing water (MW) as well as curing conditions (moist curing and steam treatment) and exposure conditions (moist air, water and atmospheric air) on the compressive strength of FAB. These cylinders were made from fly ash-sand mortars, containing 60% FAB (mix of HCOSFA+LCCFA) and 40% sand. Specimens containing 36% HCOSFA, 24% LCCFA (Σ=60%), 40% sand plus 25% MW (by solid weight) were prepared to investigate the development of FAB strength in moist air and water. The ratio between HCOSFA and LCCFA in FAB mortar specimens, designed for exposing to atmospheric air after previous moist curing, were varied from 20:80% to 80:20% so as HCOSFA content in mortars varied from 12 to 48%. MW amount ranged from 15 to 25%. Steam curing specimens for testing in atmospheric air were formed from the mortar included 24% HCOSFA, 36% LCCFA, 40% sand plus 25% MW. As a reference, the certain experiments (influence of HCOSFA content on the compressive strength and strength development in atmospheric air) were done using mortar without LCCFA. It contained 60% HCOSFA, 40% sand plus 30% MW.

Pressed paste specimen-disks of 15 mm in height and 57 mm in diameter were designed for investigating the composition and microstructure of FAB after moist air and steam curing as well as after long-term exposure of cured FAB in atmospheric air. FAB involved 60% HCOSFA and 40% LCCFA. Specimens were compacted from mixture with MW of 30%.

All test specimens were molded by pressing on a Carver laboratory hand press at the compaction pressure of 4 MPa.

Test specimens cured and were exposed to various conditions:

- air at constant temperature (20$\pm$3°C) and humidity (R.H.>96%)-moist curing and moist air exposure);
- water at a temperature between 20-23°C-water curing and water exposure;
- atmospheric air at a temperature between 18-23°C and R.H. of 35-60%-atmospheric air exposure);
- steam atmosphere (temperatures of 60°C and 90°C, duration of 8 hours)-steam curing.

The mortar specimens that were designed for testing in atmospheric air or in water previously moist cured for 3 days or 1 day correspondingly. After that they were exposed to atmospheric air (atmospheric air exposure) or were submerged in water (water exposure).After curing and exposure, the mortar specimens were crush-tested to determine compressive strength.
The paste specimens were examined by XRD analysis and Scanning Electron Microscopy (SEM) to find a composition and structure for new formations of the cementing matter of FAB. XRD analysis and SEM were also used for identifying the mineralogical composition of HCOSFA and LCCFA.

RESULTS AND DISCUSSION

Effect of High-Calcium Oil Shale Fly Ash Content and Amount of Mixing Water on Compressive Strength

The dependence of the compressive strength of 28-day moist cured specimens from HCOSFA content and MW are showed in Figure 1.

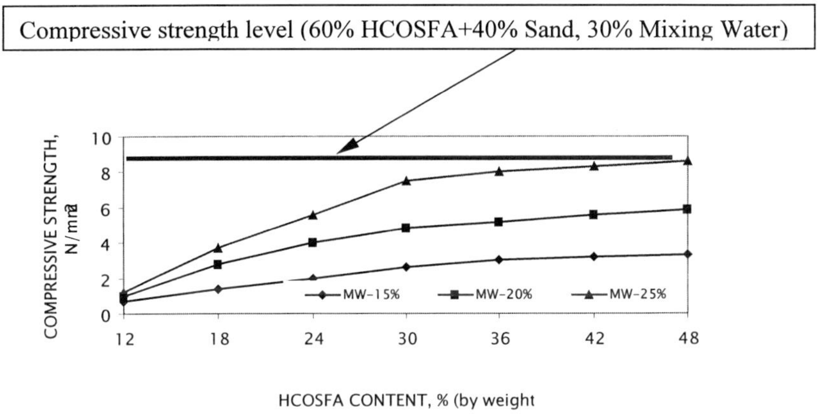

Figure 1 Effect of HCOSFA and MW content on 28-day FAB compressive strength

It can be seen that an increase of HCOSFA content and MW amount improved compressive strength of FAB. Maximum strength can be achieved by increasing HCOSFA content as well as the amount of mixing water. The relationship between HCOSFA content and compressive strength received from specimens with HCOSFA content from 12% to 30% is a straight-line characteristic.

The specimens which were made from 48% HCOSFA, 12% LCCFA and 25% MW had almost the same strength as the specimens made from 60% HCOSFA without any LCCFA. Therefore it can draw the conclusion that LCCFA is involved in the curing processes and strengthens FAB.

Effect of Curing and Exposure Conditions on Durability of Fly Ash Binders

Moist air and water

The typical curves showing changes of specimens compressive strength in moist air and in water for 36 months, are alike (Figure 2).

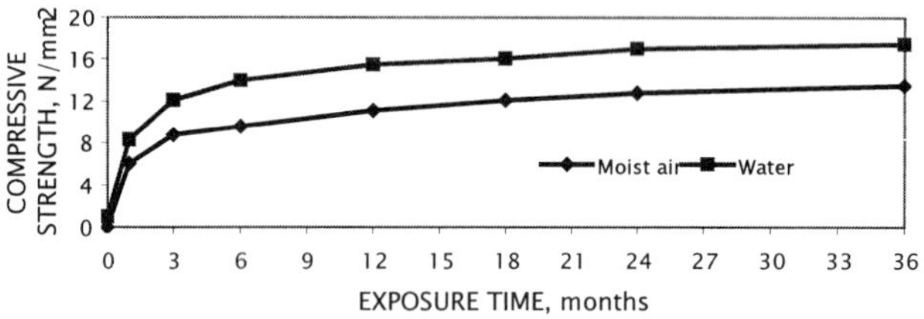

Figure 2 Development of FAB compressive strength in moist air and water

A continuous curing and increasing compressive strength of FAB were observed throughout the experiment. These processes progressed rapidly in the initial period (up to 3 months) and strength reached 60-65% of 36-month values.

Thereafter, strength growth was slowed. After 18 months strength stabilization was evident. It should be noted that curing in water was more intensive than in moist air; compressive strength in water greater was than in moist air.

Atmospheric air

In Figure 3 tests results shown an influence of HCOSFA content on the compressive strength of cured FAB in atmospheric air, are presented. Several conclusions can be drawn.

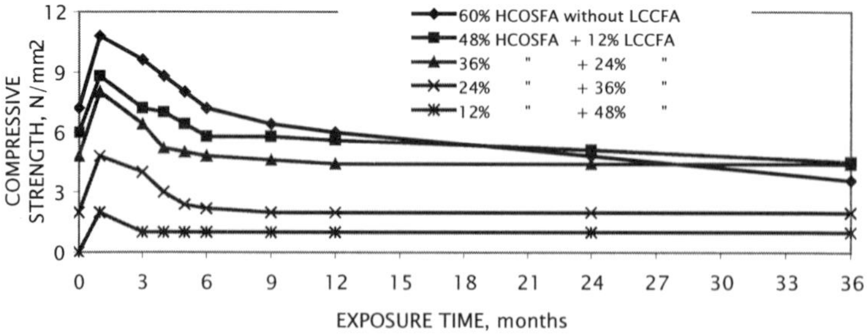

Figure 3 Development of compressive strength for 3-day moist cured specimens with several of HCOSFA content in atmospheric air

There are three clearly distinguishable parts in strength development curves. Corresponding to the first part, which generally takes one month, an increase in strength can be observed. This is caused by loss of water, drying and a gain in compressive strength of the specimens. The second part (from the second to the sixth month) is marked by a sharp drop in strength. The severity of loss depends on the initial strength (compressive strength of specimens after three days of previous moist curing), that in turn is determined by the HCOSFA content. Specimens with low initial strength (12%, 24% HCOSFA) reached the end of decrease in compressive strength sooner (during 2-4 months), while specimens with higher initial strength (36%, 48% HCOSFA) lost their compressive strength more intensively and over a longer period of time (4-6 months). Evidently this deterioration is caused by destructive processes within ettringite. Their duration depends on the initial quantity of ettringite in cured matter. The third part of strength development curves is characterized by stabilization of strength at its initial level (24%, 36% HCOSFA) or at an even higher level (12% HCOSFA). Specimens with a higher HCOSFA content (48%) do not come to full stabilization but the loss of compressive strength only slows down, similar to specimens without LCCFA which serve as reference in this study. Such change in strength points to the fact that the decay of ettringite is not yet exhausted by the end of this experiment and may continue in the future, - further studies will be needed.

Figure 4 shows an effect of curing conditions on FAB strength n atmospheric air. It can be seen that strength development curves have three parts independent from curing conditions.

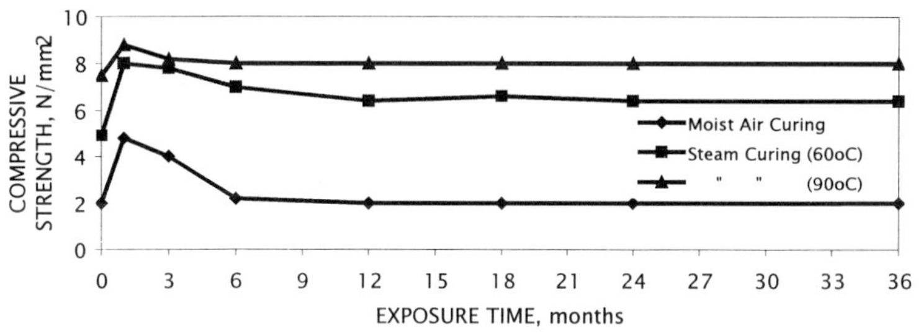

Figure 4 Effect of curing conditions on deof compressive strength of FAB in atmospheric air

The first part displays increase in strength for the first month of specimens exposure to atmospheric air. The second part reflects a loss of strength, which happens due to of destructive processes that occur in cured FAB for next 2-5 m. It is characterized by a sharp decrease in the compressive strength. Strength of specimens cured at steam treatment was reduced significantly less than that of moist cured specimens; the greater the temperature the smaller the decrease.

The last third part starting from 3-6 months shows, that negative operations in cured FAB are retarded. Between 6-12 months, they reach completion and the compressive strength of specimens is settled at a constant value.

The fastest strength stabilization (6 months of air exposure) comes for specimens cured in steam atmosphere at 90°C.

Composition and Structure of Fly Ash Binder After Curing and Exposing in Various Conditions

X-Ray Diffraction Analysis

In Figure 5 XRD patterns of FAB paste specimens after moist and steam curing (A) as well as ettringite segment of patterns after long-term exposure of cured FAB in atmospheric air (B) are presented. The presence of ettringite, quartz, calcite, mullite and fluorapatite peaks are common to all XRD patterns (Figure 5A). Differences are in the following:

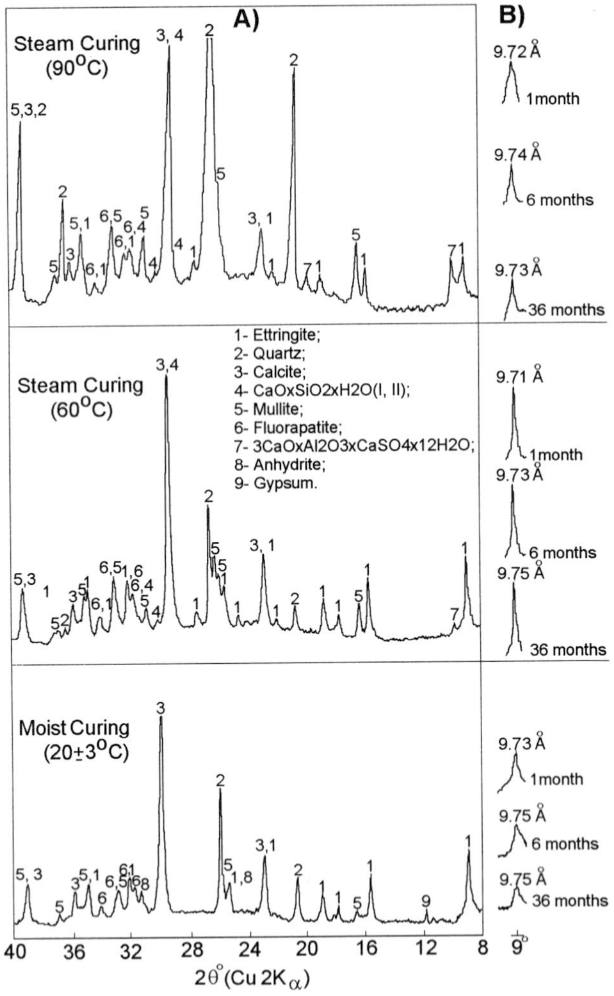

Figure 5 XRD patterns: A)- Full patterns of cured FAB; B)- Ettringite peak segments in patterns of cured FAB after atmospheric air exposure

1. Intensity of XRD pattern peaks for steam cured binder is higher than that for moist cured binder. This could be explained by greater extent of crystallization of hydration products in curing processes proceeding at higher temperature; the greater the temperature of steam curing the higher the intensity of XRD pattern peaks.

2. XRD pattern of moist cured binder has peaks of anhydrite (8) and gypsum (9). The same peaks in XRD patterns of steam cured FAB are absent.

3. XRD pattern of moist cured FAB has no peaks of calcium silicate hydrates (CSH). This may be linked to X-Ray amorphousness of CSH. However, there is circumstantial evidence of their presence. The analysis of HCOSFA content effect on strength of FAB shows that in spite of decrease of HCOSFA amount in specimens from 60% to 48-36%, i.e. on 20-40%, by substitution with LCCFA their compressive strength was close to that of specimens contained only HCOSFA (90-95%). This can be attributed only to an interaction of CaO_{free} of HCOSFA and active SiO_2 of LCCFA with producing the additional cementing matter in form of CSH. Possibility of such forming mechanism of CSH is also confirmed by the absence of lime peak in XRD patterns. The lime peak was defined in XRD pattern of row HCOSFA. The increase of curing temperature is favorable to form CSH and their crystallization. Finally, XRD patterns of FAB steam cured at 60°C and 90°C have calcium silicate hydrate-CSH(I, II) peaks with d=2.81Å and 2.97Å. The CSH strength hardened binder and can improve its stability in different conditions.

4. In addition to ettringite peaks XRD patterns of steam cured FAB have peaks with d=8.86Å and 8.93 Å. These indices belong to the crystals of calcium monosulfate hydrate ($3CaO{\cdot}Al_2O_3{\cdot}CaSO_4{\cdot}12H_20$). In all likelihood, its formation is connected to the important role that the reaction of CaO_{free} of HCOSFA and active SiO_2 of LCCFA plays in FAB curing process. It decreases the concentration of CaO_{free} in the liquid phase of cure system. A prerequisite in generating more stable low-basic calcium hydrate aluminates is created. Such course of the reaction predetermines the formation of three-sulfate form of calcium sulfoaluminate hydrate ($3CaO{\cdot}Al_2O_3{\cdot}3CaSO_4{\cdot}31H_20$) -ettringite only in the initial stage of FAB curing. In the next three-sulfate form partially grades into monosulfate form; the greater the temperature of curing the greater the quantity of $3CaO{\cdot}Al_2O_3{\cdot}CaSO_4{\cdot}12H_20$ formed. Thus, the intensity of peak monosulfate hydrate peak in XRD pattern of FAB steam cured at 90°C is higher than that at 60°C.

XRD pattern segments of open air exposed specimens of cured FAB, which belong to ettringite only (Figure 5B), indicate a marked reduction in the intensity of the 9.73-9.75 Å ettringite peaks for moist cured specimens. This is due to a decomposition of ettringite. At the same time ettringite peak intensity in XRD pattern segments of steam cured specimens hardly changed.

Microstructural Observation

The ettringite phase is the best interest between other hydrate products of FAB since its stability in various condition influents on binder durability. In Figure 6 the morphologic features of ettringite both after curing FAB and after 6- and 36-month exposure of cured FAB in atmospheric air can be seen.

Comparison of the initial images after FAB curing shows that ettringite produced at higher temperatures (60°C and 90°C) are much more crystallized and have less specific surface than those formed at 20±3°C. Therefore, it seems that atmospheric air exposure caused less reduction in strength of steam cured FAB than that of moist cured FAB (Figure 4). Ettringite structure of steam cured FAB was subjected to unessential alterations (Figure 6).

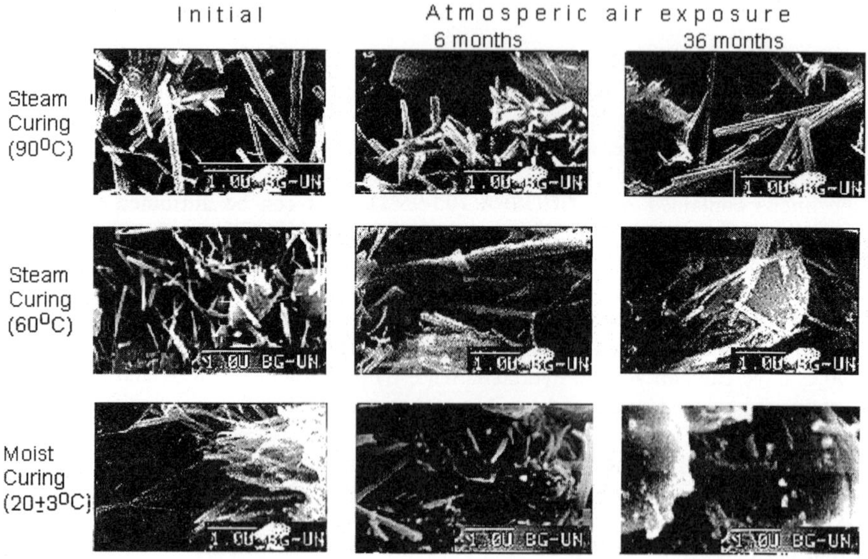

Figure 6 Scanning Electronic Microscopy observation of ettringite phases in cured FAB; 15000x magnification

CONCLUSIONS

1. The following conclusions concerning the durability of high- and low-calcium fly ash binders (FAB) could be made.

2. The durability of binders may be evaluated by changing long-term strength in various conditions.

3. The best conditions for the development of binder strength are moist air and water. A continuous curing and increase of compressive strength are observed.

4. In atmospheric air are three stages in strength change of cured FAB:
   - increase of compressive strength for 1 month
   - sharp drop after 1 month up to the 3rd-6th month
   - stabilization or slow-down of loss in strength after 3-6 months of exposure.

   Duration of the second and third stages depends on HCOSFA content and doesn't depend on curing conditions.

5. At the design of the composition of FAB the consideration should be given to the HCOSFA content. It should be remembered that HCOSFA quantity have to give required initial strength of building materials manufactured with using of FAB. Moreover, their long-term strength after the 3rd-6th month of exposure in atmospheric air must be on a level not less than initial level.

6. Combined hydration of high-calcium fly ash and low-calcium fly ash give an additional amount of high-strength and stable CSH which neutralizes the effect of decay in ettringite and improves the durability of the hardened system in atmospheric air. However a decrease in strength for a certain period of atmospheric air exposure (2-5 months) is noticed.

ACKNOWLEDGEMENTS

Author would be like to express its appreciation for the research grant made by the Jacob Blaustein Institute for Desert Research, Ben-Gurion University of the Negev (Israel) to carry out the reported work, which form part of the major research program on the development of technology of cementless building materials on the basis of fly ashes.

REFERENCES

1. KHEDAYWI, T, YEGINOBALI, A, SMADI, M AND CABRERA. Pozzolanic Activity of Jordan Oil Shale Ash. Cem. and Concr. Res., Vol. 20, 1990, pp 843-852.

2. PAPAYIANNI, J. Strength and Bond Data for Greek High-Lime Fly Ash Concrete. Proceedings of the 2nd International Conference on Use of Fly Ash, Silica Fume, Slag and Natural Pozzolans in Concrete, Madrid, 1986, pp 443-448.

3. BENTUR, A, ISH-SHALOM, M, BEN-BASSAT M AND GRINBERG, T. Properties and Application of Oil Shale Ash. Proceedings of the 2nd Second International Conference on Use of Fly Ash, Silica Fume, Slag and Natural Pozzolans in Concrete. Madrid, 1986, pp 779-802.

4. BENTUR, A, ISH-SHALOM, M, BEN-BASSAT, M AND GRINBERG, T. Cementing Properties of Oil Shale Ash: Part I, II, and III. Cem. and Concr. Rec., Vol. 10, 1980, pp. 175-182; Vol. 11, 1981, pp. 645-650; Vol. 11, 1981, pp 799-807.

5. BENTUR, A AND GRINBERG, T. Modification of the Cementing Properties of Oil Shale Ash. American Ceramic Society Bulletin, Vol. 63, No 2, 1984, pp 240-244.

6. SATAVA,V AND VEPREK, O. Thermal Decomposition of Ettringite under Hydrothermal Conditions. Journal of the American Ceramic Society, Vol. 58, No 7-8, 1975, pp 357-358.

7. MEHTA, R K. Stability of Ettringite on Heating. Journal of the American Ceram. Society, Vol. 55, No 1, 1972, pp 55-56.

8. NA, T, SUZUKI, K, ITO, S AND TAKABE, T. Decomposition of Synthesized Ettringite by Carbonation. Cem. and Conc. Res., Vol.22, 1992, pp 6-14.

9. ESTERSON, G, AIZENSHTAT, Z AND PISMEN, M. Mineral Reactions and Pyrolysis of Israeli Oil Shale. Journal of Analytical and Applied Pyrolysis. Vol. 6, 1984, pp 375-389.

10. YOFFE, O AND NATHAN, Y. The Chemistry and Mineralogy of Ashes from PAMA Demonstration Plant. 12th Israeli Conference of the Israel Association for the Advancement of Mineral Engineering, Elite, 1994, pp 270-274.

11. BENTUR, A. Application of Oil Shale Ash as Building Material. Silicates Industrials, Vol. 7/8, 1982, pp 163-168.

A TECHNOLOGY OF PRODUCTION OF SULFOGYPSUM-BASED BINDING AGENTS FROM THERMAL POWER PLANT WASTE

O S Popova

State University of Railway Communications

Russia

S P Vänninen

Tampere University of Technology

Finland

ABSTRACT. Low-temperature technology of processing the wastes into binding agents with a high content of the α-modification of calcium sulfate is proposed on the basis of studies of properties of thermal power plant sulphur - containing wastes obtained by collecting sulphur from flue gases with the help of lime mortar. Structural characteristics and other properties of binding agents obtained by this method were investigated. The obtained binding agents are suitable for the manufacturing of items of higher strength and as component of gypsum-concrete binding agents.

Keywords: Chemical gypsum, Sulfogypsum, Absorbent, Calcium sulfate α- and β-modification of calcium sulfate, Hemihydrate and dihydrate of calcium sulfate, Waste processing, Sulfogypsum-based binding agent.

Professor O S Popova is a Doctor of Technical Sciences, currently at the Department of Construction Materials and Products of State University of Railway Communications, St. Petersburg, Russia. She specialises in polymeric materials, chemical additives, and hydraulic and aerial binding agents.

Dr S P Vänninen is a Candidate of Technical Sciences, currently at the Department of Geology of Tampere University of Technology, Tampere, Finland. She specialises in chemical additives, solutions and aerial binding agents.

INTRODUCTION

Sulfur-containing wastes are obtained by collecting flue gases of thermal power plants according to the following scheme [I]:

$$SO_2 + CaCO_3 + 0.5\ H_20 \rightarrow CaHSO_3.0.5H_2O + CO_2\uparrow$$
$$CaHSO_3.0.5H_2O + O_2 + 3H_2O \rightarrow 2CaSO_4.2H_2O$$

The size of sulfogypsum crystals varies within 5–100 μM. The characteristics of chemical gypsum so obtained match those of the natural gypsum, and may even be superior with respect to gypsum purity [2].

Some estimates [3] forecast that the amount of collected sulfogypsum in the world will reach 220–280 million tons by the year 2000. The utilization of chemical gypsum may solve economical problems (obtaining new gypsum sources and relieving the land from heaps) from one side, and ecological (environmental pollution : soil, water, atmosphere) from the other side.

This paper describes a technology for sulphoxide-containing waste utilization that yields binding agents suitable for construction material industry.

EXPERIMENTAL DETAILS

Materials

Samples of sulfogypsum from thermal power plants in Finland were used. Plant quality certificate for the sulfogypsum.

Chemical composition:

SiO2	0,61% – 1,05%
Al2O3	0,23% – 0,46%
CaO	25,8% – 28,4%
SO3	19,7% – 20,2%
Humidity	10%

Experimental

All experiments were conducted in compliance with the standards ASTM C472, SFS 3167, and GOST 125-79.

Based on preliminary studies that revealed differences in the structure of natural and chemically obtained gypsum we developed a new technology of the production of calcium sulfate hemihydrate from sulfogypsum.

At the first stage of the technological process, sulfogypsum is treated with a water adsorbing agent. Sulfogypsum is mixed with the sorbent at a 22-24°C room temperature in a 1:1 proportion by weight.

The mixture is kept in a closed vessel for 24 h. At this stage, free and a part of bound adsorbed water is removed from sulfogypsum under «mild» conditions. Upon mixing sulfogypsum with the sorbent, the temperature of the mixture increases up to 42°C. It is known for example according to Flörke [4] that even below 45°C the formation of the orthorhombic modification of calcium subhydrates is already possible. According to Van Hoff [5], anhydride formation is possible below 42°C, and a rhombic syngony modification of calcium sulfate, dimorphen, was found below 45°C [6]. In our case, it is reasonable to expect that the process of mild water removal by adsorbent is endothermic. At this stage, intermediate modifications of sulfogypsum hemihydrates must already be formed. Such pretreatment of sulfogypsum permits to avoid the evaporation of excess water at the second stage of dehydration, while drying in an oven at 175°C, and so to decrease the formation of small crystals of the β-modification of calcium sulfate hemihydrate, as is generally known. The duration of the second stage is 3 h. However, further studies have shown that this time period may be shortened to 0,5 h. At this technological stage, the final dehydration of sulfogypsum is achieved yielding target binding agent. The baking is performed in a calorifier with a rapid (within 3–8 min) temperature elevation to the required level.

RESULTS AND DISCUSSION

The sulfogypsum binding agent obtained via the technological process suggested contains larger amount of particles of the α-form of calcium sulfate hemihydrate than that obtained by generally employed technology. Figures 1 and 2 show plots that represent the specific surface area of crystals, obtained (A) by direct process of dehydration in a furnace and (B) by dehydration preceded by pretreatment, as suggested here. The dispersion degree of sulfogypsum crystals is 2.5 – 3 times higher in case B than in case A. Besides, a general tendency toward the increase of crystal particles in evident in case B.

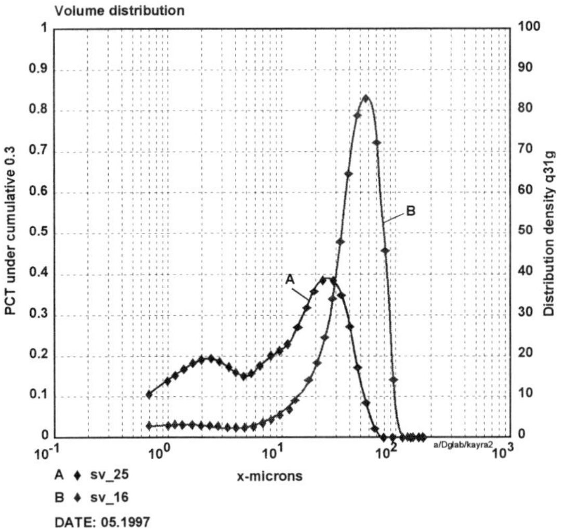

Figure 1 The specific surface area of a sulfogypsum binding agent (sulfogypsum is from Naantali). A – without dehydrating pretreatment. B – with dehydrating pretreatment

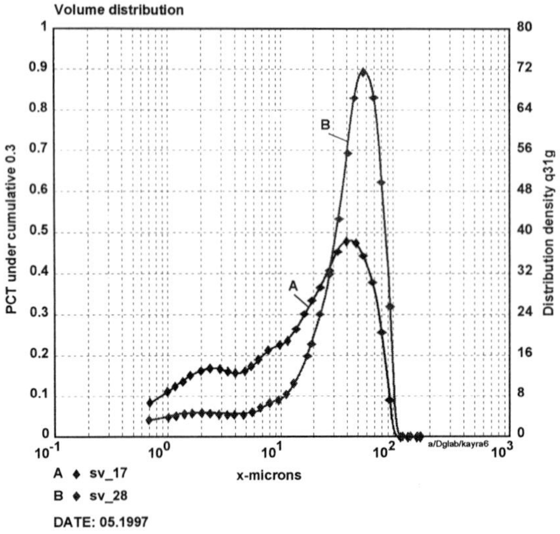

Figure 2 The specific surface area of a sulfogypsum binding agent (sulfogypsum is from Pori). A – without dehydrating pretreatment. B – with dehydrating pretreatment

Figures 3 and 4 show plots that reflect the state of water in sulfogypsum structure obtained by the method of optical combination scattering [6]. The study is fulfilled using a DFS-24 device with the neon-helium laser. Sulfogypsum transmission spectra are significantly different in cases A and B. Within the range of main transitions, total transmission area is greater in case B than in case A. This indicates that, in case B, the microscopic crystal structure is altered [7] and the mean size of sulfogypsum crystal particles is greater.

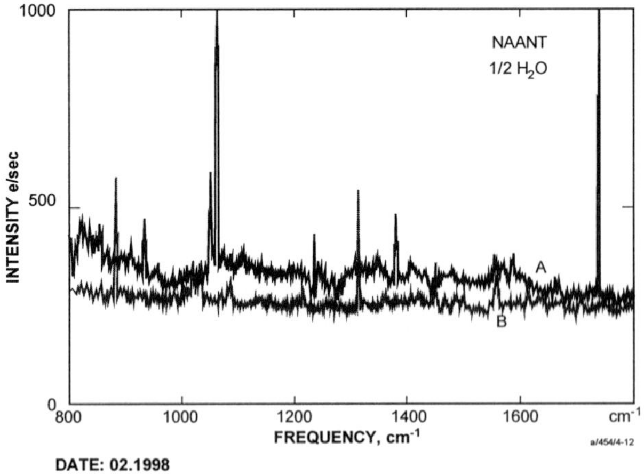

Figure 3 Transmission spectra (Ne-He) of $CaSO_4.0,5H_2O$, (sulfogypsum is from Naantari). A – without dehydrating pretreatment. B – with dehydrating pretreatment

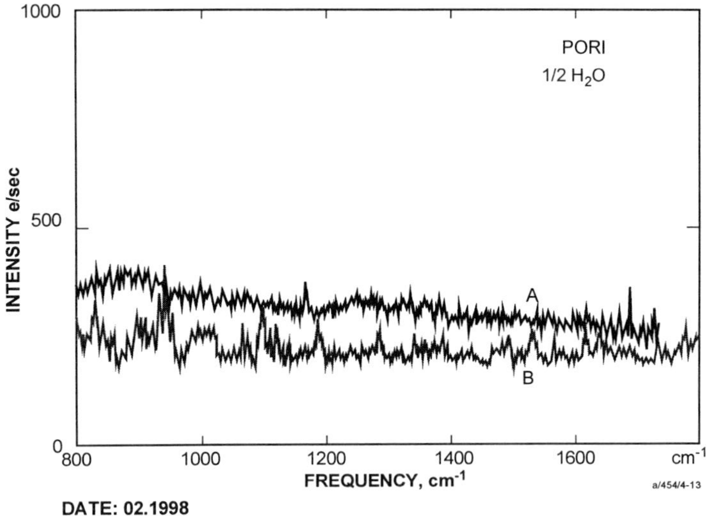

Figure 4 Transmission spectra (Ne-He) of $CaSO_4.0.5H_2O$, (sulfogypsum is from Pori).
A – without dehydrating pretreatment. B – with dehydrating pretreatment

Figure 5 shows IR spectra of water in the structure of sulfogypsum crystals.

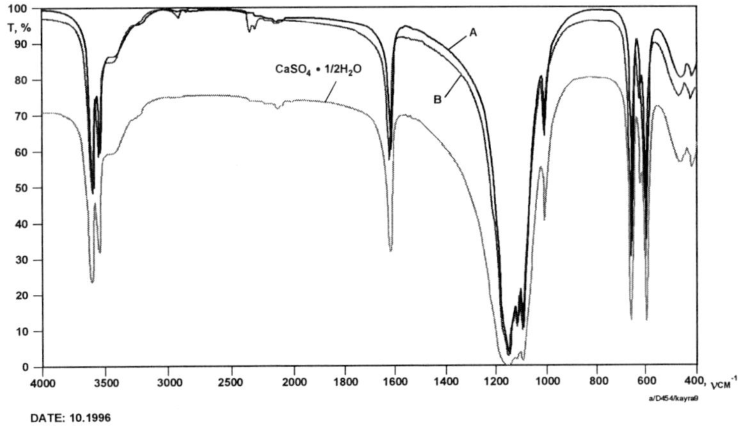

Figure 5 IR-spectrogram of $CaSO_4 \bullet 0.5H_2O$ (sulfogypsum is from Naantari).
A – without dehydrating pretreatment. B – with dehydrating pretreatment

The spectra were obtained using a Perkin-Elmer instrument with a 7900–320 cm^{-1} working range. It is known that the vibration spectrum of calcium sulfate always contains components contributed by water and sulfate groups in addition to components contributed by water bound by crystallization.

The frequencies characteristic for intramolecular structures depend on their nearest environment. Thus, the vibration spectra of molecules reflect characteristics of local crystal structure [8]. In case B, sulfogypsum spectrum displays characteristic absorption bands at 1600 cm^{-1}, an increase in multiplicity at 3600 cm^{-1}, and the disappearance of the band of completely symmetrical vibration at 100 cm^{-1}, which is characteristic for the ions of the β-modification of calcium sulfate. Thus, the sulfogypsum binding agent obtained by the two-stage processing contains less amount of β-modification crystals and larger crystals with the size increased 2.5- 3 times, which confirms the convenience of this technology for sulfogypsum utilization.

PRACTICAL IMPLICATIONS

The technology of sulfogypsum binding agent processing that includes a pretreatment stage of «mild» dehydration with water sorbents followed by baking the product in a furnace yields binding agent with a reduced content of the β-modification of calcium sulfate hemihydrate. The process of gypsum dehydration which is necessary for endowing it with binding properties is a phase transition without diffusion. As water is being removed from the crystal lattice of the dihydrate, energy conditions are being built up that cause the uncoupling of Ca^{2+} and SO_4^{2-} from water molecules. Crystallographic rearrangement occurs by shifting Ca-SO_4-Ca layers. In the elementary crystal cells of the hemihyrated gypsum crystallochemical bonds are coordinated in much the same, yet somewhat distorted, way as they are in the dihydrated gypsum, the energy of the crystal lattice of the hemihydrate increasing with dehydration [9].

The mechanism and kinetics of calcium sulfate dihydrate dehydration process depends upon the dehydration technology employed. When water is removed by evaporation, the thermal processing of the surface of dihydrate particles occurs. This results in the loosening of gypsum crystal lattice and the generation of the β-modification of calcium sulfate hemihydrate. β- hemihydrate looks like highly dispersed fine filaments and its specific area is 2,5–5 times higher than that of the α-modification. The crystals of the β-modification are small, poorly formed, and defective. When dehydration is performed under an elevated pressure, water is removed in drop-liquid state. This is not associated with crystal lattice loosening and crystal destruction, and the dihydrate forms prismatic crystals of the α-form of calcium sulfate. The crystals of α-hemihydrate are big, dense, and less defective, which provides for the strength of products manufactured.

In the suggested technology of sulfogypsum binding agent production, partial dehydration is performed by water adsorption under normal conditions, instead of using an elevated pressure. Therefore, only the remainders of adsorbed water are removed by the subsequent baking which provides for the formation of bigger crystals of the α-modification of calcium sulfate hemihydrate.

In Table 1 the values are presented of the specific surface area of sulfogypsum binding agents obtained by the suggested and generally employed technological processes. From the data it is clearly seen that the «mild» dehydrating pretreatment results in yielding bigger crystals of calcium sulfate hemihydrate.

Table 1 Sulfogypsum specific surface area (cm^2/g)
(A – sulfogypsum was not pretreated; B – sulfogypsum was dehydrated by pretreatment)

SULFOGYPSUM SOURCE	SPECIFIC SURFACE AREA OF DIHYDRATE $CaSO_4.2H_2O$		SPECIFIC SURFACE AREA OF HEMIHYDRATE $CaSO_4.0,5H_2O$	
	A	B	A	B
S. Kipso oy	1100	1800	4300	2000
Naantari	1100	1500	6200	1500
Pori	1000	1900	5000	1900

CONCLUSIONS

1. A technology is suggested for the processing of sulfoxide-containing waste into the binding agent suitable for using as a raw material for manufacture and as a component for gypsum-concrete binding agents. The technology includes the pretreatment of sulfogypsum with dehydrating absorbents followed by baking it in a furnace.

2. The sulfogypsum binding agent produced as a result of two-stage processing contains significantly less amount of β-modification of calcium sulfate hemihydrate. The specific surface area of sulfogypsum crystals so obtained is 2,5 – 5 times less than that of sulfogypsum obtained by the generally used method.

3. Crystal structure and water state in the crystals of calcium sulfate hemihydrate were investigated by IR-spectroscopy and combinational light scattering. A direct effect of dehydrating pretreatment on the state of bound water and crystal structure was found. The crystals of sulfogypsum binding agent become bigger virtually devoid of the β-hemihydrate of calcium sulfate.

REFERENCES

1. WAHLSTRÖM, M. Hilivoimalaitoiden rikinpoistojätteiden hyöttökayttö. Espoo, 1988, Kemia-Kemi, Vol. 15.

2. WIRSCHUNG, F, HULLER, R, AND OLEJINIK, R. Gypsum from flue gas desulphurisation plants: definition and legislation in the European Communities, in OECI, and in Germany. Zement Kalk Gips, 1994, 2, pp 66–69.

3. UTILIZATION OF THE PHOSFOGYPSUM PRODUCED IN THE FERTILIZERS INDUSTRY. UNIDO/4s, 533.23 Mags., 1985, pp 2, 7, 20.

4. FLÖRKE, O, W. Kristallographischen und röntgenographishen Untersuchungen in System $CaSO_4$-$CaSO_4 \cdot 2H2O$. N. Jb. Min. Abh., Stuttgart, 1952, Vol. 84, SS. pp 189-240.

5. HAND, R, J. Calcium sulfate hydrates: a review. British Ceramic Transactions, 1997, Vol. 96, N 3.

6. KUMARESON, P, AND DEVANARAYANAN, S. J. Mater. Sci (UK), 1989, Vol. 24, No. 1, pp 288.

7. Бобров В. С., Жигун И. Г., Ромашков А. В., Киселева Л. В. Известия АН СССР, сер. Нееорганич. матер. 1987, т. 23, № 8, с. 1364. (BOBROV, V, S, ZHIGUN, I, G, ROMASHKOV, A, V, KISELEVA, L, V. Izvestiya, AN SSSR, ser. Neorganich. mater., 1987, Vol. 23, No. 8, pp 1364 in Russ.).

8. HASS, M, AND SUTHERLAND, G, B. Proc. Roy. Soc. (London), Ser. A, 1956, Vol. 236, pp 427.

9. Мчедлов-Петросян О. П. Химия неорганических строительных материалов. М.: Стройиздат, 1971. (MCHEDLOV-PETROSIAN, O, P. Chemistry of inorganic construction materials. Stroyizdat, Moscow, 1971.

HYDRAULIC BINDING MATRIX WITHOUT CEMENT

I Robu

Technical University of Civil Engineering

Romania

ABSTRACT. This paper reviews my previous research on hydraulic binding matrices without cement. Pozzolanic- type reactions and various industrial wastes or local materials have been used. Among the pozzolanic materials analysed, I am listing: fly ash, furnace slag, diatomite, silica fume, while as $Ca(OH)_2$ donor materials I mention: sludge of carbide and slaked lime. Phosphogypsum was particularly used in the binding systems under research in order to accelerate the initial hardening. Specific attention was given to the binding systems with fly ash, outlining the existent possibilities for improving certain characteristics of hardening structures by using various corrective admixtures for hydraulic and granulometric purposes. The setting and hardening of the mixtures based on hydraulic admixture are adequate criteria for assessing the overall pozzolanic activity of the hydraulic admixtures. This paper presents optimal compositions, curing conditions, physical and mechanical conditions of the hardening structures, pores distribution, as well as methods for the acceleration of hardening processes in the binding systems without cement.

Keywords: Pozzolanic activity, Fly ash, Volcanic tuff, Phosphogypsum, Curing

Dr Ion Robu is a professor at the Technical University of Civil Engineering Bucharest - Romania. Currently, he is the head of the chair of Chemistry and Building Materials within the University. His research covers pozzolanic binders, possible utilisation of local materials and industrial wastes, as well as various aspects related to concrete durability.

INTRODUCTION

This paper outlines major opportunities and limits in the utilisation of some hydraulic binding matrices without cement. It is a synthesis of author's research in this area.

Hydraulic binding matrices without cement are based on pozzolanic type reactions between active acid oxides (SiO_2, Al_2O_3) present in various industrial byproducts (fly ash, furnace slag, silica fume) or local materials (volcanic tuffs, diatomite) and $Ca(OH)_2$ from slaked lime or lime substitutes (sludge of carbide [1]). These reactions give rise to cementitious calcium hydro-aluminates and silicates with hydraulic properties.

Paper [2] details on the physical chemical and technological bases for the achievement of these binding matrices without cement.

The utilisation of industrial byproducts as materials containing cementless binding matrices is of a major economic and environmental interest both in Romania and abroad. The proportion of industrial byproducts and / or local materials in the binding matrices exceeds 50% when using hydrated lime as a base reagent and even 80% when using sludge of carbide as base reagent (Table 1).

The binding compositions we have analysed are mainly fly ash binding compositions under the condition of using as base reagent either slaked lime or sludge of carbide. Other similar compositions using furnace slag, volcanic tuffs, diatomite or silica fume have also made the object of our research.

The present paper outlines the optimal compositions of some binding matrices, the key properties of the matrices within the reference binding system (fly ash – slaked lime or sludge of carbide) and proposes ways for property improvement.

EXPERIMENTAL DETAILS

Materials

The materials we used during the research process have been mainly: fly ash (activity index 0.76 as per STAS 8819-88), slaked lime(hydrated lime in powder), sludge of carbide, phosphogypsum, volcanic tuff, diatomite, silica fume, well grounded blast furnace slag, granulated slag, poligranular quartz sand (0.08 – 2 mm), glass fibre wastes (ordinary glass) and alkalis resistant glass fibres (zirconium glass).

Binding Systems

The process of obtaining hydraulic binding matrices required the analysis of several binding systems, such as: fly ash-slaked lime-water, fly ash-sludge of carbide-water, fly ash-volcanic tuff (diatomite)-slaked lime-water, fly ash-slaked lime (sludge of carbide) phosphogypsum-water. The real optimal compositions are disclosed in table 1 below.

Technological Issues

The pozzolanic capacity of hydraulic additions mixed with hydrated lime has been assessed on the basis of several physical and mechanical criteria (namely setting – hardening). The setting time was determined on normal consistency pastes using Vicat's needle device (diameter 3 mm and 1 mm), while the hardening capacity was estimated by the time resolution of mechanical strengths (bending and compressive strengths).

The pastes used in the determination of mechanical strength and of other characteristics have been stronger than normal as they contained less water (by 15 %) than normal consistency pastes; the pieces used in the determination test were 40 x 40 x 160 mm prisms or 100 mm sided cubes. The target compactness was obtained by vibration or vibro-pressing techniques, making full use of the tixotropic properties of fresh mix.

The curing conditions of test pieces after striking have varied widely: air, moist air or mix conditions. The overall maximum binding potential has been assessed based on the compressive strength after 90 – 180 days of moist air curing.

The reference binding system has been fly ash – slaked lime – water. Sludge of carbide has been used as a lime replacement, while the other hydraulic admixtures (volcanic tuffs, diatomite, silica fume, slag) have been used for enhancing some pozzolanic and granulometric fly ash properties; phosphogypsum has been equally used for increasing the early strength of the binding matrices.

Table 1 Real optimal compositions

SYSTEM COMPOUNDS (Kg / m^3)	FLY ASH-SLAKED LIME-WATER	FLY ASH–VOLCANIC TUFF – SLAKED LIME-WATER	FLY ASH-DIATOMITE-SLAKED LIME-WATER	FLY ASH-SLUDGE OF CARBIDE-WATER
Fly ash	900	726	485	806
Volcanic tuff	-	182	-	-
Diatomite	-	-	208	-
Slaked lime	180	136	208	-
Sludge of carbide (45 % humidity)	-	-	-	403
Water	457	456	549	276

RESULTS AND DISCUSSIONS

The Characteristics of the Hydraulic Binding Matrices Without Cement

The reference binding matrix has been on fly ash and slaked lime. An equivalent matrix based on fly ash has been obtained by substituting slaked lime with sludge of carbide. The core characteristics we have analysed in optimal composition in both matrices are the following: absolute and apparent density, porosity and pores distribution, water absorption capacity, bending and compressive strength, thermal conductivity, wear resistance and hardening shrinkage, mortar adhesion. Generally, the values recorded depend on the initial moist air curing period.

- The absolute density varies within 2.19 to 2.22 g/cm^3. The apparent density after striking reached 1525 Kg/m^3 for pastes made of fly ash and slaked lime and 1485 Kg/m^3 for pastes containing fly ash and sludge of carbide. However, after reaching the hygrometric equilibrium, the apparent density decreased to 1200 – 1300 Kg/m^3, respectively to 1150 – 1250 Kg/m^3 (depending on the initial curing in moist air).

- The overall porosity parameter ranged within 48 – 50 % for matrices based on ash and slaked lime and within 50 – 54 % for those based on fly ash and sludge of carbide. The submicron pores distribution was verified by means of a porosimeter with mercury, by analysing the integral and differential distribution curves; we have found that the maximum pores distribution corresponds to a 0.05 μm radius (as in case of cement pastes). However, the overall submicron porosity is 4 times higher than the normal figure for cement pastes (320 mm^3/g as compared to 80 mm^3/g), that provides also an explanation of the strength differences which exist among these types of matrices.

- Water absorption capacity depends directly on the duration of the initial moist air curing. The values obtained have been of 29.5 % (for a 90 day period) and of 38.8 % (for a 14 day period) in case of ash and slaked lime matrices and of 35.2 % and 45.9 % in case of ash and sludge of carbide matrices.

- The bending strength was assessed using 40 x 40 x 160 mm prisms. After a 28 day curing in moist air, the bending strength reached 3.3 Mpa for fly ash and slaked lime pastes and only 3.0 MPa for fly ash and sludge of carbide pastes.

- The compressive strength depends mainly on the composition and hardening conditions (Figure 1 and Figure 2). For instance, pastes containing fly ash and 25 % slaked lime and 3 % $CaCl_2$ hardened in moist air for 180 days and tested at 3 year age, have indicated compressive strengths of 40 MPa.

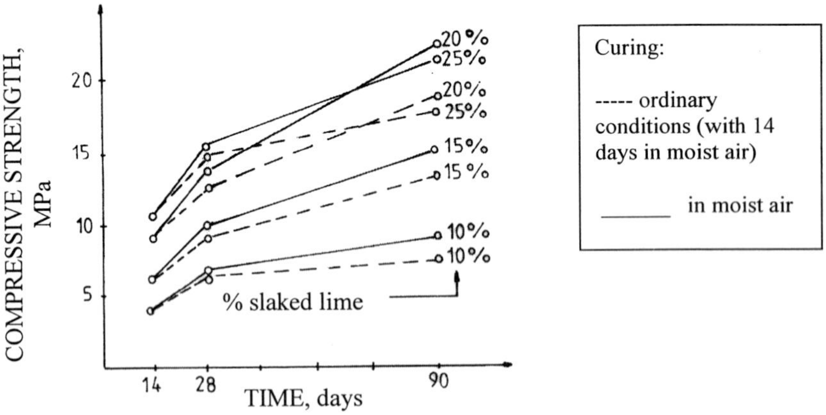

Figure 1 The time variation of compressive strength for ash and slaked lime pastes

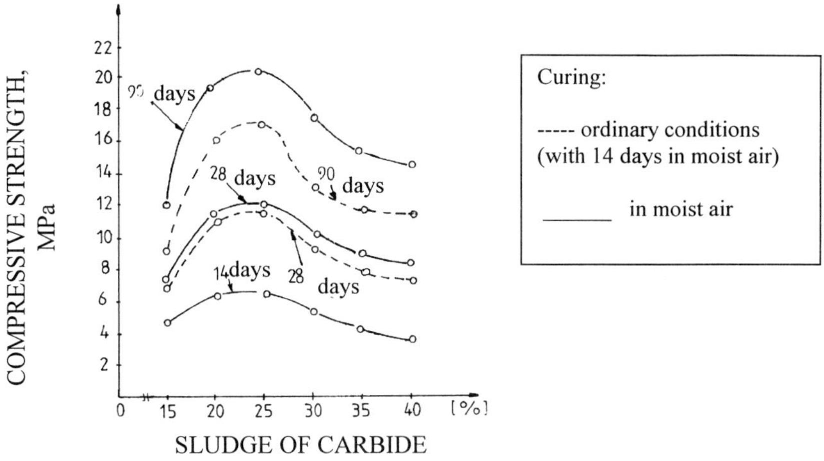

Figure 2 The dependence of compressive strength on the proportion of sludge of carbide and on the hardening time in variable conditions

- The hardening shrinkage at 90 days (including a 14 day moist air curing) was of 2.1 mm/m for fly ash and slaked lime pastes and of 2.3 mm/m for fly ash and sludge of carbide pastes. As no cracks have been noticed during shrinkage, it results a good deforming characteristic of these matrices.

- The thermal conductivity coefficient at 28 day age (including 14 in moist air) has been of 0.343 W/m.K for tests made of fly ash and slaked lime and of 0.327 for those containing fly ash and sludge of carbide; good insulation properties are thus revealed.

- The wear resistance has been measured by means of a Böhme device using 10 mm sided cubes dried at 105° C, following a 90 day curing period in moist air. The values obtained are summarised in table 2 and are in accordance with the mechanical strengths.

Table 2 Wear resistance

WEAR RESISTANCE	FLY ASH WITH 20 % SLAKED LIME	FLY ASH WITH 25 % SLUDGE OF CARBIDE
W_1 (g/cm^2)	0.31	0.49
W_2 (mm)	2.70	4.30
W_3 (cm^3)	26.10	40.40

- The capillary absorption shows values which are mainly dependent on the duration of moist air curing (Figure 3). Pastes made of fly ash and sludge of carbide present an absorption trend similar to the one illustrated in Figure 3, but with higher values for the tests cured in moist air.

- The mortar adhesion of products based on ash and slaked lime has been measured using plastering mortars M10 and masonry M25. The separation strength has been of 0.10 – 0.14 MPa in the former case and of 0.15 – 0.25 MPa in the latter. The hydraulic binding systems without cement present a crystalline–gellic structure (revealed by electronic microscopy) and contain cementitious hydrated compounds [2].

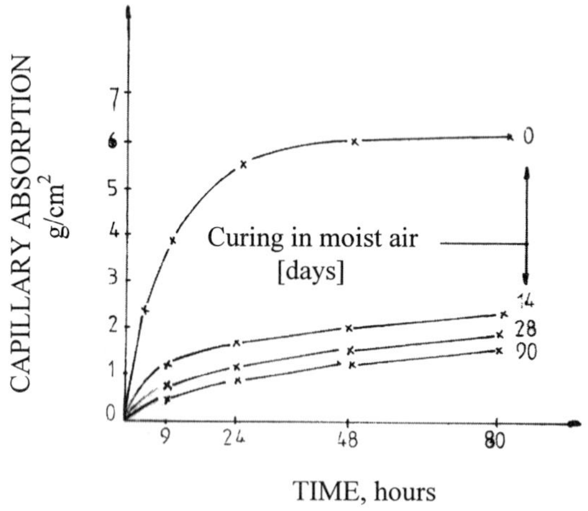

Figure 3 The influence of moist air curing period on the capillary absorption

Characteristics Improvement of Fly Ash Binding Matrices

The hydraulic binding matrices without cement containing fly ash and slaked lime (or alternatively, sludge of carbide) present a series of limitations (disadvantages) that must be taken into consideration: slow setting and hardening process, initial 10 – 14 day curing in moist air, high strain, limited binding potential and smaller tensile and impact strength compared to Portland cement.

A way to improve the mechanical strength of the matrices consist in mixing fly ash with activators of pozzolanic nature (furnace slag, volcanic tuffs, silica fume, diatomite) and slaked lime. In this way, some other properties (technological features, setting and water stability) are also improved.

The active hydraulic admixtures react consistent with the pozzolanic and granulometric effect. Figure 4 illustrates the role of volcanic tuff in the binding system fly ash-volcanic tuff-slaked lime.

The maximum strength in various binary systems (fly ash: hydraulic admixture) depends on the hydraulic activity and on the fineness of the admixture involved.

The admixture proposition in the mix containing fly ash (value relative to the maximum strength) has been of 20 % in case of volcanic tuff, 30 % in case of diatomite and 15 % in case of both silica fume and furnace slag.

Therefore, based on the data obtained and the underlying principle, we consider that structurally integrated composite matrices can be further obtained; such matrices present superior characteristics as compared to the reference system (fly ash- slaked lime) values.

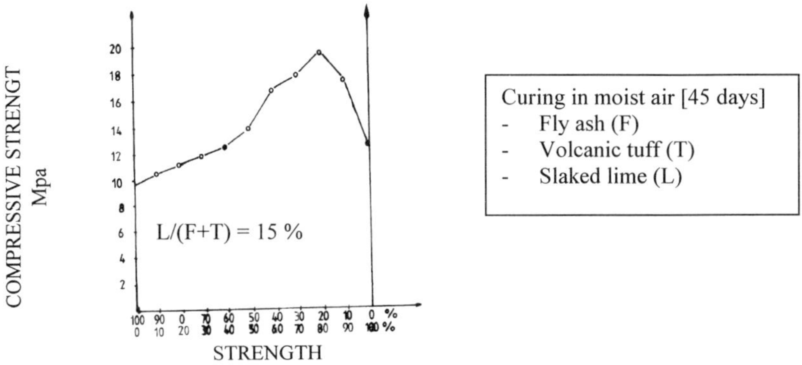

Figure 4 The influence of volcanic tuff on the compressive strength of some ash composite matrices

We have also noticed that a 10 – 15 % content of phosphogypsum may further enhance the early strength (7 days). The favourable influence of phosphogypsum can be explained by the formation of ettringite-type compound which can be revealed by X-Ray diffraction [3].

Similarly, in order to reduce the hardening shrinkage lean materials such as quartz sand or granulated slag may be used. Figure 5 illustrates the influence of 0,08 – 2 mm poligranular quartz sand on the overall shrinkage. Thus, the compressive strength of test pieces containing sand decreased by 30 – 50 % after a 90 day cure in moist air. Granulated slag has been used as an aggregate in the achievement by vibro-pressing of some experimental lots of fly ash and sludge of carbide masonry elements. As mentioned, the main limitation of the hydraulic binding matrices without cement consists in the need of an initial moist air curing period of at least 10 – 14 days. Such limitation can be overcome through thermal treatments, such as: steaming, autoclaving, greenhouse effect, high frequency current etc. [4]. However, the best results have been obtained in case of curing in high frequency field [5] or under steaming treatment [6].

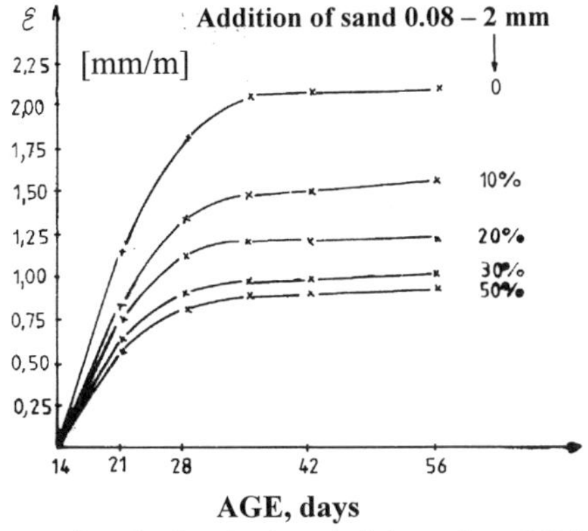

Figure 5 Decreasing the hardening shrinkage by addition of quartz sand

Within the optimal matrix (fly ash – volcanic tuff – slaked lime – water) we have aimed at increasing the tensile – bending strength and the impact strength by using ordinary glass fibre wastes and alkalis resistant glass fibres (1.5 – 3 % compared to the fresh mix and 3 – 5 cm length). Please note that significant improvements have been obtained only when using alkalis resistant glass fibres (Figure 6); in case of using ordinary glass fibres wastes, the compressive strength has been inferior to that witness, due to the alkaline attack.

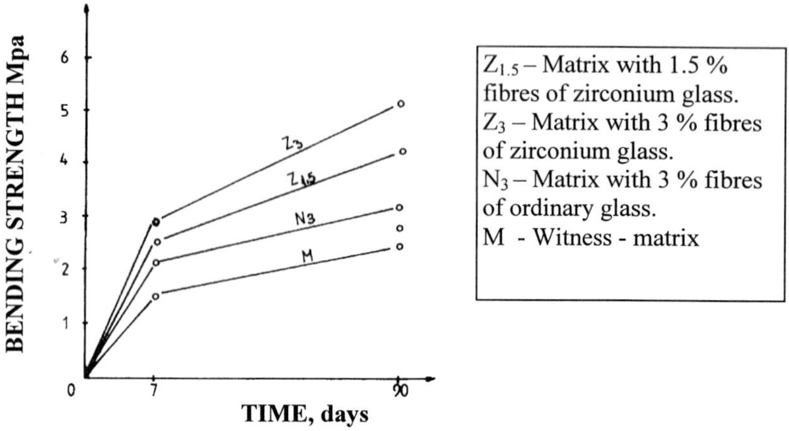

Figure 6 The bending strength of glass fibres reinforced matrices

CONCLUSIONS

Based on the above data and our related comments, several conclusions can be drawn up:

1. The hydraulic binding matrices without cement available for different hardening structures offer real opportunities for an efficient use of some industrial byproducts or local materials in the field of construction and building materials. The basic compounds are one or more pozzolanic-type materials and one $Ca(OH)_2$ donor material, such as slaked lime or sludge of carbide.

2. In order to obtain durable hardening struc4ftures with adequate features, certain conditions regarding the mix composition and the hardening process should be met. An optimal ratio between the composition with pozzolanic properties (fly ash, furnace slag, volcanic tuffs, diatomite, silica fume or a combination of these) and the base reagent (slaked lime or sludge of carbide) should exist, as well as an initial moist air curing of 10–14 days or another replacement treatment for the acceleration of hardening.

3. The sludge resulted from the acetylene manufacturing out of carbide may be used both as a lime substitute in mortars or as a base reagent in case of pozzolanic activity materials. The slaked lime or sludge of carbide content should be chosen according to technical – economic criteria and not based on maximal performance parameters, as the additional strength benefit is low after a certain threshold is met (20 % hydrated lime or 25 % sludge of carbide – the equivalent in dry powder – relative to the fly ash or mixed addition quantity).

4. The features of the reference matrices based on fly ash and slaked lime (or sludge of carbide) may be improved by adding more active pozzolanic materials (slag, silica fume, volcanic tuffs, diatomite) in ratios up to 10 – 30 % according to the hydraulic and granulometric effect. An enhance in the initial strength can be achieved by adding 10 – 15 % phosphogypsum as compared to the dried binding mix; for reducing the hardening shrinkage and for improving the technological properties either quartz sand or granulated slag can be used. In a similar manner, the parameters quantifying the bending and impact strength have been improved by using (3 % as compared to fresh mix) alkalis resistant glass fibres.

5. The hardening structures obtained based on hydraulic binding matrices without cement have proved to have features compatible to the requirements imposed by masonry elements or other building elements such as polymer coated slabs for irrigation canals.

REFERENCES

1. ROBU, I. Hardening properties of fly ash and sludge of carbide, Rev. Materiale de Constructii, 15(1), 1985, pp 51 – 53.

2. ROBU, I. Binding compositions with fly ash, Ph. D. Thesis, Civil Engineering Institute, Bucharest, 1986.

3. ROBU, I., ROBU, M. Opportunities and limits in the use of phosphogypsum, Rev. Materiale de Constructii, 1986, pp 225 –228.

4. ROBU, I. Aspects regarding the acceleration of hardening in fly ash binding systems, Rev. Materiale de Constructii, 21(2 – 3), 1991, pp 89 – 93.

5. ROBU, I. Improving characteristics of hardening structures in fly ash binding systems, Proceedings of The National Conference on Chemistry and Chemical Engineering, Bucharest, 1995, pp 362 – 366.

6. NICOLESCU, L. Fly ash in construction, Ed. CERES, Bucharest, 1978.

THE INFLUENCE OF SODIUM SALTS ON INTERACTION IN THE SYSTEM. PORTLAND CEMENT - GRANULATED BLAST FURNACE SLAG - FUEL ASH - SLAG BLEND

R F Runova **M A Kochevykh**

I I Rudenko **J L Nosovskii**

V.D. Glukhovskii Scientific Research Institute on Binders and Materials

Ukraine

ABSTRACT. Among advantages of Portland cement- based binders made with industrial wastes (multi-blends) are mentioned the following: compliance of the concrete made with multi-blends working in aggressive environments with increasingly growing engineering requirements, environmental safety and low cost. Unrestricted possibilities of such concretes may be attributed, first of all, to a great variety of their compositions, and chemical modification is of special attention. There have been developed formulations of CC with Portland cement content up to 20 % by mass and strength varying between 30… 55 MPa. These properties are reached due to the use of sodium salts which take an active part in the interactions occurring in the systems containing blast furnace slag, fuel ash-slag blends and natural glassy pozzolana. The hydration products of the cementitious materials have been examined. The developed cements may be recommended for the use in road construction in making road pavement.

Keywords: Calcium hydrosilicates (CHS), Composite cement (CC), Fuel ash- slag blend (ASB), Granulated blastfurnace slag (GBFS), Ordinary portland cement (OPC).

Professor, DrSc(Eng) Raisa F Runova is Chief Researcher of V.D.Glukhovskii Scientific Research Institute on Binders and Materials, Kyiv, Ukraine. Since 1970 she has been investigating the properties of amorphous silicates and herewith suggest of contact-condensation mechanism of hardening these systems.

Dr Marina A Kochevykh is Research Scientist of V.D.Glukhovskii Scientific Research Institute on Binders and Materials, Kyiv, Ukraine. Her research interests include creation of new materials with determined properties based on calcium hydrosilicates.

Dr Igor I Rudenko is Research Scientist of V.D.Glukhovskii Scientific Research Institute on Binders and Materials, Kyiv, Ukraine. He is specialised in the application of fuel fly ashes and slags in cement-free concretes.

Juri L Nosovskii is an Engineer of V.D.Glukhovskii Scientific Research Institute on Binders and Materials, Kyiv, Ukraine. He specialises in developing formulations of the blended cements with low content of Portland cement clinker.

INTRODUCTION

Practical experience gained from the use of blended ordinary portland cements (OPC)- based cements counts a few years and found its reflection in the European Standard ENV- 197. The role of cementitious materials which are a combination of OPC and natural and industrial pozzolanas is growing increasingly. In compliance with the norms currently in force in Ukraine DSTU B V.2.7- 46 which contain the requirements to cements intended for general application, a type of blended cement- the composite one, is introduced and specified. This allowed us to use the term «composite cements» in the present paper.

According to analysis reported in [1], the existence of blended cement is connected with the following aspects:

- technical – it is caused by a pozzolanic character of hydration which promotes the corrosion resistance enhancement to aggressive environments and the lower exothermic effect;
- ecological – is determined by compounding of the majority of existing industrial wastes of aluminosilicate composition with hydration products and decline in the OPC production due to a «greenhouse» effect;
- economical – is caused by decline in energy consumption for production of OPC, increase in lifespan of structures, decrease in expenses for repair and maintenance of concrete structures made of blended cements.

One of the way of how to effect the cement activity, first of all, of the composite one, is its modification with chemical additives.

From this point of view worth mentioning is the use of sodium and potassium salts. The known-in-the-art works on creation of alkali- containing cementitious materials- so-called «soil cements», which are known to differ considerably from the traditional ones in nature of the products of hardening (the presence of alkaline hydroaluminosilicates which are in their composition analogous to durable natural zeolitic materials [2, 3].

The present research performed in a continuation of the above works was dedicated at investigation of the interaction processes taking place in the blended cements modified with sodium salts. The objective of the study- to achieve high activity of these cements with decreased content of the OPC .

MATERIALS AND TESTING TECHNIQUE

Used as basic ingredients of the CC were the OPC with a strength class of 40 MPa and a specific surface of 400 sq m/kg (by Blaine), and natural and industrial pozzolanas as aluminosilicate constituents. The chemical composition of raw materials used in the experiments is given in Table 1.

Used as natural pozzolana was a perlite with a specific surface 420 sq m/kg (by Blaine). The mineralogical composition of the perlite is presented, mainly, by volcanic water-containing glass (96.71%) [4 , 5]. Used as industrial pozzolanas were a granulated blastfurnace slag (GBFS) and an ash- soda blend (ASB) with a specific surface 430 sq m/kg (by Blaine).

Table 1 Chemical composition of initial raw materials

NAME	OXIDE CONTENT, % by mass									
	SiO_2	AL_2O_3	Fe_2O_3	CaO	MgO	MnO	R_2O	R_2O	SO_3	LOI
GBFS	39.5	6.49	0.12	47.1	3.1	1.7	1.15	-	-	0.6
OPC clinker	51.5	42.0	3.4	0.4	0.7	-	-	1.2	0.6	0.12
ASB	53.7	17.4	12.75	3.70	2.82	0.08	3.19	1.01	0.61	4.69
Perlite	72.0	12.5	0.65	1.07	0.17	-	0.03	4.51	2.10	0.05

The GBFS may be referred to a basic type slag is is represented, mainly, by a glass phase (75…80 % by mass) [6]. Its mineralogical composition is represented by melilite (gehlenite), anorthite, sulpfides, calcite. The chemical analysis data of the ASB sample show that it may be referred to the acid ones [7, 8]. The phase- mineralogical composition of the ASB is represented by a glassy component of aluminosilicate and silicate composition with inclusions of quartz and hematite [7, 8, 9]. Used as additives in the experiments were an anhydrous sodium sulfate Na_2SO_4 and sodium carbonate Na_2CO_3 as well as their blend (SSB) in which they were contained in a ratio of 2:1. This blend is a model of an industrial by-product of chemical industry The additives were introduced together with a mixing water as a solution in a quantity of 6 % by mass of the cement mass. Used as a plasticiser (introduced in a quantity of 1 % by mass of the cement) was an additive selected from lignosulfonates. Used as fine aggregate was fine quartz sand with a gradation factor 1.42. CC was produced by mixing in appropriate proportions of the components ground to a certain specific surface. CC activity was determined on the beam- specimens (40x40x160 mm) made from a cement mortar (cement to slag ratio=1:3) of standard consistency allowed to harden for 3 days in normal conditions and 25 days in water.

EXPERIMENTAL RESULTS AND DISCUSSION

Activity

Taken as a criterion to assess the interaction processes in CC was its activity as the most important target property.

These investigations covered a wide range of CC content varying (in % by mass) from OPC (20 – 80), ASB (5- 65), perlite (5 – 65), GBFS (15- 75). The dependances between activity of CC and content of the above constituents were studied on the systems «OPC-GBFS-ASB» (Figure 1, a) and «OPC-GBFS-perlite» (Figure 1, b) (for reference).

As a result, it was found that the cements of the first system demonstrated the activity greater than 40 MPa with the ASB content less than 31 % by mass, and greater than 50 MPa - less than 5 % by mass. To maintain the 50 MPa activity with 25 mas.% ASB is possible by increasing the OPC content to 65 % by mass.

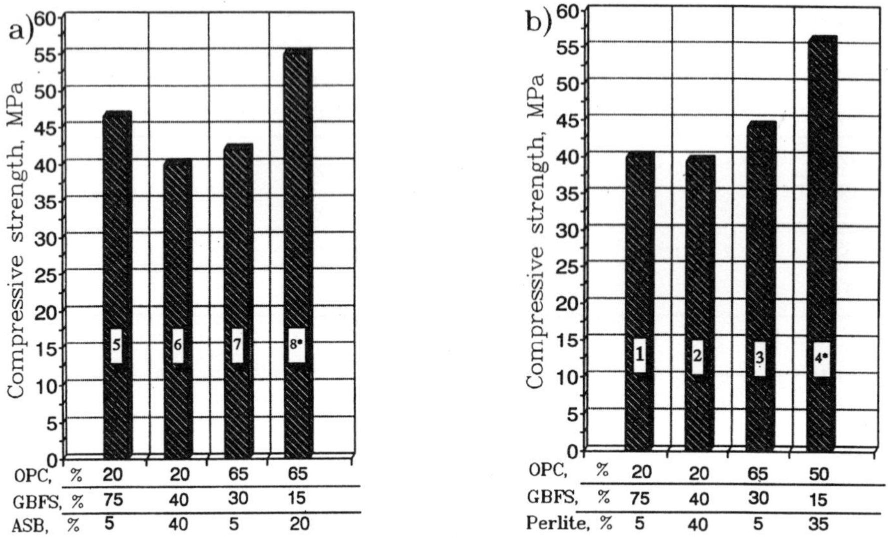

Figure 1 Activity of CC of the system 1 «OPC - GBFS - ASB» modified with SSB (a), of the system 2 «OPC - GBFS - perlite» modified with sodium sulfate (b)

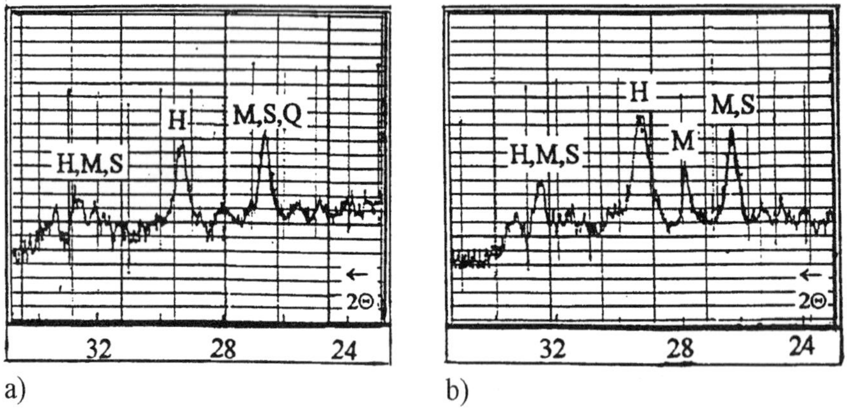

H - calcium hydrosilicates
M - melilite

S - γ-C_2S
Q - quartz

Figure 2 X-ray diffradtograms of the patterns of CC of the system 1 «OPC - GBFS - ASB» modified with SSB (a), of the system 2 «OPC - GBFS – perlite» modified with sodium sulfate (b)

To increase the content of pozzolana in the first system may be achieved by decreasing the OPC content due to activation of the aluminosilicate component by sodium carbonate [10, 11, 12].

To optimise the first system cement formulations by a criterion of activity, a simplex- grid method planning a 3-factor experiment was applied. Taken as variables in the first system with sodium carbonate were the contents of OPC, GBFS and ASB.

$$R = 56x_1+18x_2+27.28x_3$$

where, R = strength
x_1, x_2, x_3 = OPC, GBFS, ASB content (for reference)

Basing on analysis of the equation for further investigations have been selected the compositions given in Table 2.

The investigation of influence of an additive type on the cement activity showed high efficiency of the SSB additive. While investigating the interaction processes that take place in the system with 6 mas.% - content of such additive in combination with a plasticiser, some increase in activity of the cement with the increased ASB content was reported: Composition No 6 (40 mas.% ASB) is characterised by activity 40 MPa (Figure 1, a), what is by 37 % higher as that of the additive - free analog (CC activity 25 MPa) and of the cement with sodium carbonate - by 12% (35 MPa). Besides, the SSB additive affects greatly the activity of cement with ASB as compared with that containing GBFS. So, at the same OPC content (65%), Composition No 8* (20 mas. % ASB) has an activity 55 MPa, what is by 24% greater as compared with Composition N 7 (5 mas.% ASB).

Table 2 CC compositions

COMPOSITIONS Nos	COMPONENTS CONTENT, mas.%			
	OPC	GBFS	Perlite	ASB
1	20	75	5	-
2	20	40	40	-
3	65	30	5	-
4*	50	15	35	-
5	20	75	-	5
6	20	40	-	40
7	65	30	-	5
8*	65	15	-	20

Clearly followed is the influence of nature of the anion constituent of the modifying additive: the salts of sodium and weak acids are found to be the most effective in the compositions with increased calcium content (OPC) (the strong acid anion can take place in the formation of soluble calcium salts), whereas salts of sodium and strong acids - in a low-calcium compositions due to decreased activity of the last. In the last case a lack of Ca^{2+} initiates participation of the Na^{+} - ions, in the aluminosilicate glass destruction. These foundings correlate well with the comparative data on activity of the compositions with perlite containing Na^{+} [11].

Worth mentioning that it is impossible to give an objective characteristic of interaction processes in the cements containing the ASB using absolut values of activity because of variable chemical and phase composition of this product which is an industrial waste. That's why, when using the ASB containing blended cements in concrete, a extremely careful approach to analysis of raw materials and, first of all, to the ASB, and their adjustment to technology should be taken. The formulations should be chosen in which the range of ASB content would provide a required stability of properties of CC.

Character of Hydration Products

The X-ray analysis data of the hydration products of the modified cement of Composition No 6 of the first system (Figure 2, a) as compared with those of an additive-free reference cement testify to a predominant formation of low-basic calcium hydrosilicates (0.305, 0,280, 0.182 nm). The intensity of the peaks of the non-reacted ASB and GBFS in a form of residual peaks of quartz (0.335, 0.251, 0.202 nm) and coinsiding with them the peaks of melilite (0.336, 0.273, 0.220 nm) and γ-C_2S (0.336, 0.268, 0.228 nm) is by 2 times less intensive. The presence of the peaks attributed to the initial components even in the modified CC testifies to uncompleteness of the pozzolanic reaction resulted from the increased density of the aluminosilicate glass of the ASB and GBFS. This allowed to increase the OPC content up to 65% by mass. These data obtained correlate well with those of differential - thermal analysis and infra-red spectroscopy. Of interest is the comparison of these data with those obtained for the cement compositions containing perlite (the second system).

As the X-ray analysis data demonstrate (Figure 2, b) the hydration products of the second system cement compositions modified by sodium sulfate containing 20% by mass of the OPC (Composition No 2) at high perlite content are, in general, similar to hydration products of the Composition No 6 containing industrial pozzolana. Worth mentioning is the formation of large quantities of low-basic calcium hydrosilicates of CSH (1) type (d= 0.305, 0.280, 0.182 nm) accompanied by residual peaks of low intensity attributed to melilite and γ-C_2S (d= 0.427, 0.335, 0.273, 0.267 nm), testifying to a more completeness of the pozzolanic reaction. The last may be also attributed to a lower, as compared with the ASB, density of water-containing perlite glass, and its relative homogeneity. The X-ray analysis results correlate well with those obtained using the differential- thermal analysis and the infra- red spectroscopy.

The new formations composition in a form of calcium hydrosilicates of the lower basicity and the gel phases, inclusive of alkaline aluminosilicate composition, explain rather high values of activity of CC modified by sodium salts additives. The identification of such phases is possible during their crystallisation for a long- term period [10]. At the same time, the earlier reported data [12] on the character of micropores, suggested to assume that just a gel of such composition is of importance in the formation of material property. To the properties in support of this assumption refer, first of all, resistance to aggressive environments.

The results of corrosion resistance of CC compositions after they were cured for 72 days in 5% - solutions of sodium and magnesium sulfate salts are shown in Table 3.

The coefficient of resistance of all CC compositions under investigation, which is expressed as a ratio of flexural strengths of the specimens kept in solutions and those reference ones kept in water, exceeds 0.8, the coefficient of their resistance in sodium sulfate solution being somewhat greater than of those stored in magnesium sulfate solution.

Table3 Corrosion resistance of CC

COMPOSITIONS Nos (See Table 2)	FLEXURAL STRENGTH, MPa			COEFFICIENT OF RESISTANCE	
	Aggressive environments			In Na_2SO_4	In $MgSO_4$
	H_2O	Na_2SO_4	$MgSO_4$		
2	7.0	10.7	6.8	1.52	0.97
4*	6.0	8.3	8.03	1.38	1.34
6	7.5	9.8	8.2	1.31	1.09
8*	5.4	5.3	9.1	0.98	1.68

CONCLUSIONS

1. The interaction processes in the system "OPC-GBFS-ASB" are initiated by alkali metal salts.

2. The hydration products of the studied CC are found to be similar to those formed in the pozzolanic systems. A difference is the increased content of the gel-like new formations in a form of calcium hydrosilicates and the creation of conditions for possible formation of the products which in composition would be analogous to natural alkaline hydroaluminosilicates in a gel state.

3. The choice of the anion constituent of sodium salt which is used as an activator is determined by a nature of pozzolanic constituent of CC: sodium salts of weak acids are found to be the most efficient in the basic systems, whereas sodium salts of strong acids - in low-calcium systems due to the lower activity of the last, thus causing participation of the Na^+ ions, predominantey, in the destruction of glass of the aluminosilicate component of the system.

4. The new formations compositions of CC modified by additives while hydration allowed to recommend them for the use in the aggressive environments, including when laying road pavements.

REFERENCES

1. DHIR, R K AND McCARTHY, M I. Optimising Binders for Concrete Construction. World Cement, 1997, Vol.28, No. 1, pp 74- 81.

2. GLUKHOVSKII, V D. Soil Silicates, UkrSSR Gosstroiizdat, Kiev, 1959, 134 pp.

3. GLUKHOVSKII, V D. Soil Silicates- Made Articles and Structures, Budivelnik., 1967, 154 pp.

4. GORIAYNOV, K E AND GORIAYNOVA, S K. Technology of Heat Insulating Materials and Articles, Stroiizdat, Moscow, 1982, 376 pp.

5. Materials and Articles from Expanded Perlite, Edit. A A Zhukov, Stroiizdat, Moscow, 1972, 249 pp.

6. KRIVENKO, P V AND PUSHKARJEVA, E K. Durability of Slag Alkaline Concrete, Budivelnik, 1993, 224 pp.

7. SERGEEV, A M. Utilisation of Energy Industry Wastes in Construction, Budivelnik, Kiev 1984, 120 pp.

8. Energy Fuels of the USSR, Manual, Energia, Moscow, 1973, 128 pp.

9. Composition and Properties of the Ash and Slag from the Heat Power Stations, Code of Practice (Edit. V A Melent'ev, Energoatomizdat, Leningrad, 1985, 285 pp.

10. Alkaline and Alkaline - Alkali-Earth Hydraulic Binders and Concretes (Edit. V D Glukhovskii), Vyscha Shkola, Kiev, 1979, 232 pp.

11. FUNDI, Y A S. Alkaline Pozzolana Portland Cements. Proceedings of the First International Conference on Alkaline Cements and Concretes, Kiev, 1994, Vol.1, pp 181-192.

12. RUNOVA, R F, KOCHEVYKH, M A AND RUDENKO, I I. Pressed Ash Containing Concrete: Activation of Structure Formation. Proceedings of the International Conference Concrete In The Service Of Mankind, Dundee, Scotland, 1996, Vol. Concrete for Environment and Protection, pp 671-676

INVESTIGATIONS WITH SUPERSULFATED CEMENT

J Stark

U Frohburg

Bauhaus-University Weimar

Germany

ABSTRACT. The paper presented is a report on investigations made into a cement very rich in ground granulated blast-furnace slag (>75% blast-furnace slag) which is chemically activated by sulfates: supersulfated cement. Numerous investigations have been carried out to find an optimised composition of supersulfated cement especially with respect to compressive strength. The most suitable composition for supersulfated cement was found to be 85.0% ground granulated blast-furnace slag, 13.5% anhydrite and 1.5% Portland cement.

In addition to the compressive strength other parameters influencing the durability of concrete were examined, i.e. the microstructure of the concrete, gas permeability, imperviousness to water, chloride-ion penetration, depth of carbonation, air void parameters and the resistance to frost and frost-deicing salt attack and subsequent scaling. The high tendency of supersulfated cement concrete towards carbonation as well as its low resistance to frost and frost-deicing salt attack marked the main problems for this type of concrete. Various ways of improving the frost-deicing salt resistance, such as the improvement of curing, the reduction of the water-to-cement ratio, the use of a curing agent as well as the use of draining formwork material have been explored. The required improvement of the frost-deicing salt resistance could only be achieved by means of draining formwork material.

Keywords: Blast-furnace cement (BFC), Durability, Draining formwork material (DFM), Frost resistance, Frost-deicing salt resistance, Ground granulated blast-furnace slag (BFS), Ordinary Portland cement (OPC), Supersulfated cement (SSC)

Professor J Stark is Director of the F.A.Finger-Institute of Building Material Sciences at the Bauhaus-University Weimar. The main fields of research are the use of binders, the durability of concrete, structural light-weight concrete, high-performance concrete, and historic mortars.

U Frohburg is an research assistant at the F.A.Finger-Institute of Building Material Sciences, Bauhaus-University Weimar. She specialises on the durability of concrete, in particular on the frost and frost-deicing salt resistance of concrete containing cement rich in ground granulated blast-furnace slag.

INTRODUCTION

Supersulfated cement (SSC) has not been produced in Germany for approximately three decades. Today, considerable damage is observed on some SSC concrete structures. That is why this type of cement must be considered again. Because of the incompatibility with other types of cement the knowledge of the subject concerning the composition of the hardened cement stone is necessary.

For the reconstruction of a dam (in Weida/Thuringia) which had been built 45 years ago using SSC concrete a suitable SSC had to be reproduced.

OPTIMISING SUPERSULFATED CEMENT

As required by the formerly valid German standard DIN 4210, supersulfated cement had to contain at least 75% ground granulated blast-furnace slag (BFS), a minimum of 3% anhydrite and not more than 5% Portland cement, ordinary (OPC).

Precondition for the production of SSC is a highly basic BFS with a aluminium oxide content above 15% (≥ 13% according to DIN 4210).

Furthermore, DIN 4210 required: $$\frac{CaO + MgO + Al_2O_3}{SiO_2} \geq 1.6$$

In contrast to blast-furnace cement (BFC), in SSC blast-furnace slag is the sole provider of hydraulic properties. In order to achieve the optimum pH value (pH = 11.5 - 12) required for the hardening of SSC (the formation of ettringite) it contains a small amount of OPC. Along with C-S-H phases, ettringite ($C_3A{\cdot}3C\overline{S}{\cdot}H_{32}$) is an essential product of hydration. Furthermore, monosulfate ($C_3A{\cdot}C\overline{S}{\cdot}H_{12}$), C-A-H phases and $Al(OH)_3$ are formed.[1] For further investigations it was the task to choose the appropriate basic materials and to find their best composition for the production of SSC.

Basic Materials

The basic materials used for the production of supersulfated cements were anhydrite, OPC (CEM I 42.5 R according to DIN 1164-1) and nine different BFS. The BFS differed in their chemical composition (especially their content of Al_2O_3 and MgO) as well as their pH-value and their content of water-soluble ions in aqueous extracts[2]. For the process of optimising the cement composition a blast-furnace slag from the Netherlands was employed. The chemical composition of the materials used is presented in Table 1.

Characteristics of Supersulfated Cement

As the starting point for optimising the cement composition a blast-furnace slag - anhydrite - cement was chosen, originally designed as an injection mortar and foamed injection mortar, with 86.5% BFS, 8.5% anhydrite and 5.0% OPC [2][3]. This kind of cement achieved sufficient strength to produce repair mortar but for the use in concrete it had to be improved. For testing standard mortar prisms of 40×40×160 mm were made with a cement-to-sand ratio of 1:3 and a water-to-cement ratio of 0.5. The specimens were demoulded after two days and stored under water until testing. Test criterion was the compressive strength, shown in Table 2.

Furthermore, the required amount of water, setting time, tensile bending strength, volume stability and heat of hydration were determined.

Table 1 Chemical composition of the materials used

[%]	OPC (CEM I 42.5 R)	ANHYDRITE	BLAST-FURNACE SLAG (NL)	SUPERSULFATED CEMENT
GV*[1]	1.9	-	+*[3]	+*[3]
SiO_2	20.2	-	31.9	28.0
Al_2O_3	5.0	-	16.5	13.8
Fe_2O_3	2.2	-	0.43	0.4
CaO	64.5	39.9	36.6	37.1
MgO	0.9	-	10.2	8.9
K_2O	1.29	-	0.46 (0.01 wl.*[2])	0.8
Na_2O	0.20	-	0.35 (0.03 wl.)	0.32
SO_3	3.3	55.3	0.2	8.4

*[1] GV = loss due to ignition at 1000°C; *[2] wl. = water-soluble; *[3] + = increasing

Table 2 Compressive strength of the mortar prisms

AGE AT TEST	COMPRESSIVE STRENGTH RESULTS [N/mm²]				
(a) Mixture with 83% blast-furnace slag					
Anhydrite / OPC [%]	16 / 1	15 / 2	14 / 3	13 / 4	12 / 5
2 days	11.6	14.0	11.2	9.6	11.2
7 days	39.8	29.4	18.1	14.0	15.8
28days	49.2	39.5	22.2	16.9	18.6
(b) Mixture with 85% blast-furnace slag					
Anhydrite / OPC [%]	14 / 1	13 / 2	12 / 3	11 / 4	10 / 5
2 days	11.4	13.1	10.5	10.1	12.6
7 days	39.9	28.4	15.8	14.1	17.5
28days	51.6	36.5	18.6	16.6	19.7
(a) Mixture with 87% blast-furnace slag					
Anhydrite / OPC [%]	12 / 1	11 / 2	10 / 3	9 / 4	8 / 5
2 days	11.4	13.8	10.6	9.6	13.1
7 days	36.6	28.5	15.5	13.6	18.5
28days	49.6	37.1	19.4	15.8	20.8

All 15 mixtures tested fulfilled the requirements of DIN EN 196 regarding volume stability and setting time. They all could be rated as cement with low heat of hydration.

Preference was given to a cement with 85% BFS, 14% anhydrite and 1% OPC. With respect to the dispensing accuracy of mixing equipment of ±0.5% the following cement was produced in minor amounts: 85% blast-furnace slag - 13.5% anhydrite - 1.5% Portland cement. Re-testing the characteristics of the cement and the mortar made thereof did not show any relevant changes.

INVESTIGATION OF SUPERSULFATED CEMENT CONCRETE

The investigations of SSC concrete have been done with concrete without any additives. Additionally, the influence of artificially entrained air voids as well as the effect of draining formwork material (DFM) were to be studied.

Composition and Production of Concrete

Materials

The concrete tested was produced with SSC of the 85/13.5/1.5 type. The aggregates used were Rhine sand and Rhine gravel (ρ = 2,62 g/cm³) in the size fractions 0/0.25, 0.25/0.5, 0.5/1, 1/2, 2/4, 4/8, 8/16 mm. This aggregate meets the higher requirements on the resistance to frost and deicing materials according to DIN 4226. The air-entraining agent used for the production of concrete with artificial air voids was based on vinol resin. For determining the influence of a draining formwork material, a spinning-fibre material made of thermally hardened continuous polypropylene fibre was employed. In order to study possible improvements of the frost and frost-deicing salt resistance of concrete various additives were tried. Superplasticisers made from melamine resin or naphthalene sulfonate were employed to achieve the consistency required in concrete with a low water-to-cement ratio. The effect of a curing agent, an aliphatic paraffin-wax emulsion was also studied.

Mix proportions

The concrete mixtures as listed in Table 3 were included in the testing programme:

Table 3 Concrete mixes tested

SSC CONCRETE	WATER CONTENT [kg/m³]	CEMENT CONTENT [kg/m³]	AGGREGATE CONTENT [kg/m³]	AIR-ENTRAINING ADMIXTURE [ml/kg_{cem}]	CONSISTENCY (DIN 1048, p.1)	AIR VOID CONTENT [%]
Without air-entrainment	185	370	1764	—	regular	1.5
With air-entrainment	185	370	1672	2.0	flowable	5.0

The following variants were investigated:

- concrete without air-entrainment / without DFM
- concrete without air-entrainment / with DFM
- concrete with air-entrainment / without DFM
- concrete with air-entrainment / with DFM

Production and curing

All concretes were mixed in a pan-type laboratory mixer. The concrete was placed in a double mould (dimensions of one section: 500×200×500 mm). Before the fresh concrete was placed the respective area of a mould section was covered with the DFM so that its effect could be studied. After two days the concrete specimens were demoulded. Cores of the supersulfated cement concrete were drilled after seven days. The concrete specimen were protected from drying-out by covering with moist burlap and plastic sheets. Until the beginning of the tests after 28 days of curing the core samples were stored at 20°C/65%RH.
For a second test series concrete cubes of 150×150×150 mm were produced. These were stored under water until the seventh day at 20°C and subsequently 21 days at 20°C/65% RH.

Scope of the investigation

In an extensive testing programme compressive strength, microstructure of the concrete, gas permeability, imperviousness to water, chloride-ion penetration, depth of carbonation and the frost and frost-deicing salt resistance were determined, partially at certain distances from the concrete surface.

The compressive strength (according to DIN 1048, part 5) was determined at the age of 28 days on three core samples (50mmØ × 50mm). The results of the compression tests were converted in order to be comparable to 200mm test cubes.

The microstructure / pore size distribution was determined in layers by high-pressure mercury porosimetry. Starting from the moulded surface, five layers of 10mm thickness each (up to a depth of 50mm) and one layer of 10mm at a depth of 100mm were examined.

The gas permeability was measured on three specimens (50mmØ × 20mm) according to DAfStb, number 422 [4]. One side of each sample was subjected to oxygen gas pressured at 0.15 MPa to 0.35 MPa while the opposite side was kept at atmospheric pressure. The amount of oxygen gas permeating the sample was measured over time, and the rate of permeability was calculated.

Two specimens (150mmØ × 120mm) were used for the determination of the imperviousness to water according to DIN 1048, part 5. These samples were stored under water for 28 days prior to testing. During the test the samples have been subjected to a water pressure of 0.5 MPa for three days. Afterwards, each sample was split and the maximum depth of water penetration into the sample was measured.

The penetration of the chloride ions was determined in all concrete series by the capillary absorption of NaCl solution. Additionally, the water-soluble chloride proportions in the supersulfated cement concretes were determined because of their influence on corrosion. Two core samples (50mmØ × 75mm) were stored in 3% NaCl solution for 28 days. After this period of time the upper portion of 30 mm of these core samples, beginning from the moulded surface, were sawn without cooling water into segments of 5 mm thickness each, and the chloride content was measured according to the specifications of DAfStb, number 401 [5].

For the determination of the depth of carbonation the core samples were stored under water until the seventh day and afterwards their moulded surface was exposed to a 3% CO_2 atmosphere (23°C/75%RH) until the 28th day and 14 months in laboratory by 20°C/65%RH, respectively. The indicator solution phenolphthalein (0.1% in 70% ethanol) was used for measuring the depth of carbonation in the split core samples.

The air void parameters and/or the distribution of air voids was determined by image analysis of thin sections [4]. Starting from the surface, four layers (0-5mm, 5-25mm, 25-45mm and 45-65mm) of the sections were analysed.

Three and six core samples, respectively, (100mmØ × 75mm) were tested for their frost and frost-deicing salt scaling resistance by means of the CF/CDF test [6][7]. After prestorage (28 days) the samples are allowed to absorb the test solution (water and 3% NaCl solution, respectively) by capillary suction for 7 days.

The CF/CDF-test is carried out in special temperature controlled chest with a liquid cooling bath. A cycle of freezing and thawing (+20°C / -20°C) takes 12 hours and a regular CF/CDF-test takes 14 days which corresponds to 28 cycles of freezing and thawing.

For testing concrete mixes or concrete constituents in a mix it is recommended to cast and compact the test specimen on a vibrating table in 150mm cube moulds according to prEN-ISO 2736/2. For all further investigations concrete specimens were produced according to this standard.

TEST RESULTS AND DISCUSSION

The test data are listed in Table 4; see Table 3 for concrete mixes.

Compressive Strength

All tested concretes may be classified as grade B 25 concretes (at least 25N/mm² compressive strength after 28 days). Test results show that the use of the DFM has no effect on the compressive strength. Air-entrainment, however, reduces the compressive strength of the concretes.

Table 4 Test data of all investigated concretes

	SSC oA/oD	SSC oA/wD	SSC wA/oD	SSC wA/wD
Compressive strength [N/mm²] (β_{W200})	32	32	31	31
Porosity [%]				
- gel (4nm-10nm)	2.9	2.8	2.9	2.9
- capillary (10nm-100μm)	19.7	11.9	24.7	21.6
Coefficient of gas permeability [10^{-16} m²]	0.27	0.07	1.11	0.33
Depth of water penetration [mm]	15.0	9.5	13.5	11.5
Increase in mass [%] (7d absorption of 3% NaCl solution)	0.93	0.54	1.47	1.06
Chloride-ion penetration [%]				
- depth of concrete 0-5mm	2.04	1.06	2.65	1.40
- " 5-10mm	0.88	0.42	0.86	0.54
- " 10-15mm	0.52	0.38	0.56	0.49
Depth of carbonation [mm]				
- 7d water, 21d 3% CO_2-atmosphere	12.0	0.0	16.0	4.5
- 7d water, 14 months 20°C/65% RH	10.0	4.0	10.0	8.5
Air void parameters				
- air voids [%]	1.2	1.2	2.2	2.9
- air voids ≤ 0.30 mm [%]	0.6	0.5	1.1	1.3
- spacing factor [mm]	0.13	0.20	0.15	0.17
Scaling [g/m²]				
CF: test solution-water				
- cores (100 mm)	1543	64	2661	155
- cubes (150 mm)	1997	—	—	—
CDF: test solution-3% NaCl				
- cores (100 mm)	5412	531	6119	2109
- cubes (150 mm)	8321	—	—	—

* oA = without air-entrainment, wA = with air-entrainment
oD = without draining formwork material, wD = with draining formwork material

Microstructure

Table 4 lists the results for the porosity as determined for the layer 0-10 mm because this portion is relevant for the durability of concrete in general and because the major differences between the concretes tested occurred in this layer. The capillary porosity of concretes made with SSC is relatively high and must be due to the low degree of hydration at the time of testing (age of the concrete: 28 days). It is clear that the use of air-entraining admixtures in the respective concretes led to higher contents of capillary pores and air voids. A favourable effect of the DFM on the capillary porosity near the surface was established.

Gas Permeability

According to Gräf and Grube [8] the microstructure of cement concrete may be said to be dense for any permeability coefficient $K \leq 1 \cdot 10^{-16} m^2$. Concrete with a dense microstructure could be produced with SSC. Entraining artificial air voids produced a more open microstructure.

Imperviousness to Water

Regardless of the grade of concrete it may be stated that the concretes meet the requirements on the imperviousness to water as well as on a high resistance to chemical attack (depth of water penetration < 50 mm and < 30 mm, respectively).

Depth of Chloride-Ion Penetration

Considering the results obtained for the absorption of NaCl solution, it may be stated that the absorption of chloride-ions due to the increase in mass could be markedly reduced by the use of DFM.
The tests for chloride-ion penetration on the four SSC concrete series showed that the penetration of chloride-ions is considerably inhibited by the use of DFM. This applies to concretes with and without air-entraining admixtures. Especially near the surface lower chloride contents were established. This is clearly shown by the results obtained after 28 days storage in NaCl solution.

Depth of Carbonation

In general the results show that concretes with SSC have an increased depth of carbonation compared to other cement concretes, e.g. concretes with OPC or with BFC. Just as in the case of BFC concrete carbonation leads to a coarser microstructure and to the formation of metastable $CaCO_3$ modifications (aragonite, vaterite) [9] as discovered during these investigations. The rate of carbonation is, however, higher than that of blast-furnace cement. One reason is that the tendency of ettringite towards carbonation is greater than that of the C-S-H phases. Besides, there is no $Ca(OH)_2$ that could impede carbonation.[1]

By the use of the DFM the carbonation of concretes without air-entraining admixtures could be prevented and/or delayed.

Air Void Parameters

It must be stated that entraining artificial air voids in SSC concrete has proven to be difficult. Despite achieving a sufficient air void content in fresh concrete, the air void content in hardened "entrained" concrete showed no improvement compared to non-entrained concrete.

The air-entrained SSC concrete does# not meet the requirements on a minimum content of air voids ≤ 0.30 mm of 1.5% and on the air voids spacing factor ≤ 0.20 mm (according to DIN V ENV 206, Oct. 1990). It was established that in concrete made with DFM a high content of air voids occurs near the surface which decreases with the depth of the concrete.

Frost Resistance

The limiting value for the CDF test after 28 cycles of freezing and thawing is 1500 g/m² of scaled material. Experience shows that about 600 g/m² may be considered as an acceptance criterion for the CF test.[6][7]

The results of the CF test show that supersulfated cement concrete has a very low frost resistance, regardless which kind of concrete samples were tested.

One cause of the low frost resistance may be the low degree of hydration at the time of testing. The capillary porosity of the microstructure of the supersulfated cement concrete is still rather high. Furthermore, the storage under water for one week is too short to achieve better frost resistance. Certainly, carbonation which causes dusting and scaling of the concrete surface also plays an important role.[10]

Frost-Deicing Salt Scaling Resistance

Both frost resistance and frost-deicing salt resistance of SSC concrete are extremely low. As already observed with BFC concrete, SSC concrete shows a typical scaling (Figure 1). The heavy initial scaling is followed by a period of lower intensity of scaling. This course of scaling is caused by carbonation connected with the coarser microstructure and the formation of the metastable $CaCO_3$ modifications. [9][10]

Furthermore, there is a great dependence of the frost-deicing salt resistance on the size and preparation of the specimen. The samples prepared according to the standard have a larger surface subjected to carbonation and consequently heavier scaling.

WAYS OF IMPROVING THE FROST-DEICING SALT RESISTANCE

The entrainment of artificial air voids which proved so successful in OPC concrete will not lead to any significant improvement in SSC concrete. On the contrary, it had to be stated that the frost and the frost-deicing salt resistance of SSC were impaired. A way of improving the frost-deicing salt resistance is to impede carbonation as far as possible.

So, in the first place, the effect of the following changes in curing was studied:

- 23 days under water (20°C) / 5 days at 20°C/65%RH
- 23 days under water (35°C) / 5 days at 20°C/65%RH
- 23 days saturated $Ca(OH)_2$ solution / 5 days at 20°C/65%RH (according ASTM C511-93)

The courses of scaling are represented in Figure 1.

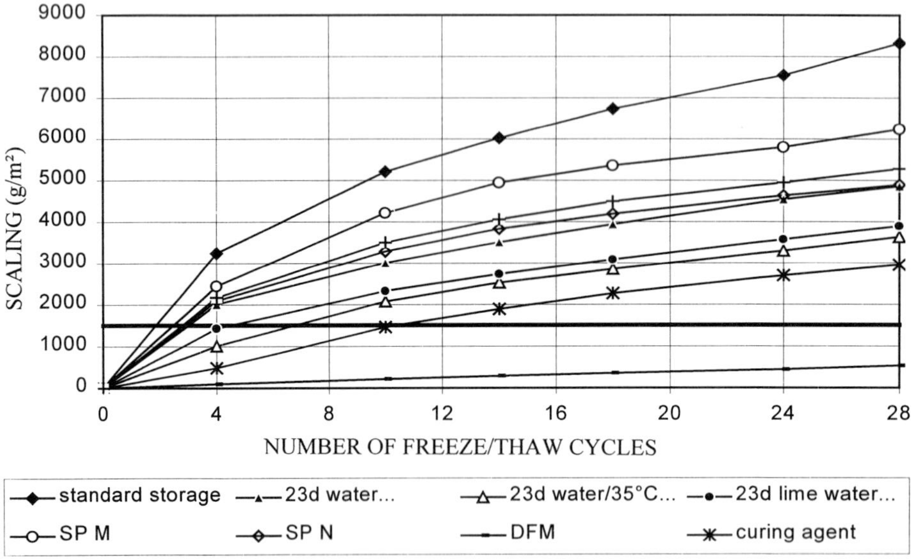

Figure 1 Scaling of concrete with SSC after different changes in curing as well as with superplasticisers made from melamine resin (M) or naphthalene sulfonate (N)/(w/c = 0.4) and with a curing agent and DFM

Curing reduced the amount of scaling, in some cases to 50% of the amount after standard storage. But a significant improvement of the microstructure could not be observed. While an improved microstructure could not be established, a reduced depth of carbonation was measured (after 28 days):

- 7 days under water (20°C) / 21 days 20°C/65%RH 4.0 mm
- 23 days under water (20°C) / 5 days 20°C/65%RH 0.5 mm
- 23 days under water (35°C) / 5 days 20°C/65%RH 1.0 mm
- 23 days saturated $Ca(OH)_2$ solution / 5 days 20°C/65%RH 0.5 mm

It has to be concluded that by changes in curing it was not possible to produce a SSC concrete with a high frost-deicing salt resistance.

Other ways of reducing the initial scaling of SSC concrete had to be considered. A denser microstructure may be achieved by a low water-to-cement ratio. As the complete hydration of SSC requires a high water content the maximum reduction of the water-to-cement ratio had to be limited to 0.4. Superplasticisers were employed to achieve the consistency required. The effect of a curing agent was also studied. Furthermore, the effect of DFM was compared to the other measures taken.

A comparison of the courses of scaling shows that the required improvement of the frost-deicing salt resistance can only be achieved by the use of the DFM. It is true that the reduction in the water-to-cement ratio and the addition of superplasticisers results in an increase in compressive strength and the formation of a slightly denser microstructure, but sufficient improvement of the frost-deicing salt resistance could not be achieved. The depth of carbonation at the beginning of the frost-deicing salt attack was 3.0mm.

The use of a curing compound could also not prevent carbonation (1.5 mm at the beginning of frost attack). Although a significant reduction in scaling has been achieved, the SSC concrete modified by these measures cannot be classified as frost-deicing salt resistant.

SUMMARY

Responding to an actual demand intensive investigations into supersulfated cement concrete have been carried out. An optimised composition of SSC was found, produced and tested. The results prove that SSC concrete of high compressive strength and durability can be produced. Nevertheless, two somehow connected problems occurred. One is the strong tendency of SSC towards carbonation, the other its rather low resistance to frost and frost-deicing salt attack. The only sound improvement concerning both problems can be offered by the employment of draining formwork material.

REFERENCES

1. STARK, J. Sulfathüttenzement. Wissenschaftliche Zeitschrift HAB Weimar, 1995, Vol.41, No.6/7. pp 7-15.

2. MIELKE, I. Beitrag zur Optimierung sulfatisch angeregter Zemente - SHZ, Working paper. Bauhaus-University Weimar, 1998, pp 18.

3. TRAUTMANN, I AND KNÖFEL, D. Entwicklung eines Hüttensand - Gips - Portland-zement - Mörtels. ZKG, 1993, Vol.46, No.2, pp 95-101.

4. DEUTSCHER AUSSCHUß FÜR STAHLBETON (DAfStb). Prüfung von Beton; Empfehlungen und Hinweise als Ergänzung zu DIN 1048, Berlin, 1991, Vol.422.

5. DEUTSCHER AUSSCHUß FÜR STAHLBETON (DAfStb). Anleitung zur Bestimmung des Chloridgehaltes von Beton, Berlin, 1989, Vol.401.

6. RILEM RECOMMENDATION (RILEM TC 117 FDC). CDF Test - Test method for the freeze-thaw resistance of concrete with sodium chloride solution. Materials and structures, 1996, Vol.29, pp 523-528.

7. RILEM TC 117 FDC. Draft Recommendation for the test method for the freeze-thaw resistance of concrete - Test with water (CF) or with sodium chloride solution (CDF). Materials and stuctures, 1995, Vol.28, pp 175-182.

8. GRÄF, H AND GRUBE, H. Einfluß der Zusammensetzung und der Nachbehandlung des Betons auf seine Gasdurchlässigkeit. Beton, 1986, Vol.12, pp 473-476.

9. LUDWIG, H-M. Zur Rolle von Phasenumwandlungen bei der Frost- und Frost-Tausalz-Belastung von Beton, PhD Thesis, Hochschule für Architektur und Bauwesen Weimar, 1996.

10. KNAACK, U AND STARK, J. Frost and frost-deicing salt resistance of supersulfated cement concrete. Proc. of the Int. RILEM Workshop on Resistance of Concrete to Freezing and Thawing With or Without De-icing Chemicals, Essen, 1997, pp 139-145.

ACTIVATION AND ACCELERATION OF PORTLAND CEMENT/GGBS BLENDS USING CEMENT KILN DUST (CKD)

R K Dhir

T D Dyer

J E Halliday

University of Dundee

United Kingdom

ABSTRACT. Cement kiln dust (CKD) is a byproduct of the manufacture of Portland cement. It contains soluble components which are capable of activating pozzolanas and slags, as well as accelerating the hydration reactions of Portland cement. This paper examines the performance of blends of Portland cement, CKD and GGBS and examines the effect of CKD on hydration chemistry.

The presence of CKD at CKD / GGBS ratios of between around 1 and 1/3 yields optimum compressive strength development behaviour, which isothermal conduction calorimetry confirms to be due to an enhanced degree of hydration. X-ray diffraction shows that the dust accelerates the hydration of both C_3S and C_3A. At 28 days the highest quantity of amorphous material is present in the PC / CKD / GGBS blend which displays optimum strength, which is attributed to the production of larger quantities of CSH gel partly due to both acceleration of PC and GGBS reactions. Whilst more research is required, the results presented in this paper suggest that it could be put to use in selective construction applications.

Keywords: Cement kiln, Ground granulated blasfurnace slag, Ternary blends, Hydration chemistry

Professor R K Dhir is the Director of the Concrete Technology Unit and Professor of Concrete Technology at the University of Dundee. He is a member of numerous national and international technical committees and has published extensively on many aspects of concrete technology, binder science, durability and construction methods.

Dr Thomas D Dyer is a research / teaching fellow working in the Concrete Technology Unit of the Department of Civil Engineering, University of Dundee. He is a materials scientist with interests in waste recycling and other aspects of sustainable concrete construction.

Judith Halliday is a postgraduate research student in the Concrete Technology Unit of the Department of Civil Engineering, University of Dundee. She is a civil engineer currently working towards a PhD on the topic of cement kiln dust and incinerator fly ash recycling in concrete.

INTRODUCTION

Cement kiln dust (CKD) is a byproduct of the manufacture of Portland cement. It generally consists of fine particles comprising raw feedstock, partially clinkered material and re-condensed volatilised compounds. Whilst a proportion of this dust can be returned to the cement kiln, the volatile components are a source of technical problems and hence this practice is restricted. The disposal of this dust therefore poses a significant problem to the cement industry, and routes for recycling the material are being sought.

The material itself contains a number of components whose properties are likely to assist in the activation of pozzolanic materials and hydraulic slags. This has been previously demonstrated with PFA. This paper investigates the performance and hydration chemistry of ternary blends containing Portland cement, CKD and ground granulated blastfurnace slag (GGBS).

EXPERIMENTAL DETAILS

Materials

The major oxide analyses, determined using XRF, of the CKD, GGBS and PC used are supplied in Table 1, whilst the mineralogical composition of the materials estimated using the Rietveld refinement method on x-ray diffraction data (see subsequent text for further details) are given in Table 2.

Table 1 Major oxide analyses of the materials used

	WEIGHT %		
	CKD	GGBS	PC
SiO_2	12.2	33.5	21.1
TiO_2	0.1	0.5	0.2
Al_2O_3	3.2	12.5	5.0
Fe_2O_3	1.8	0.1	2.7
MnO	0.0	0.2	0.1
MgO	0.9	7.1	1.6
CaO	34.8	39.3	64.9
Na_2O	1.6	0.1	0.3
K_2O	7.5	0.4	0.6
SO_3	10.6	0.7	3.3
Cl	2.8	0.0	0.03

Blends Investigated

Five blend were investigated, all with a PC content of 65%. Varying proportions of CKD and GGBS were used with the percentage of CKD in the CKD / GGBS blend being 100%, 75%, 50%, 25% and 0%. This method of describing the mix proportions is used in all subsequent Figures.

Table 2 Mineralogical composition of the materials, estimated using Rietveld refinement methods

PHASE	CHEMICAL FORMULA	CKD	GGBS	PC
Calcite	$CaCO_3$	52.8	2.1	-
Gehlenite	$Ca_2Al_2SiO_7$	-	5.9	-
Larnite	Ca_2SiO_4	-	1.6	-
Wollastonite	$CaSiO_3$	-	0.9	-
Portlandite	$Ca(OH)_2$	8.4	-	2.0
Sylvite	KCl	6.1	-	-
Anhydrite	$CaSO_4$	7.0	-	7.1
Lime	CaO	1.4	-	-
Halite (potassian)	$Na_{0.31}K_{0.69}Cl$	2.9	-	-
Quartz	SiO_2	6.9	-	-
Aphthitalite	$K_3Na(SO_4)_2$	10.2	-	-
Muskovite, 2m1	$K_2Al_4(Si_6Al_2O_{20})$	3.3	-	-
Calcium Harmotome	$Ca_2Si_{12}Al_4O_{32}(H_2O)_{12}$	0.9	-	-
C_3S	Ca_3SiO_5	-	-	48.7
C_2S (β)	Ca_2SiO_4	-	-	23.3
C_3A	$Ca_3Al_2O_6$	-	-	7.0
C_4AF	$Ca_4Al_{1.6}Fe_{2.4}O$	-	-	9.8
Syngenite	$K_2Ca(SO_4).H_20$	-	-	1.9
Amorphous		-	89.5	-

Whilst measures were taken to minimise the exposure of the samples to atmospheric CO_2 at all stages, it is inevitable that some reaction would have taken place.

Thermal Analysis: Both thermogravimetry (TG) and differential thermal analysis (DTA) were carried out simultaneously on the paste samples. A heating rate of 20°C/minute was used to heat 10mg samples to 1000°C in a flowing nitrogen atmosphere.

Powder X-ray Diffraction: Paste samples were ground finely by hand using a pestle and mortar before being analysed using x-ray diffraction. A Philips diffractometer using a Cu-K-α radiation source and fitted with a curved graphite single crystal monochromator was used. A step size of 0.05°2θ was used over an angular range of 3 to 80°2θ.

The mineralogical composition of the hydrated pastes was estimated using Rietveld refinement techniques on the x-ray diffraction traces. The computer programs X-fit and Koalariet were used [3,4,5,6,7,8].

An unknown quantity of amorphous material was present in all the samples as both CSH gel and the glassy phases of the slag. These unidentifiable phases present a problem in terms of quantification using Rietveld refinement. This problem may be overcome by the inclusion of a known quantity of an internal standard.

However, since thermogravimetric analysis was also carried out on the pastes, portlandite contents were able to be determined accurately, thus allowing the quantities of this mineral present in the samples to be used as 'pre-existing' internal standards.

Structural data for the various crystalline phases was obtained from the Inorganic Chemical Structures Database held at Daresbury, UK [9]. In the case of the calcium carboaluminate hydrate phases, no structural data could be obtained. However, since the structures of these compounds is known to be analogous with that of calcium monosulfoaluminate hydrates [10]. Therefore, when such phases were identified on the X-ray diffraction traces, several independent monosulfate phases were used in the initial Rietveld refinement. Once the calculated profiles for the monosulfate phases had been fitted to the peaks on the traces, the basal spacings of these structures could be matched to those for calcium carboaluminate phases in the literature [10], and the replacement of sulfate ions with carbon and hydroxide ions was then possible for the final stages of refinement.

Isothermal Conduction Calorimetry: Isothermal conduction calorimetry was carried out on paste samples containing 30g of binder with a water / binder ratio of 0.5 using Wexham Developments apparatus. Distilled water was used.

RESULTS

Compressive Strength

Figure 1 shows the development of compressive strength in the PC / CKD / GGBS blends. It is clear that higher strengths are consistently observed in blends containing both CKD and GGBS. Blends containing only PC and CKD generally possess the lowest strengths at a given age, demonstrating the absence of cementitious components in this material. By 28 days, the blend with a CKD / GGBS ratio of 1/3 displays the highest strength, although at earlier ages the best results are obtained with higher ratios.

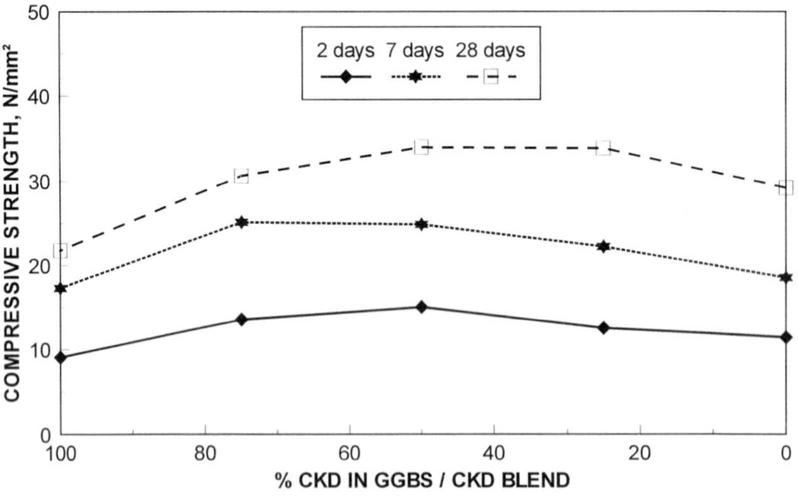

Figure 1 Compressive strength development of PC / CKD / GGBS blends

Isothermal Conduction Calorimetry

It should be noted that the quantity of heat generated by GGBS as it reacts is small and, hence, the findings obtained from isothermal conduction calorimetry only relate to PC hydration. Figure 2 shows the total heat evolved at 24, 48 and 72 hours. It is clear from these plots that by 72 hours the blends containing CKD and GGBS have generated more heat than blends containing only one of these components, a sign of enhanced activity. In general, the 72 hour results shown in Figure 2 resemble the compressive strength results previously discussed.

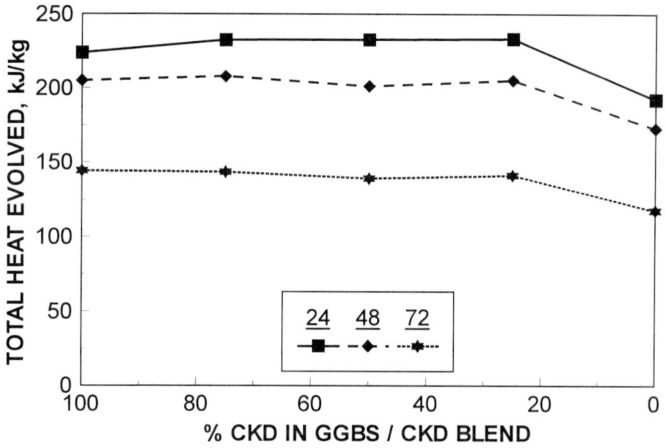

Figure 2 Total heat evolved at 24, 48 and 72 hours by CKD / GGBS / PC blends

Thermal Analysis

Thermal analysis was used in this study solely to determine the levels of portlandite present in the hydrating blends. These results, expressed as a percentage of the sample residue weight at 1000°C, are shown in Figure 3.

Portlandite levels are primarily influenced by two factors: the degree of hydration of the calcium silicate cement phases, whose reactions produce this compound, and the degree to which the GGBS has reacted, which will generally lead to a decrease in portlandite. In fact, the reactions of GGBS also generate portlandite to some extent [11], and so the observed consumption of this compound is less than that for wholly pozzolanic materials such as pulverised-fuel ash (PFA). At all ages, as the quantity of CKD in the blends increases, so does the quantity of portlandite. This can be attributed to three factors:

I. the CKD initially contained small quantities of portlandite and lime (which hydrates to form portlandite),

II. the consumption of portlandite by GGBS will increase as GGBS levels increase,

III. it is also possible that the CKD is having an acceleratory effect on the hydration of the PC calcium silicate phases.

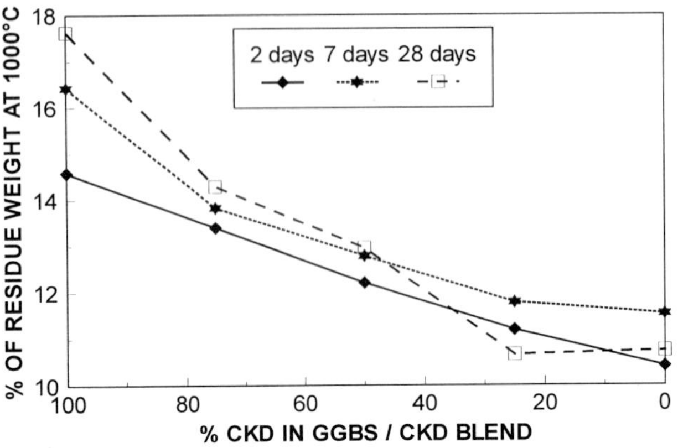

Figure 3 Portlandite levels in the PC / CKD / GGBS blends at 2, 7 and 28 days, determined using thermogravimetry

The reactions of GGBS are clearly demonstrated by the low rate of increase in quantities of this phase, in marked constrast to the blend containing only CKD and PC. This is particularly apparent in the blends containing the two highest quantities of GGBS, in which 28 day portlandite levels drop to below those at 7 days. This is most pronounced for the blend containing both CKD and GGBS, implying that CKD is accelerating the reactions of the slag

X-Ray Diffraction

Amorphous Phases: Figure 4 shows how the amorphous content of the pastes alters with time. Initially the only amorphous phases are those comprising the GGBS. By 2 days, the amorphous content of all the phases has increased greatly, due to the production of amorphous hydrates. However, beyond an age of two days, the quantity of amorphous matter changes less significantly. By 28 days, the previously straight line of amorphous content has become noticeably distorted, indicating differences in the degree and nature of hydration reactions in the pastes. Since the reactions of blastfurnace slag involve both the erosion of amorphous material and the precipitation of gel hydrates, it is impossible to precisely determine the degree to which these two processes have occurred. However, the maximum value observed in the blend with a CKD / GGBS ratio of 1/3 corresponds well with the compressive strength results, as well as with the low portlandite level observed for this blend at 28 days, implying that the high strength is at least partly the result of a suggested acceleration in the reaction of GGBS caused by the CKD, and leading to the production of higher quantities of CSH gel.

Calcium Silicates: Figure 5 shows how the C_3S content of each blend changes with time as hydration proceeds. The higher levels observed in the blends containing less CKD indicate that this material has the effect of accelerating C_3S hydration. Whilst not shown here, the hydration of C_2S proceeds at an equal rate, regardless of how much CKD is present. It is likely that the presence of potassium chloride in the CKD is the main reason for the acceleratory effect observed here [12].

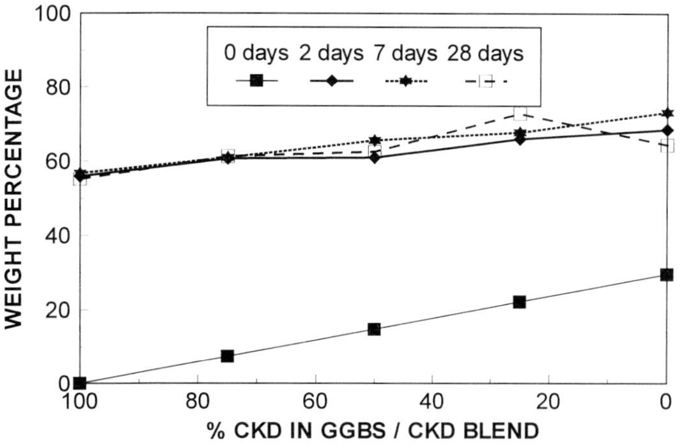

Figure 4 Quantities of amorphous material present in the hydrating blends at various ages, determined using powder x-ray diffraction

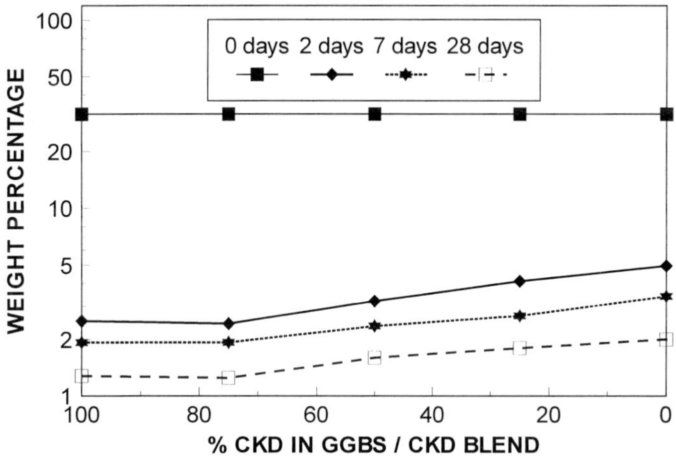

Figure 5 C_3S contents of the hydrating blends at 0, 2, 7 and 28 days, determined using powder x-ray diffraction

Calcium Aluminates: Figure 6 shows the C_3A content of the pastes at 0, 2, 7 and 28 days. At an age of 2 days it is apparent that more C_3A has reacted in the blends containing CKD, indicating that the presence of this material accelerates the hydration of this phase.

However, by 7 days, differences in the degree of C_3A hydration have disappeared. Due to the low quantities of C_4AF in the PC, meaningful qunatitative results of this phase could not be obtained.

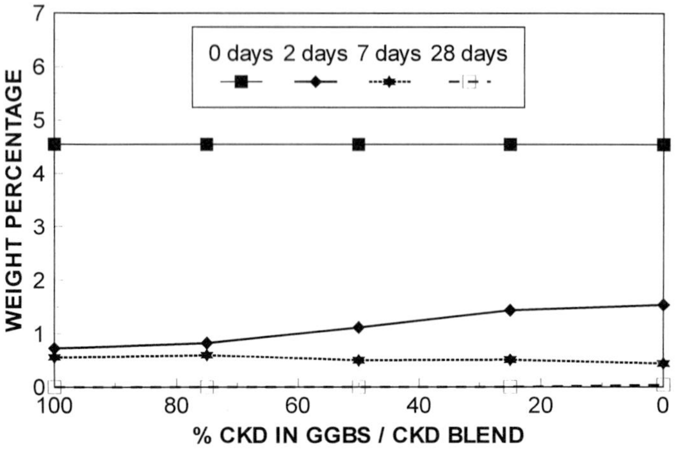

Figure 6 C_3A content of hydrating blends at 0, 2, 7 and 28 days, determined using powder x-ray diffraction

In systems containing hydrating Portland cement, the stability of calcium aluminate (and alumino-ferrite) hydrates follows the general trend ettringite ($3CaO.Al_2O_3.3CaSO_4.31H_2O$) > Friedel's salt ($3CaO.Al_2O_3.CaCl_2.10H_2O$) > monocarbonate ($3CaO.Al_2O_3.CaCO_3.11H_2O$) > monosulfate ($3CaO.Al_2O_3.CaSO_4.12H_2O$) [13]. The only discrepancy in this sequence is shown by ettringite in the presence of hydrating calcium aluminates, which, once all soluble sulfates are exhausted, and where chlorides and carbonates are abscent, will convert to monosulfate to accomodate the additional aluminates. Since CKD contains sulfates, chlorides and carbonates there is clearly much scope for the production of a wide range of calcium aluminate hydrates.

The two calcium aluminate phases observed in the highest quantities are ettringite and Friedel's salt, although other sulfate and carbonate-bearing phases also occur in some instances. The development of ettringite is shown in Figure 7 at ages of 2, 7 and 28 days. In the CKD-rich blends, the production of ettringite is significant and continual - a result of the high quantities of soluble sulfates. However, where CKD is not present, the quantities of ettringite in the blend containing no CKD can be seen to lessen as this phase converts to the monosulfate phase.

In blends containing CKD, and hence chloride compounds, the formation of Friedel's salt is also observed (Figure 8). Between 0 days and 7 days there is a clear development of a maximum in Friedel's salt levels at a CKD / GGBS ratio of 1. However, beyond 7 days there is a large increase in the quantity of Friedel’s salt produced by blends with a CKD / GGBS ratio greater than 1. The reason for this behaviour can be related back to the stability sequence discussed previously. In blends containing CKD, there are also quantities of soluble sulfates, leading to the formation of ettringite. Only after this supply of sulfates is exhausted will the less stable Friedel’s salt be permitted to form. Therefore, in blends containing lesser quantities of CKD, ettringite is produced for a brief period, followed by Friedel’s salt, whereas in high CKD blends large quantities of ettringite are produced for lengthy periods of time before Friedel’s salt can be formed.

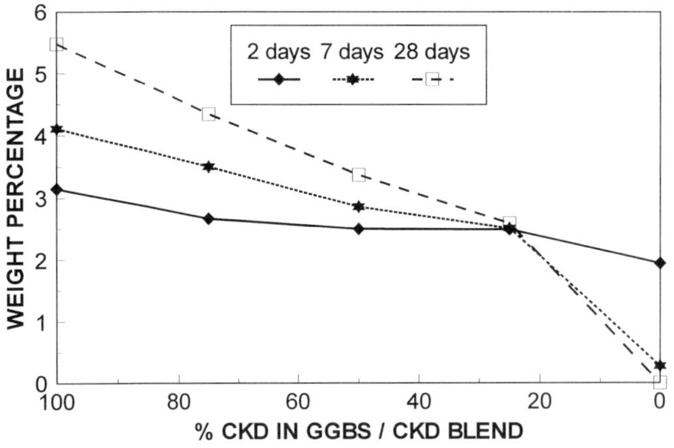

Figure 7 Ettringite content of hydrating blends at 2, 7 and 28 days, determined using powder x-ray diffraction

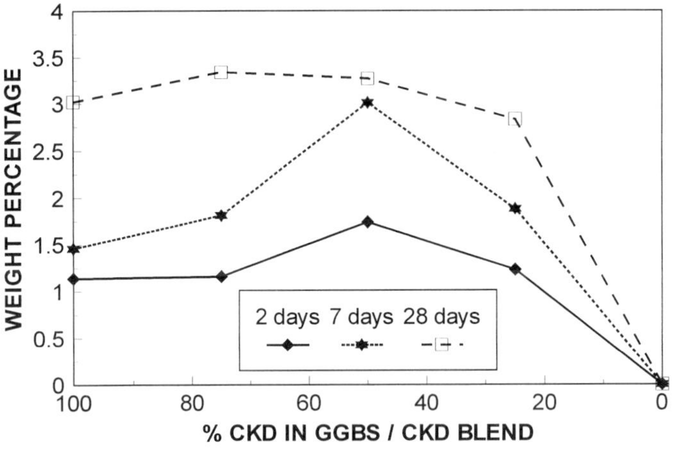

Figure 8 Friedel's salt content of hydrating blends at 2, 7 and 28 days, determined using powder x-ray diffraction

CONCLUSIONS

1. The strength development of PC / GGBS blends is enhanced by the inclusion of CKD. In this instance optimum performance was obtained with CKD / GGBS ratios between around 1 and 1/3.
2. The mechanism by which strength development is enhanced appears to be by an acceleration of the reactions of both PC and GGBS.
3. Higher quantities of ettringite are formed in blends containing CKD, as well as Friedel's salt.

The chloride content of CKD immediately excludes it from use in applications in which concrete is in contact with steel. However, there is nonetheless scope for using this material for other construction purposes. Before such uses can be considered, three points need to be addressed. Firstly the consequences of pronounced ettringite and Friedel's salt formation on the overall stability and durability of blends containing CKD must be investigated. Secondly, the presence of high quantities of alkalis in CKD means that the issue of alkali-silica reaction must be addressed. Finally, small quantities of potentially toxic trace metals are present in the material, and the long-term ability of blends containing CKD to immobilise these substances must be demonstrated.

REFERENCES

1. DYER, T.D., HALLIDAY, J.E., and DHIR, R.K., An investiagtion into the hydration chemistry of ternary blends containing cement kiln dust, Journal of Materials Science, 1999 (in press)
2. BRITISH STANDARDS INSTITUTION, BS EN 196 - 1, Methods of testing cement - Part 1: Determination of strength, 1995
3. BERGER, H., Study of the k-alpha emission-spectrum of copper, X-ray Spectrometry, Vol.15, 1986, pp241-243
4. CHEARY, R.W., AND COELHO, A.A., A fundamental parameters approach of x-ray line-profile fitting, Journal of Applied Crystallography, Vol.25, 1992, pp109-121
5. CHEARY, R.W., AND COELHO, A.A., Synthesising and fitting linear position-sensitive detector step-scanned line profiles, Journal of Applied Crystallography, Vol.27, 1994, pp673-681
6. CHEARY, R.W., AND COELHO, A.A., Theoretical model for axial divergence with primary and secondary Soller slits in x-ray line-profile analysis, 1996, in preparation
7. MARQUARDT, D.W., Journal of the Society for Industrial and Applied Mathematics, Vol.11, 1963, pp431-331
8. NASH, J.C., Compact Numerical Methods for Computers, Adam Hilger, Bristol and New York, 1990
9. FLETCHER, D.A., MCMEEKING, R.F., AND PARKIN, D., J., The United Kingdom Chemical Database Service, Journal of Chemical Information and Computer Sciences, Vol.36, 1996, pp746-749
10. LEA, F.M., The Chemistry of Cement and Concrete, 3rd Ed, Arnold, London, 1970, 727pp
11. TAYLOR, H.F.W., Cement Chemistry, Academic Press, London, pp301
12. DOUBLE, D.D., New developments in understanding the chemistry of cement hydration, Phil. Trans. R. Soc. Lond., A 310, 1983, pp53-66
13. ZHANG, F., ZHOU, Z., and LOU, Z., Solubility product and stability of ettringite, 7th International Congress on the Chemistry of Cement, Paris, Volume II, 1980, pp88-93

NEW CEMENT-FREE BINDERS AND CONCRETES

L B Svatovskaya **M N Latutova**

A P Komohov **V A Chernakov**

L L Maslennikova **O U Makarova**

A V Smyrnov

Railway University

Russia

ABSTRACT. In this work is considered the possibility of the development and characteristics of new without-roasting water- and cold-resistant materials based on the natural clays and loams of the St.Petersburg area, as powders and activating additives and water or ortophosphoric acid as mix liquids. Materials developed on the basis of clay-bearing mixtures and water are named "Clayconcretes" and with the addition of phosphoric acid – "Glinphs". Clayconcretes have a Portland cement content of not more than 5%, i.e. PC is present only as an activating additive. Developed materials harden in natural conditions and evolve considerable amounts of heat allowing work to be carried out in low temperatures.

Keywords: Glinphs, Clayconcretes, Phosphoric acid, Loam, Clays, Strength.

Professor L B Svatovskaya is the Head of Department of Engineering Chemistry, Railway University, St.Petersburg, Russia. She specializes in the chemistry of the binders.

Dr M N Latutova, Department of Engineering Chemistry, Railway University, St.Petersburg, Russia. Area of scientific interests – silicate and phosphate binders.

Dr A P Komohov, Railway University, St.Petersburg, Russia. Area of scientific interests – concretes.

Mr V A Chernakov, Department of Engineering Chemistry, Railway University, St.Petersburg, Russia. Area of scientific interests – concretes.

Dr L L Maslennikova, Department of Engineering Chemistry, Railway University, St.Petersburg, Russia. Area of scientific interests – clay materials.

Mrs O U Makarova, Department of Engineering Chemistry, Railway University, St.Petersburg, Russia. Area of scientific interests – phosphate materials.

Mr A V Smyrnov, Railway University, St.Petersburg, Russia. Area of scientific interests – ecology.

INTRODUCTION

Natural highly dispersed products (clayey minerals) which have a sufficient source base and do not require grinding have the potential to be used to as a base for an inorganic binder or be included in materials of cool (without roasting) hardening, used in construction [1].

It is known that in the composition of clays there are present minerals of kaolin $AI_2O_3.2SiO_2.2H_2O$ or montmorillonite groups of the general formula $AI_2O_3.4SiO_2.2H_2O$ as well as hydromicas [2].

Surface decomposition of clayey materials can pass in accordance with two acid-base schemes: reaction and catalytic. In the latter case the acid and base active centres of minerals of clay.

The first scheme is realised due to amphoteric nature of aluminum silicates, which are reactive and unstable both in acid, and in alkaline media. In an acid medium, aluminum hydrosalts are formed, in alkaline media- aluminosilicates of metals with base properties, than aluminum. In the first case the following general reactions take place:

$$AI_2O_3.2SiO_2.2H_2O + 6H^+ + nH_2O \rightarrow 2AI^{3+} + 2SiO_2.mH_2O \quad (1)$$

In the second case, if the cation of alkaline metal is charged, the general type of reaction is:

$$Al_2O_3.2SiO_2.2H_2O + 4OH^- + 2Me^{2+} \rightarrow MeO.AI_2O_3.SiO_2.nH_2O + MeO.SiO_2.mH_2O \quad (2)$$

Decomposition of aluminum silicates on mechanisms of acid-base catalysis by means of solid bases (calcium hydrosilicates) or solid acids (calcium sulphates anhydrites) consists in the following: if in the system, containing minerals of clay, solid acids are present, then being donor of proton, they capable to weak and break oxygen-element bonds, taking electronic pair of oxygen on vacant s-orbit of proton. Solid bases, being donor of electronic pair, capable to break aluminum-element bonds, placing its unshared electronic pair of oxygen from OH^--group on vacant p-orbit of aluminum [3,4]. It is important to note that the break of A1-O bond can be catalysed by materials, acting also on the principle of redox catalysis, as far as covalent component of the bond in this case is significant.

Prediction and provision of the physicomechanical and technical characteristics of materials on the base of aluminosilicates and phosphoric acid can be realised by accounting of particularities of electronic configurations of cations, participating in the formation of the phases, crystal-chemical particularities of cation and anion, as well as particularities of processes of the transmission of charge, occurring in the process of formation of artificial stone with the use of the phosphoric acid.

It is known that material on the base of clays, loams and phosphoric acid are not water-resistant, though possess in air-dry conditions sufficiently high strength. The absence of resistance to water is connected with formation of the aluminum hydrophosphates, which are soluble:

$$Al_2O_3.2SiO_2.2H_2O + 6H_3PO_4 + mH_2O \rightarrow 2AI(H_2PO_4)_3 + 2SiO_2.nH_2O \quad (3)$$

Solubility of new growths can be decreased by complicating of a composition both on cation, and on anion; resistance to water is possible to raise by reinforcement of the system by water-resistant products.

Materials

In this work about the use of activity of clayey minerals in alkaline and acid medium were used St.Petersburg loams and Cambrian clays. As mineral components for clayey matrixes in the alkaline medium was chosen compositions, consisting from by-products of industry: the mixture I – belite-containing – 90%, acid fluoride – 10%; the mixture II – blast furnace slag before 10%, belite-containing product – 80% and acid fluoride – 10%; in both cases over 100% was added Portland cement – 3...5%. For the acid medium (based on the 28% phosphoric acid) as activating additives to loams were used industry products, containing d^{6-10}-elements. Mixture thereby consisted of the clay-containing product and certain activator.

RESULTS AND DISCUSSION

In the Table 1 are presented results of "Glinphs" test – rasistance to water and cold of nonpressed, without roasting materials, which can be used in the form of building bricks or concrete blocks [4,5,6,7].

Table 1 Physicomechanical characteristics of "Glinphs"

ACTIVATOR (20%)	COMPRESSIVE STRENGTH AFTER 2 DAYS OF SATURATION WITH WATER, N/mm^2		SETTING PERIOD, HOURS	
	7 Days	28 Days	Beginning	End
–	–	–	20.5	58
CuO	3.0	5.0	0.25	3.15
Cu_2O	12.0	9.0	0.25	0.50
FeO	18.0	20.0	0.40	0.60

Calorimetric studies have shown that in formation of this material a heat is evaluating, which amount decreases by acid modification. Material have mainly size of pores ~10 nm and small shrinkage.

Particularity of electronic structure reveals itself in the colour of material, as far as d-elements have a distinctive colouring. The use of white clays or technical aluminum hydroxide in combination with corresponding compounds of d-elements as powdered part, make it is possible to develop coloured without roasting compositions-paints, which were used in decoration of building materials.

Table 2 Compositions for clayconcrete preparation

NUMBER	COMPOSITION		AGGREGATE PART
	Loam	Inorganic Mixture	
1	1	0.25	–
2	1	0.50	1
3	1	0.75	2
4	1	1.00	3

Table 3 Strength of clayconcrete of normal hardening, separated mixture preparation

NO	WATER BINDER RATIO	BINDING STRENGTH, N/mm^2			COMPRESSIVE STRENGTH, N/mm^2		
		14 days	28 days	56 days	14 days	28 days	56 days
1	1.10	0.09	0.23	0.47	0.48	1.80	2.60
2	0.84	1.66	3.54	4.80	4.80	13.10	16.40
3	0.67	4.23	7.10	8.60	16.20	20.85	24.60
4	0.60	4.68	7.53	9.00	17.00	21.50	26.30
1	1.30	0.14	0.28	0.50	0.48	1.42	2.90
2	0.86	0.70	2.50	3.80	4.00	10.70	14.60
3	0.71	3.26	4.61	5.70	12.3	16.60	20.10
4	0.64	3.76	5.17	6.50	13.00	17.50	22.40

It is necessary to note one more important "Glinphs" particularity – they can include oil-contaminated loams and thereby to block oil-contaminants in the stone, which can be applied in transport and industrial construction.

At the preparation of an clayconcretes two variants of technologies were used: on separate or joint scheme of mixtures preparation [5]. Separate technology includes blending in the concrete mixer apart obtained inorganic mixture and clay-containing component and their joint grinding in the ball mill. These technological particularities influence upon conditions of an heat and moisture processing and strength of clayconcrete. Herewith, essential role belongs to the homogeneity degree of mixture. The ratio of inorganic mixture to clay-containing component of clayconcrete is shown in the Table 2. On the base of these compositions were prepared and tested beams 4x4x16 cm and cubes 10x10x10 cm at the age of 7, 14, 28, 56 days; condition of hardening – normal.

In the Table 3 data are given about changing of strength of developed clayconcretes. With the increase of an aggregate contents (quartz sand, 1.6 mm) the water/binder value for mixtures with equal plasticity decreases and accordingly increase compressive and bending strength of

Table 4 Strength of clayconcrete of normal hardening, joint mixture preparation

NO	WATER BINDER RATIO	BINDING STRENGTH, N/mm^2			COMPRESSIVE STRENGTH, N/mm^2		
		14 days	28 days	56 days	14 days	28 days	56 days
1	0.90	2.40	3.10	4.50	9.50	12.50	13.70
2	0.80	4.50	6.50	7.80	12.10	16.00	19.10
3	0.65	5.26	7.13	8.30	15.00	18.60	22.40
4	0.40	5.70	7.90	9.20	16.00	19.70	24.30
1	0.97	2.10	2.90	3.40	7.60	10.30	11.50
2	0.86	3.60	5.00	6.20	15.10	18.00	21.80
3	0.71	4.70	5.50	6.80	16.60	18.60	22.50
4	0.46	5.60	6.70	8.20	18.40	20.70	24.80

clayconcretes. In the Table 4 are given characteristics of clayconcretes obtained under joint technology of clay-mixture preparation. The advantage of technology of clayconcrete preparation under the joint grind of the source mixture is obvious. Small aggregate of clayconcrete (quartz sand) can be successfully replaced with sifting of fine fractions (0.16-2.5 mm) of waste from the granite rubble production.

It is necessary to note that the water/binder ratio of clayconcrete is 1.1-0.6 that in contrast with usual cement concretes higher in 1.5 times. It is necessary to note one more circumstance – a high value of binding strength/compressive strength ratio for clayconcretes, which in average is 0.30-0.33 that is mach above the one for usual cement concretes.

The comparison of kinetics of strength growth of steamed clayconcrete shows that the strength after steaming is equal to seven-days strength of normal hardening, and that the strength of steamed under 40^{0}C clayconcrete in a day age is close to strength of this material of normal hardening in 28 days ages. Results of clayconcrete's resistance to cold tests are presented in the Table 5. Data shows that clayconcrete bears 30 cycles under -50^{0}C.

According to the derivatograph analysis data, the most intensive hydration of clay-mixture composition takes place at the temperature 40 °C. Maximum loss of chemically bound water are 11.5% and 11.8% accordingly with specified binder composition.

Elastic and plastic deformation were researched, Poisson factor, modulus of elasticity and nature of clayconcrete destruction under the stress. These results are shown in the Table 6. Poisson factor for clay-containing material is more than usual values for stone material, including products on the base of Portland cement. Starting module of elasticity (under 0.3 x ultimate strength) is 1.5 times less, than for the cement concrete with equal strength. For the given material cubic and prism strength at an age 28 days are nearly equal. Test were carried out on samples 10x10x10 cm and 10x10x40 cm. The factor of prism strength of clayconcrete is nearly 1 that witnesses on strengthening of structures of material at the destruction

THEME FOUR:

WATER-REDUCING/ SET CONTROLLING ADMIXTURES

unhydrated grain (Figure 4). These were identified as calcium chloroaluminate hydrates, *viz.*, $C_3(A,F).3CaCl_2.xH_2O$ (in our notation AFt chloride analogue, abbreviated as AFt_{Cl}) and $C_3(A,F).CaCl_2.yH_2O$ (in our notation AFm chloride analogue, abbreviated as AFm_{Cl}). The latter low chloride form is known as Friedel's salt where y=10 and it forms a series of solid solutions with calcium aluminate hydrates (like $C_4(A,F)H_{13}$). AFt_{Cl} also form solid solution with ettringite (AFt phase).

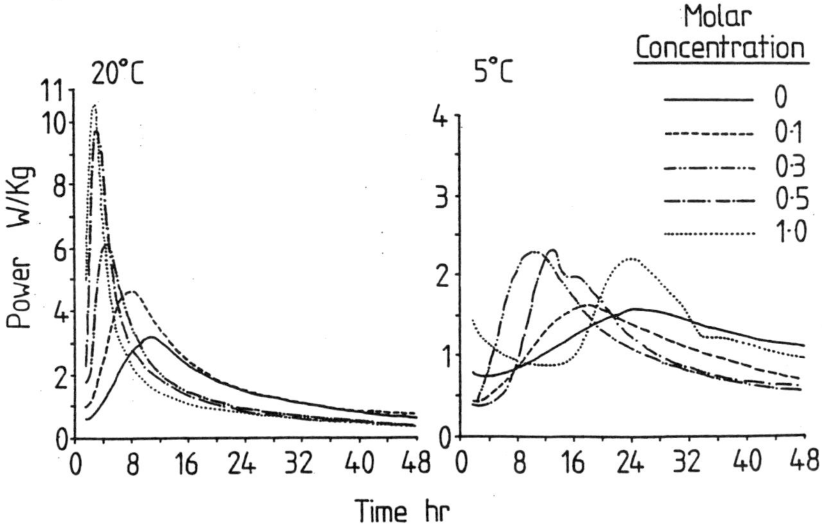

Figure 1 Rate-of-heat-evolution curves of oilwell cement with various additions of $CaCl_2$ at 20°C and 5°C

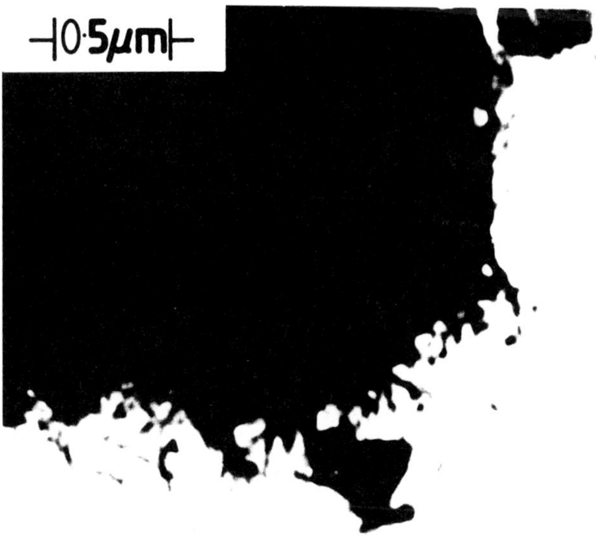

Figure 2 HVEM of oilwell cement hydrated at 20°C for 1 hour with 1.0M addition of $CaCl_2$

Figure 3 Fracture surface of oilwell cement hydrated at 20°C for 1 day with 0.5M addition of $CaCl_2$

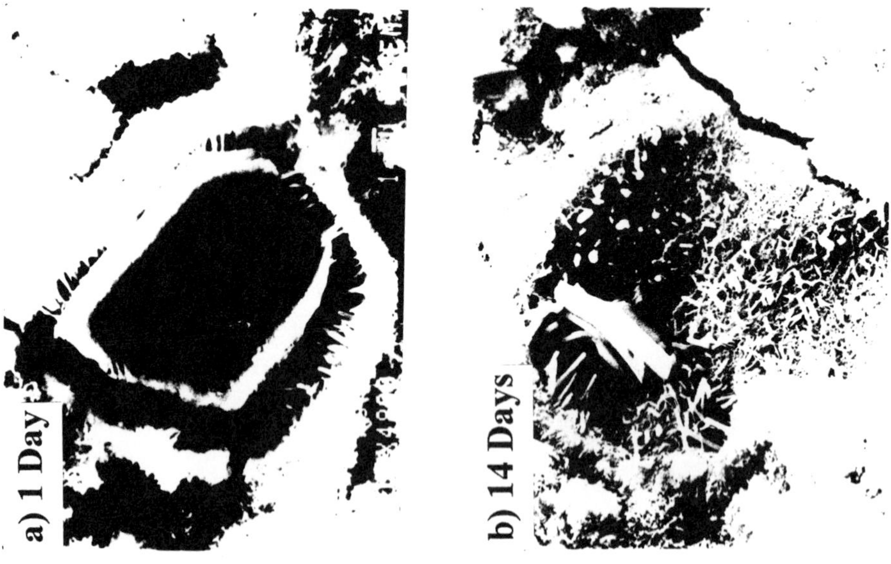

Figure 4 Fracture surface of oilwell cement hydrated at 20°C for a) 1 and b) 14 days with 0.3M addition of $CaCl_2$

X-Ray Powder Diffraction (XRD)

XRD patterns of the cement hydrated at 20°C and 5°C show a complex nature of the formation of the hydration products (Figure 5). The occurrence of ettringite, AFm_{Cl} (~11.2°2θ) and calcium aluminate hydrates (like $C_4(A,F)H_{13}$) depends upon the temperature and dosage of $CaCl_2$. With 0.3 and 1.0M addition at 20°C ettringite and AFm_{Cl} solid solution appears after 1 day and persist at 1 year. At 5°C & 0.3M addition ettringite is seen at all ages and AFm_{Cl} appear after 1 day and persist at 1 year. Also at 5°C, with 1.0M addition ettringite is detected at 1 year and AFm_{Cl} is present from 1 day onwards. The development of ettringite in the paste is similar to that in the control at 20°C and 5°C. The AFt_{Cl} is very unstable & decomposes readily at 20°C.

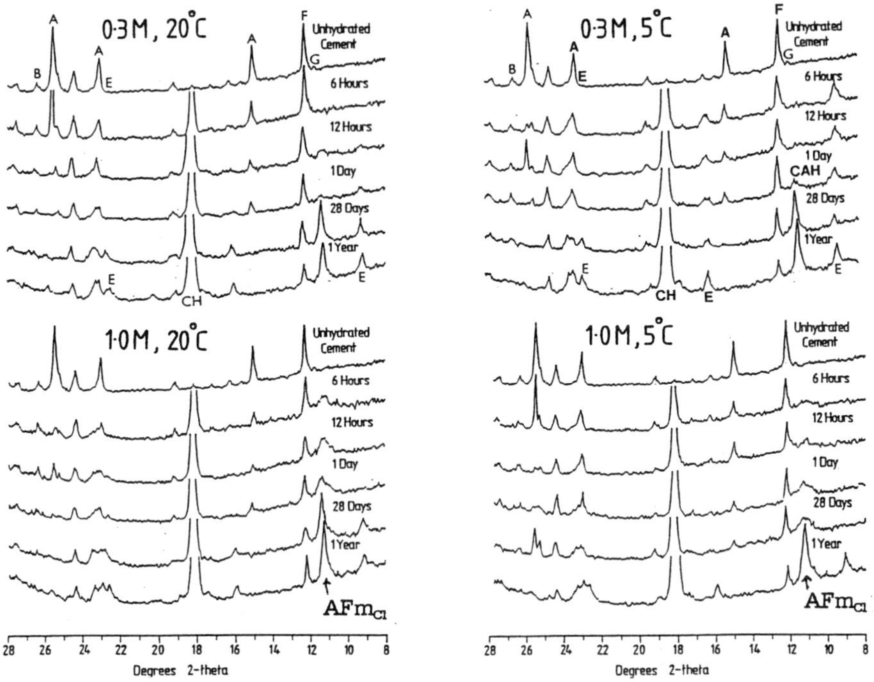

Figure 5 XRD patterns of Class G oilwell cement hydrated at 20°C and 5°C with various additions of $CaCl_2$

Infrared (IR) Spectroscopy

IR spectra of the pastes agree with XRD traces in that early formation of ettringite-AFm_{Cl} are seen. All the spectra indicate extensive CSH formation at early ages [1].

Thermal Analysis

The amount of calcium hydroxide (CH) and loss-on-ignition (LOI) in the paste hydrated at 20°C and 5°C is shown in Figures 6 and 7 respectively. At 20°C, at 3-6 hours the amount of CH formed increases with dosage. After 12 hours the most amount of CH is generally formed with 0.3M addition. A similar trend is seen at 5°C with lower amounts of CH. The LOI of the paste follows a similar pattern to that of CH. Here the LOI after 28 days with 0.3 and 1.0M additions is not that different. It appears that 0.3M addition $CaCl_2$ accelerates hydration and that the 1.0M addition accelerates early hydration to such an extent that subsequent hydration is stifled.

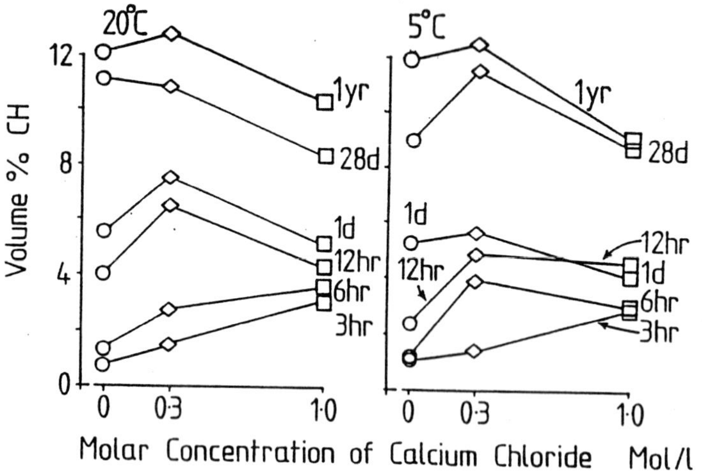

Figure 6 CH content of oilwell cement hydrated at 20°C and 5°C with various additions of $CaCl_2$

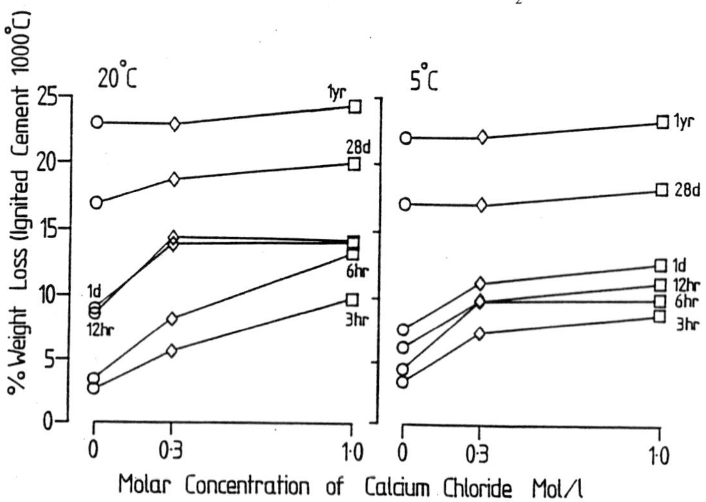

Figure 7 LOI of Class G oilwell cement hydrated at 20°C and 5°C with various additions of $CaCl_2$

Compressive Strength Development

Development of compressive strength of the cement at 20°C and 5°C is shown in Figure 8. At 20°C, generally the strength of the paste increases as the dosage of $CaCl_2$ is increased. At 28 days the difference in strength between 0.5M and 1.3M additions is not that great. In contrast at 5°C the strength of the paste at corresponding ages and dosages is always lower than at 20°C. The 0.5M addition gives the optimum increase in strength.

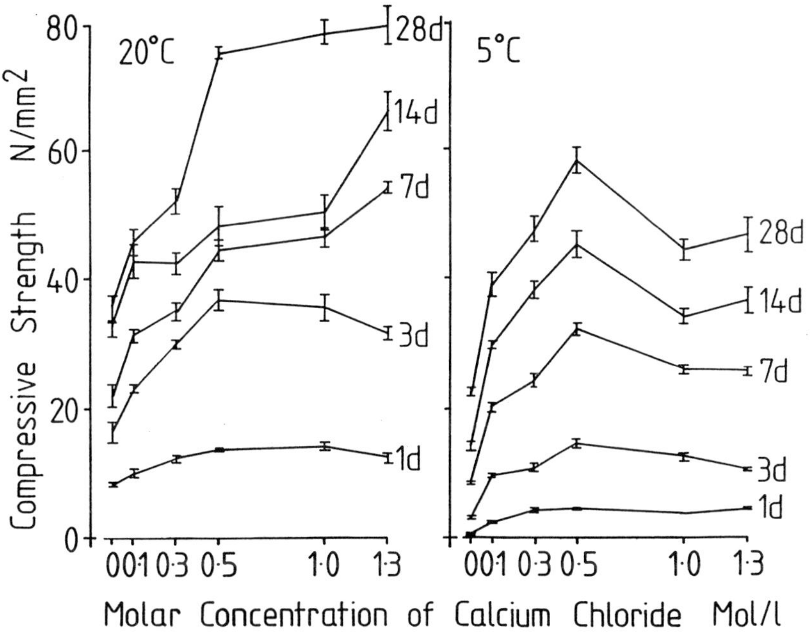

Figure 8 Compressive strength development of oilwell cement hydrated at 20°C and 5°C with various additions of $CaCl_2$

GENERAL DISCUSSION AND MECHANISM OF HYDRATION

It has been shown that the acceleratory effect of $CaCl_2$ at the two temperatures depends not only upon the temperature of curing and the dosage, but also on the technique that is being used to measure it. For example, the rate-of-heat evolution curves show that $CaCl_2$ is not as effective an accelerator at 5°C as at 20°C.

Clearly the hydration of the cement at the two temperatures is different. The effect of temperature and dosage influence the cement reactions and on the development of cement microstructure especially the CSH. Fracture and polished images of the cement paste show the presence of needle and hexagonal plate-like features.

A mechanism is hypothesised as follows:

- Initial thin film covering of hydration product forms completely around cement grain.
- This film of hydration product acts as a partial semi-permeable protective membrane, allowing the ingress of water and the migration of dissolved ions into and out of the hydration shell forming. The CSH produces a fibrous morphology in the wet state and on drying this becomes reticulated or honeycombed network.
- $CaCl_2$ dissolved in the mix water, precipitates CH and the chloride ions migrate into the hydrating cement grain. Once within the cement grain AFt_{Cl} and AFm_{Cl} form in association with ettringite (AFt) and monosulphate (AFm) respectively. The CH precipitated provides nucleation sites for CH and CSH, hence accelerating hydration.
- Eventually the hydration shell thickens and inner product forms and hydration slows down.

CONCLUSIONS

- The hydration of Class G oilwell cement in the presence of calcium chloride at 20°C is different from that at 5°C.
- Calcium chloroaluminate hydrate phases were found to occur within the hydrating cement grain boundaries and not in the space between the hydrating grains.
- The ferrite phase in the cement plays an active part in the hydration process at both temperatures.
- Chloride modifies the morphology of the AFt and AFm phases. This could be due to the $CaCl_2$ depressing the solubility of gypsum.
- A mechanism to explain the experimental observations is proposed.
- The optimum dosage of $CaCl_2$ that should be added to the paste to achieve the most acceleratory effect is strongly dependent upon which technique is used to measure it.
- The definition of "acceleration" needs more careful thought.
- There is no universal mechanism of acceleration of cement hydration since $CaCl_2$ behaves very different from sodium aluminate discussed earlier [2].

REFERENCES

1. LOTA J S, The hydration of Class G oilwell cement, PhD Thesis, Department of Materials, Imperial College of Science Technology and Medicine, University of London (1994).

2. LOTA J S, BENSTED J and PRATT P L, Effect of sodium aluminate on Class G oilwell cement hydration at low and ambient temperatures, Proceedings of the 10th International Congress on the Chemistry of Cement, Gothenburg, 3, Paper 3iii014, 8pp (1997).

LIGNOSULFONATES IN CONCRETE: EFFECT OF DOSAGE RATE ON BASIC PROPERTIES OF FRESH CEMENT-WATER SYSTEMS

J Zhor

T W Bremner

University of New Brunswick

Canada

ABSTRACT. Lignosulfonates are well known as efficient and cost effective water-reducing admixtures for concrete. In the present study, the influence of dosage rate of lignosulfonate-based water-reducing admixtures on the fundamental properties of fresh cement pastes and mortars was investigated. The addition of three different lignosulfonate samples at dosage rates between 0.25 and 2.5% (w/w, solids by mass of cement) was applied and different testing methods were used. The effect of the dosage rate on water reduction, plastification/ fluidification, setting time and air content was determined. The newly developed methodology used for this testing includes both standard (ASTM C 185, C 187 and C 191) and non-standard (torque test, modified mini-slump test) methods. The dependence of the properties of fresh cement pastes and mortars on the dosage rate of lignosulfonate-based water-reducers was confirmed. However, the effect of the admixture was not always directly proportional to the added amount. The optimum dosage rate is dependent not only on the admixture sample but also on the mixture-proportions and on the method of testing. Contrary to general belief, the highest dosage rate of lignosulfonate did not always cause the highest set retardation. Also, the positive influence of air-entrainment on plastification of cementitious systems was not observed in all cases.

Keywords: Acceleration, Admixture, Air-entrainment, Dosage rate, Fresh cement paste, Fresh cement mortar, Lignosulfonate, Plastification, Retardation, Setting time, Water-reduction.

Mr Jiri Zhor is a doctoral candidate in the Department of Civil Engineering at the University of New Brunswick in Fredericton, N.B., Canada. After ten years of experience in various areas of concrete technology, he has been engaged in research on concrete admixtures since 1986. His current research interests are the chemistry of superplasticizers for concrete and properties of fresh concrete.

Dr Theodore W Bremner is a professor of civil engineering at the University of New Brunswick in Fredericton, N.B., Canada. He is past chairman of ACI Committees 122-Energy Conservation and 213-Lightweight Aggregate and Concrete. His research interests include structural lightweight concrete, properties of fresh and hardened concrete and durability of concrete in the marine environment.

INTRODUCTION

The application of lignosulfonates as concrete admixtures has been known for more than fifty years. Since the late nineteen thirties lignosulfonates have been used as cement dispersants [1]. They have been recognised as concrete plasticizers and their water-reducing, set-retarding and air-entraining effects have been investigated [2]. In fresh cement-water systems, the reduction of water is typically considered to be the main effect of lignosulfonate-based admixtures. If water-reduction is not needed and the water/cement ratio is not decreased, lignosulfonates have a plasticizing/fluidifying influence on the cement-water mixture. Besides these two interdependent effects, two additional side effects of lignosulfonates are recognised: setting time retardation and air-entrainment. According to ASTM Standard Specification for Chemical Admixtures for Concrete - ASTM C 494 [3], lignosulfonate-based admixtures are usually classified as water-reducing and retarding admixtures and also as water-reducing, high range, and retarding admixtures [4].

The effectiveness of water-reducing admixtures is related to the amount of admixture added to the cement-water mixture before the saturation concentration is reached. Generally, it is believed that the effects of the admixture are directly proportional to its dosage. In other words, the higher the dosage the higher the water reduction (or plastification). Thus, in the absence of side effects, the admixture dosage can be optimised by determining the saturation amount, above which the workability (or water reduction) does not improve [5]. If side effects are present, however, they have to be taken into consideration [6]. In the case of lignosulfonates it has been found that the higher the dosage the higher the set retardation and air-entrainment [7].

To meet the requirements for a high range water-reducing admixture, lignosulfonates have to be added in amounts above 0.3 % (w/w, solids by mass of cement) [4]. On the other hand, this amount of lignosulfonate is in some instances considered to be unrealistic, as it can cause excessive retardation and air-entrainment [8]. During the last few decades a number of new, mostly synthetic, products competes with lignosulfonates on the market. The major advantage of lignosulfonates is usually considered to be their price. It has been documented that the higher dosage rate of lignosulfonate is economically more feasible but, because of side effects, technically questionable [9].

In this work the influence of the dosage rate of lignosulfonate-based admixtures on the properties of fresh cement pastes and mortars has been investigated. The testing methodology previously published by the authors [10] was used for the experimental evaluation. Four basic effects of three lignosulfonate samples were tested, namely water-reducing effect, plasticizing/ fluidifying effect, set-accelerating/retarding effect and air-entraining effect. The broad range of dosage rates from 0.25 to 2.5 % (w/w, solids by mass of cement) was chosen in order to study the effects of rather excessive amounts of lignosulfonates on the behaviour of fresh Portland cement-based systems.

EXPERIMENTAL PROGRAM

Materials

Ordinary Portland cement (OPC - ASTM C 150, Type I) and potable water were used. Standard silica sand graded "20-30 sand" according to ASTM C 778 was incorporated in the cement mortars. Three lignosulfonate-based water-reducing admixtures were tested. One commercial (SL1) and two experimental ALCELL® lignin-based (SL2 and SL3) formulations were applied in dosage rates ranging from 0.25 to 2.5 % (w/w, solids by mass of cement). The basic properties of admixtures are shown in Table 1.

Table 1 Basic properties of lignosulfonate samples

PROPERTY	SL1	SL2	SL3
Physical State		aqueous solution	
Appearance		dark brown liquid	
Specific Gravity	1.190	1.170	1.170
Solids Content, % w/w	45.7	38.1	38.4
Volatiles, % w/w (as water)	54.3	61.9	61.6
pH	7.2	9.3	11.3
Viscosity, cP	N/A	35	43

Test Methods in Cement Pastes

Normal consistency determination by Vicat apparatus (ASTM C 187)

Apparatus: In accordance with ASTM Standards (C 305 and C 187).

Mixture Proportions: Portland cement, Type 10 : 650 g
Water (and resulting w/c) : As required for normal consistency
Admixture: 0/0.25/0.5/0.75/1/1.5/2/2.5 % (w/w, by mass of cement)

Procedure: In accordance with ASTM C 187. The water reduction was quantified as a reduction of w/c needed to maintain a normal consistency of paste with the desired amount of an admixture, as compared to a cement paste without admixture. The data obtained were used to evaluate the water-reducing effect of the admixtures.

Time of setting determination by Vicat apparatus (ASTM C 191)

Apparatus: In accordance with ASTM Standards (C 305 and C 191).

Mixture Proportions: The proportions determined by the normal consistency test were used.

Procedure: In accordance with ASTM C 191. The data obtained were used to evaluate the set-accelerating/retarding effect of the admixtures.

Modified mini-slump test

Apparatus: Mini-slump cone as specified by Kantro [11], hand-held Rival® kitchen mixer (600 rpm) with 1 "egg-beater" paddle, toploading Sartorius balance - model 1364MP.

Mixture Proportions: Portland cement, Type 10 : 75.0 g
Water : 26 mL
W/C : 0.35
Admixture: 0/0.25/0.5/0.75/1/1.5/2/2.5 % (w/w, by mass of cement)

Procedure: The procedure proposed by Kantro was altered by using a different mixing regime and by implementing the measurement of cement paste density. The tested admixture was added at the desired dosage rate to the mixing water in a 200 mL polyethylene beaker prior to adding cement. Immediately after adding the cement, a 5 minute mixing regime (1.5 min. mix; 2 min. rest; 1.5 min. mix) was applied. The cone was placed on a Plexiglas[7] sheet located on the weighing pan of a toploading balance. After transferring the mixture into the cone in three layers (each compacted with 10 strokes of a small spatula) and levelling the surface, the cone was lifted 6.5 minutes after the start of mixing. The mass of fresh cement paste was recorded and used to determine the air content by a method derived from ASTM C 185. After hardening, the formed paste cake base area was measured using a planimeter. The data obtained from this test were used to evaluate the air-entraining and plasticizing/fluidifying effects of the admixtures.

Torque test

Apparatus: The rotational torque-meter developed by the authors was used [10].

Mixture Proportions: Portland cement, Type 10 : 1500 g
Water : 525 mL
W/C : 0.35
Admixture: 18.75 mL (5 times 3.75 mL)
(20% w/v solution) (i.e. 0.25 % (w/w, solids by mass of cement))

Procedure: The cement-water mixture was premixed by hand for 5 minutes and then mixed in the assembled mixing chamber for another 5 minutes prior to the beginning of the test. The mixing in the chamber was continuous throughout the entire test. Data collection was started ten minutes after the first contact of cement with water. After another ten minutes, 3.75 mL of a 20% w/v admixture solution (0.05% w/w, solids by mass of cement) was applied. At 10 minute intervals 3 additional doses were added, resulting in the total dosage of 0.25% (w/w, solids by mass of cement). The torque decrease between times t=10 min (immediately before the addition of the first sample increment) and t=60 min (10 minutes after the addition of the last sample increment) was used to quantify the plastification/fluidification. The time of setting determined from this test was arbitrarily chosen as the time from the beginning of the test until the torque value, after initial drop, returned to a level of 0.6 N·m. The data obtained were used to evaluate the plasticizing/fluidifying and set-accelerating/ retarding effects of the admixtures.

Test Methods in Cement Mortars

Flow test by the flow table (ASTM C 185)

Apparatus -- In accordance with ASTM Standards (C 305, C 230 and C 109).

Mixture Proportions: Portland cement, Type 10 : 350 g
20/30 standard sand: 1400 g
Water (and resulting w/c) : As required for a flow of 87.5±7.5%
Admixture: 0/0.25/0.5/0.75/1/1.5/2/2.5 % (w/w, by mass of cement)

Procedure -- In accordance with ASTM C 185. The water reduction was quantified as the reduction of w/c needed to maintain a constant flow of 87.5±7.5% of the mortar with the desired amount of an admixture, as compared to cement mortar without admixture. The data obtained were used to evaluate the water-reducing effect of the admixtures.

Air content determination by the gravimetric method (ASTM C 185)

Apparatus: In accordance with ASTM Standards (C 305, C 185, C 109 and C 1005).

Mixture Proportions: The proportions determined by the flow table test were used.

Procedure: In accordance with ASTM C 185. The data obtained were used to evaluate the air-entraining effect of the admixtures.

RESULTS AND DISCUSSION

Water-Reducing Effect

The water-reducing effect of different dosages of evaluated admixtures, determined by the normal consistency test on cement pastes, is presented in Figure 1. The saturation concentration (optimum dosage) was 1.5 % for admixtures SL2 and SL3 but was not reached within the tested range for admixture SL1. The highest dosage (2.5 %) of admixture SL3 had an adverse effect on water reduction. The water-reducing effect of different dosages of evaluated admixtures determined by the mortar flow test is shown in Figure 2. The saturation concentration (optimum dosage) was 0.25 % for admixture SL3, 1% for admixture SL2 and 1.5 % for admixture SL1. Increasing the dosage of SL3 above the saturation concentration had no effect on water reduction.

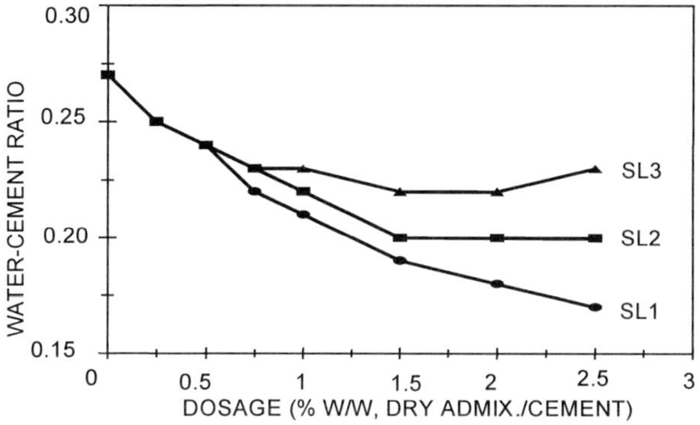

Figure 1 Influence of dosage rate on water-reducing effect of admixtures in pastes with normal consistency determined by Vicat apparatus (ASTM C 187)

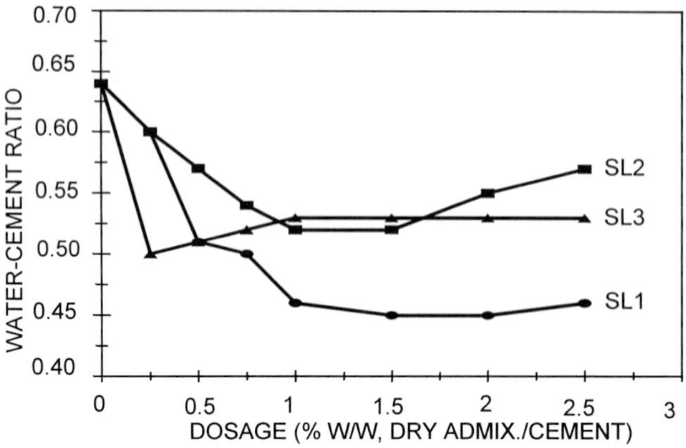

Figure 2 Influence of dosage rate on water-reducing effect of admixtures in mortars with a flow of 87.5±7.5% determined by the flow table (ASTM C 185)

Plasticizing/Fluidifying Effect

The plasticizing/fluidifying effect of different dosages of evaluated admixtures determined by the mini-slump test on cement pastes is presented in Figure 3. The saturation concentration (optimum dosage) was 1.5 % for admixtures SL1 and SL3. While the effect of admixture SL3 was moderate, admixture SL1 showed clearly the highest plasticizing/fluidifying effect. Surprisingly, admixture SL2 had no plasticizing/fluidifying effect at all even if its water-reducing effect was higher than with admixture SL3.

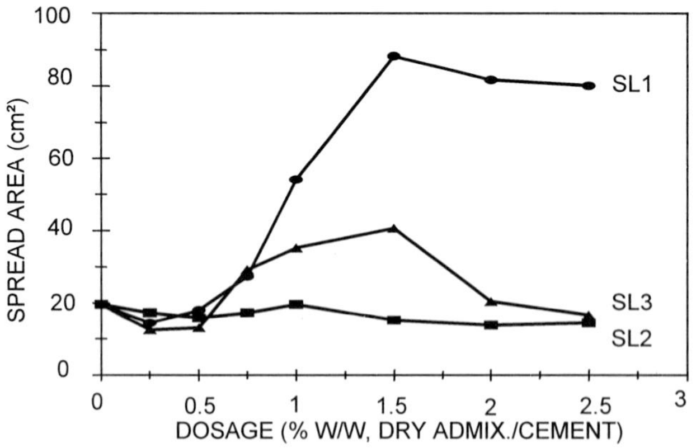

Figure 3 Influence of dosage rate on plasticizing/fluidifying effect of admixtures in pastes with w/c=0.35 determined by mini-slump test

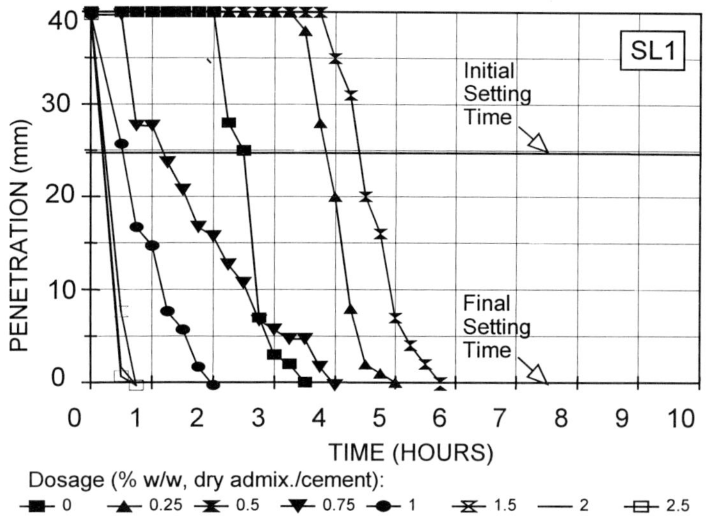

Figure 4 Influence of dosage rate of admixture SL1 on the setting time of pastes with normal consistency determined by Vicat apparatus (ASTM C 191)

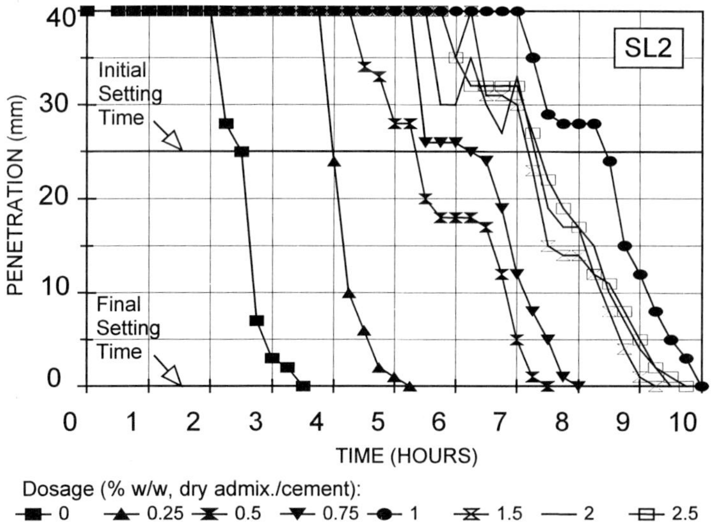

Figure 5 Influence of dosage rate of admixture SL2 on the setting time of pastes with normal consistency determined by Vicat apparatus (ASTM C 191)

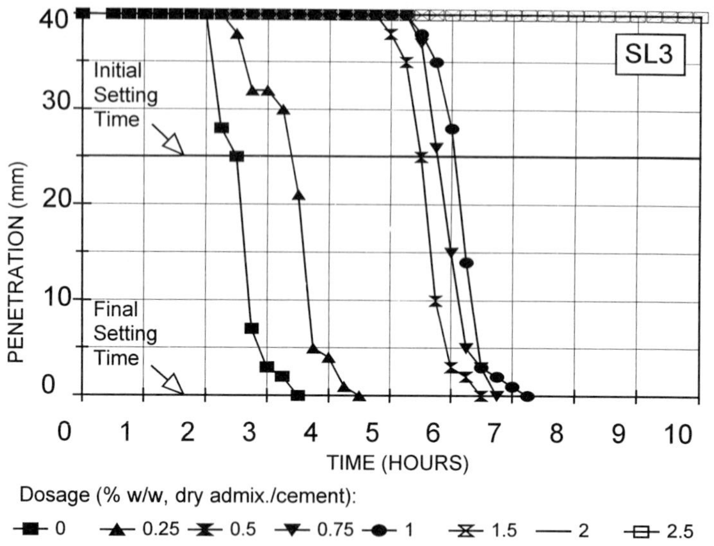

Figure 6 Influence of dosage rate of admixture SL3 on the setting time of pastes with normal consistency determined by Vicat apparatus (ASTM C 191)

Set-Accelerating/Retarding Effect

The records of the setting time measurements for different dosages of admixtures SL1, SL2 and SL3 are presented in Figure 4, Figure 5 and Figure 6, respectively. The set accelerating/retarding effect of different dosages of evaluated admixtures on the initial setting time is shown in Figure 7. The same effect on the final setting time is shown in Figure 8. The retarding effect of admixtures SL2 and SL3 increased to a saturation concentration of 1 %. Above this amount, the retardation caused by SL2 did not increase, while the retardation affected by SL3 increased beyond 12 hours and was not further recorded. Interesting behaviour was observed when admixture SL1 was used. The retardation increased up to a dosage of 0.5 % and then gradually decreased until the saturation concentration of 1.5 %. Actually, above a dosage of 0.5 % admixture SL1 had an accelerating effect on both initial and final setting time. Thus, in contrast to the general rule, the highest dosage of admixture SL1 did not cause the highest set retardation.

Plasticizing/Fluidifying and Set-Accelerating/Retarding Effects by the Torque Test

The record of the torque test results is shown in Figure 9. In this case a constant dosage of 0.25 % (w/w, solids by mass of cement) of all admixtures was applied. A comparison of the plasticizing/fluidifying and set-accelerating/retarding effects caused by 0.25% of evaluated admixtures, determined from the measurements of torque variation with time as well as other methods, is presented in Table 2. The effect of admixtures on pastes under static (mini-slump, Vicat test) and dynamic (torque test) conditions expressed in per cent is compared.

The plasticizing/fluidifying effect of admixtures SL1 and SL2 was proportional for both mini-slump and torque test values. The admixture SL3 had a much better effect under dynamic conditions. As the torque test monitors the behaviour of fresh cement paste under continuous motion, extended set retardation occurred.

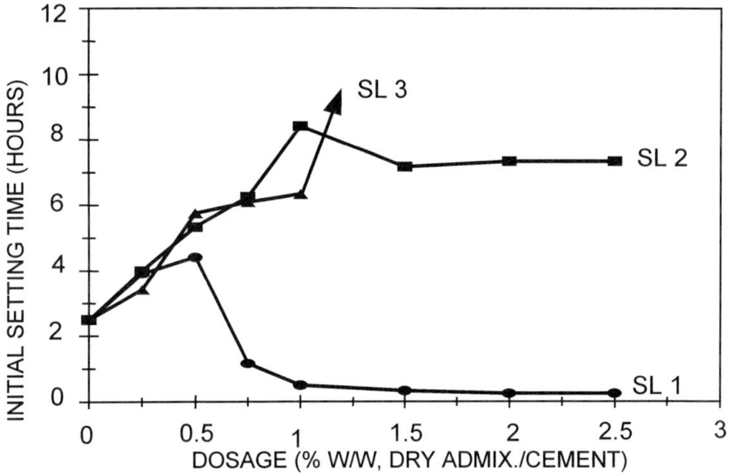

Figure 7 Influence of dosage rate on initial set-accelerating/retarding effect of admixtures in pastes with normal consistency determined by Vicat apparatus (ASTM C 191)

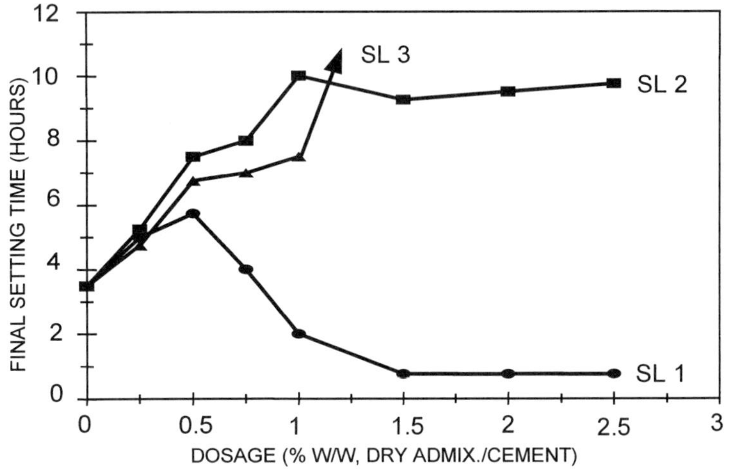

Figure 8 Influence of dosage rate on final set-accelerating/retarding effect of admixtures in pastes with normal consistency determined by Vicat apparatus (ASTM C 191)

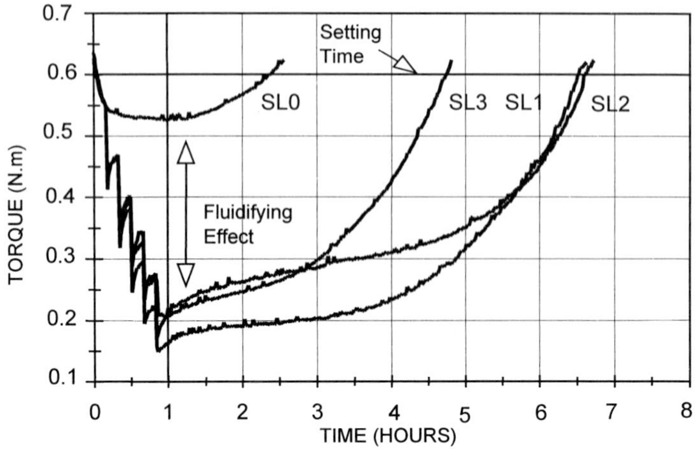

Figure 9 Record of torque test results on pastes with w/c=0.35, both with O.25% (w/w, solids by mass of cement) and without (SL0) admixtures

Table 2 Comparison of plasticizing/fluidifying and set-accelerating/retarding effects

SAMPLE CODE	PLASTICIZING/FLUIDIFYING EFFECT				SET ACCELERATING/RETARDING EFFECT					
	Mini Slump		Torque Test		Vicat-Initial		Vicat-Final		Torque Test	
	(cm^2)	(%)	(N.m)	(%)	(min)	(%)	(min)	(%)	(min)	(%)
SL1	14.5	100	0.33	100	85	100	90	100	250	100
SL2	17.4	120	0.39	118	90	106	105	117	256	102
SL3	12.6	87	0.34	103	55	65	75	83	142	57

Air-Entraining Effect

The air-entraining effect of different dosages of admixtures determined by the modified mini-slump test is presented in Figure 10. It can be seen that admixture SL3 had the highest air-entraining effect at a dosage rate of 0.75 %, while the air entraining effect of admixtures SL1 and SL2 was continuously increased with the dosage rate and the saturation concentration was not reached. The air-entraining effect determined by the gravimetric method according to ASTM C 185 on cement mortars is shown in Figure 11. In this case the air-entraining effect of all admixtures was comparable, with the maximum at 2 % of admixture SL1 and at 1.5 % of admixtures SL2 and SL3. Above this amount the actual air-entrainment of concrete mortar decreased.

This might be caused by lower stability of the air-void system at high dosages of admixtures and by the ambiguous influence of aggregate, which both supports (whipping effect) and disturbs (breaking effect) the creation of foam (related to the air-void system).

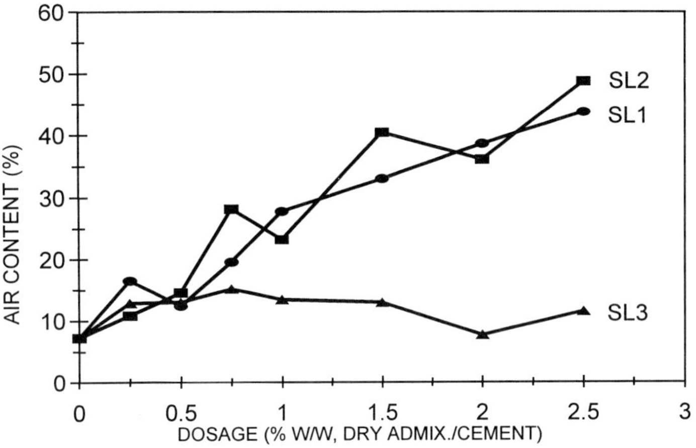

Figure 10 Influence of dosage rate on air-entraining effect of admixtures in pastes with w/c=0.35 determined by mini-slump test

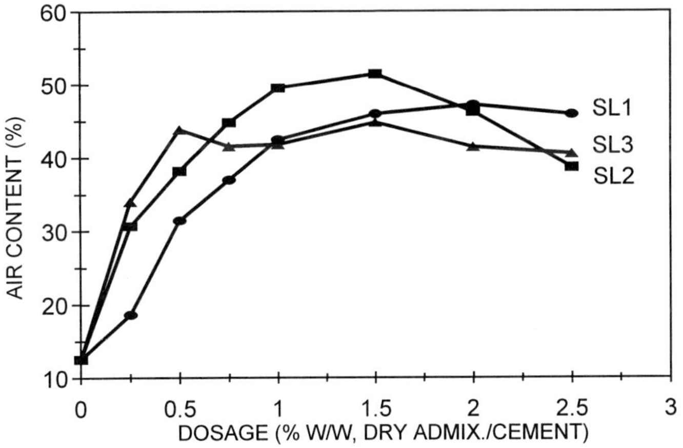

Figure 11 Influence of dosage rate on air-entraining effect of admixtures on mortars with a flow of 87.5±7.5% determined by the gravimetric method (ASTM C 185)

Correlation between Air-Entraining and Plasticizing/Fluidifying Effects

It is known that air-entrainment generally contributes to the plastification of fresh concrete mixes. The behaviour of admixtures SL1 and SL3 was somewhat in agreement with the above statement. On the other hand, no agreement was found for admixture SL2, especially in cement pastes. In this case the air-entraining effect was gradually increasing with the dosage but the plasticizing/fluidifying effect was negligible for all dosages (compare Figure 3 and 10). The more usual opposite behaviour, when the plasticizing/fluidifying effect increases with the admixture dosage while the air-entraining effect does not increase at all, was reported by the authors elsewhere [10]. Nonetheless, the air-entraining effect must always be taken into consideration when evaluating not only the plasticizing/fluidifying effect but also the water-reducing effect of concrete admixtures.

CONCLUDING REMARKS

The influence of dosage rate of lignosulfonate-based water-reducing admixtures on basic properties of fresh cement pastes and mortars was investigated. A relatively wide range of dosages from 0.25 to 2.5 % (w/w, solids by mass of cement) was used to describe the correlation among the main and side effects of different dosages of three different lignosulfonate samples on cement-water systems (both with and without aggregate). Based on the results and observations throughout the testing, the following conclusions and summarising comments can be made.

1. All studied effects: The relationship between the dosage rate of an admixture and the properties of a fresh cement-water system is dependent on the admixture sample being tested and the cement-water system (mixture proportions) being used as well as on the testing procedure employed.

2. Water-reducing effect: This effect generally increased with the dosage rate of all admixtures until the point of saturation. The optimum dosage was lower in cement mortars than in cement pastes, where even 2.5 % of SL1 was not enough to reach the saturation amount.

3. Plasticizing/fluidifying effect: Under static conditions the optimum dosage rate of SL1 and SL3 was 1.5 %; in the case of SL2 the plasticizing/fluidifying effect was not detected (but was found under dynamic conditions). Thus, the admixture with a good water-reducing effect did not always have an equally good plasticizing/fluidifying effect. Also, the relative effectiveness of admixtures was sometimes different under static (mini-slump test) and dynamic (torque test) conditions.

4. Set-accelerating/retarding effect: Three different types of behaviour were observed. First, the retardation gradually increased with the dosage rate (SL3). Second, the retardation first increased, then decreased and further remained moderate and unchanged (SL2). Third, the small retardation by low dosages changed to acceleration by higher dosages and further remained unchanged (SL1). These differences may be related to the different w/c ratios of individual mixtures, to the possible presence of an accelerator in the commercial lignosulfonate admixture and also to the lignosulfonate chemistry.

5. Air-entraining effect: In cement pastes, air-entrainment gradually increased with the dosage rate of SL1 and SL2 and the saturation amount was not reached. Sample SL3 had very low air-entraining effect at any dosage with the maximum at 0.75 %. In cement mortars, all three samples were comparable and reached the saturation amount (between 1.5 and 2 %) above which the entraining effect decreased. The positive influence of air-entrainment on plastification (water-reduction) of cement-water systems was not confirmed in all cases.

ACKNOWLEDGEMENTS

The authors wish to thank Mr. Jeff Keefe and Mr. Brian Stocker for their experimental help.

REFERENCES

1. ERNSBERGER, F.M. AND FRANCE, W.G. Portland Cement Dispersion by Adsorption of Calcium Lignosulfonate. Industrial and Engineering Chemistry,Vol.37, No.6, 1945,. pp.598-600.

2. FORBRICH, L.R. The Effect of Various Reagents on the Heat Liberation Characteristics of Portland Cement. Proceedings of the American Concrete Institute, Vol.37, 1941, pp.161-184.

3. AMERICAN SOCIETY FOR TESTING AND MATERIALS. Annual Book of ASTM Standards, Section 4: Construction, Vol.04.02: Concrete and Aggregates. ASTM, Philadelphia, PA, 1996.

4. ZHOR, J, AND BREMNER, T W. Influence of New Environmentally Beneficial Lignin-Based Superplasticizer on Concrete Performance. Radical Concrete Technology , Edited by R K Dhir and P C Hewlett E & FN Spon, London, 1996, pp. 301-311.

5. DE LARRARD, F. A Method for Proportioning High-Strength Concrete Mixtures Cement, Concrete and Aggregates, Vol.12, No.2, 1990, pp.47-52.

6. COLLEPARDI, M. Admixtures Used to Enhance Placing Characteristic of Concrete. Cement and Concrete Composites, Vol.20, No.2/3 (April/June), 1998, pp.103-112.

7. RIXOM, M R, AND MAILVAGANAM, N P. Chemical Admixtures for Concrete, 2nd edition, E&FN Spon, New York, NY, 1986

8. MASSAZZA, F. Admixtures on Concrete. Advances in Cement Technology, Edited by S N Ghosh, Pergamon Press, New York, NY 1983, pp 569

9. RIXOM, M.R. The Economic Aspects of Admixture Use. Cement and Concrete Composites, Vol.20, No.2/3 (April/June), 1998, pp.141-147.

10. ZHOR, J AND BREMNER, T W. Advances in Evaluation of Lignosulfonates as Concrete Admixtures. Proceedings, 4th International Conference on Recent Advances in Concrete Technology, Tokushima, Japan. Edited by V M Malhotra, ACI SP-179, 1998, pp 1101-1042

11. KANTRO, D.L. Influence of Water-Reducing Admixtures on Properties of Cement Paste. A Miniature Slump Test. Cement, Concrete and Aggregates, Vol.2, No.2, 1980, pp.95-102

FLUIDIZING MECHANISM OF POLYCARBOXYLATE-BASED SUPERPLASTICIZERS FOR SEVERAL BINDER MATERIALS

A Ohta

NMB Ltd

T Uomoto

University of Tokyo

Japan

ABSTRACT. We studied the dispersing components and dispersion maintaining components of polycarboxylate-based high-range water reducing AE agents. We studied the adsorption of these two types of components to various binding materials, and investigated the relationship between these components and the fluidity of cement paste. The results of our experiments showed that that as the relative surface area of a binding material increases, the adsorbed amount of dispersant per unit weight of binding material rises, however, the adsorbed amount remains constant per unit area of binding material. This results suggests that it is possible to estimate the adsorbed amount required to achieve particle dispersion for binding materials that have different relative surface areas, such as blast furnace slag particle and limestone particle. It may be possible to estimate the dosage of high-range water reducing AE agent required when binder particles of different relative surface areas are mixed together to produce high fluidity concrete.

Key words: Polycarboxylic acid, High-range water reducing AE agent, Adsorption characteristics, Adsorbed amount per unit area

Akira Ohta is a Research chemist of Central Research Laboratories, NMB Ltd. His research interest is the synthesis of various polymer dispersing agents and evaluation of their performance.

Professor Taketo Uomoto is a member of the Center for Collaborative Research, University of Tokyo. He specializes in: 1) Shape optimization of structures, 2) Nondestructive testing, 3) Durability of concrete, 4) New composite materials, 5) Computer simulation of cement hydration, 6) Evaluation of deteriorated concrete structures, 7) Analytical computer modeling phenomena and durability of shotcrete, and 8) Self compacting concrete.

INTRODUCTION

Owing to their excellent water reducing ability and good slump maintaining performance, high-range water reducing AE agent is an important component in the production of high-fluidity concrete and high-strength concrete.

In the manufacture of these kinds of concrete, it is necessary to combine various types of inorganic fine-particle binding materials such as ordinary Portland cement, belite cement, blast furnace slag powder, fly ash, and silica fume. In the production of high-fluidity concrete, these powders are combined in predetermined dosages, the concrete is mixed and fluidity and segregation resistance are tested. If there is any change to the mix proportions, then tests have to be made to determine the dosage of high-range water reducing AE agent. If different kinds of binding materials are used, or the relative surface areas of chemically identical binding materials are different, then testing to determine the dosage of high-range water reducing AE agent must be made.

We have studied the main component in high-range water reducing AE agents responsible for dispersion and have investigated their adsorption to cement particles, their zeta potential, and their relationship with the fluidity of mortar and concrete[1],[2]. In this study we investigated the dispersing components and dispersion maintaining components of polycarboxylate-based high-range water reducing AE agents. We studied the adsorption of these components onto various binding materials, the fluidity of pastes of binding materials, and the relationship between changes in the dosage of these components in concrete.

EXPERIMENTAL DETAILS

Characteristics of Binder Materials

The binding materials, their physical qualities and chemical composition are shown in Table 1.

Limestone powder (LS) is not generally included as a binding material, but in this study, six types of limestone powders, each having a different relative surface area, were used.

Other binding materials were blast furnace slag(BF), fly ash (FA) and ordinary portland cement (OPC).

Chemical Structure of Polycarboxylate

Two types of polymer dispersing agents, A-3 and A-5 were used in isolation. These are both dispersing components used in high-range water reducing AE agents. The chemical structure of the principle component of polycarboxylate-based high-range water reducing AE agents used in this study is shown in Figure. 1.

A-3 has high ratio of methacrylic acid groups and A-5 has high molar ratio of methacrylic EO-ester groups. n/m ratio: (A-3) > (A-5)

Table 1 Physical qualities and chemical composition of binding materials

BINDER TYPE	ABBREV.	DENSITY	REL. SURAFCE AREA	MgO	SO_3	SiO_2	Al_2O_3	Fe_2O_3	CaO	$CaCO_3$
Ordinary Portland Cement	OPC-1	3.15	3250	1.60	1.9	21.2	5.1	3.0	64.9	-
	OPC-2	3.15	3160	2.10	1.8	21.6	5.3	2.3	64.5	-
Blast Furance Slag	BF-40G	2.90	4110	6.60	2.1	32.1	13.2	0.2	42.5	-
	BF-60G	2.90	5810	6.50	2.1	33.1	12.6	0.2	42.7	-
	BF-80G	2.90	7960	6.10	2.0	32.1	14.2	0.1	42.9	-
	BF-100G	2.90	9800	6.60	2.0	32.6	12.9	0.2	42.6	-
Fly Ash	FA	2.27	3190	2.50	-	48.8	30.3	5.1	7.0	-
Limestone Powder	NF	2.70	5440	0.25	0.0	0.29	0.07	0.0	55.3	97.3
	LS-3500	2.76	3480	0.81	0.0	1.86	0.15	0.15	53.7	95.8
	LS-4700	2.79	4730	0.49	0.0	0.47	0.09	0.06	54.9	98.0
	LS-8900	2.79	8380	0.90	0.0	2.70	0.18	0.19	53.3	95.2
	LS-11000	2.77	10970	0.91	0.0	2.71	0.15	0.19	53.3	95.1
	LS-20000	2.77	18180	0.78	0.0	1.71	0.13	0.17	54.0	96.3

Remarks : Density: g/cm^3, Relative Surface Area : cm^2/g, Chemical Composition : %

(1) Measurement of adsorbed amount of polycarboxylate dispersant

Cement paste with a W/C ratio of 35% was prepared in an ASTM mortar mixer and mixed for three minutes after the addition of the final ingredient. After standing for a specified time, the cement paste liquid phase was separated via vacuum filtration.

The organic carbon content of the separated liquid layer was measured using a total organic carbon measuring device (Shimadzu TOC-5000). The amount of polymer adsorbed to the binding agent was calculated by subtracting the amount of carbon in the filtrate from the amount of carbon contained in the plain paste and the amount of additive.

To study the adsorbed amount of polycarboxylate dispersant in alkaline environments, we determined the adsorbed amount when aqueous sodium hydroxide solution with a pH of 12.5 was used as the mixing water and mixed with the binding material and cement. The mix proportions of the binding materials with cement were as follows.

Blast furnace slag powder : OPC = 5:5, FA : OPC = 3:7, LS : OPC = 3:7

(2) Measurement of distribution of particle size

Shimadzu laser particle size measuring devices SALAD-2000 and SALAD-3000 were used to measure the particle size distribution and average particle diameter of the binding materials.

The relative surface areas of the binding materials were calculated from the average particle diameters.

(3) Test of dispersibility using binding material pastes

The fluidity of paste of each binding material was measured and the dispersibility of the particles was assessed. An ASTM mixer was used to prepare binding material paste having a W/C ratio of 25 or 30%. The paste was poured into a cylindrical container with an internal diameter of 4 cm and a height of 7 cm. The container was then placed on top of a flow test stage and the spread of the paste was measured.

RESULTS AND DISCUSSION

Relative Specific Surface Area of Binder Particles

Figure 2 shows the relationship between the Blain value, derived from relative surface area of the binding material, and the relative surface area of the binding material, calculated from the particle diameter measured using a laser particle meter. The graph indicates that with the exception of one of the LS test samples, despite some variation, there is a correlation between Blain values and relative surface area.

Observation of the binding material particles using a scanning electron microscope (SEM) revealed that with the exception of FA, the particles were plate-like particles or extremely angular. Even when ethanol was used to disperse the particles, smaller particles adhere to the surface of larger ones. Regarding these particles to be spherical, this adhesion of particles presents problems when deriving the relative surface area based on the particle diameter distribution. Since there were problems with large relative surface areas, Blain values were used in this experiment to represent relative surface area.

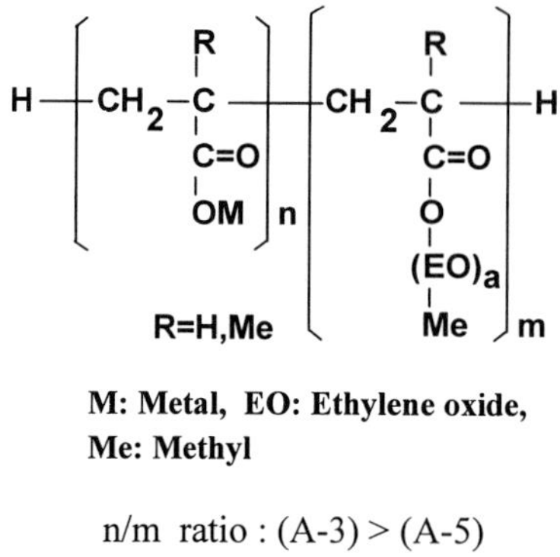

Figure 1 Chemical structure of polycarboxylate

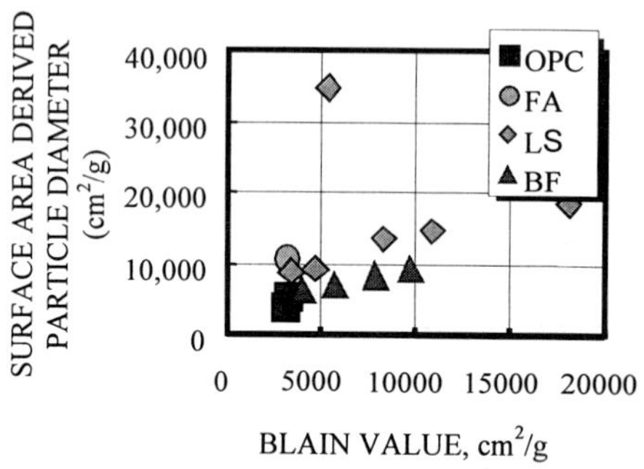

Figure 2 Relative surface area derived from particle diameter and Blain values

The Adsorbed Amount of Polycarboxylate Dispersant to Binder Particles

The two types of polycarboxylic acid (dosage: B x 0.15%) were in turn added to each type of binding material particles and the adsorbed amount was measured. The results are shown in Figures 3 and 4. The adsorbed amount of the two polycarboxylic acids varied considerably

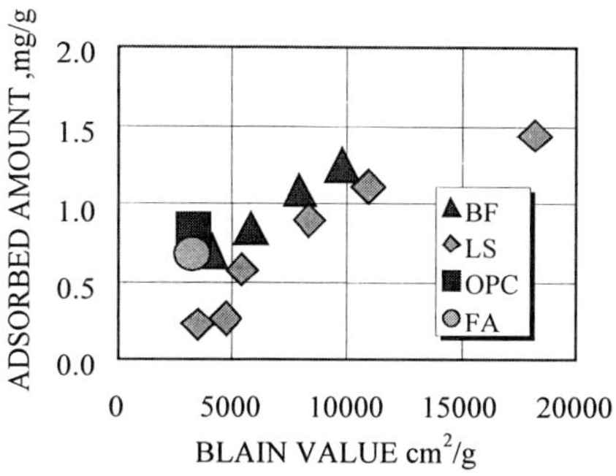

Figure 3 The adsorbed amount of A-3 polymer to binder particles

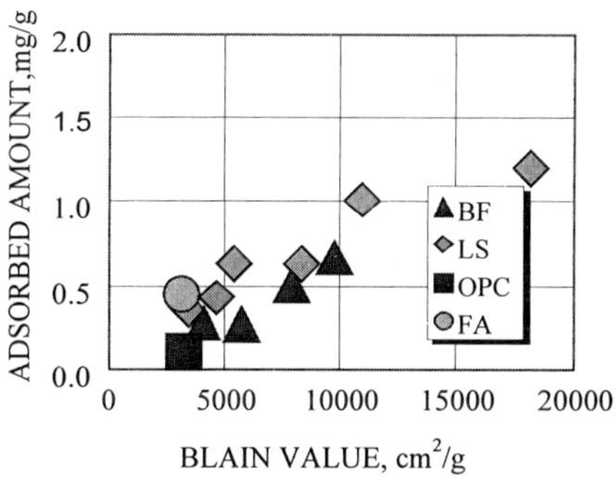

Figure 4 The adsorbed amount of A-5 polymer to binder particles

between the surface area of the binding material particles. For A-3 polymer, the binding materials can be ranked according to adsorbed amount as follows: LS < BF < FA < OPC; while for A-5 polymer the order is OPC < BF < LS < FA. For either polymer, the adsorbed amount per unit weight of binding material increased as the relative surface area of the binding material increased.

There was a large difference in the amount of A-3 and A-5 polymer adsorbed by OPC, but the difference was small in the case of LS and FA. The difference in the amount of A-3 and A-5 polymer adsorbed by BF was moderate.

Figures 3 and 4 show that the difference in amount of A-3 and A-5 polymer adsorbed by BF is not as great as the difference in amount absorbed by OPC.

The adsorbed amounts of both A-3 polymer and A-5 polymer by BF increase as the relative surface area of the particle increases. The amount of A-3 polymer adsorbed by 40G is comparably smaller than that adsorbed by OPC (Blain value of 3,250), and is virtually the same as 60G.

The Adsorbed Amount Per Unit-Area of Binder Particles

Figure 5 shows the adsorbed amount of A-3 and A-5 polymers per unit area by fine blast furnace slag particles at a dosage of B x 0.15%. The difference between the per unit area amounts of A-3 and A-5 polymers absorbed by OPC was greater than six times, but only two to three times with respect to BF fine particles. The amounts of A-3 and A-5 polymers adsorbed by BF fine particle per unit area was constant irrespective of relative surface area.

In Figures 3 and 4 no difference can be seen in the characteristics of adsorption of A-3 polymer and A-5 polymer by different grades of LS fine particles; as the relative surface area of the powder increases, the adsorbed amounts of both A-3 polymer and A-5 polymer increase. Figure 6 shows the per unit area amounts of A-3 and A-5 polymers adsorbed by LS fine particles. Like BF fine particles, the per unit area amounts of A-3 and A-5 polymers adsorbed by LS fine particle remain constant irrespective of relative surface area.

Previously, the steric repulsive effect produced by the layer of adsorbed polymer was believed to be closely related to the amount of admixture adsorbed by the cement. The finer the cement particles—in other words, the smaller their diameter—the greater the effects of the forces of steric repulsion[3]. Another characteristic of particles is that the finer the particles, the more frequent the defects in the crystal structure and the greater the surface energy. It is said that structures that have suffered a lot of damage—in other words, structures that have a high surface potential energy—adsorb well[4].

In the past, researchers have assumed that the smaller the particle diameter the greater the adsorbed amount. Our experimental results however, suggest that if we consider the adsorption per unit surface area, then polymers are well adsorbed onto particles with small diameters. There is a relationship between the size of the relative surface area and adsorbed amount, and it becomes clear that the adsorbed amount per unit surface area is constant. Depending on the type of powder, the adsorbed amount per unit surface area differs; however, as long as the binding material is the same, the adsorbed amount per unit surface area will remain constant regardless of the size of the particle diameter. Unfortunately, the types of powders tested in this study did not clarify the relationship between different adsorptions per unit surface area and the following: the chemical structure of the binding material, mineral composition, surface conditions of the particles.

Relationship Between Adsorbed Amount and Dispersibility of Binder Paste.

Figure 7 shows the relationship between flow values for the BF paste and the adsorbed amount of A-3 polymer per unit area. Dosage of A-3 polymer is BF ×0.05 0.30 %.We can see a correlation between the flow values for BF pastes and the adsorbed amount of polymer per unit area. For each of the four grades of BF particle each having a different relative surface area, a correlation was observed between the amount of polymer adsorbed and the flow value of the paste. The lines of approximation were almost the same.

Figure 8 shows the relationship between flow values for LS particles paste and the adsorbed amount of A-3 polymer per unit area The results observed above show that in the relationship between dispersibility and the adsorbed amount of polymer that achieves dispersion of particles through the steric effects, a correlation was observed for cement particles and other kind of binding materials.

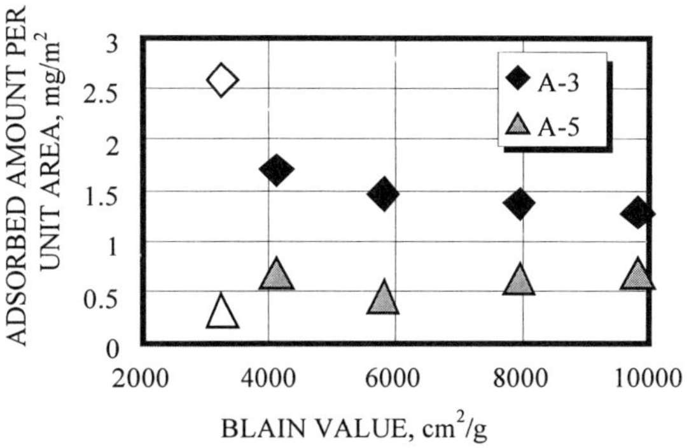

Figure 5 The adsorbed amount per unitarea of BF Particles

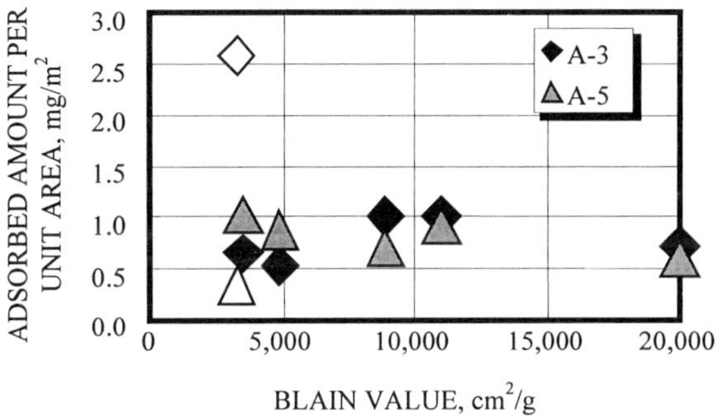

Figure 6 The adsorbed amount per unit area of LS Particles

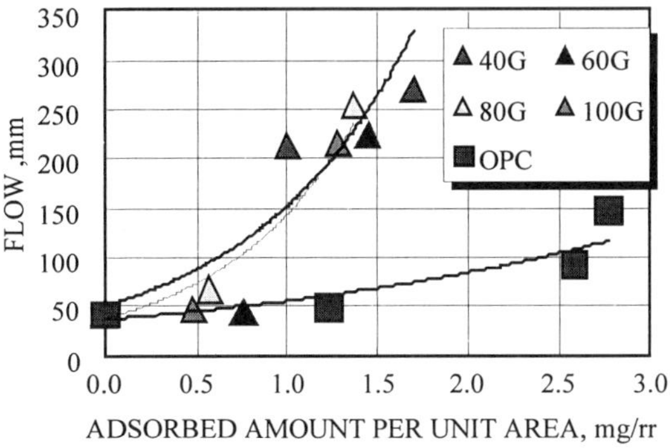

Figure 7 Relationship between adsorbed amount of A-3 on BF and dispersibility

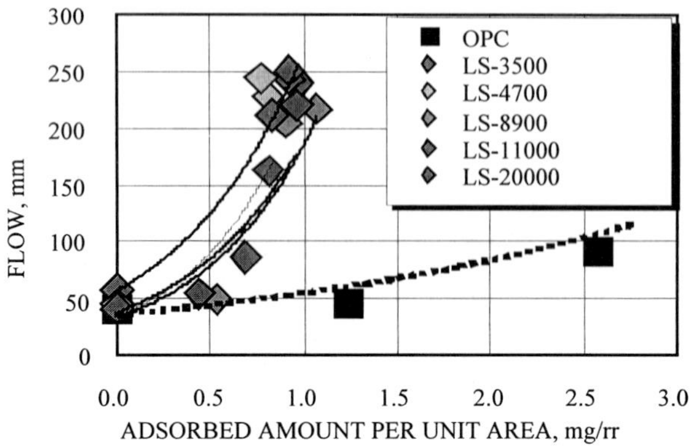

Figure 8 Relationship between adsorbed amount of A-3 on LS and Dispersibility

Furthermore, if mixed with different binding materials or combinations of different binding materials, the adsorption differs depending on the type of polymer.

Adsorbed Amount of Polymers for Several Alkali Conditions.

Figure 9 shows the adsorbed amount for per unit weight of binder particles of polymers under an alkali environment and when mixed with OPC. From Figure 9 we see that under an alkali environment the amount of A-3 polymer adsorbed on FA does not change when OPC is added. Under an alkali environment, the adsorbed amount of A-3 polymer when mixed with BF-40G declined about 35%, but remained virtually unchanged when OPC was added. The results were virtually identical when A-3 polymer when mixed with BF-100G and then combined with OPC. Under an alkali environment, the amount of A-5 adsorbed declined when combined with OPC. In particular, the amount of adsorbed dramatically declined for OPC.

There was almost no difference in the amount of A-3 and A-5 polymer adsorbed when mixed with limestone powder, however, under an alkaline environment, the difference increased when OPC was added. The results were virtually identical when A-3 polymer when mixed with BF-100G and then combined with OPC. In an alkaline environment where eluted sulfate ions were present in solution, a change in the spread of the polymer could be detected using a light scattering method[1],[2],[5]. This suggests that various factors cause the shape of the A-5 polymer to change in an aqueous solution, and this change greatly influence the adsorption of the polymer onto the surface of the binding material particles, based on the results of our researches of dispersant adsorption with the various cements, analysis of the ions eluted from the cement paste, and the change in polymer size (a light scattering analyzer was used and the root mean square radius of the polymer in an aqueous solution was calculated : RMS radius) under different sulfate ion concentration[5].

Figure 10 shows the estimated values calculated from OPC values (0.84mg/g) and each binder paste value of A-3 polymer. There was almost no difference in the amount of A-3 polymer adsorbed between calculated value and experiment value.

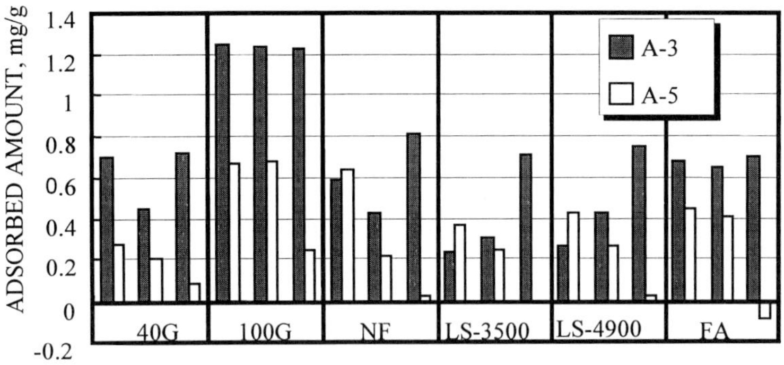

Note) From the left of each column Binder paste, Alkaline environment, With OPC

Figure 9 Adsorbed amount of alkali environment and when mixed with cement

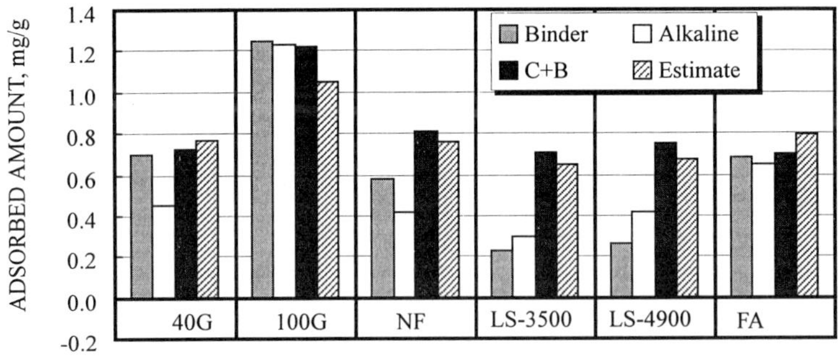

Figure 10 Adsorbed amount of A-3 polymer in alkali environment and when mixed with cement

Relationship Between Admixture Dosage and Relative Surface Area of Binding Materials in Concrete

Figure 11 shows the relationship between 1) Blain value and 2) the dosage of high-range water reducing AE agent containing A-3 and A-5 polymers in concrete mixed with cement and binding materials (relative surface area as Blain value: 3,250 to 5,810 cm^2/g) unit weight: 450 to 591 kg/m^3, with a standard flow value of 65 +- 5 cm[6],[7].

Figure 11 indicates that as the relative surface area of the binding material increases, to a certain extent the dosage of admixture increases by a constant rate.

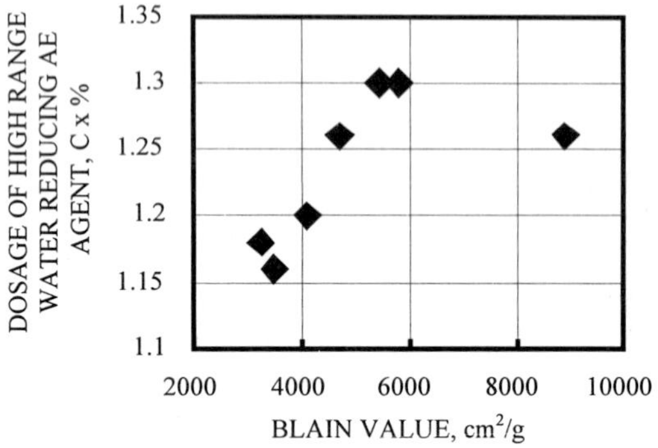

Figure 11 The relationship between relative surface area of binding materials in concrete and admixture dosage

CONCLUSIONS

1. The adsorption of dispersant and dispersion maintaining agent varies depending on the type of binding material. As the relative surface area of powder of the same type of binding material increases, the adsorption per unit weight of binding material increases, but the adsorption per unit area of binding material remains constant. This suggests that it is possible to estimate the required dosage of dispersant for powders of the same type of binding material that have different relative surface areas.

2. Under an alkaline environment the adsorption of the polymer to the binding material changes when cement is added, particularly in the case of A-5 polymer. The reason is believed to be that the adsorption of A-5 polymer to cement particles is greatly influenced by the shape of the A-5 molecules in an aqueous solution.

3. The relationship between the dosage of dispersants / dispersion maintaining components in high-range water reducing AE agents and the relative surface areas of the binding materials used in high-fluidity concrete increases at a constant rate as relative surface area increases.

4. This study investigated the dispersing effects of the principle dispersing components and dispersion maintaining components used in polycarboxylate-based high-range water reducing AE agents by measuring their adsorption with various types of binding materials containing limestone. It has been found that the adsorption of the polymer to particles is closely involved with the steric effects that are mainly responsible for the dispersing effects of polycarboxylate-based dispersants. In this study, however, we clarified the state of this adsorption under various environments.

ACKNOWLEDGMENTS

We would like to thank Mr. Sugiyama of NMB Co., Ltd. for his cooperation in the measurement of adsorption amounts, and express our gratitude for the assistance of Mr. Katsuno, a student at Shibaura Engineering College, in carrying out the experiments.

REFERENCES

1. OHTA, A., UOMOTO, T, "Basic Research into the Chemical Structure of High-Range Water Reducing AE Agents and Their Effects on the Fluidity of Mortar (1)," Seisan Kenkyu. Vol. 49, No. 12, pp. 637-640, December 1997.

2. OHTA, A. ET AL, "Fluidizing Mechanism and Application of Polycarboxylate-Based Superplasticizers," Fifth CANMET/ACI Superplasticizers and Other Chemical Admixtures in Concrete 1997, pp. 359-378.

3. NAWA, T. " Status of High-Belite-Based Cement", Concrete Journal Vol. 34, No. 12, pp. 16-25, 1996.12

4. KOISHI M. ET AL. "Technical data for Surface Improvement of Powder." Soft Technology Publishment Dept. October 1982

5. OHTA, A, UOMOTO, T, "Study on Dispersion Action Mechanism of Polycarboxylate-based Dispersant" 52th Cement Technology Conference Summary, pp.134-135. 1998

6. UEMATSU, K., KISHI T." Effect of Lime Stone Micro-Powder on Heating Process of Cement Hydration" Proceedings of the 52th Annual Conference of the Japan Society of Civil Engineers. No. 5, pp.180-181, 1997.9

7. Laboratory Report of NMB Co., Ltd.

NEW SUPERPLASTICIZERS: FROM RESEARCH TO APPLICATION

Y F Houst
R J Flatt
P Bowen H Hofmann
EPFL
Switzerland

U Maeder
J Widmer
U Sulser T A Buerge
Sika
Switzerland

ABSTRACT. Increasing use of high performance, pumping and self-levelling concrete is driving the leading global producers of chemical admixtures to develop new products. The underlying mechanisms that account for the action of superplasticisers are nowadays not well understood. The increasing variety of superplasticisers currently on the market and the complexity of cementitious systems do not help alleviate incompatibilities and unexpected problems. The above reasons have led us to initiate an European *Eureka* project with the aim to better understand the fundamental aspects of the behaviour of superplasticisers as well as their effect on concrete. The fundamental aspects studied are adsorption, surface charge, and rheology of cement paste. The rheological properties of fresh mortars with the same superplasticisers were also measured. Furthermore, the effect of a grinding aid on properties of cement paste and fresh mortar is shown.

Keywords: Superplasticisers, Dispersion mechanisms, Adsorption, Rheology, Grinding aid.

Dr Yves F Houst is a chemist and is the Head of the Cement Group at the Powder Technology Laboratory (LTP), Swiss Federal Institute of Technology (EPFL), Lausanne, Switzerland.

Robert J Flatt is a chemical engineer, and was a Ph.D. student at the LTP. He studied interactions of cement with superplasticizers.

Dr Paul Bowen earned his PhD from the University of Cambridge in physical chemistry. He is currently a research associate at the LTP.

Dr Heinrich Hofmann is a Professor at the Materials Science Department of the EPFL and Head of the LTP.

Dr Urs Maeder is Head of Concrete Admixtures Department Research and Development at Sika, Zürich.

Dr Jürg Widmer is Chief Chemist at the Department of Chemical Research and Analytical Chemistry at Sika, Zürich, Switzerland.

Ueli Sulser is a technician, specialising in polymer characterisation at Sika, Zürich.

Theodor A Buerge is a chemical Engineer, Director of Corporate Research and Development at Sika, Zürich.

INTRODUCTION

Even nowadays, the production of quality concrete seems to non-specialists a relatively simple task. The ingredients are well known: Portland cement has been produced for more than one and a half century, sand and gravel are natural products, and water as well. Different admixtures are furthermore at disposal for improving the desired properties: accelerators, retardators, (super)plasticisers, etc. Concrete technologists are only too aware that the situation is not so ideal. Even pure Portland cement can vary in its chemical and mineralogical composition.

Sulphate, C_3A, alkali content influence the properties of fresh and hardened concrete. Blended cement and addition of by-products, which are nowadays commonly used, render the situation still more complex. Aggregates also vary in their chemical and mineralogical composition, shape, particle size distribution and surface properties. Problems of compatibility cement/admixtures are known, and generally avoided by preliminary tests [1]. The precise cause of incompatibility frequently remains unknown. The development of high performance concrete still increases the risk of incompatibility.

Material scientists and chemists involved in the development of chemical admixtures have not only the task to synthesise and formulate new efficient products, indeed they should be able to understand and possibly to foresee the behaviour of the admixtures in complex cementitious systems. Fulfilling these requirements is rendered very challenging because of the complexity of concrete components and the various climatic conditions in which they are used. The influence of a grinding aid, a chemical admixture used for the cement production, will illustrate this point hereafter.

Despite very interesting studies especially carried out since 1970's [2-9], numerous questions remain unsolved. This situation led us to initiate a Eureka project with the aim to study and understand the behaviour of superplasticizers in real cementitious systems. Participants come from university, research and testing laboratory, industry, and practice as well. The scope and the aim of this project are described in more detail in [10].

MECHANISMS OF ACTION OF SUPERPLASTICISERS

Cement particles are always agglomerated in water suspensions. This leads to a viscosity increase by an apparent particle volume increase. A part of the water is entrapped in the porosity of the agglomerates and does not contribute to the flowability. The role of the superplasticizers (polymers) is essentially to break down these agglomerates by modifying the balance of interparticle forces. Four mechanisms might act for the dispersion of cement particles, as well as for other powders. The two most important are linked to polymer adsorption: steric hindrance through the adsorbed layer thickness and electrostatic repulsion through the induced electrical charge. In reality, once the particles are close enough for their adsorbed layers to overlap, both effects come into play. Understanding which mechanism is dominant and what parameters have an effect, is paramount for the design of polymers with improved performances. It should also allow a better use of common polymers. Much has been learned from the field of colloidal science, where surface forces and their role on interparticle interactions has been a subject of ongoing research for many years.

Horn [11] shows how colloidal, surface and interparticle force concepts can be applied to ceramic processing. Russel et al [12] give an extensive treatment of these effects, illustrating how these effects are the regulating factors of the rheology of colloidal dispersions.

Chemical effects are also associated with superplasticizers since they influence the hydration process [2]. For instance, there is a strong indication that sulphonated naphthalene formaldehyde polycondensates (SNFC) molecules can participate in the same type of reaction as the sulphate ions inhibit the early hydration of the aluminate phases. In general, the superplasticizers interfere with the nucleation and growth of aluminate hydrates, ettringite and C-S-H products and affect setting time. The reactivity of the cement with water has a significant influence on the superplasticiser dosage to get a given workability. The lower the reactivity, the lower the superplasticiser demand. A quartz powder requires much less superplasticiser than a C_2S rich cement, which itself requires less superplasticiser than a C_3S or a C_3A rich cement [2].

The most commonly used superplasticizers until now are certainly either SNFC (generally the Na salts) or sulphonated melamine formaldehyde polycondensates (SMFC). It is generally believed that the dispersive action of such superplasticiser types is essentially due to the electrical repulsion [5]. They are called of first generation.

Superplasticizers of the new generation are copolymers containing polyethylene glycol as graft chain and carboxylic function in the main chain. These polymers induce less negative zeta potentials (a function of the charge of the particles with adsorbed polymers) than those of the first generation. According to Sakai and Daimon [3], these polymers induce a zeta potential as small as -5 to -10 mV. In this case, the dispersion caused by these superplasticizers is attributed to steric hindrance [6]. Uchikawa et al [7] were led to the same conclusion by measuring surface forces by atomic force microscopy. Nevertheless, the distance at which interaction appears is several times larger than would be expected from the side chain length of the polymers used. Furthermore, they are less incorporated in the hydration products and disperse cement at lower concentration. This leads in general also to a smaller slump loss in time. Further elucidation of the mechanisms and more accurate modelling of the interparticle forces are needed to fully understand and apply these second generation superplasticisers.

EXPERIMENTAL

Materials

The main characteristics of model powders, cements, and silica fume, are give in Table 1. Two model powders, MgO and $Mg(OH)_2$, were used for certain measurements in this study. They have similar surfaces properties to calcium oxide, and in particular isoelectric points (zero zeta potential close to the pH=12) as cement suspensions [8]. The main advantage in using such powders is that they are "non-reactive" and the influence of hydration can be avoided.

Two Portland cements obtained from the same clinker were used. The first was ground without any grinding aid.

The second was ground in presence of triethanol amine acetate, a common commercial grinding aid. Grinding times were adjusted so as to obtain two cements with the same specific surfaces.

Table 1 Cement and silica fume characteristics

POWDERS USED	SPECIFIC SURFACE AREA [m^2/g]	
	Blaine	BET (N_2)
MgO (Martin Marietta, USA)	-	0.77
$Mg(OH)_2$ (Martin Marietta, USA)	-	11.8
CEM I 42,5 with grinding aid (Olten, CH)	0.349	0.950
CEM I 42,5 without grinding aid (Olten, CH)	0.327	0.963
S.F. Anglefort undensified (Péchiney, F)	-	35.3

Seven structurally different polymers, which are representative of most of the superplasticizers used today, have been selected for this study. They are listed in Table 2, along with their average molar mass determined by GPC. Four of these polymers, PCA-1, PCA-2, PCE-3 and SNFC-1 are commercially available. The other three, PCE-1, PCE-2, and PCA-3 are laboratory test products. Apart from the SNFC-1, these polymers are found as concentrated aqueous solutions, with a solid content of about 30-40% by mass.

Table 2 Chemical nature and molar mass of the polymers

DESIGNATION	CHEMICAL TYPE	AVERAGE MOLAR MASS [g mol^{-1}]
PCA-1	Na-polycarboxylate-polysulphonate	* 16’000 ± 2’000
PCA-2	Na-polycarboxylate-polysulphonate	* 11’500 ± 2’000
PCE-1	Na-polycarboxylate-polysulphonate, containing PEG-ester	* 9’500 ± 1’500
PCE-2	PEG-ester containing polymer, weakly ionic	* 15’000 ± 5’000
PCE-3	Na-polymethacrylic acid, containing PEG-ester, weakly ionic	≠ 10’000 ± 2’000
PCA-3	Na-polyacrylate	≠ 2’500 ± 500
SNFC-1	Sulphonated naphthalene formaldehyde polycondensate, Na-salt	≠ 6’000 ± 1’000

* Number Average Molar Mass ≠ Mass Average Molar Mass

Zeta Potential

The zeta potential was measured at 23-25° C by acoustophoresis. In this method, particles are put into movement by an acoustic wave. The field that their surface charge induces allows calculation of the zeta potential. The suspensions were prepared by dispersing 20 g of powder in 300 ml NaOH 0.01M (pH 12.0 ± 0.1 at 20°C) under vigorous agitation and sonification. The superplasticizers were progressively added as concentrated aqueous solutions with a microdispenser.

Adsorption

Adsorption measurements were performed in suspensions of 5 g of $Mg(OH)_2$ in 20 ml NaOH 0.01M (pH 12.0 ± 0.1 at 20°C), previously thermostated in a rotary bath (different amounts were taken to achieve similar volume fractions). Superplasticizers were added as concentrated aqueous solutions (10-15% solid content by mass) with a microdispenser.

After 30 min in the rotary bath, the suspensions were centrifuged. From the supernatant, 5 ml, buffered with acetic acid, were filtered and a part was analysed by reverse phase chromatography with a high performance liquid chromatograph (HPLC) [8].

The diode array detector, which only detects molecules that absorb in the UV or visible range, has limited the number of polymers whose adsorption could be studied. Polymers PCE-1 and PCA-2 were selected. They are detectable by UV, have similar molar masses, and HPLC chromatograms suggest that the molar mass distributions are also similar. Finally, they have a similar structure, but differ mainly by the insertion of some PEG-ester units into PCE-1.

Rheology

Samples were inserted into the sample holder of a coaxial cylinder device mounted on a rheometer thermostated at 25°C. The gap between both cylinders is 0.85 mm. Over this gap, the velocity gradient is assumed to be constant. More details are given in [13].

Flow Table Spread

The tests were performed according to DIN 18555 on mortars made with 1 part of cement and 3 parts of sand (0 – 4 mm) by mass.

Heat Generation

The temperature variations as a function of time were measured by means of thermocouples placed in the centre of mortars identical to those used for the flow table spread test.

RESULTS AND DISCUSSION

Zeta Potential

Electrostatic repulsion, one of the possible mechanisms involved in dispersion and water reduction, is intimately linked to zeta potential. The influence that all superplasticizers used in this study have on the zeta potential of $Mg(OH)_2$ suspensions is shown in Figure 1. Potentials are plotted as a function of superplasticiser added, given in mg of their dry mass by m^2 of $Mg(OH)_2$ in suspension. Results show that the polymers PCA-1,-2,-3 and SNFC-1 induce the more negative final potentials (around -23 mV), while the polymers PCE-1,-2,-3 induce potentials ranging from -5 to -18 mV. The main reason for these differences is that the first group of polymers are all strong electrolytes, while the others are weak or very weak electrolytes, due to the insertion of PEG-ester units. This acts as a charge spacer and decreases the charge density of the PCE-type polymers.

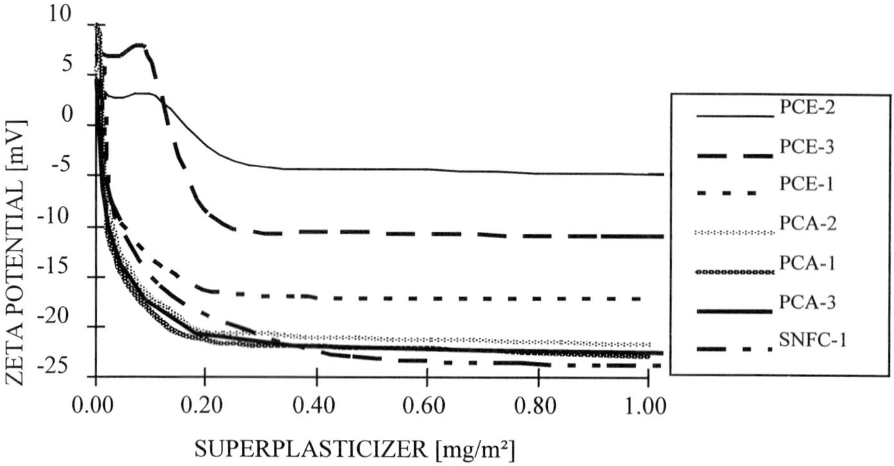

Figure 1 Zeta potential of magnesium hydroxide suspensions as a function of the added mass of superplasticizers [8]

From these results, it is concluded that dispersion by electrostatic repulsion (ER) may be much smaller for the PCEs than for the PCAs or the SNFC. However, they can all be successfully used as water reducers and superplasticizers. Dispersion by PCE-type polymers should therefore involve at least one mechanism other than ER.

Since the surface charge induced by the PCA and SNFC strong electrolytes were all found to be very similar, it is expected that dispersion due to electrostatic repulsion will be similar.

Any major differences in the dispersion efficiencies would therefore be a consequence of non-electrostatic effects.

Adsorption

For a substance in suspension at a fixed temperature, an adsorption isotherm gives the concentration of this substance at the surface of the solid phase versus its concentration in the bulk of the liquid phase. Usually, these isotherms have a horizontal asymptote (a plateau concentration), which corresponds to surface saturation of the solid phase.

Plateau concentration of adsorption isotherms of PCA-2 and PCE-1 as a function of the temperature are given in Figure 2.

It appears that the adsorption of PCE-1 on MgO increases with temperature, while the one of PCA-2 do not show significant temperature dependence. Adsorption of polymers onto surfaces is mainly linked to either polymer-surface, solvent-surface, polymer-polymer or polymer-solvent interactions. One of these factors, or a combination of them, should explain why the plateau concentration of PCE-1 adsorbed on MgO increases with temperature.

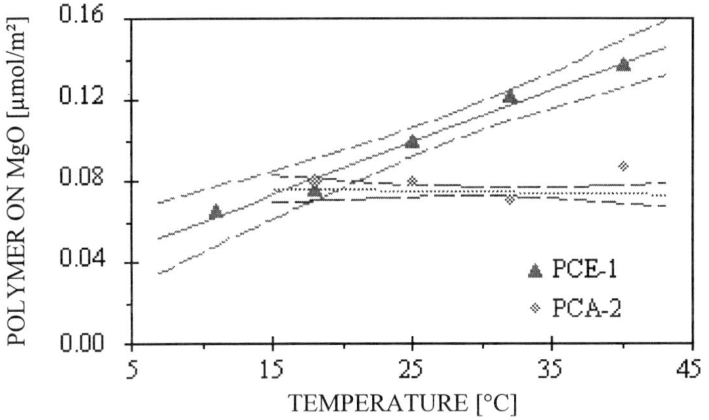

Figure 2 Temperature effect on the adsorption of PCA-2 and PCE-1 on MgO. The full straight lines are obtained by a linear regression and the dotted lines show the 95% confidence interval

The PCE-1 polymer is similar to the PCA-2, but its anionic groups are further apart, due to the insertion of PEG-ester groups. Consequently, its electrostatic attraction towards the positively charged surfaces must be smaller than it is for PCA-2. Mechanisms are discussed in more details in [8].

Such different behaviours of almost similar superplasticizers can have important practical implications. For instance, it could be assumed that the workability with PCE-1 is very much influenced by the temperature.

Rheology

It is shown in [13] that a simple model such as the two parameters Bingham model is insufficient to account for the complete shear rate-shear stress curves. Above a certain superplasticiser dosage, all curves indicate very similar flow behaviours. Increasing the shear stress allows us to identifying the yield stress. Up to a certain value of shear stress, the shear rate is equal to zero. Then, there is a sudden increase in shear rate. This limiting stress value is the yield stress.

The flow curves with the cement with and without grinding aid, with PCA-1 and PCA-2, have been measured. It has been found that optimum dispersion was obtained just before the saturation level observed in adsorption isotherms with PCA-2. This is in agreement with similar studies with model powders [9].

With PCA-1, the optimum dispersion was obtained with an amount exceeding the saturation level observed in adsorption isotherm. This can be attributed essentially to a difference in adsorption kinetics. PCA-1, however, allows us to obtain suspensions with a lower yield stress.

These properties can be related to the chemical structure and molar mass distribution of the two superplasticizers, the latter being significantly different.

The presence of a grinding aid influences the rheological properties of cement suspensions. The minimum yield stresses of the cement without grinding aid are slightly higher than those with grinding aid, but the amount of polymer required is smaller, with respect to adsorbable amounts [13], i.e. more polymer is needed when using a grinding aid, but the overall result is better.

Flow Table Spread

Numerous measurements were carried out on different mortars made with different cements, different superplasticizers and dosages, and different w/c. As examples, results obtained with the superplasticizers SNFC-1 and PCE-2 and the cement with and without grinding aid are given in Figures 3 and 4 respectively.

The mortars prepared with the cement without grinding aid have always a higher spread than those prepared with the cement containing a grinding aid.

The spread loss in time is not influenced by the grinding aid. With SNFC-1, the spread loss is high already after 30 min. The mortars with PCE-2 and the highest superplasticiser dosage (0.6%) exhibit very few spread loss.

This is certainly due to a dosage higher than the saturation in adsorption isotherm. This compensates the fraction of superplasticiser incorporated in hydration products.

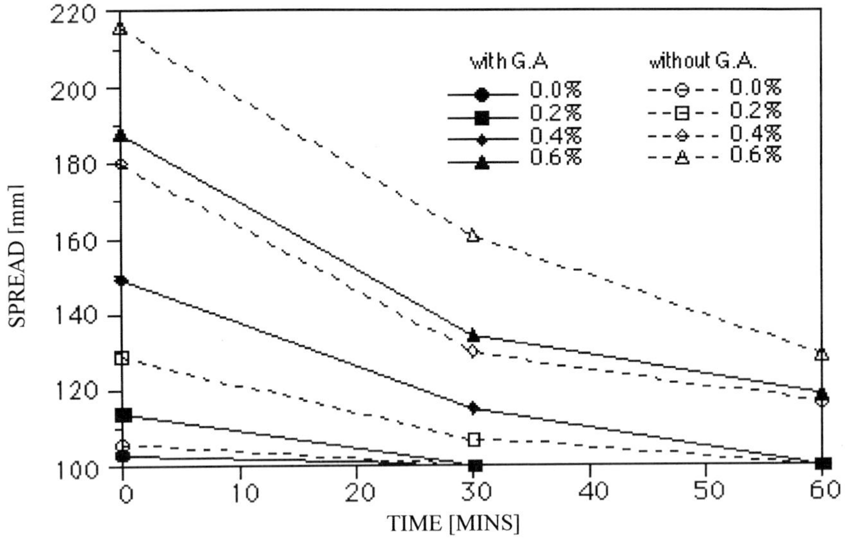

Figure 3 Flow table spread of mortars prepared with various dosage of superplasticiser SNFC-1 on a CEM I with and without grinding aid (w/c=0.36)

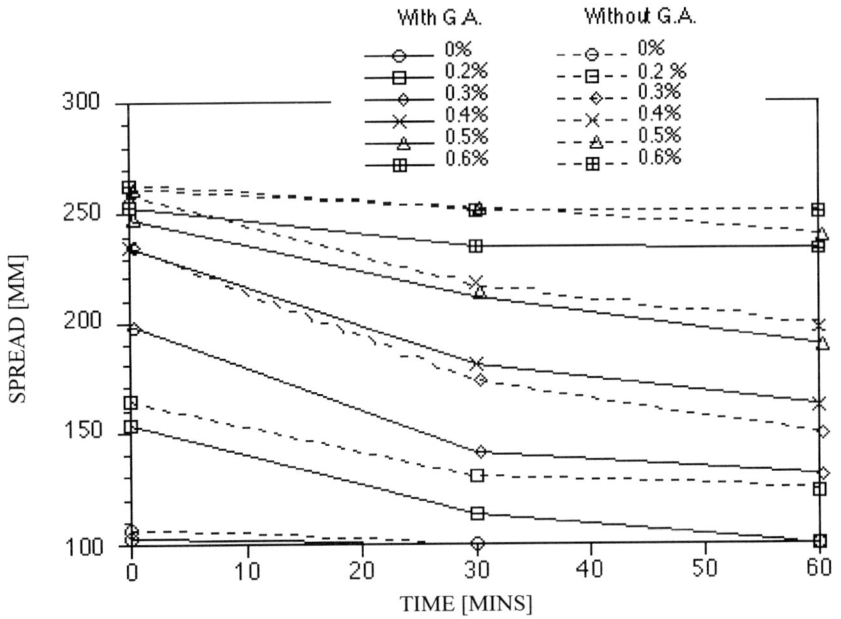

Figure 4 Flow table spread of mortars prepared with various dosage of superplasticiser PCE-2 on a CEM I with grinding aid and without grinding aid (w/c=0.36)

Heat Generation

The maximum of the heat generation is usually attributed to the crystallisation of $Ca(OH)_2$ and to the dissolution of C_3S and the formation of C-S-H. The maximum temperatures measured on mortars are reported in figure 5. The grinding aid has no influence on the reference mortars without any superplasticiser. On the other hand, all superplasticizers delay the maximum temperature, this trend being amplified by increasing superplasticiser dosage. A noticeable difference can be observed between superplasticizers PCA-1 and PCA-2 or PCE-1. The grinding aid has no effect with PCA-1, but it does have an effect with PCA-2 and PCE-1. Their effects are opposed: grinding aid increases the effect of PCA-2 and decreases the one of PCE-1.

All these effects are linked with the cement hydration and have been extensively studied with SNFC and SMFC superplasticizers. The state of the art has been recently presented [2, chap. 5]. It can be noticed that numerous examples are cited, but almost always without explanation of the fundamental reasons, which cause such behaviours. This shows avenues for new research.

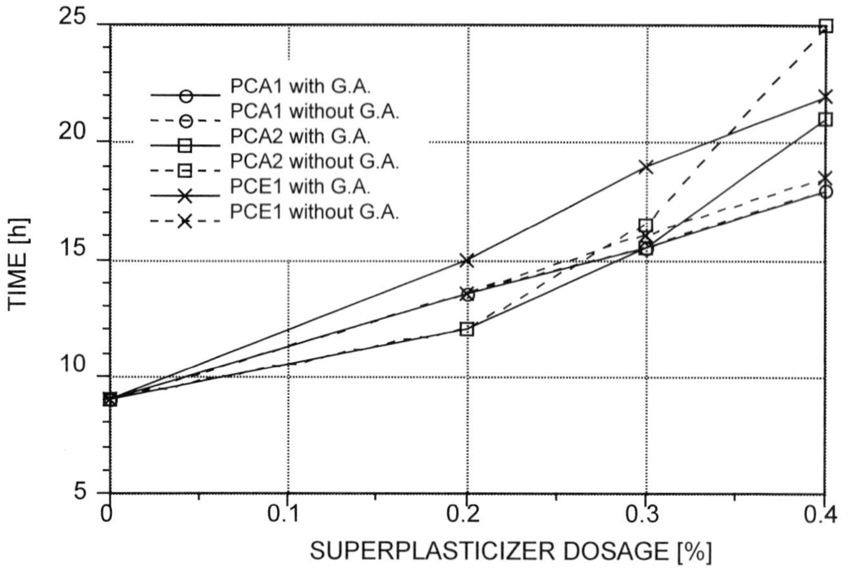

Figure 5 Influence of superplasticiser dosages on the displacement in time of the maximum of heat generation of mortars made with CEM I with and without grinding aid (w/c=0.36)

CONCLUSIONS

The seven superplasticizers studied belong either to the group of strong electrolytes or to the group of weak or very weak electrolytes. It has been shown that the polymers of the second group, which act essentially by steric repulsion, can be used with success as superplasticizers.

Adsorption isotherms characterise the interaction of superplasticizers-cement surfaces. It was found from rheological measurements that the adsorption plateaux (surface saturation) correspond roughly to the amount of superplasticiser allowing the optimum dispersion.

It has been confirmed that a superplasticiser without sulphonated functions has less spread loss than a SNFC superplasticiser belonging to the first generation of superplasticizers.

The retarding effect of superplasticizers has been confirmed. However, significant differences can be observed, depending on the couple cement/superplasticiser.

A grinding aid, commonly used nowadays in the cement production, interferes with superplasticizers and influences most of the measured properties. This effect is sometimes positive, sometimes negative. The mechanisms behind these effects remain to be elucidated.

The examples given above in this paper show why it is necessary to better understand the mechanisms through which superplasticisers act. The development, the selection and the correct use of these admixtures strongly depend on it.

ACKNOWLEDGEMENTS

This research was made within the framework of the EUREKA project 1363 ICHRWR and received financial support from the Swiss commission for Technology and Innovation (CTI).

REFERENCES

1. AÏTCIN, P C, JOLICOEUR, C, AND MACGREGOR, J G. Superplasticizers: How They Work and Why They Occasionally Don't. Concrete International, 1994, Vol. 16, No 5, pp 45-52.

2. RAMACHANDRAN, V S, MALHOTRA, V M, JOLICOEUR, C, AND SPIRATOS, N. Superplasticizers: Properties and Applications in Concrete. CANMET, Ottawa, Canada, 1998, pp 404.

3. SAKAI, E, AND DAIMON, M. Mechanisms of Superplastification. Materials Science of Concrete IV (Skalny, J and Mindess, S, Eds.). The American Ceramic Society, Westerville, USA, 1995, pp 91-111.

4. JOLICOEUR, C, NKINAMUBANZI, P C, SIMARD, M A, AND PIOTTE, M. Progress in Understanding the Functional Properties of Superplasticizers in Fresh Concrete. Proc. of the 4th CANMET/ACI International Conference on Superplasticizers and Other Chemical Admixtures (editor: Malhotra V.M.). American Concrete Institute, Detroit, 1994, SP 148, pp 63-87.

5. JOLICOEUR, C, AND SIMARD, M A. Chemical Admixture-Cement Interactions: Phenomenology and Physic-chemical Concepts. Cement and Concrete Composites, 1998, Vol 20, pp 87-101.

6. YOSHIOKA, K, SAKAI, E, DAIMON, M, AND KITAHARA, A. Role of Steric Hindrance in the Performance of Superplasticizers for Concrete. J. Am. Ceram. Soc. 1997, Vol 80, No 10, pp 2667-71.

7. UCHIKAWA, H, HANEHARA, S, AND SAWAKI, D. The Role of Steric Repulsive Force in the Dispersion of Cement Particles in Fresh Cement Paste Prepared with Organic Admixture. Cem. Concr. Res., 1997, Vol 27, No 1, pp 37-50.

8. FLATT, R J, HOUST, Y F, BOWEN, P, HOFMANN, H, WIDMER, J, SULSER, U, MAEDER, U, AND BUERGE, T A. Interaction of Superplasticizers with Model Powders in a Highly Alkaline Medium. Proceedings of the 5th Canmet/ACI International Conference on Superplasticizers and Other Chemical Admixtures in Concrete (Malhotra V.M., Ed.). American Concrete Institute, Detroit. 1997, SP-173, pp 743-762.

9. FLATT, R, J, HOUST, Y F, BOWEN, P, HOFMANN, H, WIDMER, J, SULSER, U, MAEDER, U, AND BUERGE, T A. Effect of Superplasticizers in Highly Alkaline Model Suspensions Containing Silica Fume. Proc. 6th CANMET/ACI International Conference on Fly-Ash, Silica Fume, Slag and Pozzolans in Concrete (Malhotra V.M., Ed.). American Concrete Institute, Detroit, 1998, SP 178, Vol. 2, pp 911-930.

10. HOUST, Y F, FLATT, R J, BOWEN, P, HOFMANN, H, MAEDER, U, WIDMER, J, SULSER, U, AND BUERGE, T A. Optimisation of Superplasticisers: From Research to Application. Proc. of the RILEM International Symposium on the Role of Admixture in High Performance Concrete (Rivera Villareal R. and Cabrera J.G., Eds). 21-26 March 1999, Monterrey, Mexico. Spon, London, etc., 1999, Accepted.

11. HORN, R, G. Surface forces and their Action in Ceramic Materials. J. Amer. Ceram. Soc., 1990, Vol. 73, No. 5, pp 1117-1135.

12. RUSSEL, W B, SAVILLE, D A, AND SCHOWALTER, W, R. Colloidal Dispersions. Cambridge University Press, New York, 1991.

13. FLATT, R J, HOUST, Y F, BOWEN, P, HOFMANN, H, WIDMER, J, SULSER, U, MAEDER, U, AND BUERGE, T A. Influence of Superplasticiser Adsorption on the Rheology of Cement Pastes. Intern. Symp. On the Role of Admixtures in High Performance Concrete, Monterrey, Mexico, March 21-26, 1999, Accepted.

EFFECT OF SILICA FUME ON LATEX-SUPERPLASTICISER CONCRETE

I Ray

Jadavpur University

R Chattopadhyay

Government of West Bengal

India

ABSTRACT. Studies were made on the effects of 4, 8, 12 and 16 % of silica fume (SF) by weight of cement on a concrete modified with latex-superplasticiser (LS) system (5% latex and 2% superplasticiser by weight of cement). A uniform range of flow 60 to 65%(15 strokes) was maintained for all the mixes to facilitate direct comparison.

The water demand estimated with respect to the binder (cement + SF) did not rise up to 8% of SF addition. The setting of concrete has improved up to 8% of SF. A 28 % rise in compressive strength, 100 % rise in flexural strength and 50 % increase in energy (at break-point) over LS was noticed for LS + SF concrete. Loss in strength and weight in 2% H_2SO_4 and 2% HNO_3 were significant up to 12% SF, the H_2SO_4 being more corrosive. However LS + SF performed better in 5% Na_2SO_4 solution compared to only LS system.

Keywords: Latex, Superplasticiser, Silica fume, Concrete, Strength, Energy, Density, Chemicals.

Dr Indrajit Ray is a Reader in Civil Engineering Department, Jadavpur University, Calcutta, India. His research areas are influence of admixtures and binder systems including polymeric latex, superplasticiser, silica fume on Portland cement based mortar and concrete; development of high performance repairing and overlay materials, durability of high performance concrete with special attributes. He has been in teaching and research for last fifteen years and published a number of articles on the concrete science and technology.

Raja Chattopadhyay is presently an Assistant Engineer in the Department of Agri-Irrigation, Government of West Bengal, India and is actively involved in various minor-irrigation and related construction projects. He was a formerly postgraduate research student in the Civil Engineering Department, Jadavpur University, India and was associated with the research in the field of concrete technology.

INTRODUCTION

Combined use of latex and superplasticiser in Portland cement concrete has many beneficial effects. The improvements of concrete properties occur not only due to the individual effects of latex [1] and superplasticiser but also for their synergistic actions [2,3]. Latex reduces the possible bleeding, harshness and brittleness of only superplasticised concrete. Superplasticiser, on the other hand, improves the delayed stiffening, unwanted air entrainment, compressive strength and density of the concrete. Indications are there that concrete modified with such binary admixture system (latex and superplasticiser) can further be improved by inclusion of a mineral admixture such as silica fume [4]. Silica fume have been used abundantly for the improvements of properties of concrete both in fresh and hardened state due to its excellent pozzolanic and other properties and a lot of research data are available on this [5]. There are however little or no specific information available in open literature on the compatibility of silica fume in a latex-superplasticiser modified concrete.

The present paper shall therefore examine the effect of such addition of silica fume in latex-superplasticiser modified concrete. The silica fume, superplasticiser and latex used in this study are all proprietary items so that the information from the present study may find a practical application after necessary trials.

EXPERIMENTAL INVESTIGATIONS

Materials

Ordinary Portland cement (OPC) complying with IS 8112 (Indian standard for 43 grade OPC) was used. Fine aggregate used was natural sand passing through 2.36 mm and retaining on 0.075 mm sieve. Coarse aggregate used was 10 mm graded crushed rock aggregate. Both of them conformed to IS 383 (Indian Standard specification for aggregates). Latex based on methyl methacrylate- butyl acrylate- 2 ethylhexyl acrylate polymer dispersion with solid content 55 % was used. Superplasticiser based on sulphonated melamine formaldehyde condensate with solid content of 40 % was used. The latex and superplasticiser were selected on the basis of their compatibility in a Portland cement medium [3]. Silica fume (SF) used was as follows: Fineness (nitrogen adsorption) = 22500 m^2/ kg , specific gravity = 2.3, unit weight = 250 kg/ m^3.

Mix Proportions

All the mixes were consisting of cement : fine aggregate : coarse aggregate = 1: 1: 2 (by weight), latex at 5 % (solid polymer) by weight of cement, superplasticiser at 2% (liquid) by weight of cement and a silicone based defoamer at an appropriate dosage. SF was incorporated at 4%, 8%, 12% and 16% by weight of cement as an addition. Quantities of water were varied to achieve a uniform range of flow 60 to 65% (Indian standard 1199) for all the mixes. Table 1 shows the mix variations and its fresh properties.

Experimental Procedure

For LS system, the cement and aggregates were dry mixed in a regular laboratory drum mixer for 3 minutes. A liquid system consisting of latex, superplasticiser, defoamer and

Table 1 Mix variations and its fresh properties

MIX TYPE	SF (%)	WATER/ CEMENT RATIO	WATER/ BINDER RATIO	FLOW (%)	AIR CONTENT (%)	PENETRATION RESISTANCE (N/mm^2)		
						4 Hrs	6 Hrs	8 Hrs
LS	0	0.36	0.36	65	5.35	0.4	1.5	6.2
LS+SF4	4	0.37	0.36	60	4.5	0.8	2.6	6.8
LS+SF8	8	0.39	0.36	65	3.8	0.5	1.9	6.4
LS+SF12	12	0.43	0.38	65	3.8	0.4	1.6	4.4
LS+SF16	16	0.47	0.41	60	3.5	0.2	0.9	4.0

SF = silica fume ; Binder = cement + silica fume

gauge water was prepared separately by mixing them in a planetary mixer for 2 minutes. This was added to the dry mix and a further mixing of 5 minutes were carried out. For all LS + SF systems the procedure followed was same except in those cases dry mix was performed with cement, aggregates and SF.

100 mm cube moulds were used for 3-day, 7-day, 28-day and 91-day compressive strength, bulk density and ultrasonic pulse velocity (UPV) tests and 100 x 100 x 500 mm prism moulds were used for flexural strength test (third-point loading) in accordance with Indian standard code using load control machine. A smaller size of specimen, 40 x 40 x 200 mm prisms were cast to get the complete load-displacement diagram and energy of the beam under centre-point loading at breaking point (measured as area under the diagram) with an Instron machine at cross head speed of 0.50 mm/ min.

For testing of resistance against acids and sulphates the 50 mm cube samples were oven dried at 105^0 C and immersed in 2% HNO_3, 2% H_2SO_4 and 5% Na_2SO_4 solutions for 45 days. The weight- loss and strength-loss were noted with reference to a set of undisturbed samples cured in water.

TEST RESULTS AND DISCUSSIONS

Fresh Concrete Properties

LS concrete exhibited slight bleeding and lack of cohesiveness due to the presence of sulphonated melamine formaldehyde group of superplasticiser although it was blended with latex. This phenomenon was completely eliminated when SF was added with the LS system. The mix with SF however needed more water to maintain a uniform range of flow (60 to 65%) but the water demand with respect to binder (cement + SF) did not change up to 8% SF. The demand of water measured with respect to cement however increased with the increase in SF content and reached to a maximum of 30.5% for LS + SF 16% (Table 1).

The values of air content decreased with the increase in SF content. The value dropped from 5.5% for LS to 3.5% for LS + SF 16.

At a given time after mixing (4, 6, 8 hours) the penetration resistance values have increased up to SF 4% and then decreased (Table 1). The reductions in 4,6 and 8-hour penetration value corresponding to LS + SF 12 and LS + SF 16 indicates that addition of SF in excess of 8% may interfere with the hydration process. The initial setting time of concrete was measured corresponding to a penetration resistance of 3.5 N/mm^2 (Indian standard 8142). It shows that addition of SF 4% and 8% reduced the setting value by 18 minutes and 12 minutes respectively whilst SF 12% and 16 % increased the setting by 6 minutes and 36 minutes.

Reference 5 has reported similar observations regarding water demand, cohesiveness and setting time on addition of SF in Portland-cement-only concrete.

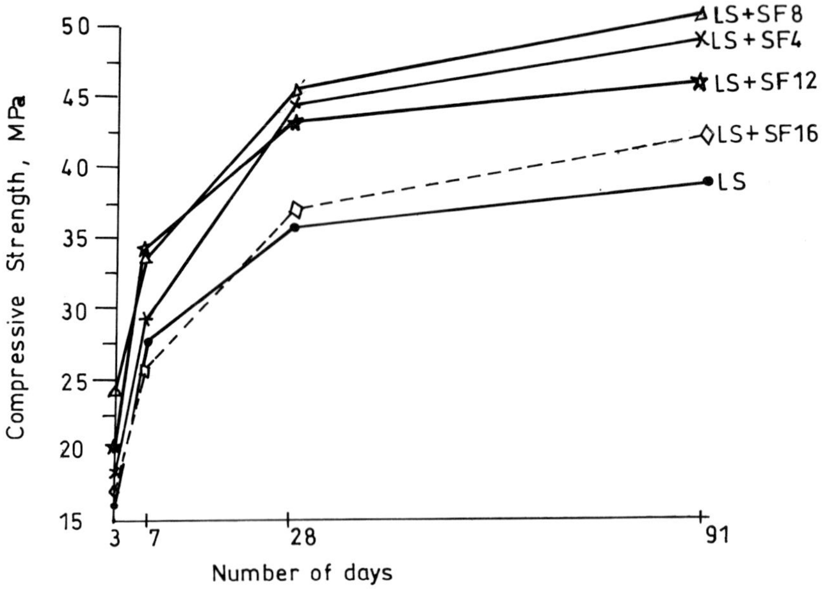

Figure 1 Compressive strength-age relationships

Table 2 Hardened properties of concrete

MIX TYPE	28 DAY COMPRESSIVE STRENGTH (MPa)	28 DAY FLEXURAL STRENGTH (MPa)	BULK DENSITY (kg/m³)	UPV (km/s)	ENERGY[1] (J/m³)
LS	35.7	4.25	23.91	4.30	1592
LS+SF4	44.5	8.50	23.34	4.50	2216
LS+SF8	45.7	7.65	22.91	4.19	2405
LS+SF12	43.4	5.38	22.80	4.00	2325
LS+SF16	37.0	5.10	22.23	3.75	1958

[1] At breaking point per unit volume

Compressive Strength and Tensile Strength

Compressive strength-age relationship (Figure 1) shows that LS + SF 8 concrete has the highest 28-day and 91-day strength followed by LS + SF 4, LS + SF 12, LS + SF 16 and LS. The LS +SF 4 concrete has developed maximum gain in strength between 7-day and 28-day. Table 2 reveals that addition of SF from 4% to 16% to LS system improved the 28-day compressive strength . However the improvement was maximum for LS + SF 8 (by 28%) and minimum for LS + SF 16 % (by 3.6 % only). Even the addition of 4% SF enhanced the compressive strength by about 25 %.

Addition of SF at all percentages improved the flexural strength. A significant rise in flexural strength (by 100 %) occurred when SF was added at 4%. Even for addition of SF by 8 % the flexural strength enhanced by 80 %. The improvement in compressive and flexural strength due to addition of SF is reported by previous authors [5, 6, 7, and 8] on only SF modified concrete.

Bulk Density and UPV

Bulk density was found to reduce marginally with the addition of SF, which caused a slight reduction in weight (Table 2). The bulk density of LS + SF 16 was reduced by 7 % over LS. The reduction was not high because SF was incorporated as addition not by replacement. UPV values show a similar trend of result. The UPV values greater than or equal to 4 km/s for SF addition up to 12% ensure a good quality of concrete. The reduction in density has increased the strength/ density ratio of all SF modified systems with respect to LS (Table 2) which is advantageous.

Load-displacement Curve

Figure 2 displays the complete load-displacement curves of three-point bending specimens. The long tail of the descending branch of LS system is indicative of its ductility in the post-peak load region, which is the characteristic of a latex-modified system. When SF was added at 4% and 8%, the post-peak branch became relatively shorter. But there was a significant rise in peak-load for both the cases. As a result there was a net gain in the value of energy (at break-point) per unit volume of the material. The addition of 4% and 8% SF to LS system has increased the energy by 39 % and 51 % respectively (Table 2). Although the peak loads of LS + SF 12 and LS + SF 16 were closer to LS, the energy (at break- point) per unit volume of the material was higher by 46 % and 23% respectively. The descending or softening branch of load -displacement curve of LS + SF 12 and LS + SF 16 were much longer than those of LS + SF 4 and LS + SF 8. The present result indicates that addition of SF in LS concrete does not necessarily reduce the ductility rather it improves the energy per unit volume of the material. This behaviour is however different from the cases where addition of SF in Portland-cement-only concrete reduces the ductility and energy absorption capacity and increases the brittleness [9,10]. Further, addition of SF in LS system has caused the ascending branch of load- displacement curve more flat indicating a gradual reduction in stiffness with the increase in SF. The indication is there that behaviour of SF in presence of latex may not be exactly same when it is used without latex.

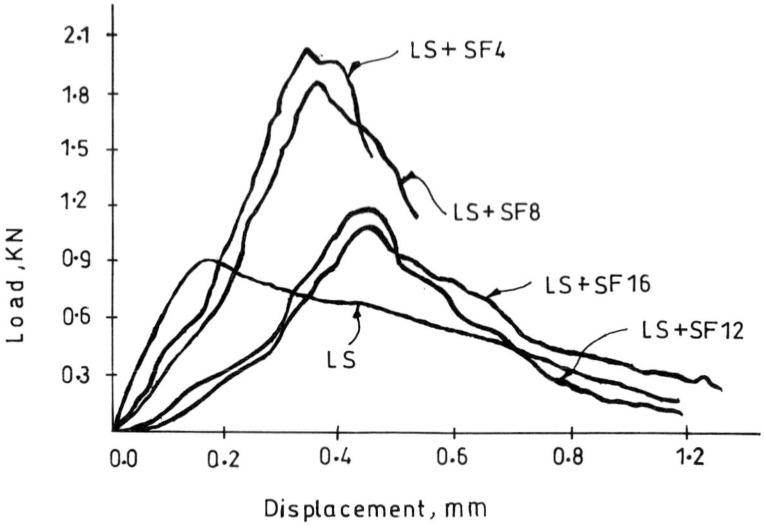

Figure 2 Load-displacement diagram for 3-point bending specimens

Resistance Against Chemicals

It is evident from the test results (Figure 3) that immersion in 2% H_2SO_4 has caused maximum loss in weight and strength, followed by HNO_3 and Na_2SO_4 .The maximum weight-loss suffered by LS + SF 8 (by about 8 %) is not very significant, however the reductions in strength for LS + SF 8 and LS + SF 12 by about 50 % are alarming. The strength-loss was significantly reduced at SF 16% which might have occurred due to reduction in capillary porosity in LS + SF 16. LS + SF systems performed much better in HNO_3 than in H_2SO_4.

The maximum strength-loss of 24 % has occurred for LS + SF 12 and minimum loss of 12% for LS + SF 4. A reduced tendency of both weight and strength-loss was noted when the SF content was increased from 12 % to 16 %. The deterioration of concrete below 10 % addition of SF and at medium strength level was reported by previous authors [11, 12] but the results were based on only SF modified concrete.

There was a slight increase in weight due to immersion of LS + SF samples in 5% Na_2SO_4 solution, however the value was negligible. This may be due to the formation of additional crystals within the concrete pores/ voids. Not only the addition of SF reduced the strength-loss in Na_2SO_4 medium but there was a marginal gain in strength also for LS + SF 12 and LS + SF 16 samples. The improvement of performance of LS + SF over only LS system in sulphate solutions may be due to formation of reduced $Ca(OH)_2$ and refined pore structures [11].

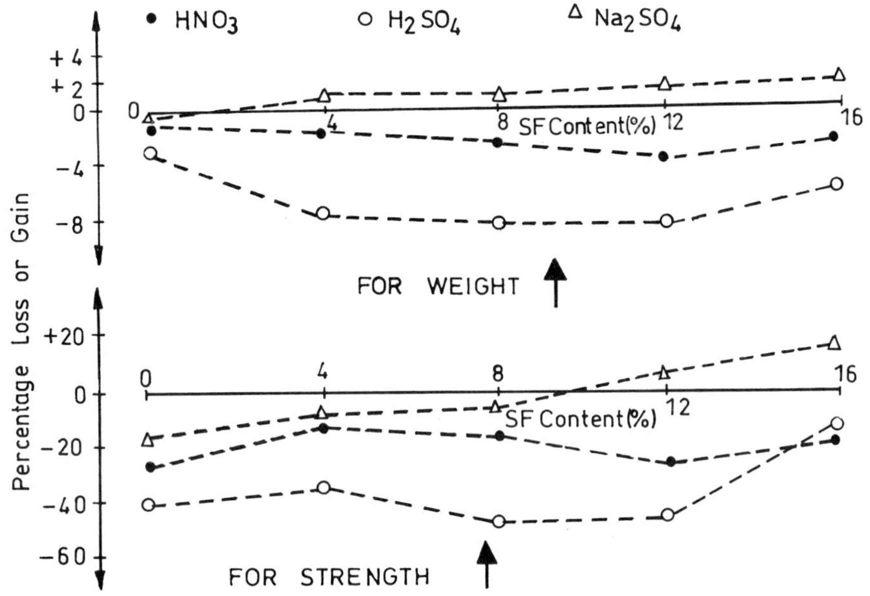

Figure 3 Percentage loss (-) or gain (+) in weight and strength for specimens immersed in H_2SO_4, HNO_3 and Na_2SO_4

PRACTICAL APPLICATION OF THE MATERIALS

Enhancement of properties of LS system by incorporating SF may be utilised in various practical fields. The cohesive mix without segregation and bleeding having a high flow value (60 to 65 %) may be utilised in repairing or as new construction material for concreting in narrow and constricted places. Out of four modified mixes, LS + SF 4, LS + SF 8 and LS + SF 12 performed well in all respect except in acid-resistance case. Good compressive strength (44.5 to 43.5 MPa), high flexural strength (8.50 MPa to 5.38 MPa), high energy absorption capacity, good ductility may be utilised in bridge and roadway overlays, flooring, ductile concrete foundation, water tank etc. after adequate trials with the local materials. A higher strength-density ratio (1.70 to 2.00 MPa-m^3/ kN) for all LS + SF materials compared to LS system (1.50 MPa-m^3/ kN) is advantageous in many respect. The better short-term resistance of LS + SF over LS against sodium sulphate solution may be beneficial in case of concrete prone to sulphate attack. The acid resistance character can be improved by using acid resistance lining or coating.

CONCLUSIONS

1. Addition of SF up to 8% did not increase the water demand when measured with respect to binder content. It reduced the air content of the mix. Addition of SF up to 8% reduced the initial setting time slightly and made the concrete highly cohesive with good flow.

2. Addition of SF 8 % enhanced compressive strength by 28 % and flexural strength by 100%.

3. Bulk density and UPV were reduced with the increase in SF content.

4. Concretes with LS + SF 12 and LS + SF 16 were more ductile in post-peak load region than LS + SF 4 and LS + SF 8. Inclusion of 8 % of SF improved the energy absorption capacity per unit volume by 51 % over only LS.

5. The concrete with SF had undergone weight-loss in both 2% H_2SO_4 and HNO_3 solutions in which the deterioration in HNO_3 was less. The performance of LS + SF in 5% Na_2SO_4 was much better compared to that in the acid solutions. A slight gain in weight and strength was noticed due to blocking of pores with crystals.

ACKNOWLEDGEMENTS

The authors gratefully acknowledge Caltech India and Sika Qualcrete Pvt Limited for supplying polymeric latex, superplasticiser and silica fume for the study.

REFERENCES

1. KARDON, J B. Polymer-modified concrete: Review. Journal of Materials in Civil Engineering, 1997, Vol 9, No 2, pp 85-92.

2. RAMACHANDRAN, V S, BEAUDOIN, J J AND SHIHUA, Z. Compatibility studies of cement pastes containing a mixture of superplasticiser and a polymer. IL Cemento, 1989, Vol 86, July-Sept., pp 185-196.

3. RAY, I, GUPTA, A P AND BISWAS, M. Effect of latex and superplasticiser on Portland cement mortar in the fresh state. Cement & Concrete Composites, 1994, Vol 16, No 4, pp 309-316.

4. ATZENI, C, MANTEGAZZA, G, MASSIDDA, L AND SANNA, U. Combined effect of superplasticiser and acrylic latex in cement mortars with and without silica fume. Proceedings 3rd international conference, Ottawa, Canada, 1989, ACI Detroit , SP 119-24, pp 457-470.

5. MEHTA P K. Condensed silica fume, 1986, Cement Replacement Materials, Surrey University Press, Ed. Swamy R N, pp 134-170.

6. SABIR, B B. Strength and fracture toughness of silica fume concrete. Magazine of Concrete Research, 1997, No 179, pp 139-146.

7. BAYASI, Z AND ZHOU, J. Properties of silica fume concrete and mortar. ACI Materials Journal, 1993, Vol 90, No 4, pp 349-356.

8. KHATRI, R P AND SIRIVIVATNANON, V. Effect of different supplementary cementitous materials on mechanical properties of high performance concrete, Cement and Concrete Research, 1995, Vol 25, No1, pp 209-220.

9. XIE, J, ELWI A E AND MACGREGOR, J G. Mechanical properties of three high-strength concretes containing silica fume, ACI Materials Journal, 1995, Vol 92, No 2, pp 135-145.

10. TASDEMIR, C, TASDEMIR, M A, LYDON, F D AND BARR, B I G. Effects of silica fume and aggregate size on the brittleness of concrete, Cement and Concrete Research, 1996, Vol 26, No 1, pp 63-68.

11. TORII, K AND KAWAMURA, M. Effects of fly ash and silica fume on the resistance of mortar to sulphuric acid and sulphate attack, Cement and Concrete Research, 1994, Vol 24, No 2, pp 361-370.

12. GANESH BABU, K AND PRAKASH, P V SURYA. Deterioration of silica fume concretes, Proc. National seminar organised by National council for cement and building materials, 1998, January 15-17, New Delhi, pp 179-185.

PROPERTIES OF FRESH AND HARDENED CONCRETES PREPARED BY NEW GENERATION SUPERPLASTICIZERS

M H Özkul

A Doğan

Istanbul Technical University

Z Çavdar **A R Sağlam**

N Parlak

Sika Deteks

Turkey

ABSTRACT. New generation superplasticizers exhibit flowing properties as well as retension of these properties for a longer time. The superplasticizer used in this study was a N-vinyl copolymer having a poly (oxyethylene) chain which is connected to the polymer via ester bonds. The effects of cement and admixture dosages on the rheological properties of mortars were investigated by using a coaxial viscometer. Workability properties of concrete were measured by slump flow test and also by a newly developed slump flow test apparatus. In-situ tests were made to find out the pumpability of superplasticized concrete and also the filling properties were determined by using a special mould which contains pipes representing reinforcing bars. The compressive strength of specimens taken from the concrete mixes and of cores cut out of the in-situ cast block were also presented.

Keywords: Self compacting concrete, Flowability, Rheological properties, Filling properties, Superplasticizer.

Professor M Hulusi Özkul is a member of Civil Engineering Faculty in Istanbul Technical University.His main research interests are chemical admixtures, waste materials and polymer concretes.

Anıl Doğan is a graduate student at the Institute of Technology and Science in Istanbul Technical University.

Zeynep Çavdar is the director of Research and Development Department in Sika Deteks.Her resarch interests include construction chemicals.

A Raif Sağlam is the chief of Concrete Laboratory in Sika Deteks.

Nazmiye Parlak is the chief of Chemistry Laboratory in Sika Deteks.

INTRODUCTION

New generation superplasticizers differ from the conventional ones not only with the ability of high range water reduction but also retention of high workability for a longer time. This admixtures allow the concrete to have a high fluidity which is characterized as flow concrete or self compacting concrete . It is possible to produce high performance concrete with this new generation admixtures at high workability levels.

The dispersibility of cement particles with these admixtures are due to the electrical repulsion as well as steric effect as a result of the long chain of polymeric molecules adsorbed on the fine particles. High resistance to segregation and having compactibility are maintained by both addition of a thickening agent and a fine powder. The amount of fine particles (smaller than 90 μm) including cement should be chosen greater than a certain limit in order to achieve segregation resistance.

There are a number of methods proposed to measure the properties of new range superplasticized fresh concrete. Rheology parameters by using Bingham model can be measured by a viscometer on both mortar and concrete [1] V-shape tunnel test [2] and box test [3,4] were improved to evaluate the segregation resistance and fluidity of concretes. Also some tests were presented to measure the filling capacity of highly flowable concrete through reinforcing bars [5,6] .

Flowability is measured as flow spread according to DIN 1048, but without jolting or by L-shaped flow meter with [7] or without reinforcement [8] . The flow velocity measured in the latter tester represents the plastic viscosity and the lateral flow and vertical slump correspondance to the slump and flow in convensional slump test, respectively.

In this study , the influence of new range superplasticizers on rheological properties of mortars was investigated by using a coaxial viscometer . Slump flow tests were carried out on concretes at different cement dosages. Flowability was measured with a modified slump test which contains reinforcement around it. Finally , the mechanical properties of concretes were presented.

EXPERIMENTAL

Materials

Cement

PC 42,5 type cement (Turkish standard 19) was used in the experiments . Blaine surface area was 3,119 cm^2/g. and the density was 3.14g/cm^3. Chemical composition percentage weight of cement was as given below:

C_3S	61.1
C_2S	12.2
C_3A	9.6
C_4AF	7.9

Aggregates

Two types of sand , natural and crushed stone, were used as fine aggregates , and the specific gravities of them were 2.56 kg /dm^3 and 2.61 kg /dm^3 respectively . The maximum size of the coarse aggregate of crushed stone was 25 mm and its density was 2.69 kg /dm^3.A lime stone powder with a Blaine surface area of 320 m^2 /kg was used as fine material for some mixtures.

Superplasticizer

The superplasticizer is based on water-soluble N-vinyl copolymers having a poly (oxyethylene) chain which is connected to the backbone of the polymer via ester bonds.[9]

Mix Design

Natural sand: Crushed stone sand : crushed stone coarse aggregate ratios were 1:0,64:1,485 for all concrete mixes. The cement dosages tested were 450,500,550 and 600 kg/m^3 . Water content was adjusted to have a slump spread of either 57 ± 3 cm or 70 ± 5 cm . Superplasticizer content was kept constant at 1,5 % in weight of cement.

As the second series, 150 kg/m^3 lime stone powder was used as fine material and the cement content was kept at 450 kg/m^3.

Mortar part of the concrete with the same proportions of fine aggregate as in concrete , but smaller in size than 2 mm , was used for measuring rheological properties.

The concrete was mixed in a pan mixer with a capacity of 50 dm^3 for 3 minutes and a small mixer with a capacity of 5 dm^3 was used for mortar mixing.

Laboratory Tests

Rheological properties of mortars were measured by a coaxial viscometer .It is programmed to have rotating speeds between 10-150 rpm. The measurements were repeated at every ½ hours up to 2 ½ h . Before each measurement the sample was remixed for 30 sec in the mixer and kept in covered position until the next test. Free flow spread was used for flowability by using DIN 1048 slump cone. Also , a newly developed slump test was applied to measure the same property as well as segregation resistance through reinforcements. The instrument contains a cylinder , opened both ends, and reinforcement bars located around it having 35 mm gaps , as shown in Fig . 1 .

After raising the cylinder to a hight of 10 cm , the concrete flows throught the bars; the time for flow to 50 cm and final flow can be recorded , as well as the final spread . After the flow is completed, the concrete remained in the middle place surrounded by bars was taken and the portion of coarse aggregate was determined by wet sieving .

15 x 15 x 15 cm^3 cubic samples were taken from the concrete for compressive strength testing and kept in water at the temperature of 22 ± 2 $^{\circ}$ C .

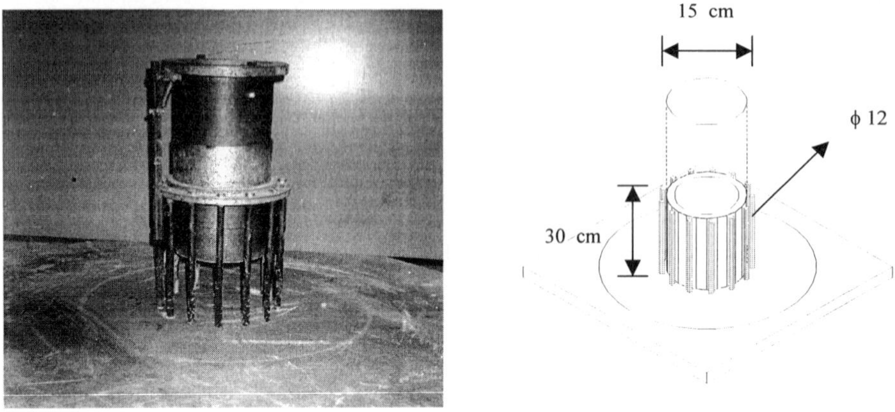

Figure 1 Developed slump flow apparatus

In-Situ Tests

$4m^3$ concrete mixtures were prepared by using the mixer with a capacity of $2m^3$ at a ready-mixed concrete plant. The cement and admixture dosages were 600 kg/m^3 and 1,5 % respectively. Concrete was loaded into a truck mixer with a capacity of $7m^3$ and after agitating for ½ h it was pumped into the mould which contains plastic pipes representing reinforcement bars as shown in Fig.2 . The concrete block left at site for 14 days after demoulding and cores (hight : diameter = 2)were taken out of it. The cores were kept in the laboratory at 22°C ± 2°C and 65 % ± 5% r.h. for further 14 days and finally tested for compression strength.

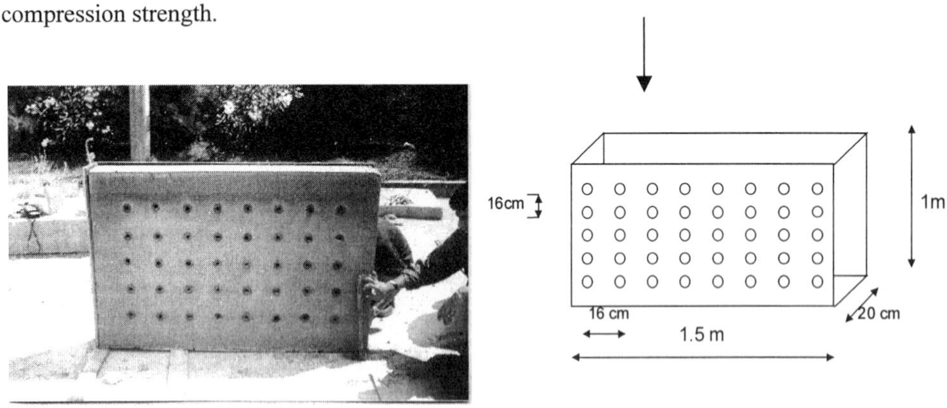

Figure 2 In-Situ concrete block

TEST RESULTS AND DISCUSSION

Rheological Properties of Mortars

Plastic viscosity and yield stress of mortars were measured by assuming the mortar as a Bingham body. For constant cement dosage of 550 kg/m^3 and superplasticizer content of 1,5 % , the variations of plastic viscosity and yield stress by time were shown in Figure 3. Both properties increase with increased W/C ratio , as expected. The increases in viscosity and yield stress by the time up to 2 ½ h are slight at all W/C ratios for the former property , and at higher W/C ratios for the latter one . However , for the lowest W/C tested, the yield stress showed a rapid increase after 1 h ; indicating that time has an effect on this property more than on viscosity , especially after 1 h.

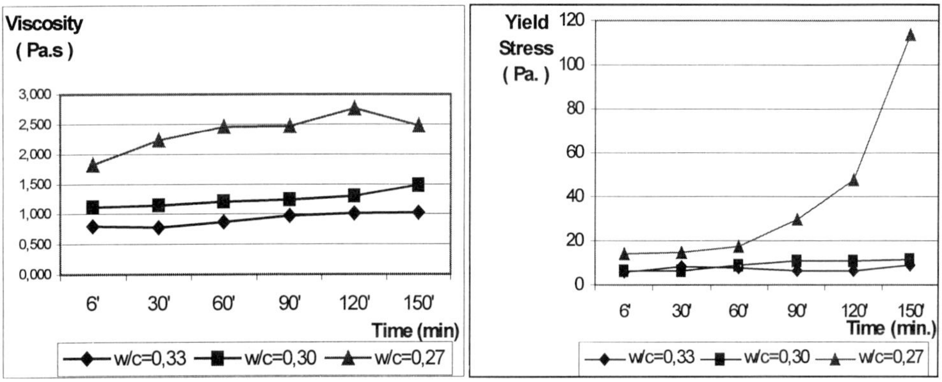

Figure 3 Variations of viscosity and yield stress by time for different w/ c ratios

The plots of viscosity and yield stress of mortars with time at constant cement content of 550 kg/m^3 and W/C ratio of 0,27 are given in Figure 4, in which each curve corresponds to a different superplasticizer dosage . For admixture content of 1,5 % and lower than this, the yield stress starts increasing after 1 h , showing that admixture contents higher than 1,5 % give better results , i.e. lower yield stress.

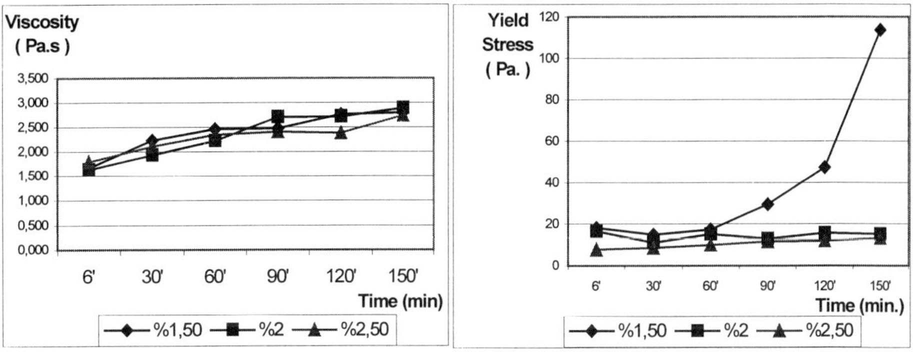

Figure 4 Variations of viscosity and yield stress by time for different admixture contents

Workability Properties of Concretes

Test results of both slump flow (with the cone of DIN 1048) and the developed slump method with reinforcements were given in Table 1. Although the latter method contains reinforcement, the spread values are close to the former because the hight and amount of concrete in the latter method is higher than that of the former . Table 1 shows that for both moderate and high spread values , the slump loss was slight up to about 1 ½ h , but for low cement contents segregation was observed . It seems that more than 500 kg/m^3 fine material is necessary to have an enough segregation resistance.

Table 3 Fresh and hardened concrete properties

CEMENT+ FINE MATERIAL (kg/m^3)	W/C	ADMIX. DOSAGE (%)	TIME	SPREAD		STRENGTH (MPa)		COMMENTS
				Free (cm)	Confined (cm)	7 days	28 days	
450+0	0,37	1,5	initial	55	-	49,0	62,4	Segregation
			30′	50	-			
			60′	52				
			90′	52				
			120′	51				
			150′	57				
500+0	0,38	1,5	initial	56	57	56,2	62,4	Slight segregation
			30′	56	57			
			60′	57	58			
			90′	58	57			
			120′	45	40			
			150′	28	30			
550+0	0,36	1,5	initial	65	65	48,2	62,0	No segregation
			30′	63	65			
			60′	64	63			
			90′	64	66			
			120′	68	65			
			150′	59	45			
600+0	0,33	1,5	initial	72	73	55.8	67.8	No segregation
			30′	71	73			
			60′	72	74			
			90′	70	71			
			120′	61	64			
			150′	42	40			
450+150	0,33	1,5	initial	67	70	46.3	53.0	No segregation
			30′	67	65			
			60′	33	39			
			90′	22	-			
			120′	-	-			
450+150	0,33	2,5	initial	73	74	45.6	58.6	No segregation
			30′	74	75			
			60′	70	65			
			90′	65	59			
			120′	49	50			
			150′	50	45			

For the second series, limestone powder was used as cement replacement , but the admixture content was needed to increase to 2,5 % , because of rapid slump loss experienced with the dosage of 1,5 % . Hence , even after 1 ½ h , this mixture remained flowable at the increased admixture dosage.

With the improved slump test, the concrete remained in the middle space surrounded by bars was taken and after wet sieving compared with the original mixture. It was found that the grading difference is slight . Similar comparison was made with respect to the water contents of concrete in the confined area and the original mixture; the water content of the former was only 5 % less than that of the latter.

In-Situ Test Results

The special mould which contains pipes representing reinforcing bars was filled with concrete from one side by inserting the pumping pipe and concrete was pumped from a hight of about 13m . The pumping pressure was the same as normal pumpable concrete. After filling the mould , no level difference was found between the pouring point and the distant point. The variation of concrete strengths obtained from the cores are given in Table 2. The results show that the change in strength in the horizontal and vertical distances from the pouring point are slight , and the mean of the strength is 51.7 MPa and standard deviation is 4.6 MPa.

Table 2 In-situ core strength results

	COMPRESSIVE STRENGTH (MPa)				
Horizontal distance (cm):	0	50	98	130	
Vertical distance (cm)					Mean
10	56,8	53,9	41,5	51,3	50,9
26	59,3	44,7	47,5	51,4	50,7
58	55,6	48,3	50,0	51,8	51,4
90	56,2	51,3	53,7	54,4	53,9
Mean :	57,0	49,6	48,2	52.2	

Hardened Concrete Properties

For the mixtures with W/C between 0.33-0.45 , the strength varied between 53 MPa and 67.8 MPa at 28 days age.Test results showed that it is possible to prepare a flowable concrete with a strength of 67.8 MPa .7 days / 28 days strength ratios remained between 78-90 %.

CONCLUSIONS

1) Rheological studies on mortars showed that the elapsed time and superplasticizer content affect yield stress more than viscosity.

2) The superplasticizer used in this study made the production of flowable concrete possible and this property was retained over 1 ½ h time

3) It is found that for segregation resistance of concrete at least 500 kg / m^3 fine material should be used

4) The developed slump method showed promising results for flowability through reinforcing bars as well as segregation resistance.

5) In- situ test results showed that the superplasticized concrete can be pumped easily and the filling capacity is perfect.The strength variation in the mould with respect to the hight and distance is slight .

6) It is possible to produce a flowable concrete at strength levels of 65 MPa at 28 days age

ACKNOWLEDGEMENTS

Authors wish to thank Nuh Ready Mixed Concrete Company and Mr Erdem Tandirli, Quality Control Director of the company, for their help in in- situ tests.

REFERENCES

1. NORNBERG J, PETERSON Ö and BİLLBERG P, Effect of new generation superplasticizers on the properties of fresh concrete, Superplasticizers and Other Chemical Admixtures in Concrete, Proc. Fifth CANMET/ACI Int. Conf, Ed.by V M Malhotra,Rome, S.P.-173,1997.pp 583-98 .

2. FUKUTE T, MORIWAKE A, SONO K and HAMASAKİ K, Development of superworkable concrete for multi functional port structures, Second CANMET/ACI Int. Symp. on Advanced in Concrete Technology, ACI, S.P.-154,1995, pp 335-56.

3. KASAI Y, HIRAISHI S, TABINAI K and OSADA K, Mixture proportion, flowability, strength and durability of flowing concrete ACI, SP148, 1994, pp 209-26

4. YAMMAMURO H, IZUMI T and MIZUNUMA T, “Study of non adsorptive viscosity agents applied to self compacting concrete, ACI, SP 173, 1997, pp 427-43.

5. YURUGI M, SAKATA N, IWAI M and SAKAI G, “Mix proportions for highly workable concrete”, Concrete 2000, Ed. By R K Dhir and M R Jones, F&FN Spon, 1993, pp 579-89.

6. OKAZAWA S, UMEZAWA K and TANAKA Y, “A new polycarboxylate based polymer: Physical properties of concrete , Concrete 2000”, Ed . by R K Dhir and M R Jones, F&FN Spon 1993, 1813-24.

7. NISHIBAYASHI S, INOUE S, YOSHINO A, KURODA T and KUME T , “ A study on the flow of highly superplasticized concrete ”,A.C.I., S.P.-148, pp 177-187.

8. MITSUI K, YONEZAWA T, KINOSHITA M. and.SHIMONO T, Application of a new superplasticizer for ultra high-strength concrete, A.C.I., S.P.-148, 1994,pp 27-45.

9. BURGE T.A., “Super high strength concrete ”,Presented in Fifth CANMET /ACI Int. Conference, Rome,1997.

THEME FIVE:

DURABILITY ENHANCING ADMIXTURES

Keynote Paper

DURABILITY ...... HOW FAR CAN IT BE EXTENDED WITH ADMIXTURES

M Corradi

R Khurana

MAC S.p.A. Modern Advanced Concrete

Italy

ABSTRACT. The durability of concrete structures was considered to be an intrinsic property of the material. It is now evident that it is very much dependent on the environmental conditions in which the structure is situated.

Some basic requirements, such as good quality concrete, which is caracterized by a low water cement ratio and hence a low permeability, adequate cover to reinforcement, proper compaction and curing of the concrete, are of fundamental importance for its durability. Besides the design and construction practice, material properties and exposure conditions are of great importance.

The causes of concrete deterioration are manifold and may be broadly classified as of physical and chemical nature. Admixtures added to the concrete mix during mixing will help to mitigate the effects of the degradation agents. Several families of admixtures are now available to enhance the durability of concrete. In this paper these admixtures presented in relation to their effectiveness against these deteriorations.

Keywords: Admixtures, Air entraining, Alkali aggregate reaction, Biological attack, Chloride diffusion, Corrosion inhibitors, Cracking, Durability, Exposure conditions, Freeze thaw resistance, Porosity, Reinforcement corrosion, Shrinkage, Superplasticizers.

Dr Mario Corradi is the Managing Director and Director of Research & Development of MAC S.p.A. (Modern Advanced Concrete), Treviso/Italy which is a part of the MBT-SKW Group. He is actively involved since 1970 in the development and use of admixtures to improve the quality of concrete structures and has authored several papers on this subject.

Ing Rabinder Khurana is the Technology Director at MAC S.p.A. (Modern Advanced Concrete), Treviso/Italy. He has been involved for more than 25 years for enhancing the role of admixtures in concrete construction. He has an active member of CEN TC104 SC3 "Admixtures for concrete" and CEN TC104 SC8 "Products and systems for the protection of concrete structures".

INTRODUCTION

Concrete is surely the most common and widely used man made material. Estimates indicate that every year one cubic meter of concrete is made for every two human beings. Its wide distribution is due to its low cost and the availability of the basic raw materials. Good quality concrete is normally considered to have durability as an intrinsic property. But the enormous amount of money being spent all over the world to repair concrete structures clearly indicates that much has to be still done to improve the durability and maintain the serviceability of concrete structures.

Factors influencing durability

The factors which determine the durability of concrete may be classified under four categories; for example design, construction practice, material properties and exposure conditions. Errors in design or carelessness in detailing may lead to stress concentrations and cracking. These should be of concern to the design engineer. The environmental conditions to which the concrete structure is exposed will determine the eventual cause of its deterioration.

These exposure conditions may vary over a wide range which include sheltered, indoor concrete, cyclic freeze thaw conditions, hot and dry desert ambient. Contact with, or the presence of certain chemicals in concrete such as sulphates, chlorides, sulphides, acids, carbon dioxide etc., will also lead to deterioration. Higher ambient temperatures will accelerate the reaction rate. Micro-organisms may also attack the concrete. Mechanical abrasion or erosion by water or wind will also effect the service life of a structure. These factors may act singly or in combination and the degradation mechanism may be either a physical effect such as shrinkage, creep, expansion, erosion, etc. or a chemical reaction like sulphate attack, carbonation, reinforcement corrosion, alkali silica reaction, etc. Examples of erosion by cavitation and severe cracking caused by delayed ettringite formation are shown in Figure 1 and Figure 2 .

Scale 1:100

Figure 1 Erosion of concrete in a stilling basin of a dam caused by cavitation

The constituent materials of concrete determine its ability to resist certain aggressions. Photo no. 3 shows concrete deterioration caused by thaumasite. Rock excavated from a highway tunnel was crushed to produce aggregates which were used to produce concrete for the tunnel lining. During some phase of the tunnel construction sulphate bearing rocks were encountered and used as aggregates.Therefore, the choice of sound and good quality materials is of prime importance. Another basic requirement for a high quality and durable concrete is its low permeability. The permeability depends upon the quality of cement or binder and aggregates, water binder ratio, degree of hydration, compaction and curing and the absence or presence of cracks. The right choice of cement type and aggregates may exclude certain kinds of aggressions such as sulphate attack, alkali silica reaction, etc. The reduction of the water cement ratio will reduce the quantity of pores and their size in the concrete. Effective compaction (related to the workability of the concrete mix) and efficient curing will determine the degree of hydration, which influences the dimensions of the pore structure. Therefore, by reducing the amount of pores and their dimensions it is possible to reduce the permeability and hence the rate of ingress of moisture borne aggressive agents.

Scale 1:5

Figure 2 A prestressed concrete railway sleeper cracked by delayed ettringite formation

Admixtures and durability

European Standard pr EN 934 part 2 defines admixtures as "Material added during the mixing process of concrete in a quantity not more than 5% by mass of the cement content, to modify the properties of the mix in the fresh and/or hardened state". Amounts exceeding 5% are considered to be "additions" according to prEN 206.

As mention earlier, concrete is subjected to numerous types of aggression. Different types of admixtures are now available to control and block their deleterious effects and therefore, enhance the durability of the concrete. Since the permeability of the cementitious matrix plays a fundamental role in the deterioration process, the water cement ratio is the crucial parameter to be controlled. To reduce the water cement ratio, either the cement content has to be increased or the water content has to be decreased. An increase in the cement content leads to an increase of materials cost, heat of hydration and shrinkage . Also, reducing the water

Scale 1:50

Figure 3 This concrete column just a few months after its construction is rendered unfit due to the formation of thaumasite. Sulphate rocks were inadvertently used for making concrete. The structure had to be demolished and remade

content to decrease the water cement ratio will cause a reduction of the workability of the concrete mix. The stiffer is the concrete mix, more the compactive effort it will require to achieve full compaction. This will require powerful compaction equipment and longer placing times. Therefore, the first category of admixtures that will be considered is the water reducing type.

Water reducing admixtures

These admixtures may be classified as water reducing/plasticizing admixtures or as high range water reducing/superplasticizing admixtures (water reduction more than 12%) which may have other functions such as set retardation, set acceleration, hardening acceleration and air entrainment. These multifunctional admixtures effect several properties of the fresh concrete at the same time. Modification of the setting properties of the concrete help avoid construction problems such as cold joints, thus improving the durability. Commercially available water reducing admixtures may be classified in polymers of five generic families:

i) Modified ligninsulfonate (MLS)

ii) Sulfonated melamine formaldehyde condensate (SMF)

iii) Sulfonated naphthalene formaldehyde condensate (SNF)

iv) Carboxylic ethers (CE)

v) Others

Typical performances of these admixtures are reported in Table 1.

As mentioned earlier, the workability of the concrete at the time of its placement is also of fundamental importance. A loss in workability of the concrete mix during the time interval from mixing to placing means that the concrete has to be placed and compacted with greater care and effort. If this is not done properly, the hardened concrete will result to be more porous and hence, less durable. If water is added to the concrete mix to restore its workability (retempering), this will lead to an increase in the water cement ratio and consequently, a higher porosity and permeability of the hardened concrete. Innovative types of superplasticizers, based on polymers of polycarboxylic ethers (CE) are now available which allow the concrete mix to retain its workability for 90 minutes and even more, if required.

Table 1 Performances of various water reducing admixtures

FAMILY	TYPICAL DOSAGE, %	WATER REDUCTION, %
Modified ligninsulfonate	0.2 - 0.6	up to 10
Sulfonated melamine formaldehyde	1 - 3	up to 25
Sulfonated napthalene formaldehyde	0.8 - 1.5	up to 25
Polycarboxlic ethers	0.5 - 1.5	up to 40

Therefore, the role of water reducing admixtures having secondary functions, such as long slump retention, are of extreme importance for reducing the water cement ratio and hence as a consequence, the porosity and permeability of the hardened concrete. All durability factors relating to the porosity and permeability such as chloride diffusion, sulphate ingress, rate of carbonation, etc. are, therefore, improved.

Air entraining admixtures

These are admixtures that allow a controlled quantity of small, uniformly distributed air bubbles to be incorporated in the concrete mix during the mixing and which remain in it after hardening (prEN 934 part 2). Air entraining admixtures are surfactants whose molecules are adsorbed at the air-water and solids-water interface. They may be of anionic or cationic nature. The commercially available air entraining admixtures are usually salts of wood resins (neutralized vinsol resin or synthetic detergents (alkyl aryl sulfonates). Other types of air entraining agents include salts of petroleum acids, proteinaceous materials, fatty and resinous acids and sulfonated hydrocarbons. Air entrainment reduces strengths of the hardened concrete. However, this reduction can be offseted by the use of water reducing admixture in the same mix.

The amount of air entrained, containing bubbles of the right size and spacing, will control the damage to concrete cause by frost and de-icing salts (see figure 4 and 5).

Scale 1:100

Figure 4 A concrete retaining wall heavily damaged by frost action

In a properly air entrained concrete, the air content should be 4-8% (depending on the maximum size of the aggregate), bubble frequency of 300 to 600 per meter, specific surface of 24-48 mm^{-1} and a spacing factor of 100 to 200 microns.

Scale 1:50

Figure 5 Deterioration of concrete guard rails caused by de-icing salts

Corrosion inhibitors

Reinforcement corrosion is one of the major cause of deterioration of concrete structures. A wide spread increase in the use of chlorides (sodium and calcium) for the removal of snow and ice from roads, pavements and bridge decks in cold climates has aggravated the situation (see figure 6).

Scale 1:50

Figure 6 Reinforcement corrosion and heavy spalling in a beam of a bridge deck caused by chloride ingress

Methods for preventing or controlling reinforcement corrosion in new or existing structures include the use of epoxy coated steel bars, waterproofing membranes to prevent chloride ion ingress, impregnation of concrete with polymers, removal of chloride ions and cathodic protection. For new structures, corrosion inhibiting admixtures may be added to the concrete mix to combat reinforcement corrosion. These admixtures are either inorganic salts or organic materials.

The inorganic type are grouped in three categories, namely anodic, cathodic or mixed types, depending upon their interaction to the corrosion process. They include calcium and sodium nitrite, sodium benzoate, sodium chromate, zinc oxide and disodium phosphate. Organic materials showing corrosion inhibiting action include hydrazine hydrate, chelating compounds, carboxylic acids, analine and petrolactam compounds. Other parameters, such as permeability, cracks, resistance to chloride ions, etc. also influence the corrosion of steel. Use of superplasticizers and supplementary cementitious materials such as fly ash, slag and silica fume are also beneficial.

Alkali aggregate reaction inhibiting admixtures

Alkali aggregate attack is an example of damage caused by chemical reactions occuring within the concrete. The expansion is caused by reaction between the alkali hydroxides, usually derived from cement, and reactive silicious or carbonate aggregates. This reaction involves a formation of a gel of variable composition. This gel absorbs water and swells. The stresses caused by this swelling may cause cracking and or localized "pop outs"(see figure 7).

Scale 1:200

Figure 7 Alkali silica reaction has caused intensive cracking in a concrete dam. Attempts to repair it by injecting with epoxy resins was made

The major cause of the alkali aggregate reaction is the alkali content in the concrete. Major source of alkalis is the cement. Recently, due to the pressure from the environmental lobby, all the effluents from the cement production are recycled back to the production process. This has led to an increase of the alkali content (sodium equivalent) of the cement and in some cases it is more than 1%. As a consequence, even in countries where alkali silica reaction was unknown, it is now quite frequent. Most international specifications limit the sodium equivalent content in cement: the ASTM C150 limit is 0.6%. Other standard specifications for example BS, EN. etc. limit the total alkali in concrete, derived from all sources, to 3 kg/m^3. Besides controlling the nature of the aggregates and the alkali content of the cement, fly ash, blast furnace slag, silica fume, rice husk ash and recently, finely ground glass have also been used as additions to prevent alkali silica reaction. Chemical admixtures that reduce or eliminate the expansion caused by the alkali silica reaction include lithium compounds (hydroxide, carbonate, nitrate and floride) sodium silica floride and alkyl alkoxy silane.

Biological attack resistance admixtures

Concrete structures made with durable concrete can suffer severe damage due to the action of bacteria, fungi and insects. The most common manifestation is in the sewage structures (see figure 8).

Scale 1:20

Figure 8 Acid corrosion caused by bacteria action in a concrete sewer

The metabolic activity of certain bacteria and fungi releases corrosive chemicals which promote the deterioration of concrete. In some cases, protection of the concrete surface may be required for resistance to acid produced by these organisms. Admixtures to be added to the concrete mix to combat the action of fungi, bacteria and termites are also available. They do not effect the properties of fresh and hardened concrete. Bactericide admixtures are polyhalogenated phenols, sodium benzoate, benzaalkonium chloride and copper compounds. Emulsified dieldrin chemical added to the concrete mix has a potent toxic effect on termites.

Shrinkage reducing admixtures

Concrete, right from its placement and finishing is subjected to volume changes such as settlement, plastic, thermal and drying shrinkage. All these shrinkages may lead to some form of cracking in the concrete structure. Cracks are preferable paths for ingress of aggressive agents into the concrete. Settlement shrinkage can best be taken care of by an appropriate mix design. Plastic shrinkage is combatted with appropriate curing and use of synthetic fibers.

Thermal shrinkage is related to construction techniques and is normally minimized by controlling the thermal gradient along the cross section of the concrete pour. Drying shrinkage is caused by the loss of moisture from the cementitious paste and takes place over the life span of the structure. To reduce the drying shrinkage and thus reducing the possibility of crack formation, shrinkage reducing admixtures are now available. These admixtures include oligoalkylen glycols, polyalchols and derivatives. Their mechanism is based on the reduction of the surface tension of the pore water and therefore lowering the stresses caused by drying which could lead to cracks.

CONCLUSIONS

The durability of a structure depends upon the quality of the concrete and the exposure condition to which it is subjected. The right choice of the constituent materials of concrete make it less vulnerable to certain aggressions. Correct construction techniques and adequate specifications also enhance the durability. The role of water reducing admixtures and superplasticizers is of fundamental importance as they reduce the porosity of the concrete, thereby making it less penetrable to aggressive agents which may undermine its durability. Air entraining admixtures, incorporating the right quality and quantity of air voids in the hardened concrete, renders the structure resistant to freeze thaw cycles and deicing salts. Corrosion inhibitors provide extra protection against reinforcement corrosion. Specific admixtures, along with additions like fly ash, slag and silica fume ensure protection against alkali aggregate reaction. The tendency of concrete to crack due to drying shrinkage is reduced if shrinkage reducing admixtures are incorporated into the concrete mix. Even attack by microorganisms and insects can be counteracted by appropriate admixtures. For certain aggressions such as abrasion, acid attack and specific chemicals, concrete must be protected with coatings or linings. Admixtures that help improve construction practices such as underwater concreting, construction in cold climates, pumping aids, shotcreting, improving bond, damp proofing and water resisting, etc. are also available but are not directly related to the durability of the concrete structure. Durability is enhanced if the existing knowledge is applied and ensured by adequate quality control and quality assurance systems. Admixtures for known aggressions are readily available and new developments are always happening. Admixtures play a fundamental role for extending the durability of concrete structures so that they may arrive intact into the fourth millennium.

REFERENCES

1. RAMACHANDRAN, V.S., Concrete Admixtures Handbook, Noyes Pubblications, Second Edition, 1995.

2. MAILVAGANAM, N.P., Repair and Protection of Concrete Structures, CRC Press, 1991.

3. MEHTA, P.K., Concrete Structure, Properties and Materials, Prentice Hall, 1986.

4. NEVILLE, A.M., Properties of Concrete, Pitman Publishing, Fourth Edition, 1995.

5. MALHOTRA, V.M., Superplasticizers: A Global Review with Emphasis on Durability and Innovative Concretes, American Concrete Institute SP-119, 1989.

DAMAGE AND ITS RESTRAINT OF HIGH STRENGTH CONCRETE UNDER THE ACTION OF DOUBLE DAMAGE FACTORS

H D Yan **W Sun**

Y M Zhang **R Mu**

A Yan

Southeast University

China

ABSTRACT. In this paper, the process and regularity of damage and its restraint of high strength concrete (HSC) under the action of double factors which were load and freezing-thawing cycles were studied systematically. The test results indicated that the damage process and inferior extent of HSC performance were accelerated in the process of the freezing-thawing cycles when the HSC were bearing different stress ratio(0.5) loads. And, the higher the stress ratio is, the larger the damage extent is. Because of the crack resistance effect of steel fibers or the pressure relief effect of air-entraining agents, or especially of the composite effects of both, the damage process and inferior extent of HSC performance were restrained effectively. The frost resistance of HSC was enhanced obviously under the action of load.

Keywords: High strength concrete, Damage, Damage restraint, Frost resistance, Steel fiber, Air-entraining agent, Durability, Freezing-thawing cycles.

Dr Han D Yan is a Lecturer in the Department of Civil Engineering, HuaQiao University and a Ph.D. student at his post in Southeast University. His main research interests are basic theories of high performance concrete.

Professor Wei Sun is a Ph.D. advisor of the Department of Material Science and Engineering, Southeast University, P.R.China. She specializes in high performance fiber reinforced cement-based composites and basic theories of high performance concrete.

Dr Ya M Zhang is a Lecturer in the Department of Material Science and Engineering, Southeast University. Her main research interests are fiber reinforced cement-based composites.

R Mu is carrying out research at the Southeast University.

A Yan is carrying out research at the Southeast University.

INTRODUCTION

It has been a significant scientific and technical problem to improve the durability and to prolong the service life of concrete in the world [1-6]. It is well known that frost resistance is of great influence to the durability of concrete. But for half a century, studies on the durability of concrete have been confined to the influence of single factor-the action of freezing-thawing cycles. In practical engineering, such as highway pavements, airport pavements, bridge decks and dams in hydraulic engineering, the damage may be caused by double action of load and freezing-thawing cycles, especially in cold area [6]. In this paper, the damage processes of different strength grade concrete under the simultaneous action of load and freezing-thawing cycles are demonstrated. The damage restraint processes, due to the crack resistance effect of steel fibers or the pressure relief effect of air- entraining agents or the composite effect of both steel fibers and air-entraining agents are also demonstrated. Surely these results are of great actual significance for critically assessing the durability of concrete under comprehensive conditions.

MATERIALS AND EXPERIMENTS

Materials

A 525# portland cement made by Jiangnan cement plant was used, its specific surface area being 3700cm^2/g. The fineness modulus of fine aggregate is 2.1. The coarse aggregate used is a crushed stone with a maximum diameter of 10mm. Superplasticizer was supplied by Shanghai Xinpu Chemical plant. Air-entraining agents was provided by the Materials College of Tongji University. The steel fiber is square straight fiber cut low carbon steel plate. The length of steel fiber is 20mm, and the aspect ratio is 60, the tensile strength is 750MPa. The mixture proportions of concrete are given in Table 1.

Table 1 The mixture proportions of different strength grade concrete

SERIES		V_f (%)	W/C	SAND RATIO (%)	CEMENT (kg/m^3)	WATER (kg/m^3)	FINE AGGREGATE (kg/m^3)	COARSE AGGREGATE (kg/m^3)
C50	PC	0	0.38	48	398	151	734	1198
	SFRC	1.5	0.38	45	398	151	849	1038
C60	PC	0	0.32	35	440	142	666	1237
	SFRC	1.5	0.32	44	440	142	817	1039
C80	PC	0	0.32	33	477	124	700	1299
	SFRC	1.5	0.32	42	477	124	787	1133

The air-entraining agent was incorporated into plain concrete (PC) and steel fiber reinforced concrete (SFRC). Air entrained concrete was given a token APC and ASFRC, respectively. The air content was controlled at 4%~6%.

Experiments

The specimens were 40×40×160mm for flexure strength and frost resistance as well as 40×40×40mm for compressive strength.

A special loading device was designed (Figure. 1) for the simultaneous action of load and freezing-thawing cycles. The load was applied by adjusting the deformation of springs. The loading device has nearly no stress relaxation in test through measurement and calculation. The stress ratio (P/P_0) adopted 0, 0.1, 0.25 and 0.5 respectively (P=actual flexural strength, P_0=maximum flexural strength). The specimens were cured for 24 days at 20±2□ and more than 95% relative humidity , and then soaked into water for 4 days. The freezing-thawing cycles experiment under the action of load was conducted in a TDR1 type fast freezing-thawing test machine for concrete according to ASTM666. The dynamic elastic modulus was tested at the end of every twenty freezing-thawing cycles.

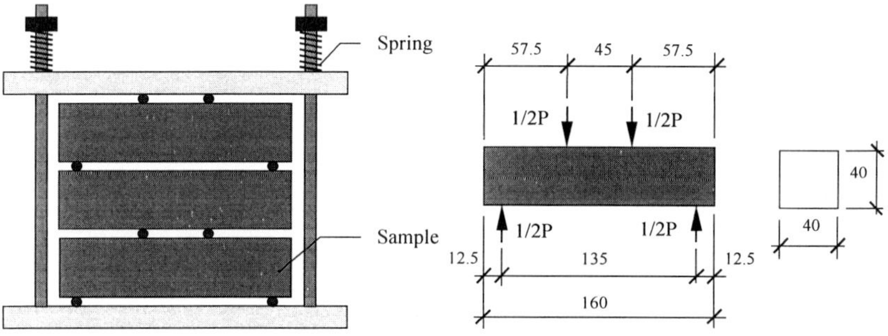

Figure 1 Schematic description of the loading device

RESULTS AND DISSCUSIONS

Damage Regularity of PC under Double Action of Load and Freezing-thawing Cycles

Figure 2(a) to (d) show the changes of dynamic elastic modulus of different grade concrete under the simultaneous actions of load and freezing-thawing cycles at the stress ratio of 0, 0.1, 0.25 and 0.5. It can be seen that, for plain concrete, strength grade, stress ratio and freezing-thawing cycles are important factors that determine the properties of concrete under the double damage actions. At the same stress ratio, concrete having higher strength can undertake more freezing-thawing cycles, and the dynamic elastic modulus decreases slower with freezing-thawing cycles. Obviously, the damage caused by the double actions at high stress ratio is the most serious. When stress ratio equals to 0.5, PC80 can endure only 150 cycles of freezing-thawing which is much lower than it can do at stress ratio of 0.25. When compared with the test results under freezing-thawing cycles without loading, not only the damage extent is increased, also is the damage rate accelerated.

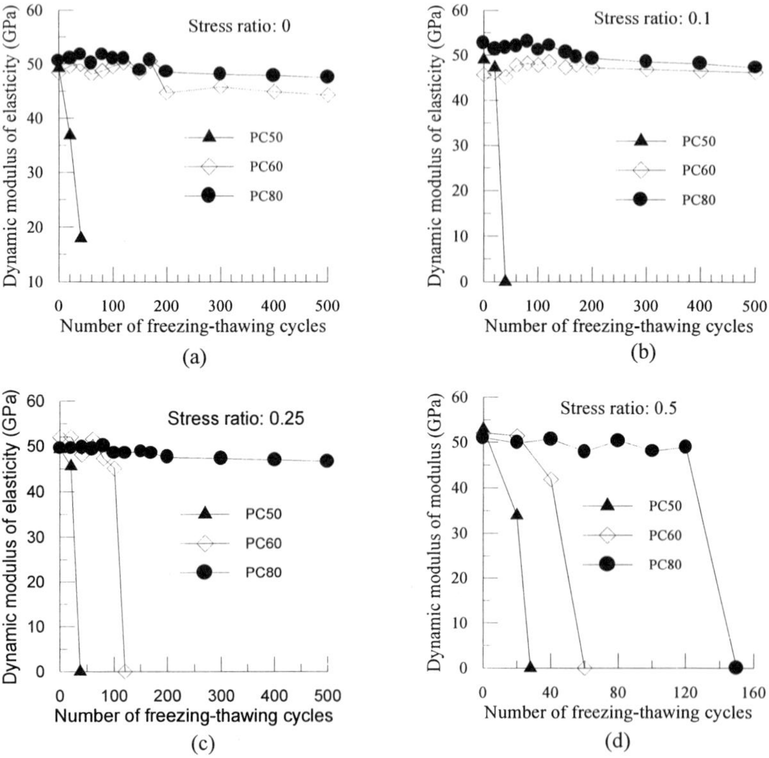

Figure 2 The changes of dynamic modulus of elasticity of PC under the double action of different stress ratio and freezing-thawing cycles

The Effect of Air-Entraining Agents on the Damage Regularity of APC under the Double Action of Load and Freezing-thawing Cycles

Figures 3(a)~(d) show the changes of dynamic elasticity modulus of different strength grade APC under the double action of load and freezing-thawing cycles at the stress ratio of 0, 0.1, 0.25, 0.50.The air content of APC60 and APC80 was 6.7% and 4.2% respectively. It can be seen that, at stress ratio of 0.25, the PC60 can bear only 120 cycles of freezing-thawing, but the APC60 can endure 500 cycles of freezing-thawing and the loss of dynamic elasticity modulus is very small. It is able to take 400 cycles of freezing-thawing at stress ratio of 0.5 which is much more than that of PC60. The APC80 has already had the capability of enduring more than 500 cycles of freezing-thawing at stress ratio of 0.5. Lots of air bubbles well-distributed into the concrete matrix have the action of releasing the pressure caused by the double damage factors, the damage extent of PC is greatly restrained because of incorporating air-entraining agents into the PC.

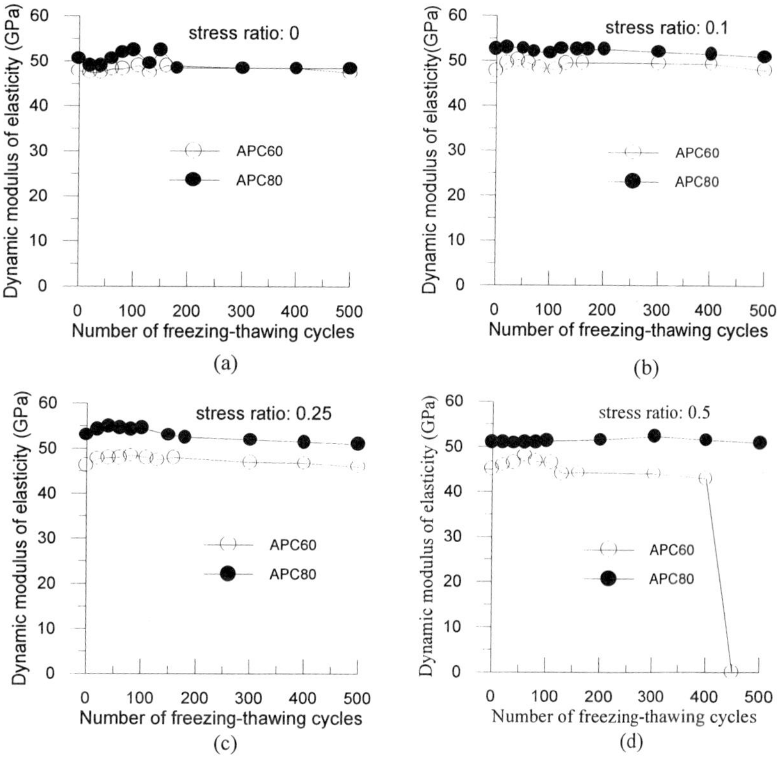

Figure 3 The changes of dynamic modulus of elasticity of APC under the double action of load and freezing-thawing cycles

The Effect of Steel Fibers on the Damage Regularity of SFRC under the Double Action of Load and Freezing-thawing Cycles

Sun etc. [7~9] have already indicated that the steel fiber can effectively restrain the initiation and propagation of crack in the process of forming and failing of HSC structure, mitigates the stress concentration at the tip of crack. So, the effect of steel fibers on the damage regularity of SFRC under the double action of load and freezing-thawing cycle was studied. The results are shown in Fig. 4(a)~(d). It can be seen that the ability of SFRC to resist the damage caused by double damage factors is greatly improved. For the same grade concrete, SFRC can undertake more freezing-thawing cycles than PC can do at the same stress ratio. At the stress ratio of 0.1, SFRC50, SFRC60 and SFRC80 can endure 500 cycles of freezing-thawing without failure of concrete. At the stress ratio of 0.5, SFRC80 suffers little after 500 cycles of freezing-thawing. But compared with APC, the damage extent of SFRC under the double damage factors is greater than that of APC at high stress ratio. The effect of restraining damage that provided by the crack resistance of steel fiber is not so good as the pressure relief of air-entraining agent.

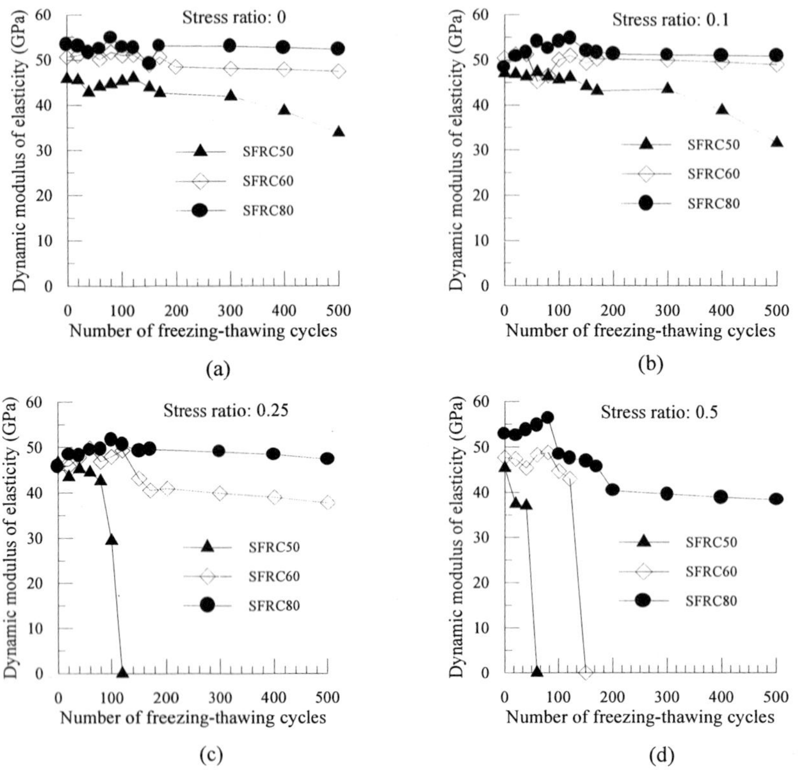

Figure 4 The changes of dynamic modulus of elasticity of SFRC under the double action of load and freezing-thawing cycles

The Effect of the Composite of Steel Fibers and Air-Entraining Agents on the Damage Regularity of ASFRC under the Double Action of Load and Freezing-thawing Cycles

Furthermore, the composite of steel fibers and air-entraining agents added to concrete can further improve the ability of inhibiting damage under the simultaneous actions of load and freezing-thawing cycles (Figure 5). At the stress ratio of 0.25, the air-entrained steel fiber reinforced concrete SFRC60 shows little change in dynamic elastic modulus after 500 freezing-thawing cycles. Even if at the stress ratio of 0.5, the loss of dynamic elastic modulus is a little.

For SFRC80 concrete, the dynamic elastic modulus keeps almost unchanged. The results show that the ability of frost resistance of air entrained steel fiber reinforced concrete under different stress ratios might be enhanced further because of the crack resisting effect of steel fibers and the pressure releasing effect of air entraining agents.

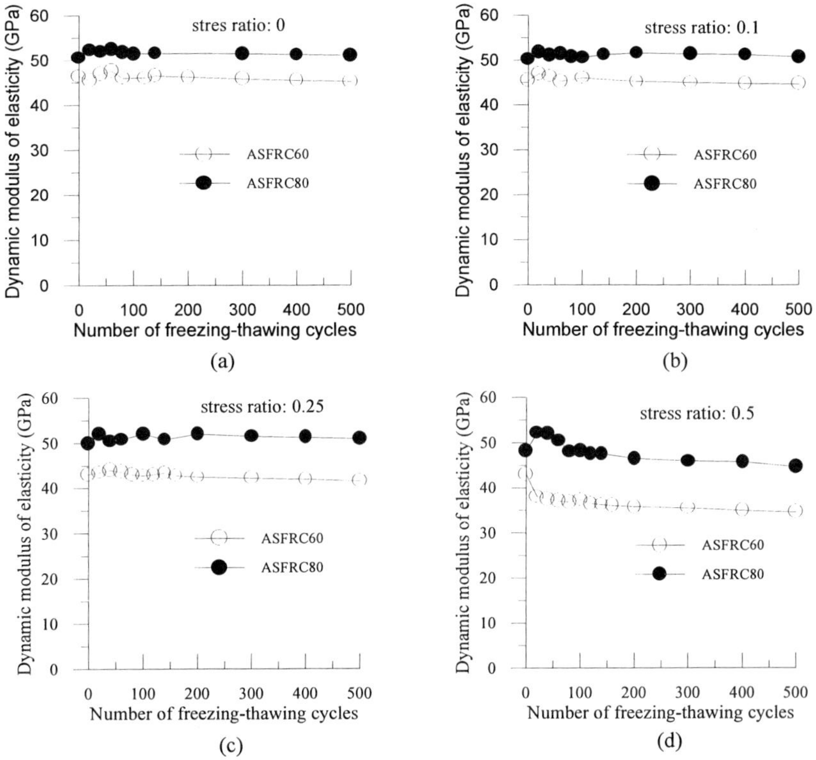

Figure 5 The changes of dynamic modulus of elasticity of ASFRC under the double action of load and freezing-thawing cycles

Discussion

When the HSC only bear the frost action (stress ratio equals to zero), the damage extent of HSC is not serious, even PC60 can endure 500 cycles of freezing-thawing. But PC80 can only hold 150 cycles of freezing-thawing under the simultaneous action of load (stress ratio equals to 0.5) and freezing-thawing cycles. These results fully indicated that the damage extent of HSC under the double damage factors of load and freezing-thawing cycles is much greater than that of HSC under single damage factor of freezing- thawing cycles. At the same double damage factors, APC60 and SFRC60 and ASFRC60 can respectively endure 400 and 160 and 500 cycles of freezing-thawing, furthermore, APC80 and SFRC80 and ASFRC80 can all endure more than 500 cycles of freezing-thawing. These results demonstrated that air-entraining agents and steel fibers have effect of restraining damage in different level. The composite of air-entraining agents and steel fibers is of the best effect to restrain the damage caused by the double damage factors of load and freezing-thawing cycles.

CONCLUSIONS

1. Stress ratio is an important factor that influences the inhibiting capacity of concrete on damage resulted from the simultaneous actions of load and freezing-thawing cycles. The higher the stress ratio, the more the damage extent.

2. Under the simultaneous actions of load and freezing-thawing cycles, the concrete strength have somewhat influence on the damage extent. The higher the concrete strength, the less the damage extent.

3. The crack resistance effect of steel fiber and the pressure relief effect of air-entraining agents can inhibit the damage extent of HSC in varying level.

4. The composite of air-entraining agents and steel fibers has the best effect to restrain the damage extent and rate under the double damage factors of load and freezing-thawing cycles.

REFERENCES

1. VENIAMIN D.LEINER, An Evaluation of Frost Actions on Concrete. Concr. Int., Design and Structure, 1996, Vol.18, No.3, pp 42-43

2. MARZOUK H. AND JIANG D. J., Effect of Freezing-thawing on the Tension Properties of High Strength Concrete. ACI Mater. J., 1994,Vol. 91, No.6, pp 577-586

3. JACOBSEN S., HANS CHRISTIAN GRAN, High Strength Concrete Freeze/Thaw Test and Cracking. Cem.and Concr. Res., 1995, Vol. 25, No. 8, pp1775-1780

4. MENASHI D. COHEN, ZHOU Y. X., Non-air-entrained High Strength Concrete-Is It Frost Resistant? ACI Mater. J., 1992,Vol. 89, No.2, pp406-415

5. LI JINYU, CHAO JIANGUO, XU WENYU, The Study on the Mechanism of Concrete Frost Thaw Damage, Concr. and Cem. Product, 1997, pp58-69 (in Chinese)

6. ZHOU Y. X., MENASHI D. COHEN, WILLIAM L. DOLCH, Effect of External loads on the Frost-Resistant Properties of Mortar with and without Silica Fume. ACI, Mater. J., 1994, Vol. 91, No.8, pp595-601

7. SUN W., GAO J. M., Study on the Fatigue Performance and Damage Mechanism of Steel Fiber Reinforced Concrete. ACI, Materials Journal, 1996, Vol. 93, No. 3, pp206-212

8. SUN W., YAN Y., Mechanical Behaviour and Interfacial Performance of Steel Fiber Reinforced Silica Fume High Strength Concrete, Science in China. 1996, Vol. 35, No. 5, pp607-617

9. SUN W., TIAN Q., Chloride Corrosion Resistance of Steel Fiber Reinforced High Strength Concrete. Proc. of Fiber Reinforced Concrete, Eds. Zhao G.F. Guangdong Science and Technology Press. 1997, pp119-126.

PROTECTION OF STEEL REBARS IN LIGHTWEIGHT CONCRETE WITH THE USE OF CORROSION INHIBITORS

G Batis

G Grigoriadis

National Technical University of Athens

C A Meletiou

TITAN Cement Co S.A.

Greece

ABSTRACT. The use of corrosion inhibitors (CI) is a well-established method for protecting steel rebars against corrosion at almost all types of constructions, made of reinforced concrete. The scope of this study was to investigate the effect of CI on the durability of lightweight concrete made with pumice. Two types of inhibitors were used, based on calcium nitrite and aminoalcohols, respectively. Specimens made with the use of Cement Kiln Dust (CKD) as an additive, were also tested and their performance was compared to the respective one obtained with the use of inhibitors. It has been shown in previous investigations that, conventional concrete made with cement containing 6% CKD exhibited a reduced corrosion of steel rebars. All specimens were exposed to a severe, corrosive environment containing chlorides. The results indicated that, corrosion and carbonation rates were significantly reduced in all specimens treated with the two inhibitors. They also indicated that CKD performed remarkably well under highly corrosive conditions and, in some cases, provided better protection of rebars than the CI's. This last indication provides new potential for further utilization of CKD in cement and concrete, which may lead to considerable economic and environmental benefits.

Keywords: Concrete, Durability, Corrosion, Carbonation, Inhibitors, Protection, Calcium nitrite, Aminoalcohols, Cement kiln Dust (CKD), Steel rebars

Dr George Batis is Associate Professor of National Technical University of Athens, Department of Chemical Engineering, Section of Materials Science and Engineering, Greece.

Grigorios Grigoriadis is a PhD Student at Section of Materials Science and Engineering, Department of Chemical Engineering, NTUA, Greece.

Dr C A Meletiou is a Research Fellow in Concrete Technology Laboratory, Division of Research & Technology, Department of R&D, TITAN Cement Co., S.A., Greece.

INTRODUCTION

Lightweight concrete (LC) is a widely known type of concrete with many applications in modern construction technology, such as insulating materials, prefabricated structural elements, wall panels, grouts, etc. The use of this type of concrete is often quite suitable, because reduces significantly the overall construction cost while, at the same time, provides satisfactory mechanical properties. The main disadvantage of LC is that performs poorly against corrosion. The use of porous aggregates (e.g. volcanic pumice) for the production of LC, results in a dramatic increase of the final porosity. Thus, the diffusion of both atmospheric O_2 and CO_2 through the open pores is increased, along with the absorption of other corrosive factors, mainly chlorides, due to capillary action. Consequently, reinforced LC exhibits higher corrosion and carbonation rates than normal concrete [1].

The most widely used method to protect steel rebars against corrosion, is the application of corrosion inhibitors (CI). Since their first use in 1978, their utilization in the construction of reinforced concrete is being continuously increasing. Nowadays, their use is considered to be one of the most preferable methods of corrosion protection, because it is easy to apply, it costs less than all the other methods with no maintenance required and increases significantly the service life of a typical construction - in some cases even up to 100%. In addition, treatment with CI, normally, does not affect the mechanical properties of the concrete, while loss of bonding between rebars and the concrete is avoided - a problem which frequently occurs when rebars are treated with epoxy coatings, Zn or PVC [2].

In this investigation, the effectiveness of using CI in reinforced LC made with volcanic pumice is evaluated. Two types of inhibitors were used, an aqueous solution of calcium nitrite and an organic, nitrogenous solution based on aminoalcohols. Specimens were prepared by using type II cement and volcanic pumice at a high W/C ratio, in order to increase the already high porosity of the concrete and, thus, to accelerate the corrosion of the rebars. The experimental tests were carried out under a severe, corrosive environment containing chlorides at, approximately, sea-water concentration. For comparative reasons, the performance of Cement Kiln Dust (CKD) as a protective additive, was also investigated. This is a by-product of low-alkali cement manufacturing process, mainly containing calcium carbonate. Previous investigations [3, 4] have shown that CKD, can improve the durability of concrete by stabilizing pH at higher values (>11.9), due to its relatively high alkalinity, thus preventing the dissolution of the protective passive film surrounding the steel.

Table 1 Chemical composition of cement, CKD and pumice, %

COMPONENT	CEMENT II35	CKD	PUMICE
SiO_2	27.38	13.68	70.55
Al_2O_3	9.10	4.36	12.24
Fe_2O_3	5.65	2.30	0.89
CaO	45.39	42.59	2.36
MgO	2.73	1.23	0.10
K_2O	0.94	0.79	4.21
Na_2O	0.56	0.28	3.49
SO_3	2.71	0.10	0.03
Others	7.71	34.67	6.13

MATERIALS AND EXPERIMENTAL PROCEDURE

The materials used in this investigation were: cement (type II35), volcanic pumice from Yali island in SE Greece, as an aggregate and CKD. The chemical composition of these materials is shown in Table 1.

Two types of corrosion inhibitors were used, one based on calcium nitrite (CN) and the other based on aminoalcohols (AM). Inhibitor AM was applied, both as an admixture (AM1) and as an impregnation (AM3) by surface spraying. The mix proportions for the preparation of all series of test specimens, are shown in Table 2.

Each series was consisted of three prismatic specimens (80 mm $\times$ 80 mm $\times$ 100 mm) with four cylindrical steel bars (100 mm $\times$ 12 mm) embedded in each one, as shown in Fig. 1. On each steel bar was properly attached a copper wire. The top surface of all specimens and the part of steel bars which protrudes over the concrete were coated with an epoxy glue to protect them from atmospheric corrosion.

The specimens were partially immersed in a 3% wt NaCl solution, up to a height of 25mm. After immersion, the following measurements were carried out at specified time intervals:

- Corrrosion half-cell potential, versus Saturated Calomel Electrode (SCE).
- Gravimetric mass loss of the rebars, according to ISO/DIS 8407.3 method.
- Mean depth of carbonation, usingg phenolpthalein indicator (RILEM CPC-18 method).
- Porosity after 9 months of exposure, using Hg-intrusion porosimetry method (MIP).

Table 2 Mix proportions of all series of test specimens

MATERIALS	SERIES CODES						
	RF	BD	CN3	CN6	AM1	AM3	AM3/1**
II35 (g)	500	500	500	500	500	500	500
Pumice (kg)	1.5	1.5	1.5	1.5	1.5	1.5	1.5
Water (ml)	500	500	400	300	485	500	500
CKD (g)	---	30	---	---	---	---	---
Inhibitor CN (ml)	---	---	100	200	---	---	---
Inhibitor AM (ml)	---	---	---	---	15	---	---
Inhibitor AM (ml)*	---	---	---	---	---	15	15

(*) Used externally, as impregnation by surface spraying, before immersing in NaCl solution. (**) The specimens were treated with inhibitor AM, after 1 month of exposure.

MEASUREMENTS AND RESULTS

Corrosion Potential Measurements

Figures 2-5 present the measurements of corrosion potential versus time of exposure, for all series of specimens.

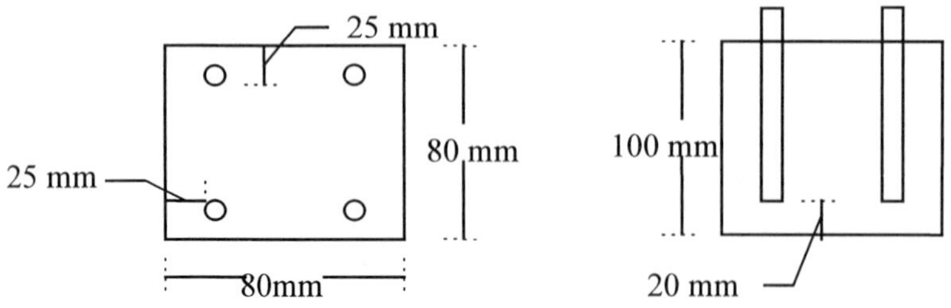

Figure 1 Specimens' shape and dimensions

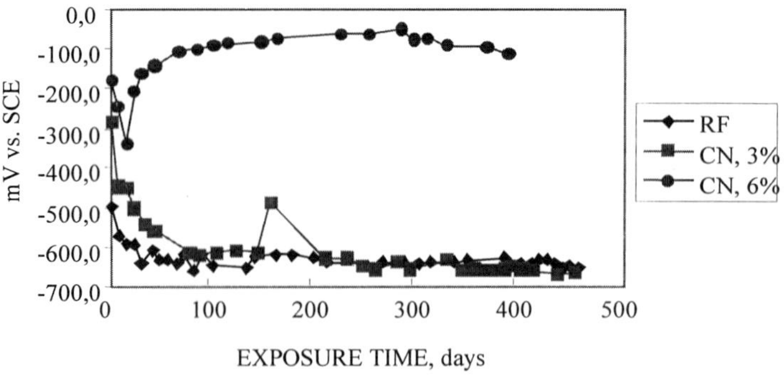

Figure 2 Corrosion potential of CN specimens

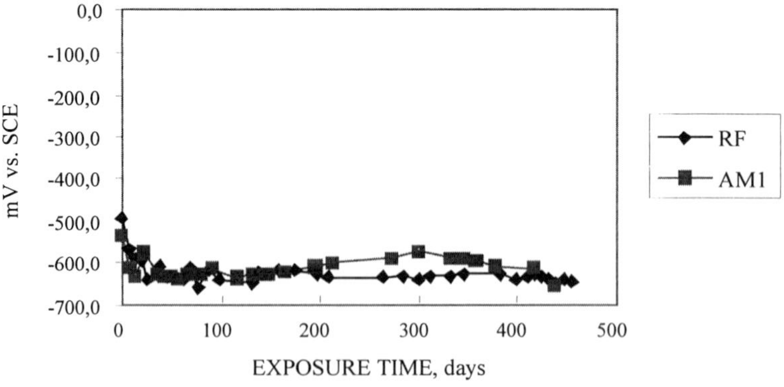

Figure 3 Corrosion potential of AM1 specimens

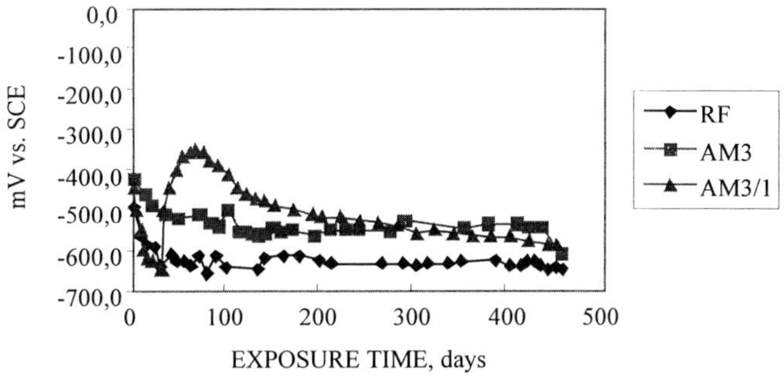

Figure 4 Corrosion potential of AM3 specimens

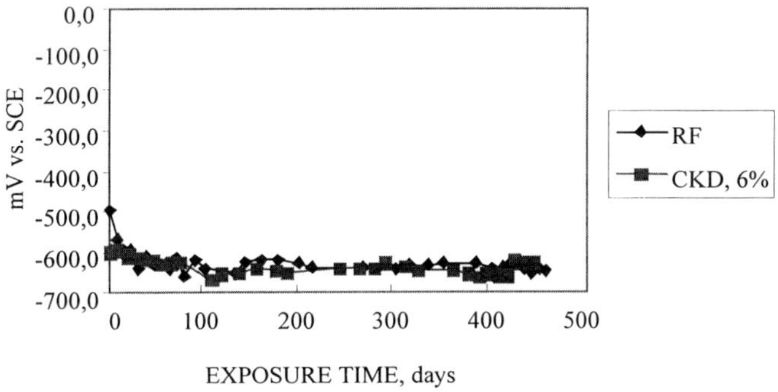

Figure 5 Corrosion potential of CKD specimens

From the potential measurements shown in Fig. 2, it is clear that the reference exhibits a rapid decrease of potential to a level of -650 mV, which takes place within about 30 days of exposure. With the addition of the CN inhibitor at 3% wt of cement in the mixture, the time required to obtain this level of potential is extended to about 100 days, thus indicating the protecting effect which is provided. In the case where this inhibitor is added at a 6% wt of cement in the mixture, the reinforcement of steel remains in the passive state (> -350 mV) for up to 400 days as the results show in Fig. 2. It is important to note that, this extraordinary protection against corrosion obtained by the use of CN inhibitor at that % level, took place in spite of the fact that a number of cracks were present in the test specimens. These cracks may have been caused by the accelerating effect produced by the excessive level of NO_2^- in concrete (the recommended maximum addition level of CN inhibitor is about 3.5% wt of cement).

The application of inhibitor AM, either as an admixture, or as surface impregnation, drives the corrosion potential to more electropositive values, as shown in Fig. 3, 4. This effect is more clearly shown in the case where AM is applied as impregnation by surface spraying 30 days after initial exposure (specimens AM3/1). The initial corrosion potential before the spraying is -650 mV, while under application the potential goes to -350 mV within the next 30 days, indicating the inhibiting effect.

The addition of 6% CKD in the mixture, presented similar performance with the reference, as far as corrosion potential is concerned. This performance can be seen in Figure 5.

Mass Loss And Carbonation Results

From the test measurements obtained after the initial 9 months of exposure, it can be seen that, relative to the reference, either the CI or CKD resulted in lower % mass loss of steel and depth of carbonation, as shown in Figures 6-7. It is noted that, significantly lower % mass loss and depth of carbonation were obtained with the addition of CN inhibitor at 6% wt. of cement in the mixture. However, the lowest depth of carbonation was obtained in the specimens made with the addition of 6% CKD, probably due to the relatively higher alkalinity, as compared to the other additions.

The above performance was also observed throughout the total test duration of 15 months. It is important to note that, the specimens made with the addition of CKD exhibited the best overall performance throughout the total test duration.

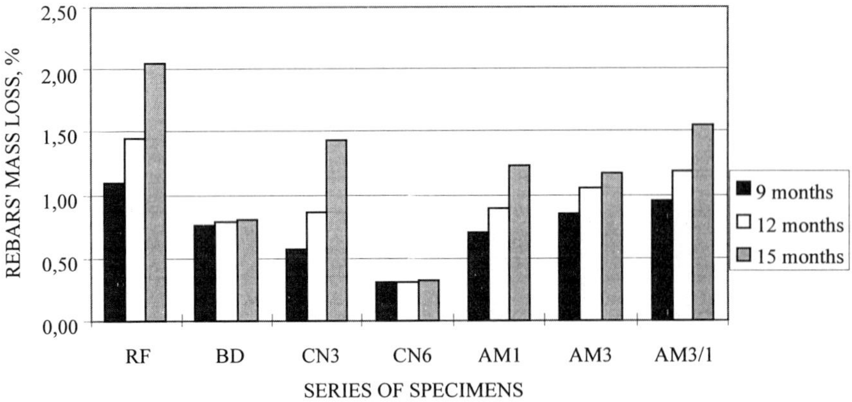

Figure 6 Mass loss of the rebars, %

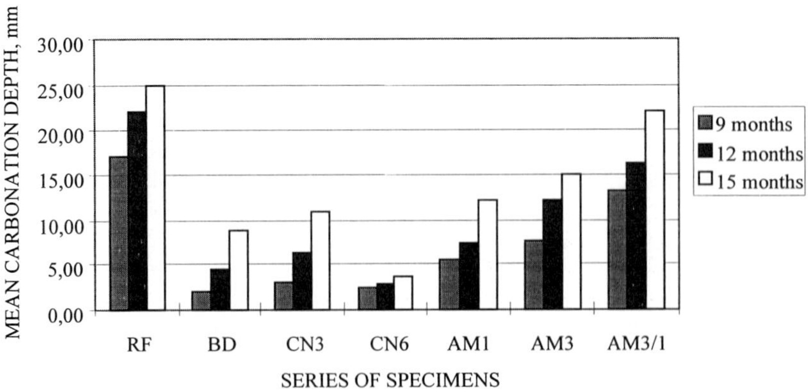

Figure 7 Carbonation depth of all specimens

Porosity Measurements

Porosity measurements show that all series of test specimens have high porosity values, between 34.0 to 38.0.

DISCUSSION

The experiments indicated that all the additions used in this investigation provided protection against corrosion in all specimens made of lightweight concrete with pumice. The use of anodic inhibitors (such as calcium nitrite) and anodic-cathodic ones (such as aminoalcohols) at a rate of 3% appears to provide about the same degree of protection. At this rate of addition, the CI exhibited a level of corrosion, similar to that normally obtained in conventional reinforced concrete made without the use of any inhibitors for corrosion protection. The use of CN inhibitor at the increased rate of 6% is prohibited, due to the observed crack formation.

Comments On CKD Anti-corrosion Action

On the other hand, the addition of 6% CKD, with the given composition, surprisingly exhibited the best protection against corrosion, although this material is not classified as an inhibitor. The protective action against corrosion of the CKD addition, is mainly attributed to its fineness and relatively higher alkalinity, which cause a reduction of total porosity and carbonation, respectively, as shown in previous investigations [3, 4].

The effect of nitrite ions

The inhibiting effect of nitrite ions on steel corrosion induced by chlorides has been reported by other investigations. Several hypotheses have been formulated, concerning steel passivation by nitrite ions. One among them, suggests that nitrite ions are easily oxidized in alkaline environments, consuming oxygen according to the following total reaction [5]:

$$2NO_2^- + O_2 \longrightarrow 2NO_3$$

Another hypothesis [6, 7] suggests that nitrite ions are sacrificially reduced to nitrogen with a simultaneous oxidation of ferrous to ferric ion, according to the following total reaction:

$$6Fe(OH)_2 + 2NO_2^- + 3H_2O \longrightarrow 6Fe(OH)_3 + N_2 + OH^-$$

The formation of iron oxyhydroxide or other similar products, would decrease the formation of ferrous or ferric chlorocomplexes and, consequently, would supress iron anodic dissolution.

According to the most universally accepted hypothesis [8, 12], a stable passive layer could be formed on the reinforcement, avoiding all the metastable intermediate forms, according to the following reaction:

$$2Fe^{2+} + 2OH^- + 2NO_2^- \longrightarrow 2NO\uparrow + 2\gamma\text{-}FeOOH$$

Thus, the nitrite ions compete with chlorides for the ferrous ions at the anode. Nitrite ions must be present in sufficient quantities and come in immediate contact, by diffusion, with the structure in order to provide protection.

The Use Of Amino-alcohols As Corrosion Inhibitors

Amines and alkanolamines and their salts with organic and inorganic acids, have been described and patented for various applications, such as for the protection of steel in cementitious matrices [5, 9]. The effectiveness of this class of compounds, has been shown by means of various experiments [6, 7, 10]. The new technology of spraying CI can be used in lightweight concrete. Earlier experiments [8, 11] have shown that aminoalcohols chemisorbs from aqueous solutions onto oxidized steel surfaces. The thickness and the composition of the absorbate phase, depend upon the composition and the concentration of the aminoalcohols solutions. The absorbate phase is formed even if chlorides are present. The corrosion inhibition effect is explained by the fact that aminoalcohols displaces, due to its strong bonding, ionic species from the oxidized steel surface, in particular chlorides which cause corrosion and forms a durable passivating film [11].

CONCLUSIONS

Based on all the measurements and results obtained in this investigation, the following main conclusions can be drawn:

- Corrosion inhibitors based on CN and AM can be effectively used for corrosion protection of lightweight concrete made with pumice
- Inhibitors applied as admixture exhibited better corrosion protection than as an impregnation by surface spraying on lightweight concrete
- The use of 6% CKD as an addition in lightweight concrete, provides a significant level of protection against chloride-induced corrosion
- The possibility of using CKD for protecting against corrosion constructions made of lightweight concrete, might have significant economic and environmental consequences.

REFERENCES

1. BATIS, G., AIDINI, A., LOUVARIS, G., NICOLAIDES, A., «Reinforcement Corrosion in Pumice Lightweight Concrete», Concrete 2000, Dundee U.K., 1993, pp153-162.

2. NUERBERGER, U., «Korrosionsschutz im Massivebau», Expert Verlag, Boebligen, 1992, pp 93-102.

3. BATIS, G., KATSIAMBOULAS, A., MELETIOU, C.A., CHANIOTAKIS, E., «Durability of Reinforced Concrete Made with Composite Cement Containing Kiln Dust», Concrete in the Service of the Mankind, Vol. 1, Edited R.K. Dhir and T.D. Dyer, 1996, pp 67-72.

4. TASIOS, TH., ALIGIZAKI, K., «Durability of Reinforced Concrete», Athens, 1993.

5. JUSTNES, H., NYGAARD, E.C., «The Influence of Technical Calcium Nitrate Additions on the Chloride Binding Capacity of Cement and the Rate of Chloride Induced Corrosion of Steel Embedded in Mortars», Corrosion and Corrosion Protection of Steel in Concrete, Edited R.N. Swamy, Vol II, 1994, pp 491-502.

6. SAGOE-GRENTSIL, K.K., YILMAZ, V.T., GLASSER, F.P., «Properties of Inorganic Corrosion Inhibition Admixtures in Steel-Containing OPC Mortars. Part 1: Chemical Properties», Advances in Cement Research, Vol. 4, No 15, 1991/1992, pp 91-96

7. YILMAZ, V.T., SAGOE-CRENTSIL, K.K., GLASSER, F.P., «Properties of Inorganic Corrosion Inhibition Admixtures in Steel Containing OPC Mortars. Part 2: Electrochemical Properties», ibit, pp 97-102.

8. PROWELL, B.D., WEYERS, R.E., AL-QUADI, I.L., «Evaluation of Corrosion Inhibitors for the Rehabilitation of RC Structures», Concrete 2000, Edited R.K. Dhir, M.R. Jones, Dundee, UK, 1993, pp1223-1238.

9. MAEDER U., «A New Class of Corrosion Inhibitors», Corrosion and Corrosio Protection of Steel in Concrete, Vol II, Edited R.N. Swamy, 1994, pp851-864.

10. MAEDER, U., «A New Class of Corrosion Inhibitors for Reinforced Concrete», Third CAMNET/ACI Int. Conf., Edited Malhotra, St. Andrews, Canada,1996, pp 215-232.

11. WELLE, A., LIAO, J.D., KAIZER, K., GRUNZE, M., MAEDER, U., BLANK, N., «Interactions of N,N'-dimethylaminoethanol with Steel Surfaces in Alkaline and Chlorine Containing Solutions», Applied Surface Science, 119, 1997, pp 185-198.

12. BERKE, N.S., HICKS, M.C., "Protection Mechanism of Calcium Nitrite", Corrosion Conference on Understanding Corrosion Mechanism in Concrete: A Key to Improving Infrastructure Durability – MIT-Cambridge University, Massachusetts, USA, 27-31 July 1997.

UTILIZATION OF INHIBITORS TO PREVENT AND TO CONTROL THE CORROSION OF THE STEEL REINFORCEMENT

E P Figueredo

C Estefani S Leão

Federal University of Goiás

Brazil

P Castro

Centre of Research and Advanced Studies

Mexico

ABSTRACT. The aim of this paper is to evaluate the use of inhibitors in order to prevent and control steel reinforcement corrosion as well as their influence on some properties of fresh and hardened concrete. Commercial products (amino-based organic) and sodium nitrite were evaluated and no changes on fresh and hardened concrete properties were observed when they were added to the mixing water despite showing plasticizer behavior. Just the nitrite based inhibitor had an influence in the corrosion process under this condition and it was due to the external chloride action. No considerable changes on corrosion parameters (corrosion potential and polarization resistance) were found when they were applied on the surface of chloride-contaminated concrete structures after a year of testing.

Keywords: Corrosion, Inhibitors, Electrochemical techniques, Chloride, Concrete, Amino based, Nitrite.

Dr Enio Pazini Figueiredo is a full professor at the Federal University of Goiás in Brazil and researcher of the Nacional Council of Scientific and Technological Developing (CNPq). His main interest is the concrete durability, repair materials and electrochemical techniques for corrosion monitoring in concrete structures.

Cinthya Estefani is an engineer's student at the Federal University of Goiás in Brazil and training researcher of the Nacional Council of Scientific and Technological Developing (CNPq).

Sérgio Leão is an engineer's student at the Federal University of Goiás in Brazil and training researcher of the Nacional Council of Scientific and Technological Developing (CNPq).

Dr Pedro Castro is a researcher in the Center of Research and Advance Studies, Merida, Mexico. His main interests include the rehabilitation of corrosion-damaged infrastructure and the effects of aggressive marine environments on the durability of concrete structures.

INTRODUCTION

Corrosion of rebars in concrete are a worldwide topic of study due to two main reasons. The first one is of economical nature due to maintenance costs related to corrosion. The second one is related to technical reasons, which include a lack of understanding about the corrosion phenomena, its causes and prevention and repair procedures.

Several authors have investigated about the economical expenses that represent the corrosion and the rehabilitation of corrosion damaged infrastructure. The economical loses for developing or developed countries (1) are between 1.25% and 3.5% of the Gross National Product (GNP). In the United States the estimated expenses for repairing the bridges affected by corrosion are about 100 billion dollars (2). In Spain, the expenses for repairing and substituting corrosion damaged infrastructure are about 1.5 billion dollars (3). According to the Hoar report, which was prepared in England in 1981 (4), the expenses due to corrosion were about 3.5% of the GNP with the construction sector being responsible of 0.6% GNP. In Brazil there is not still a deep study about the expenses due to corrosion in public and private constructions, but, according to the world tendencies, the annual expenses by corrosion in the construction sector are about 1.8 billion dollars (5). In Yucatán, México, some studies in several economic sectors as the automobiles, the construction, and the industry of chemicals, meat and metals have showed repairing and maintenance corrosion expenses in the order of 200,000 dollars annually (6). These data were obtained surveying a 7% of the big sectors in Yucatán.

While a great amount of financial resources are spent on repairing and substituting structural elements, others efforts are done to promote more efficient solutions for these problems. Each building or structure has its own particularities and, therefore, its own form to behave against the aggressives and the repairing systems effects. This is the reason why several techniques and materials are being developed worldwide in order to repair or to fight against corrosion of reinforcing steel.

This work proposes the study of inhibitors as corrosion prevention and controlling method as well as their influence on the fresh and hardened concrete. The amino and nitrite based inhibitors were used in the mixing water as a prevention method and in the hardened concrete surface as a repair method.

EXPERIMENTAL

Beams (15 x 15 x 50 cm) were constructed to evaluate the inhibitors efficiency. They have 4 longitudinal steel bars (0.8 cm diameter) with 5 stirrups (0.42 cm diameter) and a concrete cover of 2 cm. As seen in Figure 1, one of the longitudinal bars was extended 2 cm out of the beams in order to perform the electrochemical measurements. The cable for the working electrode was connected to this bar. Concrete of 18 MPa with Portland cement (CP32-IIF), a water/cement ratio of 0.60, a slump of 7 ± 1 cm, river sand and granite gravel was prepared according to proprietary methods of the EEC/UFG Materials Laboratory.

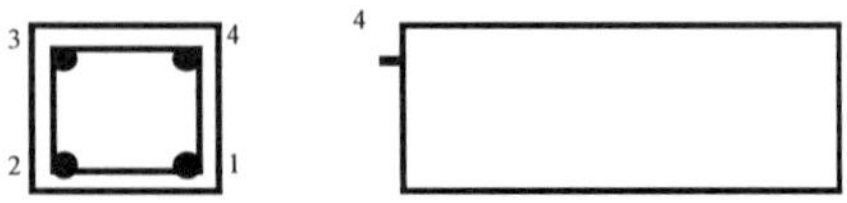

Figure 1 Sketch of the beam used in this research

Inhibitors Applied in the Mixing Water

Three corrosion inhibitors were evaluated in this work: two organic and aminoalcohol based (*A* and *B*) and one anodic which is $NaNO_2$ based (C). They were added to the concrete according to the manufacturers specifications: *A*= 0.7 l/m^3 of concrete, *B* and *C* = 3% by mass of cement being the *C* proportion from previous works (7). The inhibitors *A* and *C* were mixed and added to the mixing water without having influence on the final w/c ratio. A considerable volume of inhibitor B was added to the mixing water so that the final w/c ratio was that of design but including the inhibitor.

There were prepared specimens for testing concrete quality at 3, 28 and 91 days after casting. The workability of the mixture was tested through slump test measurements during the beams casting. Eight specimens were prepared (two for each variable) which were tested by absorption at 63 days. The beams were stored in a humid chamber after setting so they were demolded 24 hs later and conserved once more the humid chamber until the corrosion potential became stable. Then, they were subjected to wetting and drying cycles in 3% NaCl solution.

The wetting and drying cycles were done in order to accelerate, in natural form, the penetration of the chloride ions through the concrete cover. The evaluation of the inhibitors consisted on comparing of the time to initiate corrosion of the reference beams against those with the inhibitors.

Inhibitors Applied on the Surface of the Hardened Concrete

The corrosion was induced in a different manner: calcium chloride was used as an additive to accelerate the concrete hardening. This additive was used two decades ago in the Brazilian Central-West region as well as other regions and it is the cause of chloride-contaminated buildings, which are now corroding. The beams were conserved in a humid chamber at 95% of RH in order to assure the corrosion initiation.

The beams were maintained in the humid chamber for 4.5 months monitoring their corrosion potential (Ecorr). The polarization resistance technique was used at the end of this period to measure the corrosion rate (icorr). The inhibitors were applied in several stages through two different transport mechanisms: diffusion and absorption. The diffusion process was facilitated by the pore saturation grade provided by the environment of the humid chamber. The first application by diffusion of the inhibitors was after 4.5 months of testing.

In the case of the amino-based inhibitors, they were re-applied after 21 and 60 days of their first application. The beams were withdrawn from the humid chamber after 115 days of the first inhibitors application by diffusion. They were maintained at room temperature in the Laboratory and the inhibitors were re-applied again after 47 days in this condition (40%-60%). Under these circumstances, the pore saturation grade had decreased so the penetration mechanism was supposed to be by absorption. Another application of the inhibitors by absorption was made 40 days after the first one. After 350 days of testing the beams were changed to the humidity chamber again.

The inhibitor C was applied using a solution with 120 g of $NaNO_2$ and 150 ml of water.

This was done taking into account the $NaNO_2$ solubility product as well as the total volume of solution that could be able of penetrating the porous net. This last value was chosen according to empiric experience with the application of amino-based inhibitors (8). The first application was after 155 days of accelerated corrosion, therefore by diffusion. After 31 days of this application, the beams were withdrawn from the humidity chamber and stored in the Laboratory at environmental conditions (40%-60% of RH). The inhibitors were applied again, now by absorption, after 47 and 87 days at 40%-60% of RH. After 356 days of testing the beams were changed to the humidity chamber again.

Corrosion Monitoring

Corrosion monitoring was done through the Ecorr, american code, ASTM-C-876-91, and the corrosion rate with a corrosimeter GECOR6. The corrosion potential was monitored along all the test while the polarization resistance was measured in pre-determined lapses. Table 1 shows the relation between the Ecorr and the corrosion probability. The criteria for corrosion classification in terms of the icorr are shown in Table 2.

Table 1 Probability of corrosion according to the Ecorr value (ASTM C-876)

E_{corr} (mV, $Cu/CuSO_4$)	PROBABILITY OF CORROSION
> -200	5%
Between –200 and -350	50%
< -350	95%

Table 2 Corrosion classification according to the icorr value (9)

i_{corr} ($\mu A/cm^2$)	CLASSIFICATION
<0,1-0,2	Negligible corrosion
>0,1-0,2	Active corrosion
~1	Significant corrosion but no severe
~10	Severe corrosion attack

RESULTS

Influence of the Corrosion Inhibitors Added to the Mixing Water in the Fresh Concrete Properties

The influence of the inhibitors on the fresh concrete characteristics was obtained through the slump test, brazilian code, NBR 7223, 1992. According to the Table 3, the inhibitors increased the concrete plasticity. These results justify the observed behavior during the beams casting. The inhibitor *A* showed to be an efficient air-entraining admixture as observed during the vibration when several bubbles ascended to the concrete surface.

A similar behavior was observed for inhibitor *B*, although the effect was smaller. The inhibitor *C* did not change significantly the initial consistency, but the workability decreased abruptly after 30 minutes of mixing because of the setting acceleration effect.

Table 3 Results of the workability tests

SAMPLE	REFERENCE	INHIBITOR A	INHIBITOR B	INHIBITOR C
SLUMP	7 cm	11 cm	10 cm	8,5 cm

Influence of the Corrosion Inhibitors Added to the Mixing Water in the Concrete Water Absorption

The water absorption test was done according to brazilian code, NBR 9778, 63 days after the beams were cast. A small variation in the water absorption was measure to the concretes with *A* and *B* inhibitors as shown in Table 4. The amino based inhibitors (*A* and *B*) produced a small increase in the absorption value. On the other hand, a decrease of about 15% was observed with inhibitor *C* compared to the reference. It can be deduced that sodium nitrite could improve the steel protection against corrosion due to not only the formation of a passive layer but also making more hard the penetration of water and oxygen which are the necessary agents for corrosion propagation. Despite the changes in the water absorption when using the inhibitors *A* and *B*, no values close to 8% were observed that is characteristic of concretes with good durability (10).

Table 4 Results of the absorption tests

SAMPLE	REFERENCE	INHIBITOR A	INHIBITOR B	INHIBITOR C
ABSORPTION	4,50%	4,95%	4,82%	3,80%

Influence of the Corrosion Inhibitors Added to the Mixing Water in the Compressive Strength

The compressive strength tests were done at 3, 28 and 91 days by means of the brazilian code, NBR 5739,1987 method. According to the Figure 2, the compressive strength had slow variations with the addition of the inhibitors. The highest difference was observed at 28 days where a decrease in the compressive strength was observed for all the concretes containing inhibitors. However, at 91 days, all the concretes had similar strengths including that of the reference. The small differences in strength could be attributed to the small variation in the w/c ratio when using inhibitors and to the ordinary internal variation of the compressive strength testing.

The influence of the inhibitors can be analyzed in terms of increasing strength with time. The reference increased its strength 37.6% from 3 to 28 days and 6.5% from 28 to 91 days. The concrete with the inhibitor *A* increased its strength 28.6% from 3 to 28 days and 16.7% from

28 to 91 days. The concrete with inhibitor *B* increased its strength 31.7% from 3 to 28 days and 8.1% from 28 to 91 days. The concrete with inhibitor *C* increased its strength 27% from 3 to 28 days and 12% from 28 to 91 days. It can be deduced that inhibitors make concrete to reach its final strength later, this means that they are probably acting as hardening retarders.

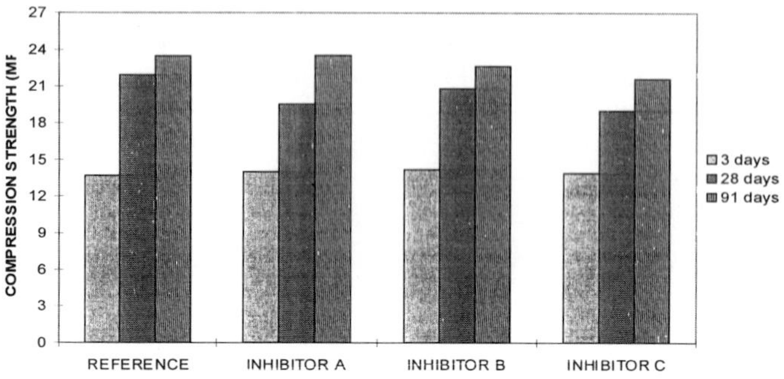

Figure 2 Influence of the inhibitors in the concrete compressive strength

Influence of the Corrosion Inhibitors Added to the Mixing Water in the Electrochemical Behavior

The Ecorr was monitored continuously since the concrete casting. The Figure 3 shows negative Ecorr values during the first week and for all the beams. The passivant layer was formed during this stage. After this stage the potential became more positive, to become fixing in a characteristic potential for each type of inhibitor. This characteristic potential to inhibitor C, B, A and reference are, respecting, -230mV, -180mV, -150mV and -130 mV. The reference group showed the most positive potentials during this period.

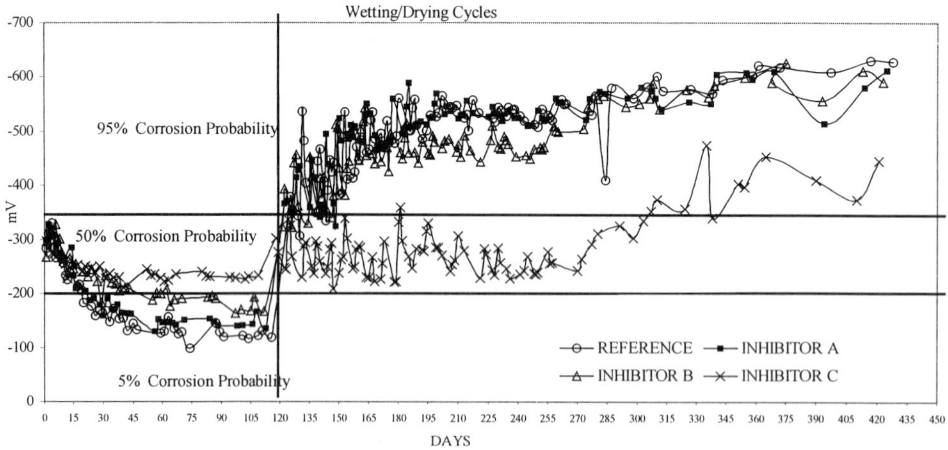

Figure 3 Corrosion potential for all the studied groups when inhibitors were added to the mixing water

The electrochemical behavior changed after starting the wetting and drying cycles. The Figure 3 shows that the reference beams depassivated first than the others with inhibitors, and they were followed by the specimens with inhibitors *A* and *B*. The beams with the inhibitor *C* ($NaNO_2$) showed the best behavior until 28 cycles of wetting and drying cycles in which present the most passive potential. The inhibitors *A* and *B*, organic compounds which are amino based, supposedly form a monomolecular layer that remains adsorbed in the metal surface (11). This layer is different from that passivant layer formed in absence of inhibitors. Depending on this layer thickness, the chloride ions have more or less difficulty to access the steel rebar (11).

The initial values of corrosion potential in the specimens with the inhibitor *C* shows the formation of the passivant layer together with those of the reference specimens. A stabilization period for both type of specimens followed with a difference of 100 mV between them (-130 mV for the reference beams and –230 mV for the beams with inhibitor *C*). This fact can be explained according to the nature of the passivant layer.

In absence of inhibitors, the passivant layer is formed in the steel surface and it is generally composed by $FeO.(\gamma H_2O)_x$, which is water soluble. In presence of $NaNO_2$, the passivant layer has a more stable chemical composition and it is formed by Fe_2O_3 (7,11,12). This can be understood observing the potential of the specimens with inhibitor *C* which remained in the uncertain zone (50% corrosion probability) during 28 wetting and drying cycles while previously the potential of the reference specimens changed to the zone with 95% of corrosion probability. Only after 28 cycles the specimens with inhibitor C changed to this zone (95% of corrosion probability).

The Figure 4 shows intensity corrosion values when inhibitors were added to the mixing water. The measures were made only after 270 days from the casting, corresponding 20 wetting/drying cycles, because of the equipment availability. This icorr behavior agrees with Ecorr values shows in the Figure 3. The best performance is related to the inhibitor C that showed negligible icorr until 25 cycles. In this moment the depassivating was recorded by Ecorr and icorr values.

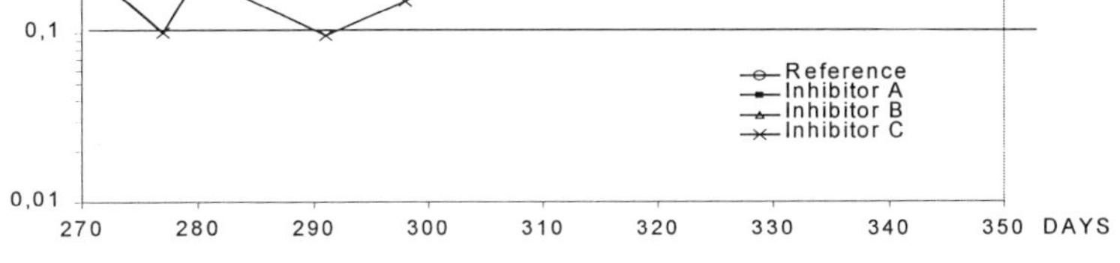

Figure 4 Corrosion rate variations for all the studied groups when inhibitors were added to the mixing water

In the technical catalogues is write that amino bases inhibitor delay the corrosion initiation time and decreasing the corrosion rate after depassivating. The Figure 3 agree with the first one, but in the Figure 4 is possible to observe that the icorr magnitude is the same for reference, inhibitor A and B.

Influence of the Corrosion Inhibitors Applied on the Concrete Surface

The Ecorr and the icorr were used to evaluate the influence of the corrosion inhibitors applied on the concrete surface. The Figure 5 shows that the Ecorr does not distinguish different behaviors among the reference and the inhibitors. This means that no influence of inhibitors *A* and *B* on the corrosion process when chloride ions are inside of the concrete.

According to these results and after 8 months of using the inhibitors through the mechanisms of diffusion and absorption, no clear corrosion control was obtained. However, the time of evaluation is not enough large to conclude inhibitor inefficiency but a good technical information was obtained. The inhibitors *A* and *B* are different commercial recipes from different companies and recommended in catalogues for application on the concrete surface. Both products have been supported by investigations (12,13) that demonstrated their better diffusion through the concrete porous net than other products. These investigations (14,15) showed by means of radioactive techniques that their products penetrated 3 to 5 cm in one month the concrete, which could be of a great efficiency against corrosion in terms of civil engineering if this penetration will be associate to decreasing of the icorr. This point could take to think that after one month the corrosion will be controlled, that is a wrong according to results of this present study. Nevertheless, in a real case of chloride contamination, as showed in this work and after 8 months of the first application, no satisfactory results are still obtained. The time at which the inhibitor reaches the rebar at an enough concentration for restoring the passivation of the steel and decreasing the corrosion rate should be determined by other techniques. However, the inhibitor effect can be determined through electrochemical techniques as was done here. The results of this study can be of help to technicians in order to determine the quantity of chloride that these inhibitors could manage in order to obtain satisfactory results.

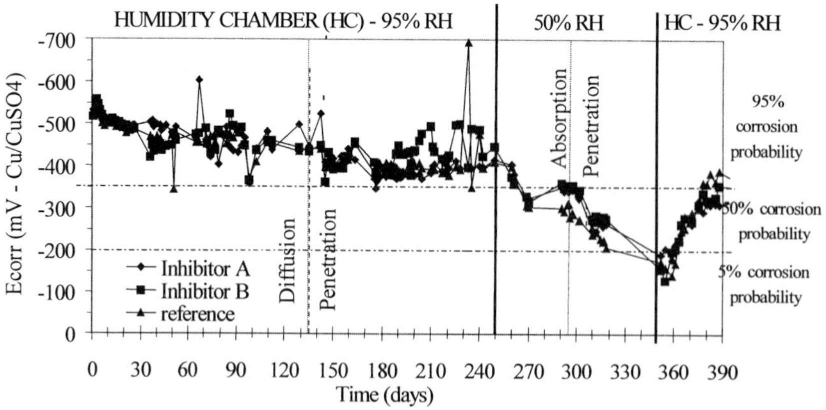

Figure 5 Corrosion potential variation due to the application of the amino based inhibitors on the concrete surface

Figure 6 shows the Ecorr and icorr behavior along time for the beam with the inhibitor *C* and the reference one. As observed, the measurements did not start in the same point due to the fact that the reference beams were cast before that with inhibitor *C*. However, they can be compared since the sodium nitrite was applied after six months of the beams casting, which is a time more than necessary to obtain the corrosion potential stabilization.

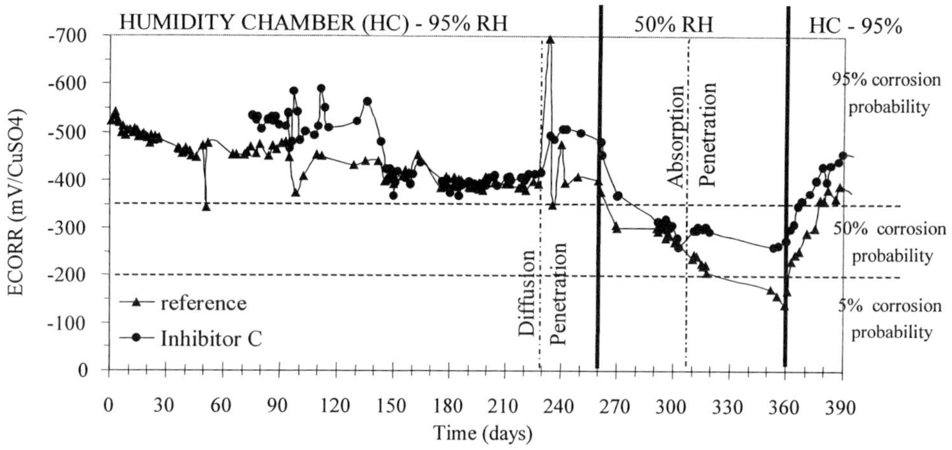

Figure 6 Corrosion potential variation due to the sodium nitrite

The beams with inhibitor *C* presented more negative potentials than the beams of reference. The reference beams and beams with inhibitor C increased their corrosion potentials during the periods of low humidity due to the decreasing of water in the concrete porosity and their consequent increase of the resistivity.

CONCLUSIONS

The methodology applied to survey the inhibitors efficiency was suitable. The values of water absorption indicate that using inhibitors could have influence on the quantity of chlorides penetrating the concrete. Under these circumstances the time for the reinforcement depassivation does not seem to be determinanted due to this factor.

The corrosion inhibitors used in this work (amino and nitrite based) do not modify the characteristics of fresh and hardened concrete. Although the strength had been reduced at 28 days using inhibitors, that of 3 and 91 days remained similar to that of the reference group. Besides this, the plastificant effect of the inhibitors allows a reduction of the w/c ratio as a mean to compensate the loss of strength.

The Ecorr results after 48 wetting and drying cycles showed that the time for reinforcement depassivation changes according to the type of inhibitor. The reference beams depassivated after 3 cycles, the beams with the *A* and *B* inhibitors in 6 cycles and the beams with inhibitor *C* depassivated after 28 cycles.

Organic inhibitors amino-based and sodium nitrite based do not play an important role controlling corrosion when they are applied on concrete surface by absorption or diffusion under the conditions of this study. It is necessary to wait for larger times in order to observe if the products can accumulate in a threshold quantity that would allow the re-passivation of the steel. It is also important to investigate if these products improve their efficiency when applied by absorption or diffusion on beams with a small chloride content.

REFERENCES

1. ANDRADE, C.; GONZÁLEZ, J.A. Tendencias Actuales en la Investigación sobre Corrosión de Armaduras. Madrid, Revista Informes de la Construcción, v.40, nº 398, nov-dic, 1988, p. 7-14.

2. KILARESKI,, W.P. Corrosion Induced Deterioration of Reinforced Concrete. Na Overview, Materials Performance. 1980. P. 48-50.

3. SEGUI, V. Estratégias de Reparación: Caso del Aluminoso en el casco Urbano de Barcelona. In: Curso de Reparación de Obras de Hormigón. Bilbao. LABEIN ed. 30-31 Enero. 1992, 16 p.

4. HOAR, T. P. Report of the Committee on Corrosion and Protection. (Coord. T. P. Hoar). Dept. of Trade and Industry. HMSO. Londres. 1981

5. PAZINI FIGUEIREDO, E.J. Avaliação do Desempenho de Revestimentos para Proteção da Armadura contra a Corrosão Através de Técnicas Eletroquímicas – Contribuição ao Estudo de Reparo de Estruturas de Concreto Armado. São Paulo, 1994. 423p. Tese (Doutorado em Engenharia). Escola Politécnica da Universidade de São Paulo.

6. CASTRO, P. ; OROZCO, J. "The cost of corrosion in Yucatán, México", Monography, 1998: 60 pp.

7. LIMA, M. G. Inibidores de Corrosão – Avaliação da Eficiência Frente a Corrosão de Armaduras Provocadas por Cloretos. São Paulo, 1996, 174p. Tese (Doutorado) – Escola Politécnica da Universidade de São Paulo

8. LEÃO, S.; PAZINI FIGUEIREDO, E. J.; CASTRO, P. "Efficiency evaluation of aminoalcohol based corrosion inhibitors applied on reinforced concrete surface", International Materials Research Congress, Cancún, México, September 1-4, (1997): p. 33

9. ALONSO, C.; ANDRADE, C. Effect of Nitrite as a Corrosion Inhibitor in Contaminated and Chloride-free Carbonated Mortars. ACI Materials Journal, v. 87, nº 2, p. 130-137, Mar/April, 1990.

10. ANDRADE, C.; FELIU, S. at al. Manual- Inspeccion de obras dañadas por corrosión de armaduras. Madrid, 1989.

11. GONZALEZ, J.A. Control de la corrosion y Medida por Tecnicas Electroquímicas. Madrid, Espanha, 1989.

12. ROSENBERG,A.M. et al. Corrosion inhibitor formulated with Calcium Nitrite for use in reinforced concrete. In: Choride Corrosion of Steel in Concrete. Chicago, 1976. Proceedings. Baltimore, 1977, p.89-99. Special Technical Publication STP-629.

13. ALONSO, C. Inhibición de la Corrosión de Armaduras Provocada por Amaso de Hormigón com Agua de Mar. Zaragoza, Espanha, 1983. 53p. Tesina. Faculdad de Ciencias Quimicas. Universidad de Zaragoza.

14. EYDELNANT, A.; OSTROVSKI, A. B.; DEMIDOV, A. M. Analysis of Diffusion Rate of Migration Corrosion Inhibitor MCI-2000 in concrete Using Radioactive Isotope Tagging Techniques. Moscow-Russia. September, 1993. 12p.

15. BENTZ, JWG; GOSCHNICK, J. Investigation of the depth of distribution of DMEA in concrete after different application condition. Institute fur Radiochemie Forschungszentrum Karlsruhe. July, 1994.

THE PERFORMANCE OF CALCIUM NITRITE AND GGBS CONCRETES

B F Perry

Grace Construction Products Limited

M D Connell

The Appleby Group

United Kingdom

ABSTRACT. The corrosion of embedded reinforcing steel is one of the major reasons why reinforced concrete structures fail prematurely when subjected to severe exposure conditions, particularly where chloride ions are present. Corrosion protection starts with the use of concretes with low water to cementitious ratios designed to offer durability and reduce ingress of chloride ions. Latent hydraulic materials such as Ground granulated blastfurnace slag (GGBS) and pozzolans such as Pulverised-fuel ash (PFA), Silica fume (SF) and Metakaolin (MK) can also be added to provide further decreases in chloride ingress. There is increasing use in corrosion inhibitors such as calcium nitrite to raise the level of chloride needed to initiate corrosion. The enhanced performance of low permeability concretes and calcium nitrite is well documented. There is limited performance data, however, on the combined use of calcium nitrite and GGBS concretes.

In this paper, the results of laboratory tests carried out to evaluate the effect of calcium nitrite based corrosion inhibitors on the properties of concretes based on cementitious blends of Portland cement and GGBS are discussed. Such properties include workability, setting time at various temperatures, permeability to chloride ions and compressive strength development.

Keywords: Corrosion, Chloride, Pozzolans, Ground granulated blastfurnace slag (GGBS), Calcium nitrite, Corrosion inhibitors.

Mr Bruce Perry is a member of the Engineering Services Group of Grace Construction Products Ltd. His responsibilities include promoting the use of admixtures and systems that enhance the durability of concrete structures, particularly where reinforced concrete comes into contact with chloride laden environments.

Mr Mike Connell is the Group Technical Manager for The Appleby Group, producers of GGBS. His responsibilities include providing technical support to the production and marketing functions of the company. He is also the President of the Institute of Concrete Technology.

INTRODUCTION

The problem of chloride induced, steel reinforcement corrosion has been covered extensively over the last 10 to 15 years with many papers having been written about its causes, mechanisms and consequences.

Also well documented are many protection measures that are available to clients, design engineers and contractors to combat this whole problem. These protection measures take many forms but can be essentially broken down into three main areas;

- reduce the risk of chloride ions penetrating the concrete and reaching the embedded steel
- reduce the risk of chloride ions causing corrosion once they reach the embedded steel
- when reinforcement corrosion starts, apply measures which reduce rate of corrosion

These three ways of addressing the problem require many different solutions ranging from simply raising the grade of concrete specified, the use of high range water reducers to significantly reduce water to cementitious ratios, use of alternative reinforcements to sophisticated cathodic protection techniques.

As already mentioned, one of the ways of achieving corrosion protection is to reduce the risk of the chloride ions penetrating the concrete and reaching the embedded steel. To do this, either a physical or chemical barrier can be applied to the surface of a concrete structure to guard against chloride ingress such as bridge deck waterproofing membranes or silane/siloxane treatments respectively.

Another way of reducing chloride ingress is to make the concrete covering the reinforcing steel more impermeable to the chloride ion. This can be achieved by either lowering the free water to cementitious ratio of the concrete thus reducing the number and diameter of capillaries in the cement paste of the concrete and/or using a cementitious binder that reduces the movement of chlorides through the cement paste in comparison to a cement paste based entirely on Portland cement. This method of corrosion protection increasingly requires the specification of high performance concretes (HPC) such as those included within the ACI definition. HPC can be defined as concrete meeting special combinations of performance and uniformity requirements that cannot always be achieved routinely using conventional constituents and normal mixing, placing, and curing practices.

These cementitious binders with reduced chloride permeability are produced by either replacing a proportion of the Portland cement with a material such as pulverised-fuel ash (PFA) and ground granulated blastfurnace slag (GGBS) or by adding a ‘super pozzolan’ such as silica fume (SF) or metakaolin (MK).

Reducing the risk of chloride ions causing corrosion once they have reached the embedded steel can also combat reinforcement corrosion. This can be achieved by either coating the embedded steel in a protective coating , i.e. fusion bonded epoxy coated rebar, the use of stainless steel reinforcement or by using corrosion inhibiting admixtures that chemically inhibit the chemical reactions associated with the dissolution of iron, i.e. corrosion.

These corrosion inhibiting admixtures can either be added to the concrete at the batching stage thus ensuring complete dispersion throughout the load of concrete or by post application of solely 'vapour phase' corrosion inhibitors to existing structures that are showing evidence of corrosion.

Perhaps the most accepted and well documented corrosion inhibitors are those based on calcium nitrite which have a history of use world-wide of over 20 years with no reported failures to date. Their use in concretes based entirely on Portland cement is well documented although their use in concretes based on 'blended' cements, particularly those containing GGBS, is less well documented.

This paper examines the compatibility of two such corrosion protections systems when used together. It discusses the effect on both plastic and hardened properties the addition of a single dosage of 15 litres per cubic meter of two, 30% weight by weight calcium nitrite based corrosion inhibiting admixtures have on concretes based on both Portland cement and also on two cementitious blends with GGBS, namely 50% and 70% replacement levels.

EXPERIMENTAL DETAILS

Materials

Portland cement complying with the requirements of BS12: 1996 and GGBS complying with the requirements of BS6699: 1992 were used. The aggregates used were 20mm and 10mm single sized crushed limestone and a land-based, medium sand.

A water reducing admixture (WRA) complying to BS5075: 1985 and based on a modified lignosulphonate was used in some of the concrete mixes at a fixed dosage of 0.18% volume by weight of total cementitious material.

Two corrosion inhibiting admixtures based on 30% weight by weight calcium nitrite were used; one normally used in cold weather environments (DCI® Corrosion Inhibitor) and one normally used in warmer or hot weather environments (DCI® S Corrosion Inhibitor). This latter product is especially formulated to minimise the effect on concrete setting time and exhibits almost neutral-set characteristics when the accelerating properties of the first product may be undesirable.

Mix Proportions

The mix proportions are summarised in Table 1.

Method for Mix Preparation and Tests on Plastic Concrete

In each batch of concrete produced, the following method was adopted;

The aggregates were oven dried and then cooled to the required mixing temperature prior to use. They were then added to a forced action mixer sufficient to produce 15 litres of concrete. 1.0 litre of distilled water, also at the required mixing temperature, was then mixed into the dry aggregates for 1 minute which were then covered and allowed to stand for a further 10 minutes to allow for water absorption.

Table 1 Concrete mix details

MATERIAL	MIX PROPORTIONS, kg/m^3					
	0.45 water:cement ratio concrete			0.40 water:cement ratio concrete		
	100% PC	50:50 PC:GGBS	30:70 PC:GGBS	100% PC	50:50 PC:GGBS	30:70 PC:GGBS
PC	380	190	115	420	210	125
GGBS	-	190	265	-	210	295
Sand*	690	690	690	635	635	635
10mm limestone*	345	345	345	350	350	350
20mm limestone*	810	810	810	825	825	825
Fines:coarse ratio		37.4%			35.1%	
Agg:cement ratio		4.85			4.31	

* aggregate batch weights are expressed in a saturated surface dry condition.

The cementitious materials were mixed into the wet aggregates followed by a further 1.0 litre of distilled water. Where the water reducing admixture was being added, it was dispersed in this 1.0 litre aliquot of water. The corrosion inhibiting admixture under test was then added directly to the concrete followed by more water until the desired workability was reached. The water content of the corrosion inhibiting admixture was taken into account in the calculation of the free water to cementitious ratio of the concrete. Mixing of the concrete continued for a total of 3 minutes.

The concrete was covered and allowed to stand for a period of 10 minutes after which it was re-mixed for 30 seconds. The concrete temperature was measured, sampled and tested to determine workability and plastic density. 100mm cubes and 100mm diameter cylinders were also cast, all following the methods given in the relevant section of BS1881. The remaining concrete was then sieved through a 5mm sieve to remove the coarse aggregates and the mortar fraction compacted into a tin and the setting time determined using the Procter needle apparatus. The tins were stored at the batch mixing temperature and the setting time of the mortar fraction was recorded at a resistance to penetration of 1.72 N/mm^2. The results of the tests on plastic concrete are given in Tables 2 and 3.

Table 2 Concretes with a free water to cementitious ratio of 0.45

CEMENT TYPE	ADMIXTURE	FREE WATER, LITRES/m3	SLUMP, mm	PLASTIC DENSITY, kg/m3	SETTING TIME, HOURS:MINS. 20°C	5°C	40°C
100% PC	WRA	168	95	2360	3:21	-	-
	WRA + DCI	171	110	2415	2:29	-	-
	DCI S	172	90	2435	3:16	-	-
50:50 PC:GGBS	WRA	170	110	2390	4:25	-	-
	WRA + DCI	169	95	2385	2:53	-	-
	DCI S	169	120	2395	3:42	-	-
30:70 PC:GGBS	WRA	168	95	2370	5:01	-	-
	WRA + DCI	168	110	2375	3:45	-	-
	DCI S	170	105	2400	4:17	-	-

Table 3 Concretes with a free water to cementitious ratio of 0.40

CEMENT TYPE	ADMIXTURE	FREE WATER, LITRES/m3	SLUMP, mm	PLASTIC DENSITY, kg/m3	SETTING TIME, HOURS:MINS. 20°C	5°C	40°C
100% PC	WRA	171	90	2385	3:37	-	-
	WRA + DCI	169	90	2405	2:30	-	-
	DCI S	169	90	2405	2:52	-	-
50:50 PC:GGBS	WRA	168	90	2385	4:13	-	-
	WRA + DCI	166	90	2400	2:32	-	-
	DCI S	166	110	2405	3:23	-	-
30:70 PC:GGBS	WRA	168	95	2385	4:21	12:30*	2:38#
	WRA + DCI	166	100	2395	3:02	10:06*	1:58#
	DCI S	165	100	2390	4:06	11:31*	2:28#

* concrete temperature after mixing; 14°C.
# concrete temperature after mixing; 30 °C

Method for Determining Hardened Concrete Properties

The cube specimens for testing at 1 day were cured at the concrete mixing temperature. The remaining cube and cylinder specimens were cured at 20°C. following the method given in BS1881. The cubes were tested to determine compressive strength at 1, 3, 7, 28 and 56 days following the method given in BS 1881.

The 100mm diameter concrete cylinders were tested to determine rapid chloride ion permeability at 28 and 90 days following the method given in AASHTO T277. The specimens from the 100%PC, 0.40 water cement ratio concretes were not tested due to equipment availability. The results of the tests on hardened concrete are given in Table 4 and Table 5.

RESULTS AND DISCUSSION

Effect of the Corrosion Inhibitors on Water Demand

All of the concretes produced, irrespective of cementitious content or type, had a similar water demand of 165 - 172 litres/m^3 at a constant workability of 90 - 120mm slump.

The corrosion inhibitor, DCI S, contains some retarding and superplasticising components in addition to the 30% w/w calcium nitrite corrosion inhibiting material. Concretes containing this inhibitor did not contain any water reducing admixture whereas the 'control' concretes (i.e. concretes without corrosion inhibitor) and the concretes containing DCI, contained the water reducing admixture at a constant dosage of 0.18% v/w of cement.

It can be concluded, therefore, that at a dosage of 15 litres/m^3, DCI S can reduce the water demand of concrete by an equivalent amount as a normal, lignosulphonate based water reducing admixture when dosed at 0.18% v/w of cement whereas DCI also at a dosage of 15 litres/m^3 shows no water reducing capabilities. This assumes, however, that the water component in the corrosion inhibitors is included in the calculation of the free water demand of the concrete.

Effect of the Corrosion Inhibitors on Plastic Density

The plastic density of all of the concretes produced did not vary significantly either between different cementitious types, different cementitious contents or whether the concretes contained a corrosion inhibitor or not. This indicates the use of corrosion inhibitors does not cause entrainment of air.

Effect of the Corrosion Inhibitors on Setting Time

As can be seen from Figure 1 and from Table 2, the same trend in setting times is observed within each cementitious type but as the level of replacement increases, setting times predictably also increase.

One of the corrosion inhibitors, DCI, clearly accelerates the setting of concretes independent of cementitious type compared to the 'control' concretes containing the WRA. This effect is to be expected since calcium nitrite is an effective set accelerator. The other corrosion inhibitor, DCI S, clearly shows a tendency to impart more neutral set characteristics because of the components which are added to the formulation to counteract the accelerating effect of calcium nitrite. This "neutral set" behaviour becomes even more evident taking into account the fact that the 'control' concrete contains a WRA (expected to induce some set retardation), whereas concrete containing DCI S does not. In other words, a "true" control mix for DCI S concrete should not contain any WRA and its setting time is more likely to be that of the DCI S concrete.

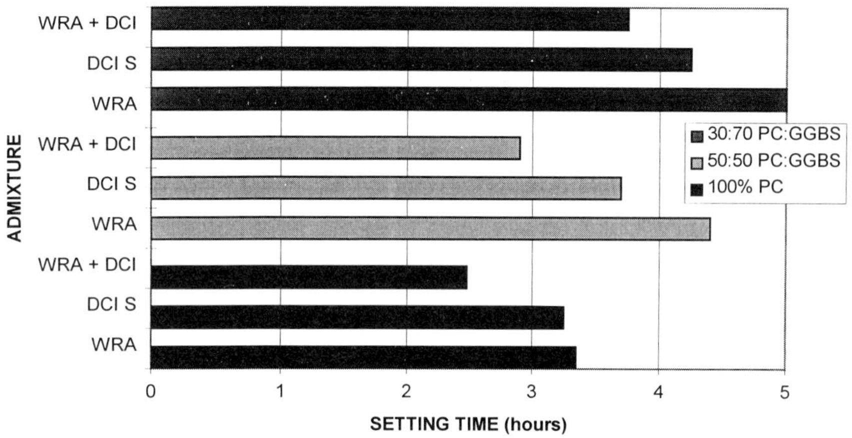

Figure 1 Setting times of 0.45 water to cementitious ratio concretes

The accelerating effect of DCI Corrosion Inhibitor can be used to advantage, particularly in colder curing temperatures. This effect was examined along with the effect at higher temperature curing conditions in one type of concrete; 30:70 PC:GGBS at a free water to cementitious ratio of 0.40, the results of which are shown in Table 3 and in Figures 2 and 3;

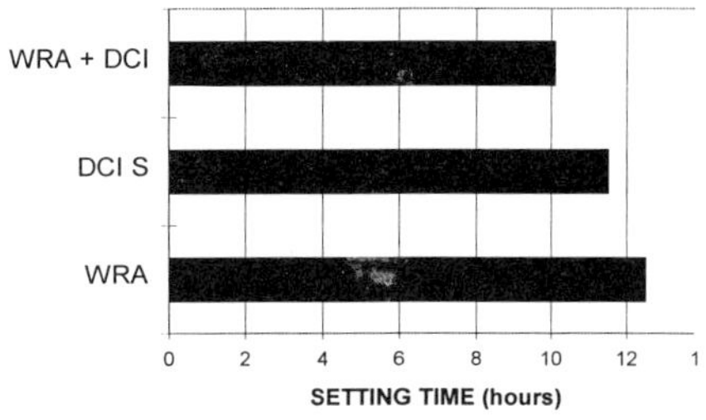

Figure 2 Setting times at 5 °C

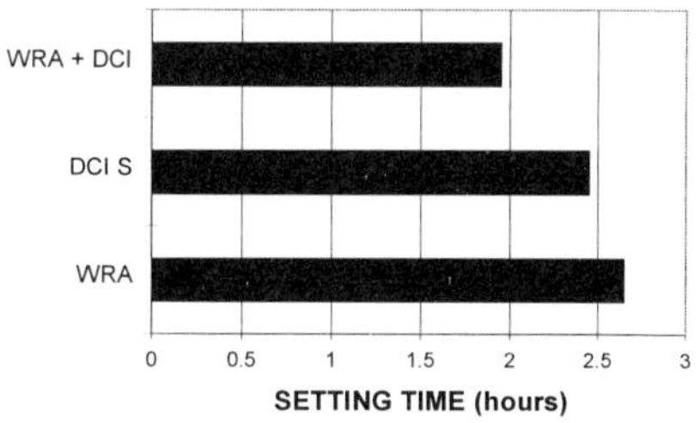

Figure 3 Setting times at 40 °C

It can be concluded, therefore, that DCI Corrosion Inhibitor accelerates the setting times of concretes when cured at 20 °C. irrespective of cementitious type.

Tests on the Concretes in the Hardened State

The following results were recorded on concretes with designed free water to cementitious ratios of 0.45 and 0.40 when tested for compressive strength at 1, 3, 7, 28 and 56 days and when subjected to the rapid chloride permeability test;

Table 4 Concretes with a free water to cementitious ratio of 0.45

CEMENT TYPE	ADMIXTURE	COMPRESSIVE STRENGTH, N/mm2					6 HR. COULOMB VALUE	
		1 day	3 days	7 days	28 days	56 days	28 days	90 days
100% PC	WRA	17.7	36.9	45.2	56.0	62.4	2645	1987
	WRA + DCI	23.2	46.5	58.3	71.2	82.0	3575	3132
	DCI S	22.2	47.5	62.2	76.1	81.8	3023	3089
50:50 PC:GGBS	WRA	9.1	23.7	43.9	66.2	75.9	909	605
	WRA + DCI	9.8	26.7	48.0	68.6	77.4	959	691
	DCI S	9.2	26.7	47.6	64.9	75.6	965	691
30:70 PC:GGBS	WRA	5.3	18.3	36.4	56.3	62.4	648	454
	WRA + DCI	4.5	20.8	38.9	59.1	66.1	494	410
	DCI S	5.8	22.7	40.6	59.1	67.4	513	389

Table 5 Concretes with a free water to cementitious ratio of 0.40

CEMENT TYPE	ADMIXTURE	COMPRESSIVE STRENGTH, N/mm2					6 HR. COULOMB VALUE	
		1 day	3 days	7 days	28 days	56 days	28 days	90 days
100% PC	WRA	25.4	44.4	56.1	68.2	71.0	-	-
	WRA + DCI	27.5	50.4	65.0	74.2	84.2	-	-
	DCI S	29.9	52.3	66.0	80.5	86.4	-	-
50:50 PC:GGBS	WRA	11.0	28.9	48.9	69.8	81.0	888	670
	WRA + DCI	13.8	33.3	52.3	73.4	75.7	958	691
	DCI S	12.1	33.9	55.3	75.8	79.8	940	713
30:70 PC:GGBS	WRA	5.9	23.3	44.4	65.7	75.4	644	389
	WRA + DCI	7.7	29.8	46.9	65.9	73.0	562	432
	DCI S	8.5	29.6	48.9	66.5	72.5	489	454

Effect of the Corrosion Inhibitors on Compressive Strength

The effect on the strength of the concretes when measured at 1 day is shown in Figure 4. It appears that the 1 day strength of concretes based entirely on 100% PC can be significantly enhanced by the addition of either corrosion inhibitor, DCI or DCI S, by 8 - 31%.

Although less significant, both corrosion inhibitors also enhance the 1 day strength of concretes based on the two blended cements; 50:50 PC:GGBS and 30:70 PC:GGBS, particularly in the higher grade, 0.40 water to cementitious ratio concretes.

It can be concluded, therefore, that the two corrosion inhibitors, DCI and DCI S, enhance the early strength development of concretes, particularly those based on 100% PC.

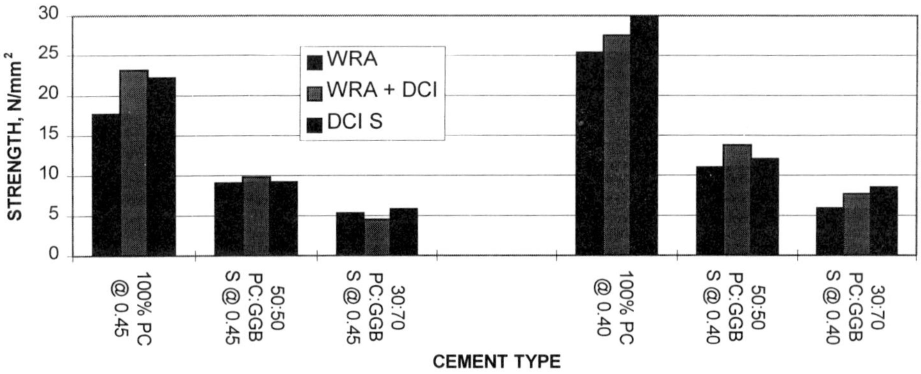

Figure 4 1 day strengths of all concretes cured at 20 °C

With respect to later strength development, even at a constant free water to cementitious ratio the addition of both DCI and DCI S appears to continue to enhance the 3, 7, 28 and 56 day strengths of concretes irrespective of cementitious type although the strengths of the concretes based on 100% PC are again significantly enhanced.

Preliminary conclusions from another, unpublished work, indicate that the strength increase appears to result from a combination of increased cement hydration and other non-thermal factors (i.e. possibly microstructural). The 28 day strengths of all of the concretes are summarised in Figure 5.

It can be concluded, therefore, that the addition of either corrosion inhibitor, DCI or DCI S, will have no detrimental effect on the compressive development of concretes irrespective of cementitious type or content but will particularly enhance those based on 100% PC.

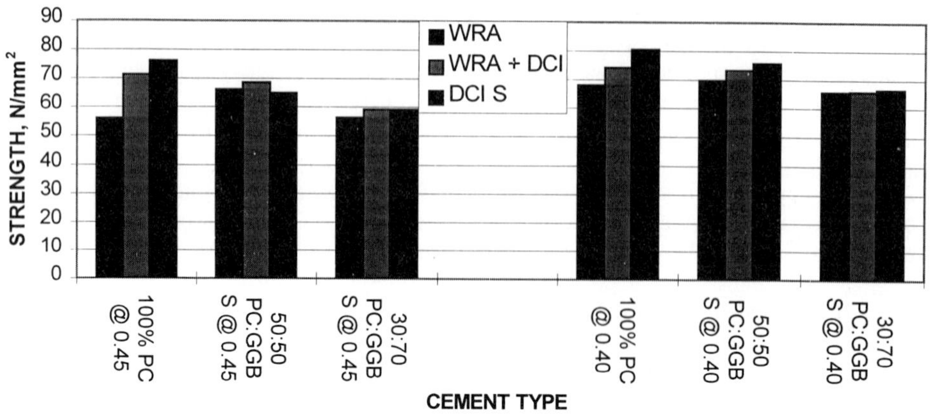

Figure 5 28 day strengths of all concretes

Effect of Cementitious Type and the Corrosion Inhibitors on Chloride Ion Permeability

The AASHTO T 277 test method involves measuring a quantity of electrical charge, i.e. coulombs, that are transported though a concrete specimen during a 6 hour period under a potential difference across the concrete specimen of 60 volts DC. This provides an indication of the resistance of the concrete to the chloride ion and has proved a useful screening test for comparing mix designs.

The 6 hour Coulomb values recorded in Tables 4 and 5 show three trends. Firstly, as the level of PC replacement with GGBS increases from 0% to 50% and then 70%, the coulomb values reduce thus indicating a reduced permeability to chloride ions, particularly when going from 100% PC to 50:50 PC:GGBS. This effect in reality is due to the chloride ion 'binding effect' and reduced permeability that cement pastes exhibit when GGBS is used as a cement replacement.

The 6 hour Coulomb values at both 28 and 90 days for all the mixes containing GGBS were below the maximum 1,000 Coulombs often specified for severe environments such as the Middle East while the concretes tested at both ages containing PC produced values significantly above that value.

The Coulomb values of all concretes measured also reduces, as does the permeability to chloride ions, as the concretes mature from 28 days to 90 days. This effect can be explained by continued hydration of cementitious material thus reducing porosity.

Finally, in the case of the concretes based on 100% PC at a free water to cement ratio of 0.45, the 6 hour Coulomb values significantly increase as a result of the addition of both corrosion inhibitors, DCI and DCI S. This might erroneously indicate that the addition of such inhibitors would increase the susceptibility of this type of concrete to chloride ion ingress.

An understanding of the test method can, however, offer an explanation of such behaviour. The main vehicle for such charge transfer is the negatively charged chloride ions which are provided on one side of the concrete specimen in the form of a 1M solution of sodium chloride. If, however, the concrete itself contains negatively charged ions in the form of nitrite ions, then these too can carry charge across the specimen thus increasing coulomb values. This test method is, therefore, not suitable for measuring the permeability to chloride ions of concretes containing calcium nitrite based corrosion inhibitors. It is suitable, however, for measuring the permeability to chloride ions of concretes made with different cement types and contents.

If a nitrite based inhibitor is to be used in the concrete, the AASHTO T277 test should be undertaken on samples both with and without corrosion inhibitor present to understand the impact on Coulomb values. It should also be reiterated that although higher values may be measured, long-term data on the use of calcium nitrite indicates no increase in chloride ingress when it is present.

It can be concluded, therefore, that concretes containing GGBS as a cement replacement material significantly reduce the permeability of concretes to chloride ion ingress.

CONCLUSIONS

1. Calcium nitrite-based corrosion inhibitors have been shown to be perfectly compatible with Ground granulated blastfurnace slag (GGBS) cement concretes and these two corrosion protection systems have now been shown to complement each other in various ways.

2. The addition of 15 litres/m^3 of a 30% weight by weight calcium nitrite based corrosion inhibitor containing no other constituents (DCI) does not affect the water demand of concretes based on either 100% PC, 50:50 PC:GGBS or 30:70 PC:GGBS assuming, however, that the water component in the corrosion inhibitor is included in the calculation of the free water demand of the concrete.

3. The addition of 15 litres/m^3 of a modified 30% weight by weight calcium nitrite based corrosion inhibitor (DCI S) does reduce the water demand of these concretes by an amount equivalent to the addition of a normal, lignosulphonate based water reducing admixture when dosed at 0.18% v/w of cement.

4. The addition of either corrosion inhibitor does not affect the plastic density of concrete and as such, does not cause any air to be entrained into the concrete.

5. Setting time comparisons have shown that concretes containing GGBS and the modified calcium nitrite based corrosion inhibitor (DCI S) will set similarly to 100% PC concretes containing no inhibitor whereas the addition of the calcium nitrite based corrosion inhibitor (DCI) accelerates the setting times of concretes irrespective of curing temperature, cementitious type or content.

6. The addition of both corrosion inhibitors, DCI and DCI S, enhances the early compressive strength development of all concretes irrespective of cementitious type or cementitious content but particularly those based on 100% PC.

7. The addition of both corrosion inhibitors, DCI and DCI S, has no detrimental effect on long term compressive strength development irrespective of cementitious type or cementitious content.

8. The replacement of PC with GGBS significantly reduces the permeability of blended cement concretes to chloride ions. Concretes containing GGBS at the 50% and 70% replacement levels produced 7 day Coulomb values below the maximum 1,000 normally specified for severe conditions. This reduced permeability significantly delays the rate of chloride ion build-up at the rebar.

9. Calcium nitrite-based corrosion inhibitors have been shown to be fully compatible with GGBS and a combination of these two corrosion protection systems is a viable means to achieving long-term durability.

A PRELIMINARY INVESTIGATION ON THE POSSIBILITIES OF PRODUCING SELF CURE CONCRETE USING HIGHLY ABSORBANT POLYMERS

S Kolias

T P Tassios **M Katsakou**

National Technical University of Athens

E Xatzi

Patzetakis SA

Greece

ABSTRACT. This report contains the first results of a study aiming at examining the feasibility of producing self-cure concrete using hydrophilic polymers with high water absoptivity. The effectiveness of a number of polymers is investigated by carrying out evaporation measurements, compressive strength tests and air permeability measurements on mortar specimens. Ordinary Portland cement is used for the main investigation but the effect of pozzolanic cements is also studied. The results are promising and the fields of further investigation are indicated.

Keywords: Water retainers, Curing, Durability, Absorbent polymers, Evaporation.

S Kolias is Associate Professor and director of the Pavements Laboratory at the National Technical University of Athens. He teaches pavement design and pavement materials. His main research activities are cemented materials, concrete, bituminous material, recycling and pavement design.

T P Tassios is Professor at the National Technical University of Athens. Research activities: Concrete technology, force-transfer constitutive laws in R.C. elements, seismic behaviour of concrete structures, redesign for repair and strengthening.

M Katsakou is a Ph.D student and research assistant at the Pavements Laboratory of NTUA. Main research activities include concrete and cement bound materials

Dr E Xatzi obtained her Diploma and Ph. D. from the Dept. of Chemical Emgineering of the National Technical University of Athens. Currently, she is working for Petzetakis S.A., a large company of Plastic Processing.

INTRODUCTION

It is generally recognized that curing is a very important parameter for concrete durability and strength development. However, it is also generally recognized [1, 2] that "adequate curing", specified by codes of practice and specifications, is in practice rarely achieved to the desirable degree. The conventional curing methods consist of either introduction of additional water (ponding, spraying or using suitable coverings kept wet) or prevention of evaporation of water from the concrete surface (plastic sheeting, curing compounds). The practical difficulties in ensuring that these methods are satisfactorily applied, are coupled by the fact that in the majority of the cases, it is not possible to apply the same method to all concrete surfaces of the same structure (vertical faces vs free horizontal surfaces). Therefore, different properties in the outer concrete layers may be developed between various concrete members of the same structure. Selfdesiccation and the resulting shrinkage is another factor which the conventional methods of curing cannot control adequately.

It would therefore be very interesting to produce a curing system whereby a chemical agent is introduced in the concrete during mixing so as to prevent it from drying (i.e. render it self-cure). This would be appealing not only in terms of concrete quality but also in terms of cost as it would enhance productivity and reduce the cost of applying conventional curing.

Dhir et al [3,4] investigated the feasibility of formulating self-cure concrete by adding water-soluble polymers during mixing that reduce water evaporation in the set concrete. They found a definite improvement not only in water retention, but also in the hydration degree, the concrete surface quality, as judged by ISAT measurements, as well as in the compressive strength. One of the polymers used, enhanced hydration further than simply by means of water retention and this was attributed to the modification of the hydration products. In another investigation Dhir et al [5] attributed the water retention improvement to three mechanisms: lowering of vapour pressure, reduction of bleeding and reduction of permeability.

The aim of this investigation is to examine the feasibility of adding during mixing hydrophilic polymers with high water absorbtivity. After the concrete is set, these polymers give back the water absorbed little by little and may thus provide the necessary water for curing.

RESEARCH PROGRAMME

The basic requirements for the hydrophilic polymers to be used as concrete admixtures were first defined as follows:

1. they should not impair the development of concrete strength
2. they should increase water retention
3. they should improve physical properties of concrete and especially permeability.

Since there was no previous knowledge in this field, a number of hydrophilic polymers were examined in order to find those which seemed to be the most promising for concrete admixture use. The basic polymers were 4 and are designated A, B, C and D.

They were in form of small grains (0 5-2mm) and were added to the mix during the mixing process in quantities calculated as percentage of the solid materials mass (aggregates and cement). Polymer D could also be premixed with another polymer (polyethylenoglycol, PEG) in fluid state in order to control the rate of water given back. A number of polymer D and PEG combinations were also examined and thus the total number of the chemicals examined were 28. The grains of polymer A were larger than those of the other polymers and it was felt that this could contribute to inhomogeneity of the mixes. After the initial stage of the programme polymer A was not used in form of grains but in form of gel with water 98. 5% by mass.

Henceforth, for reasons of brevity the polymers will be designated by the letter A, B, C or D followed by a parenthesis in which a number will indicate the polymer percentage by mass of the solid mix constituents. In the case of polymer D the parenthesis will include a second number indicating the percentage of PEG. Ordinary Portland cement CEN 1/42 5R was used throughout the investigation but in the final step (Step 5) pozzolanic cement CEN IVa/32.5R was used as well.

Since the total number of chemicals to be examined was quite large, it was decided to follow a step by step elimination procedure by performing evaporation tests on cement pastes or mortars in order to select the most promising ones. This step by step procedure is summarised below, giving the most important results. The procedure consisted of the following steps:

Step 1: Evaporation Tests on Solutions

Evaporation tests were carried out on solutions prepared with a fixed quantity of polymer placed in plastic containers in which water was added until the polymer was fully saturated. The plastic containers were then put into two different environments (20°C, RH 60% or ventilated oven with 30°C). These tests were carried out only in order to obtain some understanding of the function of the polymers A, B, C and D since it is well known that their absorption and desorption characteristics change drastically according to the chemical environment.

Step 2: Evaporation Measurements from Cement Pastes

Evaporation measurements were carried out from cement pastes prepared with various proportions of polymers in order to select those that gave better results in terms of water retention. A single W/C ratio was selected for all mixes so that the same cement and initial water quantity would be present in all of them. This requirement, however, resulted in different workabilities in the mixes due to the different absorption in the mixes. This was in turn attributed to the various types and percentages of the polymers. Moreover, the W/C ratio selected should ensure that the control mix (the mix without any polymer) should exhibit the lowest possible bleeding in order to avoid differences which could be attributed to excess of bleed water. The evaporation measurements were carried out in two different environments 25°C, 60% RH and 25°C, 45% RH.

As an example of the data collected in this step, Figure 1 shows the weight loss caused by evaporation for polymer A with the indicated mix proportions. Weight loss is defined as the difference between the initial weight of the specimen at the time of demoulding from the specimen weight at any time, divided by the initial weight.

This step enabled the elimination of a number of polymers (such as B) and combinations of polymer D with PEG because of their limited water retention ability.

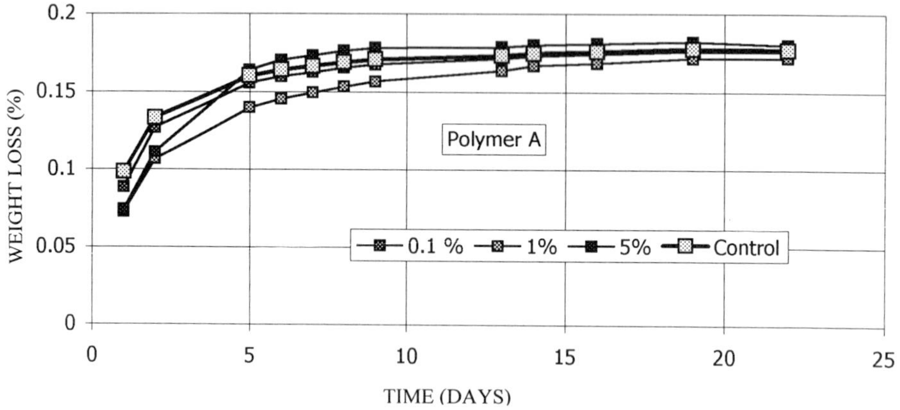

Figure 1 Weight loss from pastes with various contents of polymer A, air at 20°C/60% RH

Step 3: Evaporation and Strength Measurements on Mortars

In this step 70mm mortar cubes were prepared from a mix consisting of 323 lit/m^3 water, 380 kg/m cement (CEN I/42 5R) and crushed limestone sand (0-4 76mm) 1440 kg/m^3 W/C=0 85. Weight loss of mortar cubes with various percentages of polymer A. The following day the specimens were demoulded and kept in an environment of 20°C and 60% RH; their weight loss was measured every day until testing at 7 days. All evaporation and strength data reported are mean values from three specimens.

Some of the evaporation measurements are shown in Figure 2 and Figure 3 for polymers A and C respectively.

The histograms of Figure 4 and Figure 5 compare the 7-day compressive strength of all the mixes tested in this step with that of the control specimens, i.e. the specimens without polymer, cured under the same conditions.

As anticipated, considerable strength reductions are observed when a high polymer content or a high PEG content is used. However, a significant strength increase (21%) is observed in the case of polymer A (0.05%). In the other cases such as C (0.2) C(0 5) and D (0.5,0) a slight but insignificant strength increase or decrease [D(0.1,0), D(0.1,0.1), D(0.1,0.5), D(0.5,0.5), A(0.15)] is observed.

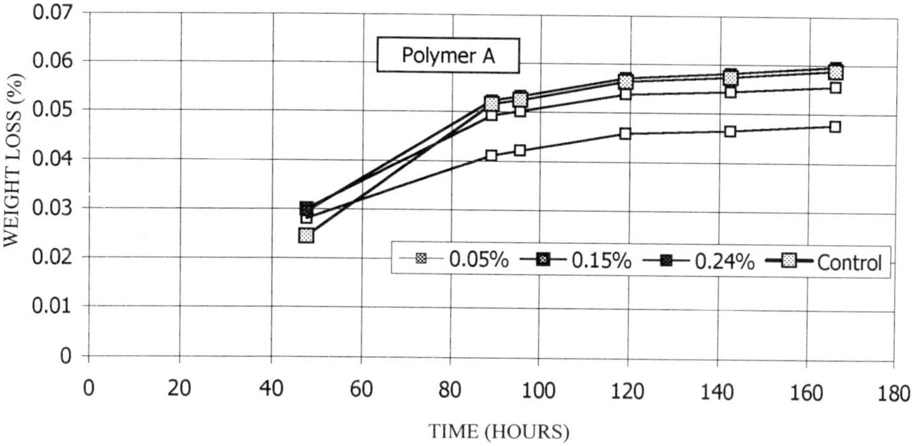

Figure 2 Weight loss of mortar cubes with various percentages of polymer A

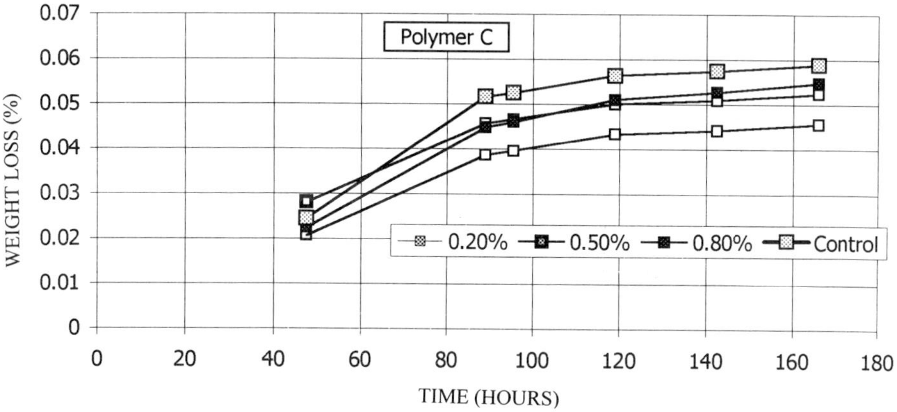

Figure 3 Weight loss of mortart cubes with various percentages of polymer C

With the results of this stage in mind, it was decided to examine in an additional step the behaviour of the following polymers:

Polymer A: A(0.02), A(0 05), and A(O. 1)

Polymer C: C(0.02), C(0 05), and C(O. 1)

Polymer D with the combinations with PEG shown in Table 1

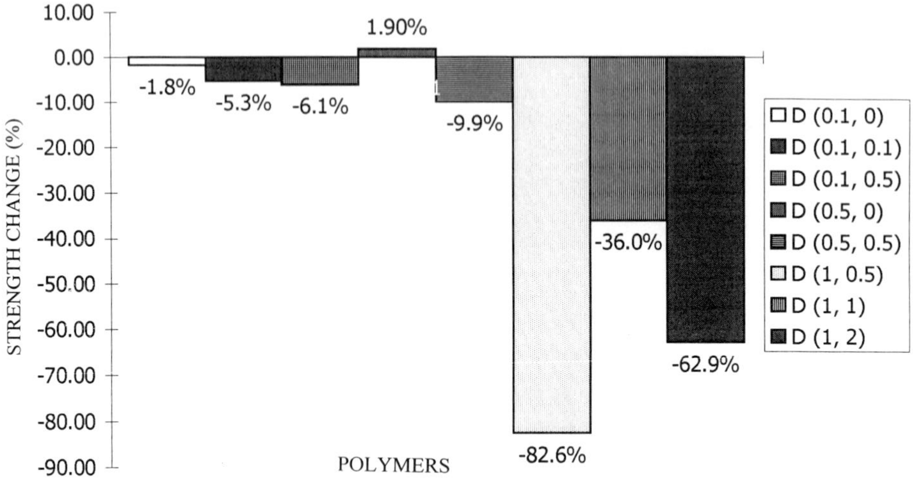

Figure 4 7 day strength changes expressed as percentage of the control specimen strength, mortars made with polymer D

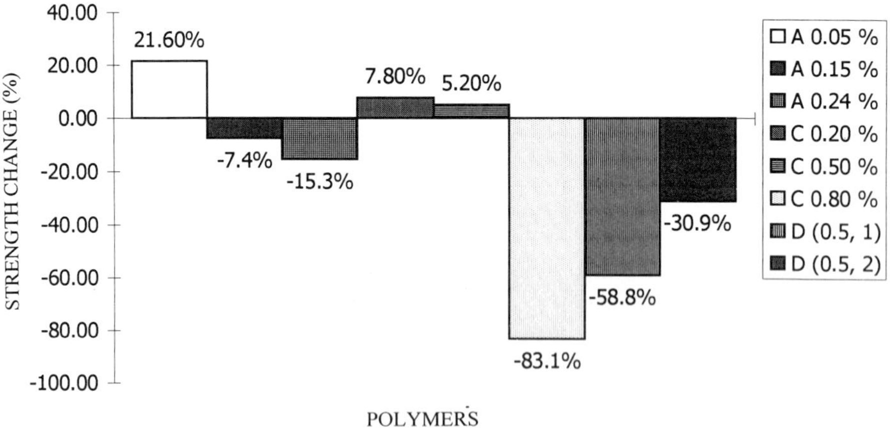

Figure 5 7 day strength changes expressed as percentage of the control specimen strength, mortars made with polymers A, C and D

Table 1 Combinations of PEG with polymer D

%OF POLYMER D	% OF PEG
0.05	0
0.05	0.05
0.05	1
0.1	0
0.1	0.2
0.3	0.5
0.3	1
0.5	1

Step 4: Evaporation, Strength and Air Permeability Measurements

This step included evaporation measurements, 70mm cube compressive strengths tests and air permeability measurements on 100 mm cubes according to Figg test [6]. Six measurements of air permeability were carried out on each cube.

Figure 6 and Figure 7 show that polymers D(0 05,0 05), D(0 05,1), D(0 3,0 5) and D(0 3,1) had little water retention capabilities while D(0 05,0) and D(0 5,1) were satisfactory. Although it is not shown here, all dosages of polymers A and C examined in this step showed good water retention.

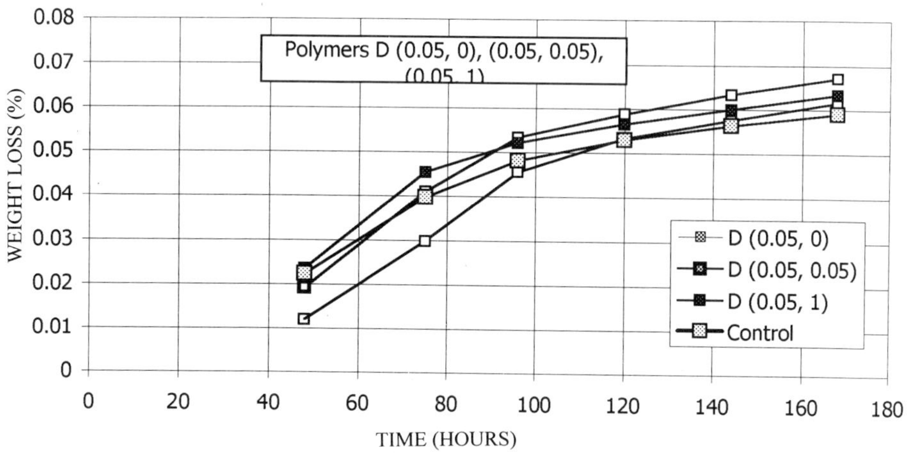

Figure 6 Weight loss of mortar cubes prepared with polymer D (0.05, 0), 0.05, 0.05) (0.05, 1)

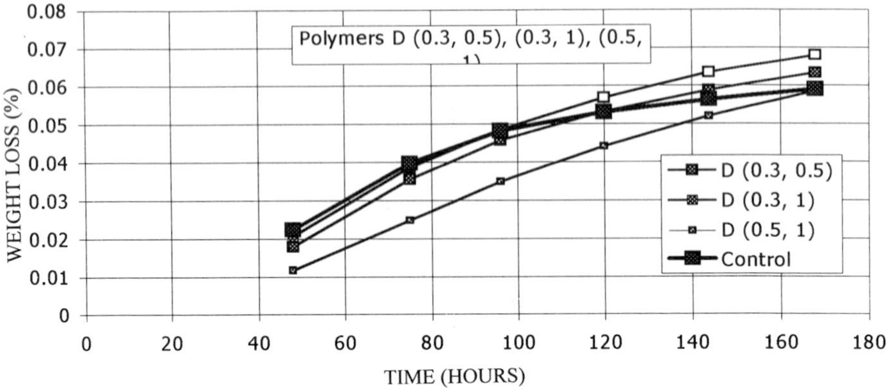

Figure 7 Weight loss of mortar cubes prepared with polymer D (0.3, 0.5), (0.3, 1), (0.5, 1)

Figures 8, 9 and 10 show the percentage differences of the 7-day strengths from those of the control specimens. It can be seen that considerable strength increases are obtained with certain polymers (such as A(0 02), D(0.1,0)7 D(0 1,0 2), C(0 02), C(O 1), C(0.05)).

However with other polymers strength decreases (D(0.5,1), D(0.3,1)), or marginal insignificant changes (D(0.05,0)7 D(0.057O.05)7 D(0.0571)7 D(0.37O.5)) were obtained.

The comparison would have been more complete if strength data from control specimens with standard curing (water or fog room curing) had been available.

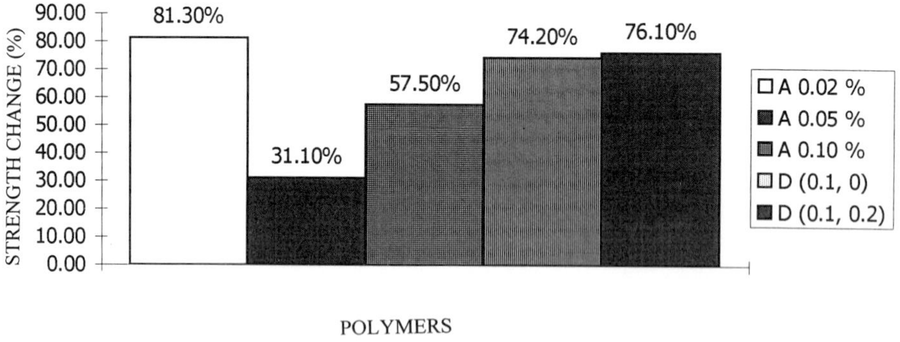

Figure 8 7 day strength changes expressed as percentage of the control specimen strength, mortars made with polymers A and D

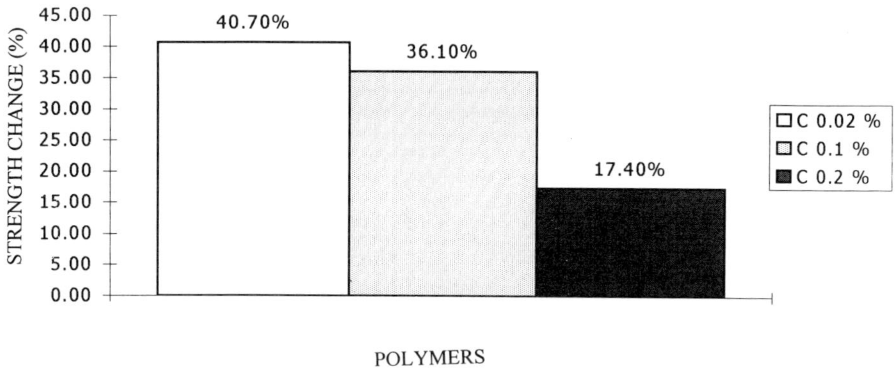

Figure 9 7 day strength changes expressed as percentage of the control specimen strength, mortars made with polymer C

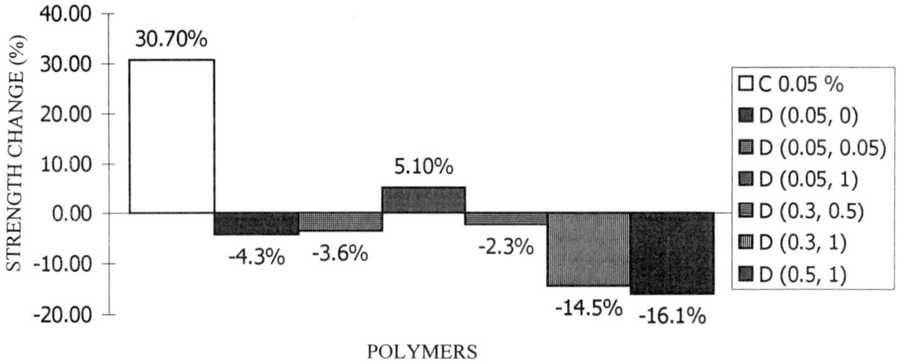

Figure 10 7 day strength changes expressed as percentage of the control specimen strength, mortars made with polymers C and D

Unfortunately due to the large number of polymers to be compared and the large number of specimens needed, it was practically impossible to include in this step standard curing as an additional parameter. However, the magnitude of the strength increase is such that it is safe to infer that the strength of those mixes would not be lower or in certain cases would exceed the strength of the control specimens with standard curing.

The results of the air permeability measurements are shown in Figure 11, 12 and 13 in terms of the time needed for the air pressure in the drilled holes (diameter lOmm, depth 40mm) to rise from -55 kPa to -50 kPa. It can be seen that 2 polymers showed lower [D(0 5,0), D(0 05,0 05)] and 2 equal [C(O 1), C(0 2)] permeability times, compared to those of the control specimen. All the other polymers showed considerable higher times.

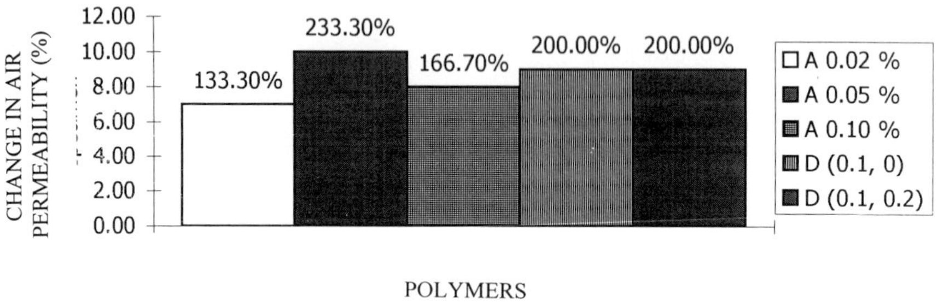

Figure 11 Change in air permeability of mortars as a result of adding polymers A and D

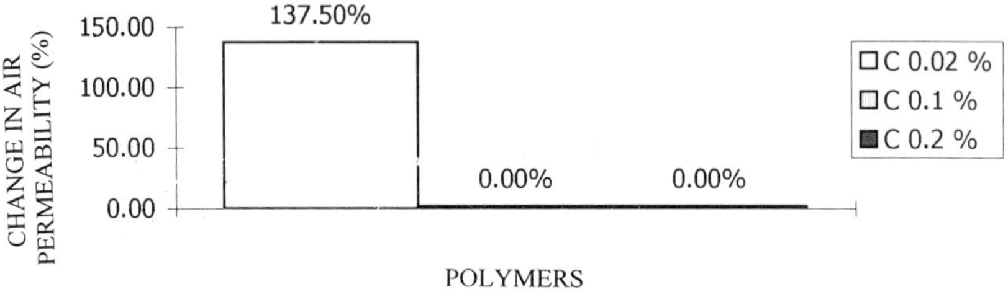

Figure 12 Change in air permeability of mortars as a result of adding polymer C

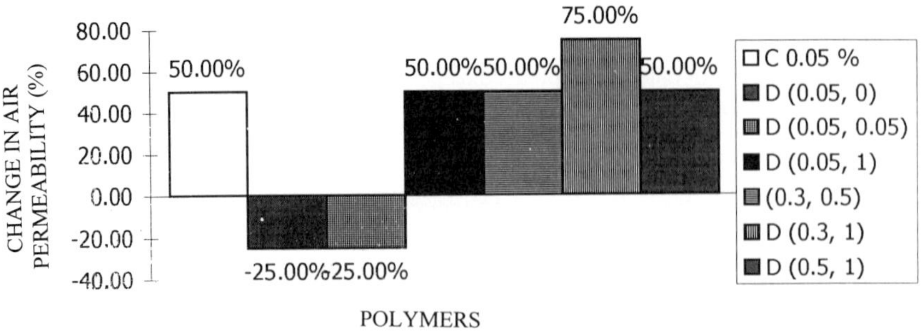

Figure 13 Change in air permeability of mortars as a result of adding polymers C and D

The permeability of mixes with polymers C(0.1) and C(0 2) is not consistent with the strength increase obtained with these polymers (Figure 9). However, due to the well known variability of the permeability measurements, these results should be considered as indicative only, since they are based on 6 measurements on one only cube specimen. More tests are needed in order to reach definite conclusions.

The conclusion to be drawn from this step is that polymers A(0 02), C(0 02) and D(O 1,0) are the most promising from those initially examined.

Step 5: Final Evaluation - Influence of the Cement Type

This step included evaporation, 70mm cube strength and air permeability measurements on mortars made with ordinary portland cement CEN I/42 5R and portland pozzolanic cement (with pozzolana or fly ash up to 20%) CEN IVa/32.5R. The relatively low polymer content allowed a more realistic W/C ratio equal to 0.76.

The evaporation measurements were extended up to 28 days and the strength tests as well as the air permeability measurements included both 7-day and 28-day results.

The results are given in Figure 14 and 15 in terms of water retention ability.

Figures 16, 17, 18 and 19 show the results in terms of 70mm cube strength at an age of 7 and 28 days

Figures 20, 21, 22, and 23 shows the results in terms of air permeability.

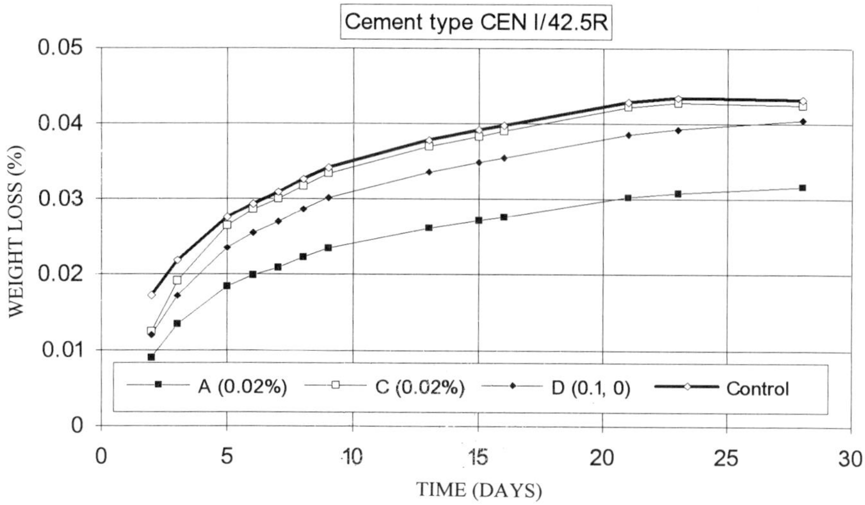

Figure 14 Weight loss from mortar with polymers A,C and D

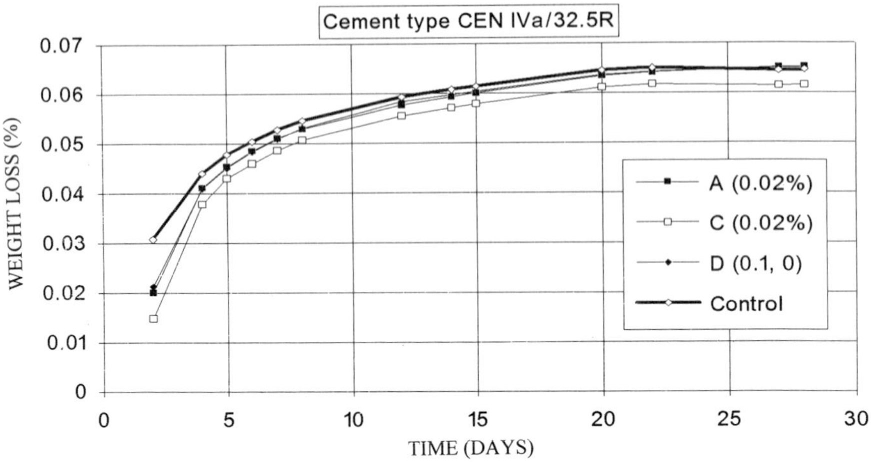

Figure 15 Weight loss from mortar with polymers A, C and D

It can be seen that the type and strength of cement is reflected in the evaporation tests since there is a significant difference in water loss between the control specimens prepared with cement I/42 5R and those prepared with cement IVa/32.5R. It can also be seen that the water retention ability of the polymers changes according to the type of cement. The most effective polymer in mortars with cement I/42 5R is A(0 02) and the least effective is C(0 02), while the ranking is reversed in case of mortars prepared with cement IVa/32.5R.

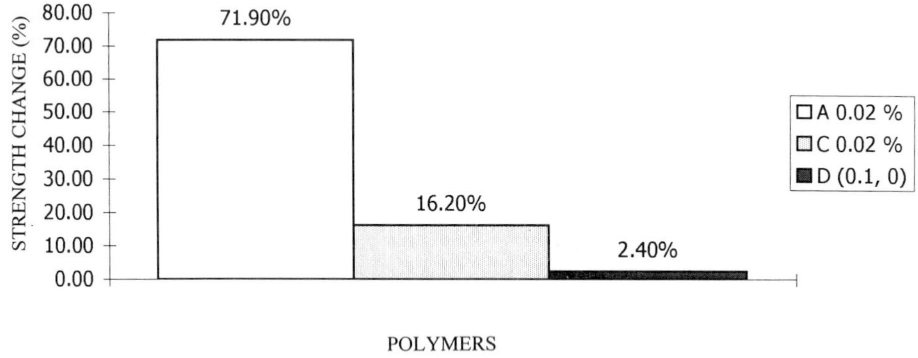

Figure 16 Strength changes expressed as percentage of the control specimen strength

Figures 16, 17, 18 and 19 show that, in general, the use of the polymers A, C and D at the indicated percentages resulted in higher strengths than those of the control specimens. These strength increases range from highly significant [A(0 02) for 7 and 28 days and cement 1/42 5R], to marginal [D(O 1,0) for 7-day mortars with cement I 42 5R and IVa/32.5R].

The influence of the cement type and strength is also apparent since smaller increases are obtained in general with cement IVa/32.5R, especially at 7 days. It should also be pointed out however, that the strength results of this step are not quite consistent with those of the previous step (4). A comparison between Figures 8 and 9 to Figure 16 reveals that while the strength increase for polymer A(0 02) is practically the same (81% compared to 72%), the strength increase for polymers C(0 02) and D(O. 1,0) show considerable difference (40 7%,and 74.2% compared to 16 2%and 2.4% for C and D respectively). No apparent reason for this discrepancy can be offered and clearly more tests are needed to clarify this subject.

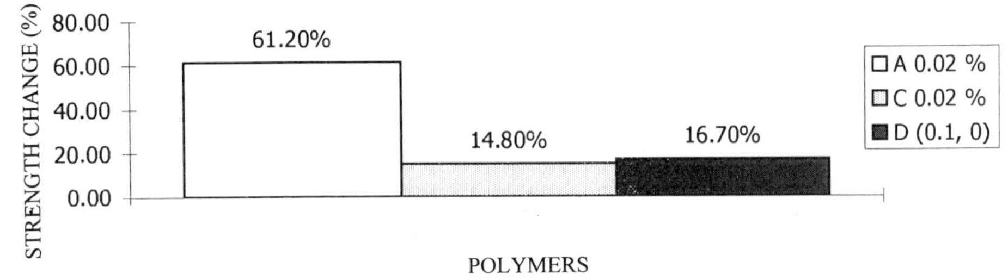

Figure 17 28 day dtrength change expressed as a percentage of the control specimen strength

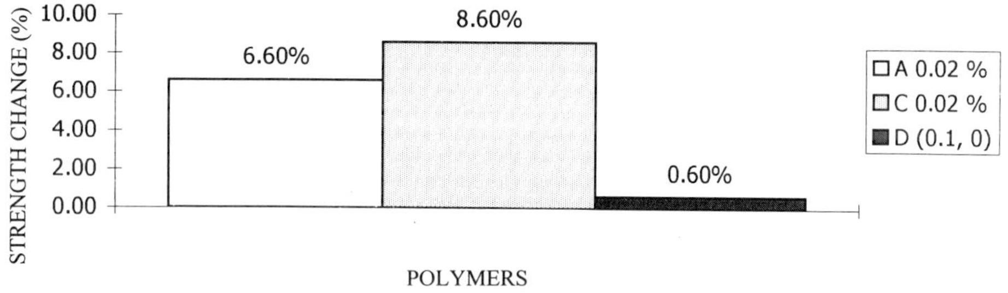

Figure 18 7 day strength changes expressed as a percentage of the control specimen strength

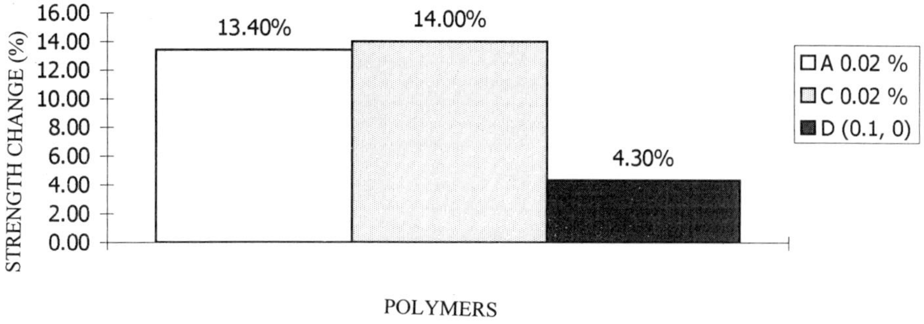

Figure 19 Strength changes expressed as a percentage of the control specimen strength

The air permeability results (Figures 20 to 23) show that all polymers considerably reduced air permeability in the case of mortars with cement CEN 1/42 5R especially at 7 days, while in the case of cement type CEN IVa/32.5R the reduction of the air permeability is smaller whereas that of polymer D(O. 1,0) is insignificant. Bearing in mind the variability of permeability tests, the results are considered similar to those of the previous step. In general all the tests showed longer permeability times in specimens at 7 days than in specimens at 28 days. This could be attributed to the fact that at 7 days the moisture content of the specimens was higher than that of the specimens at 28 days due to drying. The tests were carried out without preconditioning the specimens by oven drying. Therefore, it was not possible to estimate the hydration degree and the moisture content contribution to the improved permeability characteristics. This is clearly a matter that needs further investigation.

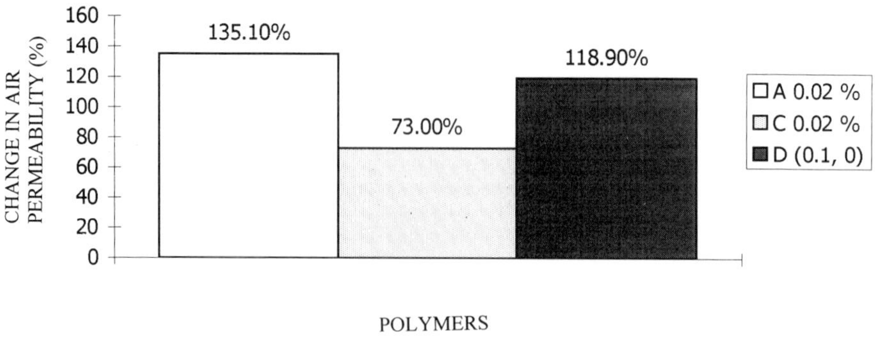

Figure 20 Change in air permeability of mortars as a result of adding polymers A, C and D, 7 days, cement CEN I/42.5R

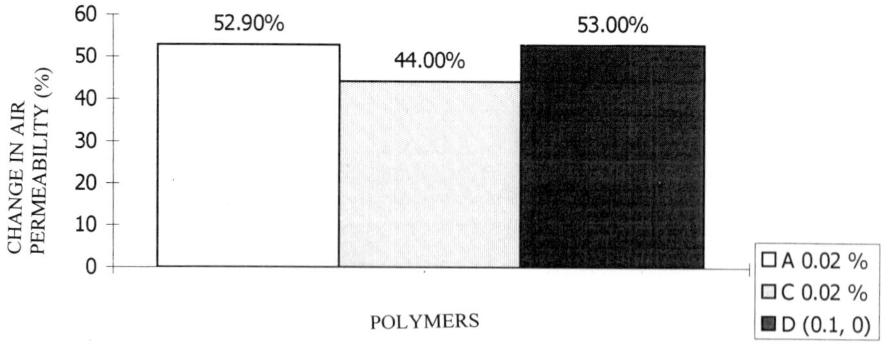

Figure 21 Change in air permeability of mortars as a result of adding polymers A, C and D 7 days, cement CEN I/42.5R

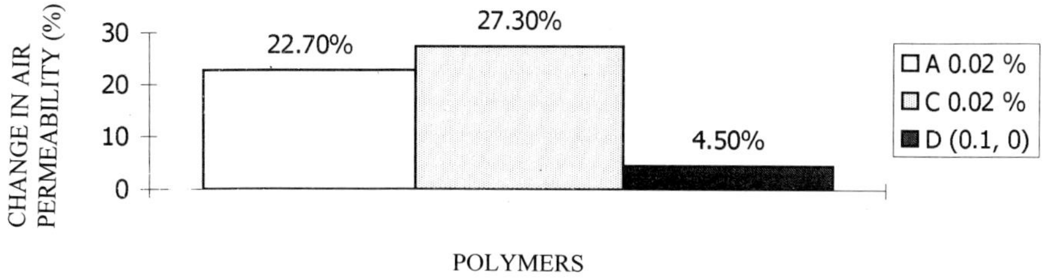

Figure 22 Change in air permeability of mortars as a result of adding polymers A, C and D, age 7 days, cement CEN IVa/32.5R

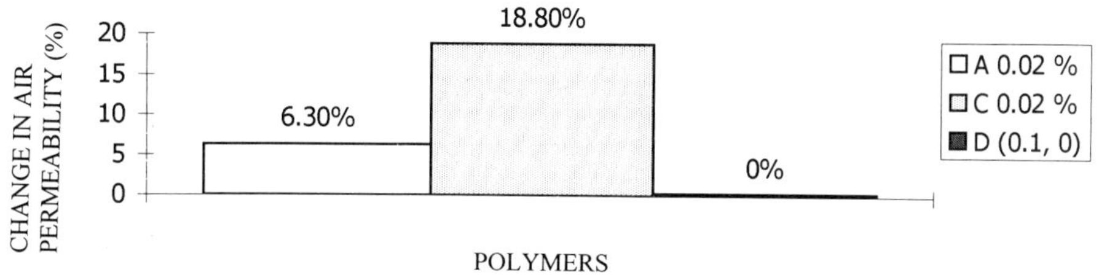

Figure 23 Change in air permeability of mortars as a result of adding polymers A, C and D, age 28 days, cement CEN IVa/32.5R

It should also be pointed out that the life of this type of polymers inside concrete is not known. This should be another subject for investigation since these polymers remaining inside the concrete pores, will definitely alter the absorption and desorption characteristics of the material.

CONCLUSIONS

The main conclusions from this preliminary research is that certain types of hydrophilic polymers in certain quantities increase water retention, do not impair but even increase strength and may also decrease air permeability. The cement type seems to influence the polymer effectiveness.

The results are promising but considerable research is needed before these chemicals can be used as concrete admixtures for producing self-cure concrete

ACKNOWLEDGEMENTS

The authors would like to express their gratitude to Prof. Andreopoulos and Prof. Bordado not only for providing the necessary polymers but also for their valuable advice during the entire investigation. The work was financed by a research programme of the Ministry of Development, Industry and Trade, General Secretariat for Research and Technology. The authors would like to express their appreciation for this financial support.

REFERENCES

1. CATHER,R., "Curing: the true story?" Guest editorial comment Magazine of Concrete Research, 1994, Vol 46, No 168, Sept., pp 157-161

2. TASSIOS T.P., KOLIAS S., AND ALIGIZAKI K., "In-situ and in-lab testing of skin concrete", Testing during concrete construction, Proc. of an Int. RILEM Workshop, Mainz, March 1990, pp. 354-366.

3. DHIR,R K, HEWLETT, P.C., LOTA,J.S., DYER,T D, "An investigation into the feasibility of formulating "self-cure" concrete". Materials and Structures, 1994, Vol 27, No 174, Dec., pp 606-615

4. DHIR,R.K., HEWLETT, P.C., DYER,T.D., "Mechanisms of water retention in cement pastes containing a self-curing agent". Magazine of Concrete Research, 1998,Vol 50, Nol, Mar.,pp 85-90.

5. DHIR,R.K., HEWLETT, P.C., DYER,T.D., "Influence of Microstructure on the Physical Properties of self-curing concrete". ACI Materials Journal Vol 93, No5, Sep.-Oct. 1996, pp 465-471

6. CATHER R., FIGG J.W., MARSDEN A.F., AND O'BRIEN T.P., "Improvements to the Figg method for determining the air permeability of concrete", Magazine of Concrete Research, 36, No 129, Dec. 1984, pp. 241-245

THEME SIX:

DESIGNER MIX METHODOLOGY

Keynote Paper

OPTIMISED DESIGN OF CONCRETE AND ITS COMPONENTS

J D Dewar

J D Dewar Consultancy

United Kingdom

ABSTRACT. The paper demonstrates ways in which computerised modelling enables fast assessment of the options for optimising the designs of concrete and its components.

Scope has been shown for the use of ultrafine cementitious materials, finer sands and intermediate sized materials and confirmed the well known benefits of introducing materials of more advantageous shape or particle size distribution. Plasticising admixtures are important means of further reducing water content and may be essential when ultrafine materials are used, in order to offset agglomeration.

There are often artificial restrictions in standards, e.g. percent of material passing 75 μm in sands, which may bear re-examination in order not to hamper design unreasonably.

Keywords: Additions, Admixtures, Aggregates, Cements, Computers, Concrete, Fillers, Mixture design, Optimisation, Water demand.

Dr J D Dewar Principal of JD Dewar Consultancy, is a civil engineer specialising in concrete technology, in particular computerised modelling of concrete and its components. He was for a number of years Director of the British Ready Mixed Concrete Association and chairman of the BSI committee for concrete standardisation.

INTRODUCTION

Materials designed for general-purpose application are unlikely to lead to optimum designs of concrete for every specialised application. Similarly, materials produced for specific application are unlikely to lead to optimum designs for general application. As a result there has been a growing trend for concrete producers to increase the number of components to improve their opportunities for optimisation of their products. This optimisation process can be further aided by the materials producers to the benefit of themselves, their clients and the future prospects generally for concrete. Survivors are those who benefit from changes.

Computerised modelling techniques have reached the stage where concretes and their components can be purpose-designed and adjusted during continuous production to obtain economic benefits for both high performance situations and for normal use.

The paper concentrates mainly on optimisation of water demand of concrete but the method described can operate on other parameters as desired, e.g. strength; economy.

REVIEW OF COMPUTERISED MODELLING TECHNIQUES

The key features of suitable methods are that they utilise readily available input data, produce output data in an acceptable format and have been proven over their working range. Preferably, optimisation systems should be built into the programmes. Progressively, more systems will become available meeting some or all of these criteria and assisting design of cements, additions, admixtures, fillers, aggregates and concretes.

Methods known to the author which have progressed significantly in this direction, and successfully combine theory and practice, include those described in other studies [1, 2, 3, 4, 5, 6, 7]. To this restricted list must be added [8] for the development and publication of very many essential modelling relationships.

This present paper introduces some aspects of the author's recent research [6] but without attempting to provide any of the principles, formulae and inner workings of computer programmes that are published elsewhere or the very detailed experimental evidence on which the method is based. This method has been adopted because, not surprisingly, it is the one with which the author is most familiar.

This method contains two separate techniques: -

1. The updated original system for designing concretes now with up to 3 cementitious materials, 3 fine aggregates and 3 coarse aggregates together with 2 admixtures which may be air entraining or plasticising;

2. A technique for use independently, or in conjunction with the first technique, for more refined optimisation purposes by combining as many as 14 particulate components into a single material

EXAMPLES OF THEORETICAL OPTIMISATION OF MATERIALS AND CONCRETES

Voids ratio is a fundamental property of a particulate material and is calculated as the ratio of voids to solids. In the various examples, values have been assumed for voids ratios of the individual size fractions. They should not be considered to be more than guidance values and would need to be determined by Vicat test or loose bulk density test for each size fraction of the particular material. The examples concentrate on optimisation for minimum water demand of concrete and do not consider any effects on other properties caused by changes to the particle size distribution.

Optimisation of Particle Size Distribution of the Cementitious Material for Minimum Water Demand of Concrete at Different Cement Contents

A process of optimisation developed bythe author [6] is summarised by way of an example commencing in Table 1, taken from a computer spreadsheet programme.

The material is considered in terms of convenient size fractions in columns 2 and 3, for which the voids ratios and relative densities, together with the proportions by mass, are listed in columns 4 to 6. The mean size of each size fraction in column 7 has been calculated on a logarithmic basis and the volumetric proportions are shown in column 8.

Material 1 is combined mathematically with material 2, using the author's Theory of Particle Mixtures, [4, 5, 6]. The resultant voids ratio, 1.35, is recorded in column 10. Similarly the mean size and relative density of this combination (omitted for clarity) are calculated. This combination is then combined with Material 3 and so on until all size fractions have been combined. It will be seen that the voids ratio progressively decreases as each size fraction is added. The final properties of the combination are shown on the right in bold type at the foot of the Table.

Table 1 Properties of individual size fractions and resultant combination for initial cement.

1	2	3	4	5	6	7	8	9	10	11
								COMBINATION		
MATERIAL	MIN SIZE (mm)	MAX SIZE (mm)	VOID RATIO U	REL DENS. RD	% BY MASS	MEAN SIZE D (mm)	% BY VOL	MEAN SIZE (mm)	VOID RATIO U	RD
1	0.0025	0.005	1.8	3.2	15	0.004	15	-	1.80	-
2	0.005	0.01	1.5	3.2	25	0.007	25	-	1.35	-
3	0.01	0.02	1.3	3.2	25	0.014	25	-	1.11	-
4	0.02	0.04	1.2	3.2	35	0.028	35	-	0.91	-
5	0.04	0.08	1.1	3.2	0	0.057	0	-	0.91	-
6	0.08	0.15	1.0	3.2	0	0.110	0	0.012	0.91	3.20

To optimise the proportions in column 6, the resulting properties for the cementitious material are fed into the computerised concrete design programme and the water demand is estimated for selected slump, cement content and a particular set of aggregates. The proportions of the aggregates are automatically optimised for cohesion. The values in column 6 are then allowed to vary systematically and automatically within the computer programme until a minimum concrete water demand is established. The results for one selected cement content are summarised in Table 2, from which it may be seen that, on average, optimisation suggests that a coarser cementitious material is to be preferred at this cement content, with a reduced voids ratio in consequence, compared with the finer cement selected initially.

Additional constraints can be applied in the optimisation process. In this example, the proportion passing 0.005 mm (5 μm) was restricted to a maximum of 15% by mass, it was assumed that no material passed 0.0025 mm (2.5 μm) and the maximum size was restricted to 0.08 mm (80 μm). Thus, material 6 was not used in this particular exercise.

Table 2 Properties of the optimised cement for minimum water demand of concrete at a cement content of 400 kg/m^3

1	2	3	4	5	6	7	8	9	10	11
								COMBINATION		
MATERIAL	MIN SIZE (mm)	MAX SIZE (mm)	VOID RATIO U	REL DENS. RD	% BY MASS	MEAN SIZE D (mm)	% BY VOL	MEAN SIZE (mm)	VOID RATIO U	RD
1	0.0025	0.005	1.8	3.2	14	0.004	14	-	1.80	-
2	0.005	0.01	1.5	3.2	9	0.007	9	-	1.47	-
3	0.01	0.02	1.3	3.2	21	0.014	21	-	1.09	-
4	0.02	0.04	1.2	3.2	40	0.028	40	-	0.84	-
5	0.04	0.08	1.1	3.2	16	0.057	16	-	0.76	-
6	0.08	0.15	1.0	3.2	0	0.110	0	0.018	0.76	3.20

The results of optimising the cementitious material at different cement contents are summarised in Table 3 from which it may be seen that the initial cement has properties closer to those for the optimised materials over the lower part of the cement content range.

At high cement contents a coarser cementitious material is to be preferred with regard to water demand. Of course, other properties, e.g. bleeding, strength, economy may determine whether this is feasible.

Over the wide range of cement contents examined, the optimum proportion of the finest material (5 - 2.5 μm) did not vary significantly from the maximum constraint value of 15%, implying that a higher percentage could be beneficial and that material finer than 2.5 μm might also be beneficial. The average optimised percentage of the coarsest material (40 - 80 μm) was 18% and ranged from 0% at the lowest cement content to 35% for the highest cement content.

The overall range for the particle size distributions for the optimised cements is shown in Figure 1.

Table 3 Optimised cement properties for concrete water demand at various cement contents

CEMENT (kg/m^3)	CEMENT PROPERTIES		
	Mean Size	Void Ratio	Relative Density
(a) Intial Cement			
All	0.012	0.091	3.20
(b) Optimised Cements			
150	0.008	1.19	3.20
200	0.017	0.79	3.20
250	0.017	0.77	3.20
300	0.013	0.88	3.20
350	0.017	0.78	3.20
400	0.018	0.76	3.20
450	0.023	0.72	3.20
500	0.020	0.74	3.20
550	0.023	0.72	3.20

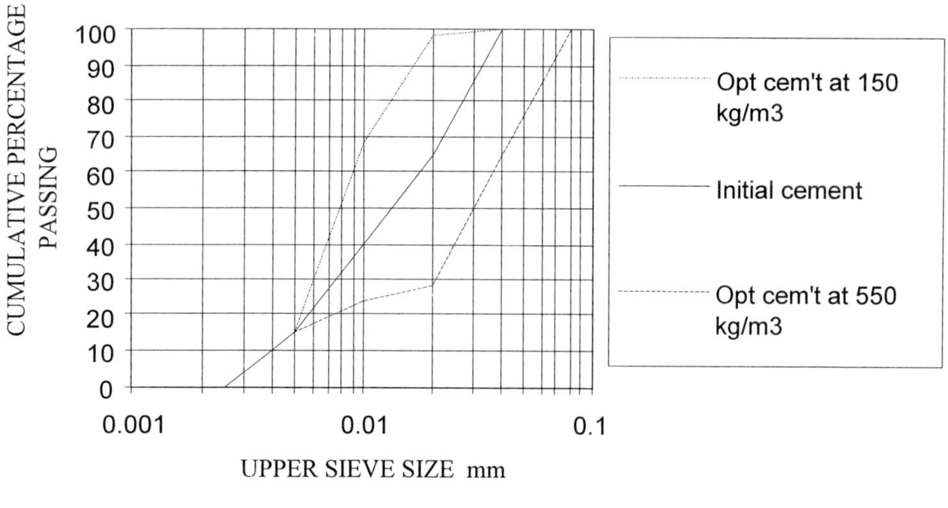

Figure 1 Range of particle size distributions optimsed for minimum water demand of concrete

Examination of Figure 2 indicates that, for this example, the initial cement was effectively optimised for particle size distribution with respect to water demand over the lower part of the working range of cement content but was less well optimised for higher cement contents. There is a potential reduction in water demand of up to 15 l/m^3 to be achieved by adopting a coarser psd for the cementitious material at high cement contents. How such optimisation might be achieved is considered later.

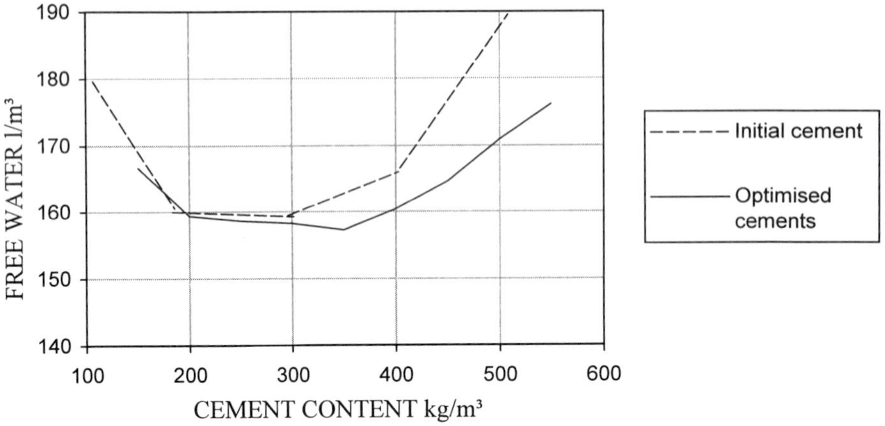

Figure 2 Influence of optimsation of particle size distribution of cement on the water demand of concrete

These findings are consistent with the often-disappointing results of using more finely ground cements, which have resulted in higher water demands of high cement content concretes. The usual reasons are that finer size fractions have greater voidage and also that the benefit of reduced particle interference has been offset by the narrow psd which has resulted in a higher voids ratio and thus increased water demand. Conversely, low cement content concretes benefit more from the use of finer cements because of the reduced particle interference and may, as a consequence enable slightly reduced water demands.

Optimisation of Particle Size Distribution of Fine Aggregate for Minimum Water Demand of Concrete at Different Cement Contents

The optimisation of fine aggregate proceeds as in Table 4, for an initial fine aggregate which lies centrally in the range of permitted gradings in BS 882: 1992. The cement used for this exercise is that adopted initially in the previous exercise.

To optimise the proportions in column 6, the resulting properties for the fine aggregate are fed into the computerised concrete design programme and the water demand is estimated. The values in Col 6 are then allowed to vary systematically and automatically within the computer programme until a minimum concrete water demand is established at the selected cement content. As shown in Table 5 a finer grading is to be preferred, despite an increased voids ratio, compared with that for the fine aggregate selected initially. The maximum size was restricted to 5 mm, the proportion passing 0.08 mm (80 μm) was restricted to a maximum of 4% by mass and it was assumed that no material passed 0.04 mm (40 μm)

The results of optimising the fine aggregate at different cement contents are summarised in Table 6 from which it may be seen that, over the complete range of cement contents from 150-550 kg/m^3, a significantly finer grading was also to be preferred compared with that of the initial fine aggregate, despite the increased voids ratio.).

Table 4 Properties of individual size fractions and of the resultant combination for the initial fine aggregate

1	2	3	4	5	6	7	8	9	10	11
								COMBINATION		
MATERIAL	MIN SIZE (mm)	MAX SIZE (mm)	VOID RATIO U	REL DENS. RD	% BY MASS	MEAN SIZE D (mm)	% BY VOL	MEAN SIZE (mm)	VOID RATIO U	RD
5	0.04	0.08	1.1	2.6	2	0.057	2	-	1.10	-
6	0.08	0.15	1.0	2.6	8	0.110	8	-	0.95	-
7	0.15	0.3	0.9	2.6	15	0.212	15	-	0.79	-
8	0.3	0.6	0.7	2.6	20	0.424	20	-	0.63	-
9	0.6	1.2	0.7	2.6	25	0.849	25	-	0.54	-
10	1.2	2.5	0.7	2.6	20	1.732	20		0.47	
11	2.5	5.0	0.7	2.6	10	3.536	10	0.64	0.44	2.60

Table 5 Properties of individual size fractions and of the resultant combination for the initial fine aggregate

1	2	3	4	5	6	7	8	9	10	11
								COMBINATION		
MATERIAL	MIN SIZE (mm)	MAX SIZE (mm)	VOID RATIO U	REL DENS. RD	% BY MASS	MEAN SIZE D (mm)	% BY VOL	MEAN SIZE (mm)	VOID RATIO U	RD
5	0.04	0.08	1.1	2.6	4	0.057	4	-	1.10	-
6	0.08	0.15	1.0	2.6	39	0.110	39	-	0.98	-
7	0.15	0.3	0.9	2.6	13	0.212	13	-	0.89	-
8	0.3	0.6	0.7	2.6	15	0.424	15	-	0.76	-
9	0.6	1.2	0.7	2.6	18	0.849	18	-	0.64	-
10	1.2	2.5	0.7	2.6	11	1.732	11		0.59	
11	2.5	5.0	0.7	2.6	0	3.536	0	0.28	0.59	2.60

Table 6 Properties of sands optimised for use at different cement contents

CEMENT (kg/m³)	FINE AGGREGATES		
	Mean Size	Void Ratio	Relative Density
(a) Intial Sand			
All	0.64	0.44	2.60
(b) Optimised Sands			
150	0.33	0.54	2.60
200	0.33	0.54	2.60
250	0.24	0.63	2.60
300	0.22	0.37	2.60
350	0.22	0.66	2.60
400	0.28	0.59	2.60
450	0.22	0.67	2.60
500	0.21	0.67	2.60
550	0.17	0.75	2.60

The gradings for the optimised fine aggregates are summarised in Figure 3 in comparison with that for the initial fine aggregate. The range of the optimised gradings is relatively narrow and does not overlap the original grading.

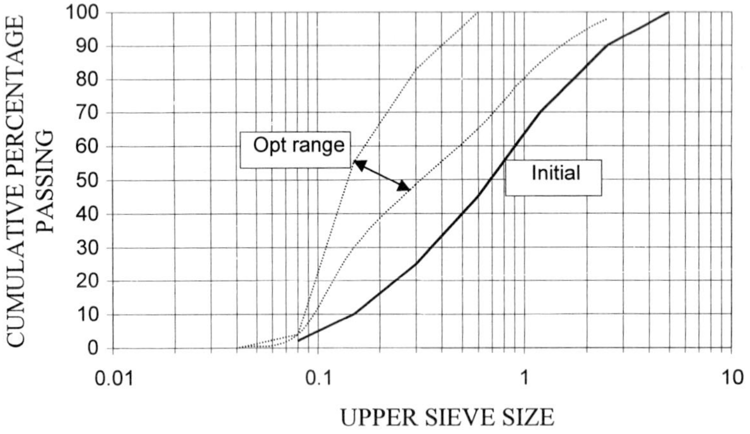

Figure 3 Optimised fine aggregate gradings for minimum water demand of concrete

In this particular example, optimising the fine aggregate grading enables a substantial reduction of water demand, in the order of 15-35 l/m^3, over the full working range of cement content, as shown in Figure 4.

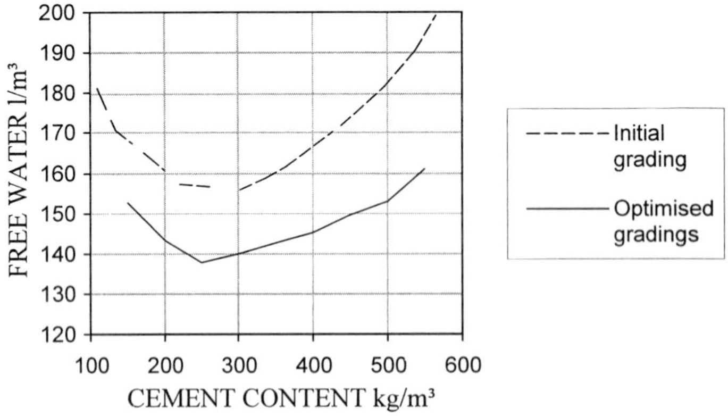

Figure 4 Effect on concrete water demand of optimising fine aggregate grading

Optimisation of Particle Size Distribution of the Coarse Aggregate for Minimum Water Demand of Concrete at Different Cement Contents.

With regard to grading of the coarse aggregate, when two sizes, 20 - 10 mm and 10 - 5 mm are available, both having the same voids ratio, computer analysis indicates that with medium sands there is a benefit to water demand in using less than 50% of the 10 mm size. For fine sands, 30 - 50% of 10 mm was optimal for the complete range of cement content, as may be seen in Figure 5.

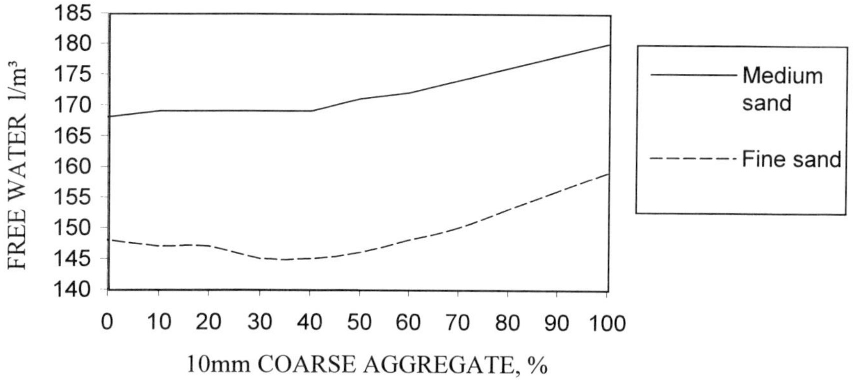

Figure 5 Example of influence on water demand of the percentage of 10 mm material in 20 mm coarse aggregate for concrete of 400 kg/m^3 cement content

If the 10 mm voids ratio is higher than that for the 20 mm material, the optimum proportion would be reduced further.

OPTIONS FOR FURTHER OPTIMISATION OF MATERIALS

Of course, the cement and aggregate industries have long experience in the successful optimisation of their materials with regard to minimising production costs, improving properties and developing market benefits for their products, both with regard to general purpose aggregates and also products for the larger niche markets. However, cements are restricted by the need to meet other requirements, e.g. strength, and aggregates are significantly restricted by the constituents, shape and grading provided in nature or as a result of processing. In addition, specifications may place restrictions upon the grading of aggregates, in particular on the proportion of very fine material, especially if it likely to contain clay rather than silt or dust.

However, there always remains considerable scope for significant general improvements to voids ratio and mean size through modifications to psd or grading, as well as modifications to meet the specialised needs of smaller niche markets.

For cements in particular, it would be beneficial to gain improvements in both voids ratio and psd at the higher cement contents.

Methods by which improvements may be obtained include any of the following taken separately or in combination

- Improvements in the voids ratios and/or gradings of the materials
- Blending with materials having more favourable shape, surface characteristics or enabling a more favourable grading.
- Addition of very fine material to extend the size distribution at the lower end, e.g. mineral filler or other powder.
- Addition of material intermediate in size between the cement and fine aggregate.

With regard to each of these, it will be observed that in the examples already described, constraints were deliberately applied, so that advantage was not taken of some of these possibilities. In subsequent sections, some of these options and possible benefits will be considered in more detail.

It will be appreciated that the opportunities for making improvements are not restricted to a single party. For example, modification of grading at the interface between cement and fine aggregate could result from efforts by cement producer, addition supplier, aggregate supplier or concrete producer.

Inclusion of Mineral Fillers in the Range of Sizes between Cement and Sand

Cement producers normally try to minimise the coarse material content in cement because of the benefit of finer sizes for strength, particularly early strength. Aggregate producers are constrained by BS 882 to a maximum of 4% silt or fine dust in a gravel sand and 16% in a crushed rock sand.

However, the analysis so far suggests that a benefit could be obtained by appropriate inclusion of material in the size range between cement and fine aggregate. Of course, if it was intended to introduce a filler of say 80 - 40 μm, say, it would be necessary to take account of any cement above 40 μm and any fine aggregate below 80 μm.

Further analyses have been made to investigate the optimum percentage of such a filler in the concretes considered earlier using the initial cement and a fine sand. The results are summarised in Figure 6 from which it may be seen that reductions of about 5 l/m^3 are possible by use of 23 % filler (80-40μm) with a sand which contained only the permitted maximum 4% passing 80μm.

Naturally, if the cement also contained a significant proportion of material coarser than 40 μm, this would reduce the benefit of including fillers.

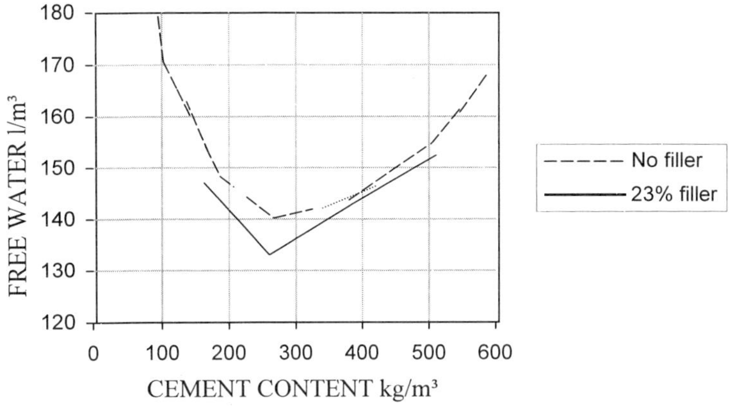

Figure 6 Influence on concrete water demand of the inclusion of 23% mineral filler with a particular fine aggregate

Inclusion of Cementitious Additions or Mineral Fillers in the Range of Cement Sizes or Finer Sizes

It is well documented that the inclusion of fine or very fine materials can often reduce the water demands of concretes and provide other benefits, e.g. economy; strength; thermal properties; durability. Some materials, due to advantageous shape and surface texture may assist in reducing the voids ratio of the cementitious material generally. Others may assist in extending the range of sizes with a similar benefit.

As recorded earlier, the optimum percentage of the finest cement size used, 5 - 2.5 μm equalled the constraint value of max 15%, implying benefit of using at least 15% below 5 μm and possibly a percentage of material below 2.5 μm.

Obviously, if the particular material had a very high voids ratio, its benefit might be limited despite the benefit of the smaller size in reducing particle interference. In addition, the small size might be accompanied by a tendency for agglomeration such that the effective size of agglomerated particles was much larger than the size of individual particles, and they might resist separation during mixing. Furthermore, the presence of this powder when mixed with cement particles might transmit the agglomeration effect to the cement as well. Of course there may be benefits to be taken into account, e.g. greater reactivity and strength.

Additional analyses were made to indicate whether experience would be confirmed by theory. This also provides the opportunity to display the input and output data involved in the author's system of concrete simulation for both fresh and hardened concrete.

Table 7 and Table 8 shows the materials input data for 9 particulate materials of which 7 have been selected for use in the example. Table 7 shows a very fine cementitious material having a high voids ratio, a medium material with a low voids ratio and the initial cement used previously. The proportions have been selected at common levels. The resultant properties of the combination are shown in bold. The adjusted voids ratio allows for the effect of a superplasticiser.

Table 7 Input data for computerised simulation of concrete using 7 of 9 particulate materials and a superplasticiser

MATERIALS	Input	MEAN SIZE D (mm) Input	VOID RATIO U Input	VOID RATIO U Adjusted	REL DENSITY RD Input
Cem 1	Addn1				
Mass %	8	0.0015	2.100	-	2.20
Volume%	10.3				
Cem 2	Addn2				
Mass %	25	0.010	0.700	-	2.30
Volume%	30.7				
Cem 3	Addn3				
Mass %	67	0.0120	0.910	-	3.20
Volume %	59.1				
Cement and Additions		**0.092**	**0.743**	**0.632**	**2.82**
C.Agg1	10mm				
Mass %	30	7.07	0.80	-	2.60
Volume %	30.0				
C.Agg1	20mm				
Mass %	70	14.14	0.80	-	2.60
Volume %	70.0				
C.Agg1	40mm				
Mass %	0	26.5	0.80	-	2.60
Volume %	0.0				
Coarse Aggregate		**11.49**	**0.726**	**0.617**	**2.60**
Fine Ag1	Filler				
Mass %	23	0.06	1.100	-	2.60
Volume %	23.0				
Fine Ag2	Fine Sand				
Mass %	77	0.24	0.630	-	2.60
Volume %	77.0				
Fine Ag3	Med Sand				
Mass %	0	0.64	0.440	-	2.60
Volume %	0.0				
Fine Agg and Mnrl. Addn		**0.174**	**0.623**	**0.529**	**2.60**

The Table also shows a mineral filler and fine sand selected at the percentages found to produce minimum concrete water demand. The medium sand is not used for this exercise. Table 7 shows the 10mm and 20mm coarse aggregates selected at percentages in the range found to produce minimum concrete water demand. The 40mm aggregate is not used for this exercise.

Table 8 provides for a number of important additional items of data about the materials and the concrete, including the air content, required slump, admixture and strength factors. In this case, a superplasticiser has been included.

Table 8 Additional input data for materials, air content, slump, admixture and strength factor

CONDITIONS				INPUT
Cohesion Factor	0.5 to 2	normally 1		1
		Air cohesion factor		1
Air Content (%)	Total	normally 0.1, 1, 5.5		1
	Entrapped	assume 0.5-1.5		1
	Entrained			0
Mortar	Input Y for mortar mixtures			N
Slump (mm)	0 to 250	normally 75 or 50		50
Admixture	Code			Sprplsr
	Rate	litres/100kg cement		0.75
	Factor	0.5 to 1. Use 1 if water redn or no admixture used		0.85
Cement Strength	28 days			55
	Strength factor	28 day factor		1
		7 day factor		1.15
	Aggreg factor			0.95
	Cemaddn Fr	Addn 1	Strength Efficiency Factor 0 to 3	2
		Addn 2	Strength Efficiency Factor 0 to 3	0.95
		PC	Strength Efficiency Factor 0 to 3	1

Specification	Input	cem + adn kg/m3
Min cement kg/m3	300	300
Max free w/c	0.35	294
Max Aggregate/cement	6	333
Char strength N/mm2	65	
Plant sd N/mm2	4	
Margin factor	2.00	
Margin N/mm2	8	
Design mean strength N/mm2	73	342
Total air %	1	
Slump	50	

Per cent of material permitted by the specification to be called cement

	For min cement	For max a/c	For w/c
Addn 1	100	100	100
Addn 2	100	100	100
PC	100	100	100

Selected concrete to meet specification			Output
Cem+Cem Addn		kg	342
Addn 1		kg	27
Addn 2		kg	85
PC		kg	229
Free Water incl admix		kg	103
Admixture	Sprplasr	litres	2.6
Filler	ssd	kg	146
Fine sand	ssd	kg	490
Med Sand	ssd	kg	0
10 mm	ssd	kg	407
20 mmSS	ssd	kg	949
40 mmSS	ssd	kg	0
Plastic density		kg/m3	2436
Free w/(c+adn)			0.30
Aggregate/(cement + adn)			5.82
% fines incl min adn			32
Total air		%	1
Expected mean str		N/mm2	73
Materials Cost		£/m3	

Figure 7 Output of computerised simulation of concrete using 7 of 9 particulate materials and a superplasticiser to meet a particular specification

An example specification and output data from the simulation are provided in Figure 7. In this example, the strength is the key requirement determining cement content. The concrete quantities and properties from the simulation process are listed on the right.

Figure 8 demonstrates the benefits to water demand of successive inclusion of more materials, including a superplasticiser.

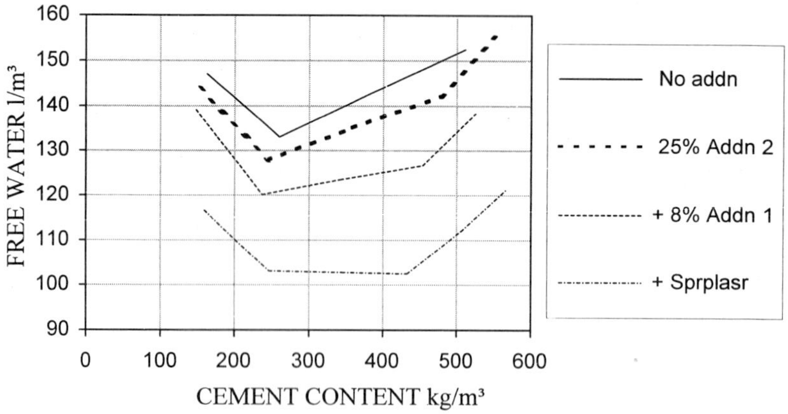

Figure 8 Comparison between the water demands for concretes containing progressively more components

INTEGRATION OF DESIGN AND CONTROL

It has become commonplace for the integration of strength control and concrete selection to take account continuously of observed changes in mean level and uniformity of strength.

It is becoming more feasible for adjustments of all aspects of design of concrete mixtures to be subject to modification to take account automatically of observed changes in materials qualities as recommended in [2, 5, 6, 9].

APPLICATION OF MODELLING TO RESEARCH AND DEVELOPMENT

Modelling techniques significantly increase the speed with which designs may be developed for new materials and products, enabling developers to immediately avoid unprofitable avenues and to home in on the areas of the most promising economic solutions.

CONCLUSIONS

The paper demonstrates the ways in which computerised modelling enables consideration of the options for optimising the designs of concrete and its components.

CONCLUSIONS

The paper demonstrates the ways in which computerised modelling enables consideration of the options for optimising the designs of concrete and its components.

Scope has been shown for the use of ultrafine cementitious materials, finer sands and intermediate sized materials and confirmed the well known benefits of introducing materials of more advantageous shape or particle size distribution or grading.

There are often artificial restrictions in standards, e.g. percent of material passing 75 μm in sands, which may bear re-examination in order not to hamper design and economy unreasonably.

REFERENCES

1. ANDERSEN, J, AND JOHANSEN, V. Computer-aided simulation of particle packing. A tool for proportioning cement based materials. GM Idorn Consult A/S, Denmark, c1989.

2. DAY, K W. Concrete mix design, quality control and specification. E & FN Spon, 1995, pp 350.

3. DE LARRARD, F AND SEDRAN, T. Innovative approach in mixture-proportioning thanks to mathematical mixture modelling and computer technology. ACI Fall Convention, Montreal, Nov 1995, pp 1-6.

4. DEWAR, J D. Computerized simulation of aggregate, mortar and concrete mixtures. ERMCO congress, London, May 1983.

5. DEWAR, J D. The structure of fresh concrete- A new solution to an old problem. First Sir Frederick Lea Memorial lecture, Institute of Concrete Technology, 1986. pp 1- 23. Reprinted by the BRMCA.

6. DEWAR, J D. Development of a Theory of Particle Mixtures and its application to Aggregates Mortars and Concretes. PhD thesis, Dept of Civil Eng, City University, London, December 1997.

7. ROY, D M, SCHEETZ, B E, MALEK, R I A, AND SHI, D. Concrete components packing handbook. Strategic Highway Research Program. NRC Washington DC. 1993. pp 1-161.

8. POPOVICS S. Concrete-making materials. Hemisphere Pub Co, 1979.

9. SEAR, L K A. Computer methods- Control systems. Inst of Concrete Technology, 26th Annual Convention, Market Bosworth, UK, April 1998. pp 1-11.

COMPUTER-SIMULATION OF PARTICLE SECTION PATTERNS FROM SIEVE CURVES FOR SPHERICAL AGGREGATE

J Zheng

Northern Jiaotong University Beijing

China

P Stroeven

Delft University of Technology

Netherlands

ABSTRACT. With the development of high-strength (-performance) concrete, it is gradually recognised that particle packing is of paramount importance. On the one hand, the particle packing has to be optimised, i.e. the sand to gravel ratio is chosen to maximize density, and to minimize void ratio. On the other hand, binder density is increased by fine-particle bleedings; particularly the packing in the particle-matrix interphase plays a major role in improving strength. The study deals with the computer-simulation for particle packing on meso-level. The formulae of the cumulative distribution functions (CDFS) for aggregate particles and intersection circles are systematically given. Based on those formulae, a numerical method for the generation and distribution of aggregate particles and intersection circles is presented. Finally, several examples are evaluated for an aggregate with particle size range between 2 mm and 32 mm, which verify the effectiveness of the computer-simulation method proposed in this paper.

Keywords: Concrete, Computer-simulation, Aggregate particle, Intersection circle, Cumulative distribution function (CDF), Meso-level.

Professor J Zheng is Vice-Director of the Bridge and Structure Research Institute, Northern Jiaotong University, Beijing, P. R. China. His main interests include the computational methods in solid mechanics, the mean value theorems in solid mechanics, the interaction between structures and media and concrete mechanics. He has published widely and serves on two National Technical Committees.

Professor P Stroeven is Head of the Group on Materials Engineering and Geosciences, Delft University of Technology, The Netherlands. He has published widely on subjects in concrete technology relevant for industrialised as well as developing countries, such as micromechanics of damage evolution, bond fibre reinforcement, cement blending. He serves as a member in RILEM committee TC162-TDF. He is a member of the Advisory Board of two International Journals.

INTRODUCTION

In the study on modern concrete technology, macro-mechanics, micro-mechanics and statistical mechanics may all be used to analyse the mechanical behaviour of concrete [1]. In phenomenological method, concrete is often assumed as a uniform material with nonlinear properties. Although the theoretical calculation based on this method can be in good agreement with the experimental results of concrete in some cases, it is very difficult for us to relate the micro-structures of concrete, such as defects, cracks and inclusions, to its macro-properties. In addition, due to the development of micro-mechanics, it has been revealed that some simple relationships between micro-structures and macro-properties exist in heterogeneous materials, but it is still considerably difficult how to take into account the complicated interactions between crack and crack, aggregate and aggregate, and crack and aggregate. Therefore, the computer-simulation of concrete becomes an effective means. On meso-level, concrete may be considered as a two-phase material consisting of matrix and aggregate particles or intersection circles. Some models based on statistical mechanics, such as lattice type model [2], have been developed to simulate the mechanical behaviour of concrete. The success of lattice type model in simulating the uniaxial tension of plain concrete is mainly ascribed to the reasonable packing of aggregate particles and the application of finite element method, in which two important physical effects, anisotropy and a cutoff at small length scales, are accounted for. Since the aggregate-matrix interfaces are determined by the distribution of aggregate particles and the macro-crack propagates along the interfaces, the generation and distribution of aggregate particles will become a key step toward computer-simulation of concrete and they directly influence the fracture laws of concrete. At present, an approximate formula given by Walraven [3] has been used for the generation of the intersection circles in two-dimensional simulation of concrete, but it is only suitable for the concrete with Fuller-type aggregate. On the basis of the research work by Walraven [3], Stroeven [4] and Zheng and Stroeven [5, 6 ,7], this paper summarises the CDFS for aggregate particles and intersection circles and presents a computer-simulation method for the generation and distribution of aggregate particles.

BASIC FORMULAE

The computer-simulation for the micro-structures of concrete is based on the CDFS for the distributions of aggregate particles in three-dimensional space and intersection circles in two-dimensional plane. On the basis of the research work by Walraven [3], Stroeven [4] and Zheng and Stroeven [5, 6, 7], the CDFS for the diameters of aggregate particles ($P_{3D}(D)$), for the diameters of intersection circles ($P_{2D}(d)$) and for the areas of intersection circles ($P_{2S}(d)$) are summarised as follows:

- For the concrete with equal-volume aggregate,

$$P_{3D}(D) = M^3 D_0^3\left(1/D_0^3 - 1/D^3\right)/(M^3 - 1) \quad (1)$$

$$P_{2d}(d) = 1 - 2D_0^2 M^2\left(1 - d^2/D_m^2\right)^{1.5} / \left[3(M^2 - 1)d^2\right] \quad (2)$$

$$P_{2S}(d) = 1 - \frac{1}{\ln(D_m / D_0)}\left[\ln\frac{\sqrt{1 + d/D_m} + \sqrt{1 - d/D_m}}{\sqrt{1 + d/D_m} - \sqrt{1 - d/D_m}} - \sqrt{1 - \frac{d^2}{D_m^2}\left(\frac{5}{6} + \frac{d^2}{6D_m^2}\right)}\right] \quad (3)$$

where D_O and D_m are the smallest and largest diameters of aggregate particles, respectively, and the magnification $M = D_m / D_o$

- For the concrete with Fuller-type aggregate,

$$P_{3D}(D) = M^{2.5} D_0^{2.5} \left(1/D_0^{2.5} - 1/D^{2.5}\right)/(M^{2.5} - 1) \tag{4}$$

$$P_{2d}(d) = 1 - 1.5 M^{1.5} D_0^{1.5} S_1(\vartheta_m)/\left[(M^{1.5} - 1)d^{1.5}\right] \tag{5}$$

$$P_{2S}(d) = 1 - \sqrt{d}\left[\overline{S_1}(\vartheta_m) + 0.5 S_1(\vartheta_m)\right]/\left[2(\sqrt{D_m} - \sqrt{D_0})\right] \tag{6}$$

where the definitions of S_1 (ϑ_m) and S_1 (ϑ_m) see [7].

- For the concrete with general aggregate, it is usually very difficult to obtain the exact solutions for $P_{3D}(D)$, $P_{2D}(d)$ and $P_{2s}(d)$. In the circumstances, suitable numerical techniques become important. In the cubic spline method [7], the interval $[D_o, D_m]$ is divided into $(L-1)$ subintervals $[D_i, D_{(i+1)}]$ $(i = 1,2,....,L-1)$. According to the given CDF's values $P_{3v,i}$ $(i=1,2....,L)$ at the discretion points D_i $(i= 1,2,....,L)$, the CDF, P_{3V} (D), for the volumes of aggregate particles can be written as

$$P_{3V}(D) = \sum_{i=1}^{L-1} P_{3V,i}(D)[< D - D_i >_+^0 - < D - D_{(i+1)} >_+^0] \tag{7}$$

in which $P_{3V,i}(D)$ and $< D - D_i >_+^0$ are defined as

$$P_{3V,i}(D) = a\widetilde{P}_{3V,(i+1)} + (1-a)\widetilde{P}_{3V,i} + t_i^2\{(a^3 - a)\mu_{(i+1)} + [(1-a)^3 - (1-a)]\mu_i\} \tag{8}$$

$$< D - D_i >_+^0 = \begin{cases} 1, & D \geq D_i \\ 0, & D < D_i \end{cases} \tag{9}$$

where $t_i = [D_{(i+1)} - D_i]$, $a = (D - D_i)/[D_{(i+1)} - D_i]$ and μ_i may be determined by $\widetilde{P}_{3V,i}$ [7]. Thus, $P_{3D}(D)$, $P_{2D}(d)$ and $P_{2S}(d)$ read, respectively

$$P_{3D}(D) = \overline{D^3} \sum_{i=1}^{L-1} P_{3D,i}(D) \tag{10}$$

$$P_{2d}(d) = 1 - \frac{\overline{D^3}}{\overline{D}} \sum_{i=1}^{L-1} P_{2d,i}(d) \tag{11}$$

$$P_{2S}(d) = 1 - \sum_{i=1}^{L-1} P_{2S,i}(d) \tag{12}$$

with

$$P_{3D,i}(D) = 0 \qquad \text{for } D \leq D_i \tag{13a}$$

$$P_{3D,i}(D) = A_{1i} \ln\left(\frac{D}{D_i}\right) + A_{2i}\left(\frac{1}{D_i} - \frac{1}{D}\right) + \frac{A_{3i}}{2}\left(\frac{1}{D_i^2} - \frac{1}{D^2}\right) \qquad \text{for } D_i < D < D_{(i+1)} \tag{13b}$$

with

$$P_{3D,i}(D) = A_{1i}\ln\left(\frac{D_{(i+1)}}{D_i}\right) + A_{2i}\left(\frac{1}{D_i} - \frac{1}{D_{(i+1)}}\right) + \frac{A_{3i}}{2}\left(\frac{1}{D_i^2} - \frac{1}{D_{(i+1)}^2}\right) \quad \text{for } D \geq D_{(i+1)} \quad (13c)$$

$$P_{2d,i}(d) = 0 \quad \text{for } d \geq D_{(i+1)} \quad (14a)$$

$$P_{2d,i}(d) = \left(A_{1i}d - \frac{A_{3i}}{2d}\right)\left[\arcsin\left(\frac{d}{D_{(i+1)}}\right) - \frac{\pi}{2}\right] + A_{2i}\ln\left[\frac{D_{(i+1)}}{d} + \sqrt{\left(\frac{D_{(i+1)}}{d}\right)^2 - 1}\right]$$

$$+\sqrt{D_{(i+1)}^2 - d^2}\left[A_{1i} - \frac{A_{2i}}{D_{(i+1)}} - \frac{A_{3i}}{2D_{(i+1)}^2}\right] \quad \text{for } D_i < d < D_{(i+1)} \quad (14b)$$

$$P_{2d,i}(d) = \left(A_{1i}d - \frac{A_{3i}}{2d}\right)\left[\arcsin\left(\frac{d}{D_{(i+1)}}\right) - \arcsin\left(\frac{d}{D_i}\right)\right] + A_{2i}\ln\left(\frac{D_{(i+1)} + \sqrt{D_{(i+1)}^2 - d^2}}{D_i + \sqrt{D_i^2 - d^2}}\right)$$

$$+\sqrt{D_{(i+1)}^2 - d^2}\left(A_{1i} - \frac{A_{2i}}{D_{(i+1)}} - \frac{A_{3i}}{2D_{(i+1)}^2}\right) - \sqrt{D_i^2 - d^2}\left(A_{1i} - \frac{A_{2i}}{D_i} - \frac{A_{3i}}{2D_i^2}\right) \quad \text{for } d < D_i \quad (14c)$$

$$P_{2S,i}(d) = 0 \quad \text{for } d \geq D_{(i+1)} \quad (15a)$$

$$P_{2S,i}(d) = A_{1i}\left[\frac{\sqrt{(D_{(i+1)}^2 - d^2)^3}}{3} + \frac{d^2\sqrt{D_{(i+1)}^2 - d^2}}{2} + \frac{d^3}{2}\left(\arcsin\left(\frac{d}{D_{(i+1)}}\right) - \frac{\pi}{2}\right)\right]$$

$$+A_{3i}\left[\sqrt{D_{(i+1)}^2 - d^2}\left(1 - \frac{d^2}{4D_{(i+1)}^2}\right) + \frac{3d}{4}\left(\arcsin\left(\frac{d}{D_{(i+1)}}\right) - \frac{\pi}{2}\right)\right]$$

$$+A_{2i}\sqrt{D_{(i+1)}^2 - d^2}\left[\frac{D_{(i+1)}}{2} - \frac{d^2}{2D_{(i+1)}}\right] \quad \text{for } D_i < d < D_{(i+1)} \quad (15b)$$

$$P_{2S,i}(d) = A_{1i}\left[\frac{\sqrt{(D_{(i+1)}^2 - d^2)^3}}{3} - \frac{\sqrt{(D_i^2 - d^2)^3}}{3} + \frac{d^2}{2}\left(\sqrt{D_{(i+1)}^2 - d^2} - \sqrt{D_i^2 - d^2}\right)\right]$$

$$+\frac{A_{1i}d^3}{2}\left[\arcsin\left(\frac{d}{D_{(i+1)}}\right) - \arcsin\left(\frac{d}{D_i}\right)\right] + A_{2i}\left[\frac{D_{(i+1)}\sqrt{D_{(i+1)}^2 - d^2}}{2} - \frac{D_i\sqrt{D_i^2 - d^2}}{2}\right]$$

$$-\frac{A_{2i}d^2}{2}\left[\frac{\sqrt{D_{(i+1)}^2 - d^2}}{D_{(i+1)}} - \frac{\sqrt{D_{(i+1)}^2 - d^2}}{D_i}\right] + \frac{3dA_{3i}}{4}\left[\arcsin\left(\frac{d}{D_{(i+1)}}\right) - \arcsin\left(\frac{d}{D_i}\right)\right]$$

$$+A_{3i}\left[\sqrt{D_{(i+1)}^2 - d^2} - \sqrt{D_i^2 - d^2} - \frac{d^2}{4}\left(\frac{\sqrt{D_{(i+1)}^2 - d^2}}{D_{(i+1)}^2} - \frac{\sqrt{D_i^2 - d^2}}{D_i}\right)\right] \quad \text{for } d \leq D_i \quad (15c)$$

where *D, D3, A~i, A2i* and A3i see [7].

In order to investigate the accuracy of the numerical method, we consider the concrete with Fuller-type aggregate of D_o = 2 mm and D_m = 32 mm. The values $P_{3v,i}$ at the discretion points of D_i = 4 mm, 8 mm, 16 mm, 24 mm are given by the following formula [6]

$$\widetilde{P}_{3V,i} = \left(\sqrt{D_i} - \sqrt{D_0}\right) / \left(\sqrt{D_m} - \sqrt{D_0}\right) \tag{16}$$

Figures 1 to 3 depict the calculation results, in which ES and NS represent the exact and numerical solutions, respectively. From these diagrams, it can be concluded that the numerical method is of high accuracy and may be applied to computer-simulation.

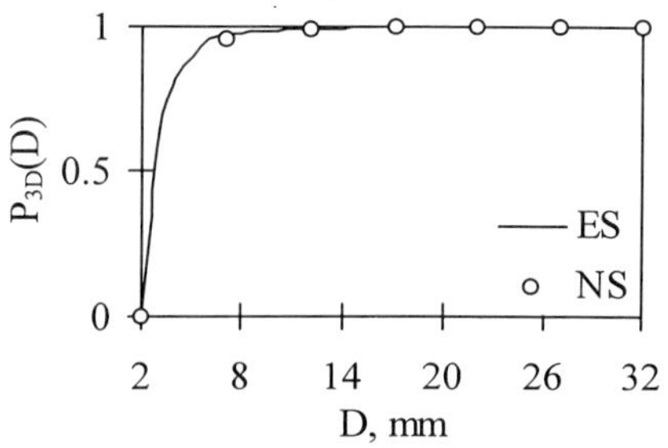

Figure 1 CDF for the diameters of aggregate particles

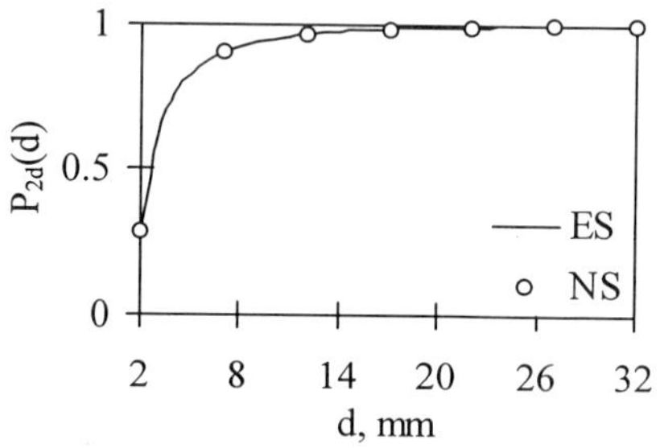

Figure 2 CDF for the diameters of intersection circles

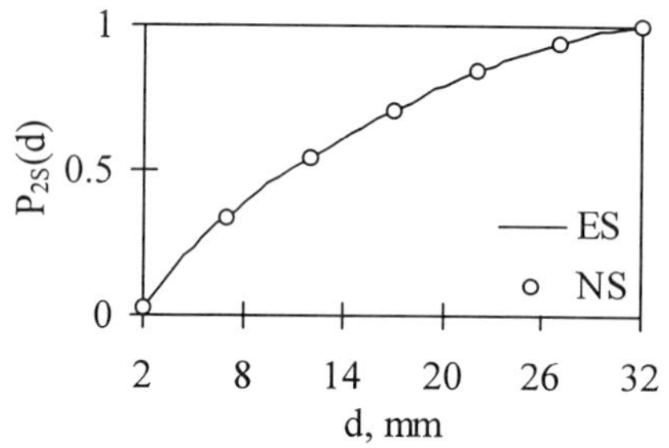

Figure 3 CDF for the areas of intersection circles

GENERATION OF AGGREGATE PARTICLES AND INTERSECTION CIRCLES

Since three- and two-dimensional computer-simulations are similar, we only discuss the generation and distribution of intersection circles in two-dimensional plane. The algorithm for the generation of intersection circles is described as follows:

Step 1: Calculate A_A and A_C

$$A_A = [1 - P_{2S}(D_0)]V_V \tag{17}$$

$$A_c = A_A A_0 \tag{18}$$

where V_V, A_A, A_O and A_C are the volume percentage of aggregate particles, the area percentage of intersection circles, the cross-sectional area of concrete and the total area of intersection circles, respectively.

Step 2: Generate a random number z_i in the interval $[P_{2d}(D_O), 1]$ with uniform distribution.

Step 3: Solve the following equation

$$P_{2d}(d) = z_i \tag{19}$$

and obtain an intersection circle of diameter d_i

Step 4: Sum the areas of those intersection circles that have been generated so far

$$A_i = A_{(i-1)} + \pi(d_i/2)^2 \tag{20}$$

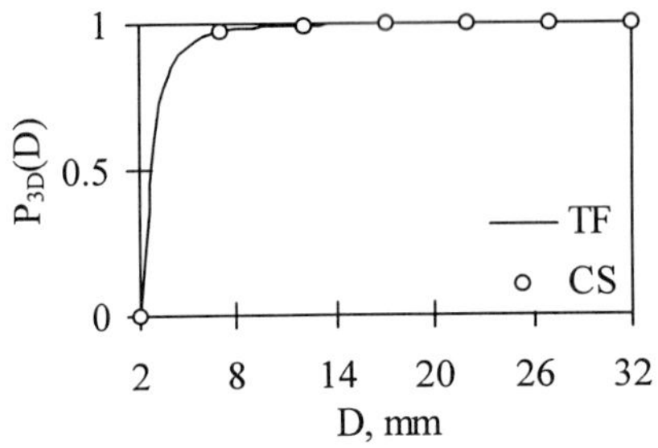

Figure 4 CDF for the diameters of aggregate particles

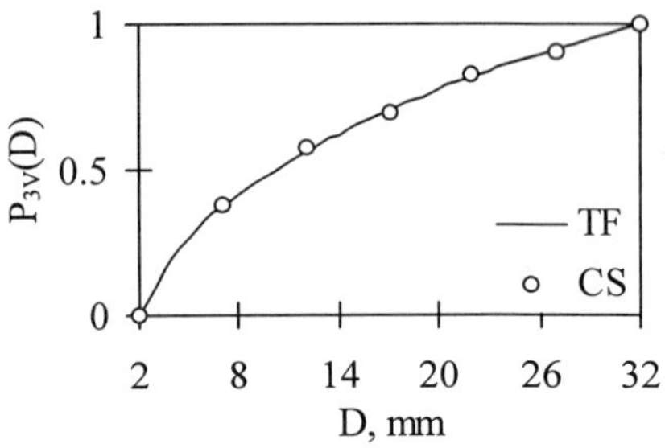

Figure 5 CDF for the volumes of aggregate particles

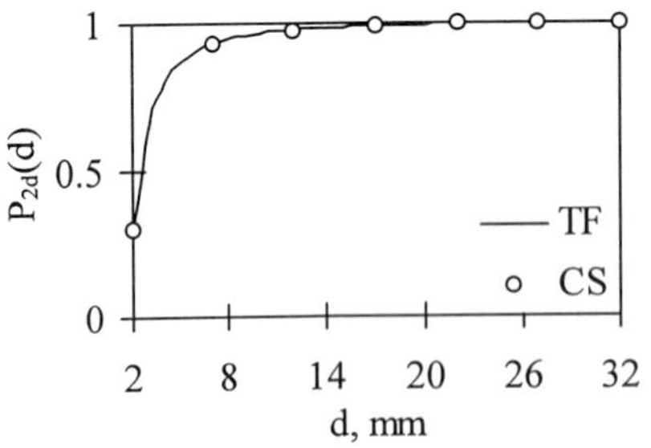

Figure 6 CDF for the diameters of intersection circles

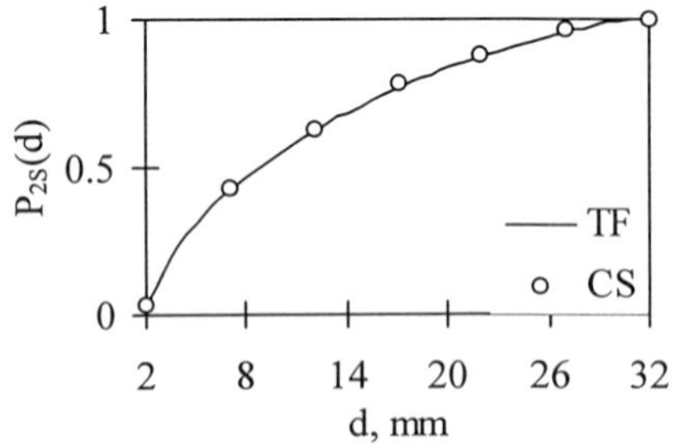

Figure 7 CDF for the areas of intersection circles

Step 5: Compare A_i with A_C. If $A_i \geq A_C$, we stop calculation. Otherwise, we return to **Step 2** and repeat the procedure from **Step 2** to **Step 5**.

For investigating the algorithm, we set D_o = 2 mm, D_m = 32 mm and V_v = 075. The values $P_{3v,\ i}$ at the discretion points of D_i = 4mm, 8mm, 16mm, 24mm are assigned as 0.1940, 0.4167, 0.6797 and 0.8588, respectively. For three-dimensional simulation, the side of concrete cube is equal to 120 mm. In the case of two-dimensional one, a square with side 500 mm is considered. The computer-simulation (CS) is compared with the theoretical formula (TF) in Figures 4 to 7. These diagrams demonstrate that the computer-simulation is in good agreement with the theoretical prediction.

DISTRIBUTION OF AGGREGATE PARTICLES AND INTERSECTION CIRCLES

For a cross-section Ω of concrete as shown in Figure 8, the distribution of intersection circles may be conducted as following procedure:

Step 1: Constitute an imaginary rectangle Ω with sides 1_x and I_y and make Ω cover the cross-section Ω of concrete.

Step 2: Rearrange the order of the generated intersection circles from the largest circle to the smallest one.

Step 3: Generate two random numbers x_i and y_i in the intervals $[0, 1_x]$ and $[0, 1_y]$ with uniform distribution, respectively, as the central coordinates of the i-th intersection circle.

Step 4: If the i-th intersection circle is involved in the Ω and doesn't overlap with the preceding (i-1) circles, we repeat **Step 3** and **Step 4** to distribute the next circle until all those intersection circles are distributed in the domain Ω. Otherwise, return to **Step 3** and re-generate a pair of random numbers x_i and y_i for the i-th intersection circle.

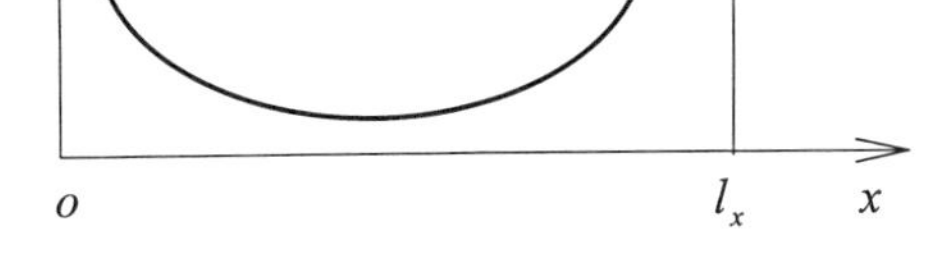

Figure 8 Cross-section of concrete and its imaginary rectangle

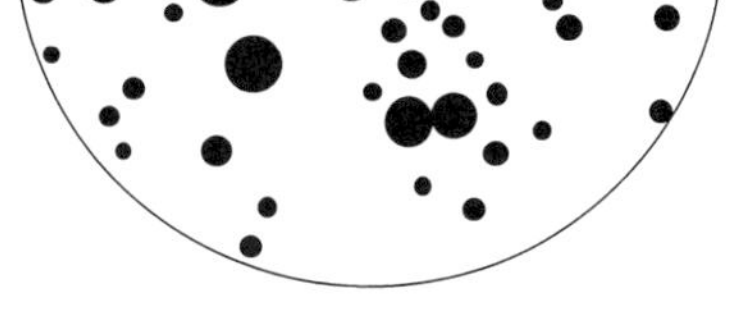

Figure 9 Distribution of intersection circles for A_A = 0.15

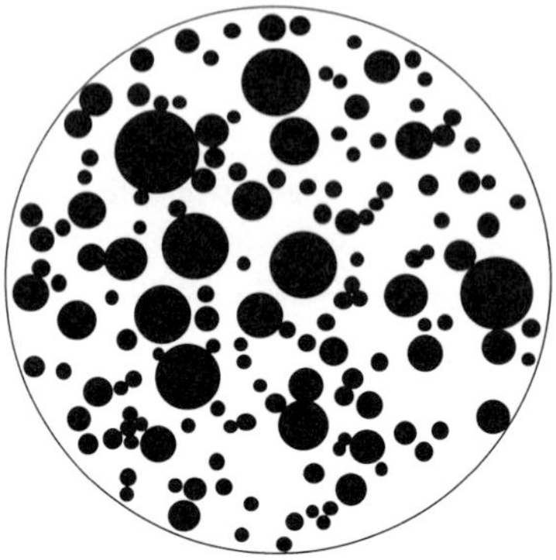

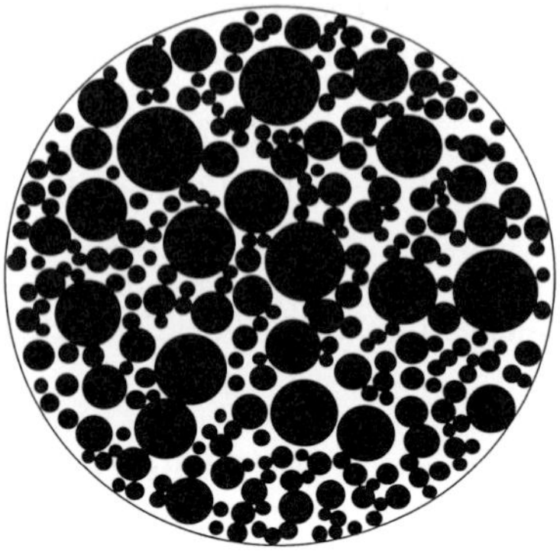

Figure 10 Distribution of intersection circles for $A_A = 0.30$

Figure 11 Distribution of intersection circles for $A_A = 0.60$

As an application, we consider a circular concrete cross-section with diameter 10cm. The aggregate particles are of Fuller-type distribution and their largest and smallest diameters are 16 mm and 2 mm, respectively. Figures 9 to 11 depict the distributions of intersection circles for $A_A = 0.15$, $A_A = 0\ 30$ and $A_A = 0.60$, respectively.

REFERENCES

1. KRAJCINOVIC, D. Damage Mechanics, Elsevier, 1996, pp 1-20.

2. SCHLANGEN, E. Experimental and Numerical Analysis of Fracture Process in Concrete, PhD Thesis, Delft University, 1993, pp 66-69.

3. STROEVEN, P. Structural Modeling of Plain and Fibre-Reinforced Concrete. J. Composites, Vol. 16, No. 2, 1982, pp 129-139.

4. WALRAVEN, J C. Aggregate Interlock: A Theoretical and Experimental Analysis, PhD Thesis, Delft University ,1980, pp 67-74.

5. ZHENG, J AND STROEVEN, P. Solutions for the Intersection Circles Distribution in Concrete with Equi-Volume of Spherical Aggregate Particles. Proceedings of the 4th Beijing International Symposium on Cement and Concrete, Vol. 2, Beijing, 1998, pp 378-382.

6. ZHENG, J AND STROEVEN, P. On the Intersection Circle Distribution in Concrete with Fuller-Type Aggregate Particles. (To be Published).

7. ZHENG, J AND STROEVEN, P. A Numerical Method for Intersection Circles Distribution in Concrete with General Aggregate Particles. (To be Published).

MANIPULATING THE WATER CONTENT AND MICROSTRUCTURE OF HIGH PERFORMANCE CONCRETE USING AUTOGENEOUS CURING

S Weber

Bundesverband der Deutschen Zementindustrie eV

H W Reinhardt

University of Stuttgart

Germany

ABSTRACT. Due to hydration and drying a shortage of water occurs even in case of high performance concrete. Because of the dense structure the traditional curing methods fail and not enough water from outside reaches the interior of the concrete.

In order to optimise the hydration a manipulation of the water content and the microstructure is proposed, using autogeneous curing. The mean idea consists in the substitution of a part of the normal weight aggregates by wet lightweight aggregates, which become an internal water supply on the disposal of the cement paste. During drying and hardening water from the lightweight aggregates is absorbed by the cement paste by capillary suction and condensation, allowing the hydration to continue. As the hydration proceeds newly built hydration products fill microcracks, improving the density of the cement paste and thus the properties of the concrete.

The procedure and the effect of internal curing are shown. The theoretical consideration, the optimisation procedure and modelling the water transport within the cement paste and from the lightweight aggregates to the cement paste are given. In order to show the improved properties of the concrete macroscopical and microscopical aspects are discussed.

Keywords: High strength concrete, Lightweight aggregates, Hydration, Microstructure, Modelling water transport, Improved mechanical properties

Dr Silvia Weber is consulting at Bundesverband der Deutschen Zementindustrie eV. Her interest concerns mainly the different types of high performance concrete, their optimisation and application in practice.

Professor Hans W Reinhardt is head of the Institute of Construction Materials, University of Stuttgart, and of the Otto-Graf Institute, Stuttgart, Germany. His interest is focussed on the mechanical and physical properties of cementitious materials.

INTRODUCTION

An essential way to achieve the designed properties of concrete is proper curing. This means avoiding water evaporation at the surface of the concrete members and supplying water from the exterior. But the traditional ways of curing often failed in practice. Even when meticulously performed only water evaporation can be reduced, but the water supply on the surfaces is still a technical problem. The time allocated for curing causes a stagnation of the construction process increasing costs and perturbing the efficiency of modern technology. And after all the detrimental perturberances the risk of damages caused by improper curing is not eliminated. In case of high strength concrete (HSC) the problems increase. Because of the high compressive strength HSC is commonly used for vertical structural elements. Due to the low water cement ratio combined with a high cement content and the addition of silica fume the concrete shows a high rate of hydration and the well known high compressive strength at early age, which would make an earlier form stripping possible, reducing construction costs. The high self-desiccation was a decisive reason for using this concrete for self-desiccating slabs [1].

The opinions and literature data about curing high strength concrete are contradictory. The very dense structure of the concrete leads to the assumption that water evaporation is low and water from the surrounding can penetrate only very slowly and/or in sufficient amount to achieve the interior of the structural members. So, curing might not be necessary. On the other hand when exposing to air water evaporation occurs [2] resulting in a considerable reduction of the compressive strength [3] and microcracks appear [4]. Wet curing for a longer time did not improve the mechanical properties [5].

Therefore the idea came up to create a water supply in the concrete independent of the environment, able to offer water to the unhydrated cement paste to continue the hydration process. Lightweight aggregate (LWA) generally can store a high amount of water. Their uniform distribution in the cement paste (CP) combined with a continuous transport of the stored water from the LWA to the CP transform the grain into a water supply for the concrete. The optimisation of concrete mixes can assure the uniform distribution of LWA in the CP. The water transport in concrete is depending on the microstructure of concrete and on the existent humidity gradient. As soon as hardening of concrete occurs, in the CP a system of pores develops. Silica fume concrete shows a refined pore structure with smaller radii than in normal concrete. Theoretically the radii of these pores are obviously smaller than the radii of the pores in the LWA so that the „conditio sine qua non“ for water transport is finally the existence of a humidity gradient. During hardening and/or drying of concrete humidity gradients on both, macroscopical and microscopical level are distinguished.

On microscopical level a shortage of water occurs due to chemical reactions. Theoretically the capillary forces of the CP are high enough to absorb the water from the LWA grain and transport it to the „drier“ CP, where a reaction with unhydrated cement occurs. By the chemical reaction calciumhydroxide is formed and reacting with silica fume new hydration products of type calciumsilicate hydrate (C-S-H) are built. They grow into the capillary pores or the available space, which can be microcracks as well, making the structure of the CP more dense. As the C-S-H have a higher strength than calciumhydroxide has, an increase in compressive strength is to be expected. Considering that the suction forces in the capillary pores are inversely proportional to the radius, the smaller the capillary pores of the cement paste get, the higher the suction forces in the smaller pores become. Thus the difference in pressure as a condition for water transport from the LWA to the CP is well given. The transport is time depending. With increasing density of the structure (later age) the transport of water and later vapour slows down and stops when the relative humidity in the LWA grain and in the hardened CP are in equilibrium.

On the macroscopical level the humidity gradient on the surfaces of concrete members exposed to the environment is considered. The lower the relative humidity of the environment is, the steeper is the gradient between the surface layer and more distant layer. Due to water evaporation this gradient increases and at the surface layer the moisture from the LWA will be transported to the CP faster than in the interior of the considered element. Because the water from the LWA is chemically bound, the structure on the surface is more dense, reducing the water evaporation and the diffusion process becomes slower. The surface of concrete with LWA will be more dense than of traditional HSC.

RESEARCH PROGRAM

Optimisation Procedure and Material Properties

During the optimisation in the reference mix shown in Table 1 the normal weight aggregate (NWA) was partially substituted by several types of LWA, replacing completely a certain fraction, while keeping the cement and water content constant. The LWA differed in grain size, strength, density, porosity and water absorption properties. The finally obtained mix containing LWA is shown in Table 1. For a theoretically complete hydration of Portland cement a water cement ratio of 0.42 is required, being equal to a water content of 189 l/m³ concrete. By the performed replacement a water supply of about 47 l was created by the LWA, which are made out of expanded clay, have a strength of 118 kN according to DIN 4226, Part 3 [6] and a total porosity of 50% and thus achieving a water absorption of 20% of their dry mass. The NWA are dry rounded sand and gravel from the upper Rhine valley. A rapid hardening Portland cement type CEM I 42,5 according to [7] was used. Silica fume was added as a slurry having a specific gravity of about 1.4 kg/l, 50% of the mass being water. A sulphonated naphthalene-formaldehyde condensate was used as superplasticizer. The LWA were introduced wet to the mix. The LWA has moisture stored inside the pores which can be determined by drying and an unknown amount of water adhering on the surface, which has to be considered. The CM-Method commonly used to determine the moisture content of sand turned out to be a reproducible adequate method for measuring the adhering water. For wetting the LWA were submersed 24 hours. Figure 1 shows the pore size distribution [9] of the used LWA. The absorption curve of the dry and desorption curve of the pre-saturated lightweight aggregates in Figure 2 show a high water absorption during the storage of 24 hours followed by an increased water evaporation when exposed to lower relative humidity. The considered relative humidity of 70% appears relatively fast in the interior of high strength concrete [8].

The dry density of the HSC with LWA is 2150 kg/m³. The workability measured on the flow table was ten minutes after mixing a_{10} >50 cm and the compressive strength $ß_{c,100,28}$ =104 MPa was determined on 100 mm cubes at the age of concrete of 28 days. After demoulding, the cubes were submersed in water and then stored in air at T= 20°±1C and 65% RH for 21 days.

Table 1 Mix proportions of reference high strength concrete and HSC with LWA

MATERIALS	AMOUNT PER m³ CONCRETE	
	HSC (reference)	HSC with LWA
Cement CEM I, 42.5 R	450.00 kg	450.00 kg
Water (W)	150.00 l	150.00 l
Dry Mass of Silica Fume (SF)	30.00 kg	45.00 kg
Superplasticizer (SP)	13.60 l	13.60 l
Retarder (R)	1.75 l	1.75 l
Fraction 0/4 NWA	607.00 kg	731.00 kg
Fraction 4/8 NWA	399.00 kg	
Fraction 4/8 LWA		234,00 kg
Fraction 8/16 NWA	729.00 kg	567.00 kg

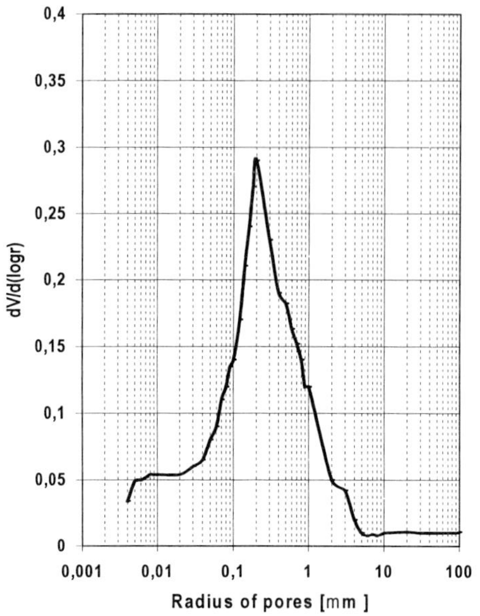

Figure 1 Pore size distribution of LWA [9]

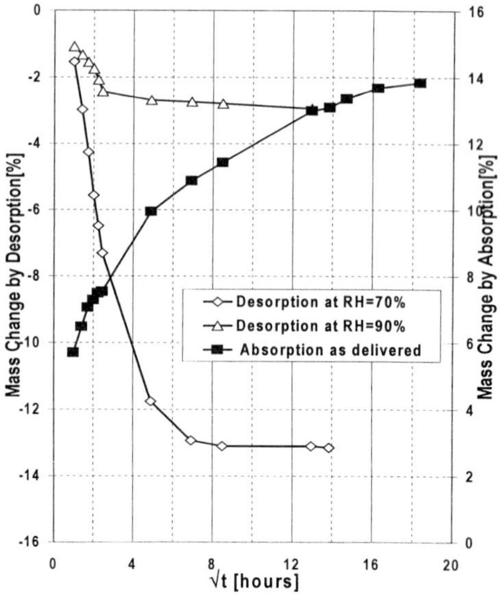

Figure 2 Water aborption and desorption of LWA

Experimental results

Figure 3 shows the monitored mass change versus time and the ratio ß/ ß $_{c,100,28}$ ß $_{c,100,28}$ where ß is the determined compressive strength of cubes 100 x 100 x 100 mm made of HSC with LWA and stored as follows: KK= in air, T= 15°C - 25°C, RH= 40%-45%; KR= in air, T= 15°C - 25°C, RH=65%; KL= sealed in aluminium and plastic foils; NK=6 days submersed in water, then in air T= 20°±1C, RH= 65%. From fragments of the crushed cubes stored in the most unfavourable curing condition KK the physically bound water was eliminated by storage in an exsiccator. After crushing the fragments, particles containing only CP and sand were selected, serving to determine the pore size distribution with mercury porosimetry shown in Figure 4. The other part of the particles was hand milled and sieved through a sieve with mesh < 0.063 mm, obtaining samples rich in hydrated cement paste. From the simultaneously performed thermal gravimetry (TG) and differential thermal analysis (DTA) on these samples the amount of chemically bound water and thus the degree of hydration was calculated as shown in Table 3. Additionally, on samples rich in hydrated cement paste X-ray diffraction was carried out. The obtained chemical composition is shown in Figure 5 and the results from the TG and DTA in Figure 6.

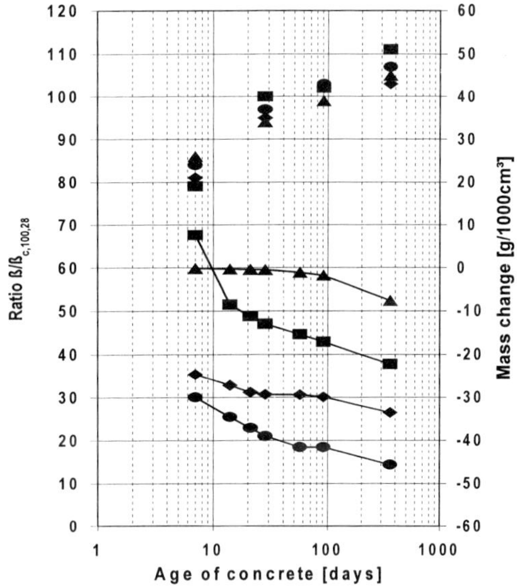

Figure 3 Mass loss and compressive strength of HSC with LWA

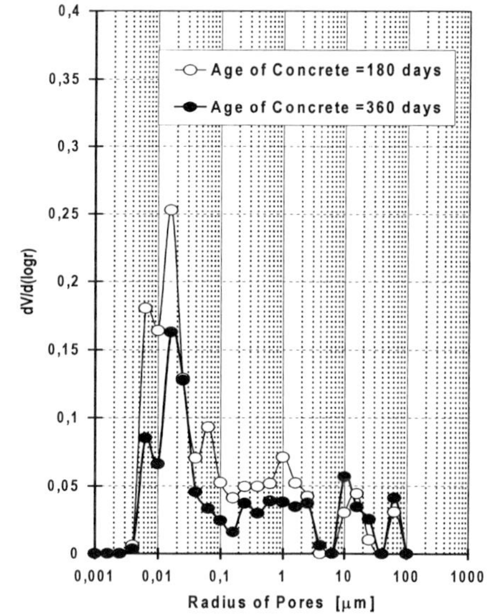

Figure 4 Pore size distribution of hydrated cement paste from HSC with LWA

Table 3 Calculation of the degree of hydration

AMOUNT OF WATER CONSIDERED [l/m³ concrete]	AGE OF CONCRETE	
	180 Days	360 Days
Initial water in concrete at **t=0**	150	150
Chemically bound water when **α=1**	121	121
Total chemically bound water determined by TG and DTA	82	100
Degree of hydration α	0.67	0.82

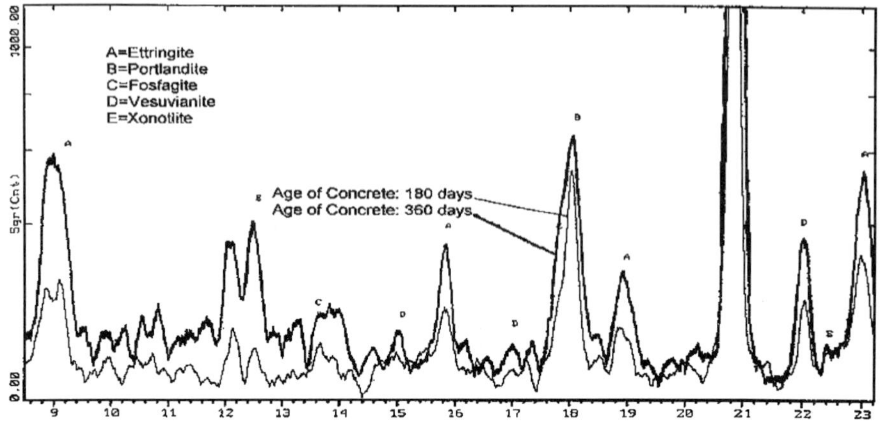

Figure 5 Spectrum of X-ray diffraction

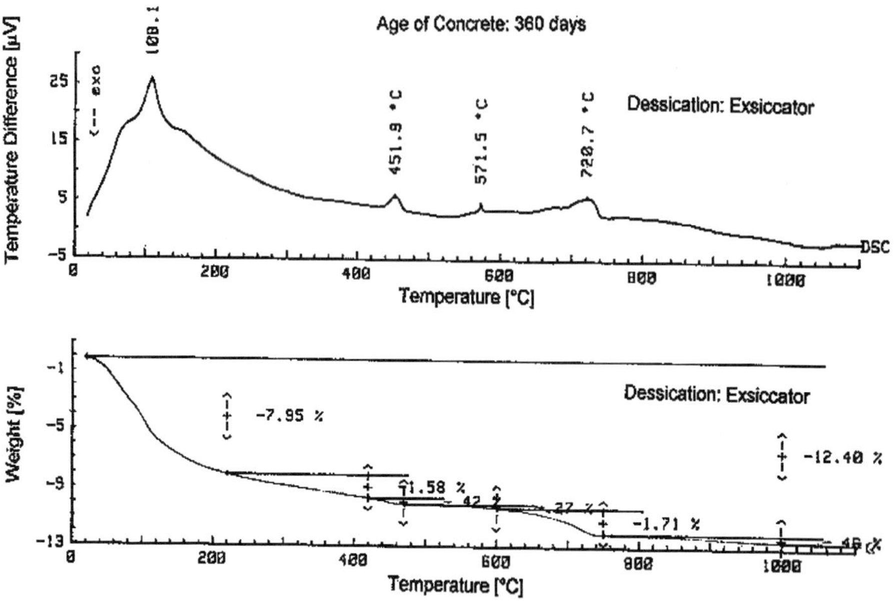

Figure 6 Results from TG and DTA

MODELLING THE WATER TRANSPORT

Water Transport within the Cement Paste

The results of the mercury intrusion are usually given as a differential distribution of the pore volume versus pore radius, using logarithmic scales i.e. Figure 4. By plotting dV/d log r versus log r the area below the distribution curves corresponds to pore volume. Using these curves the pore system of the CP can be defined by the equivalent pore radius [10], which is relatively correct for concrete with pores of smaller radius.

$$r^2_{eq} = \frac{\int_{r_{min}}^{r_{max}} \frac{dV}{d\log r} \cdot r^2 \cdot d\log r}{\int_{r_{min}}^{r_{max}} \frac{dV}{d\log r} \cdot d\log r} \tag{1}$$

Using the transport process in a single capillary pore of radius r_{eq} the water transport of the pore system can be modelled. Assuming a variation of the radii in the pore system with the tendency towards a greater number of pores with smaller radii, r_{eq} is time depending $r_{eq} = f(t)$, being smaller with increasing age of concrete.

When hardening of concrete starts the equivalent pore radius existing at early age of concrete can be considered the greatest equivalent pore radius $r_{eq,max}$ while at the age of concrete when all the water to be transported is used up, the smallest equivalent pore radius $r_{eq,min}$ dominates. The number of pores is considered to be constant. At the beginning of hydration all capillary pores described by $r_{eq,max}$ are filled with water and a hydrostatic equilibrium is established due to capillary pressure and the tension developed at the surface of the meniscus of the fluid. When hydration and/or desiccation occurs the pores become smaller achieving at a certain moment *t* a pore radius $r_{eq}(t) < r_{eq,max}$, leading to a difference in capillary pressure. At any time *t* the capillary pressure in the pores of radius $r_{eq}(t)$ is given by:

$$P_K(t) = \frac{2 \cdot \sigma \cdot \cos\theta}{r_{eq}(t)} \tag{2}$$

with σ =surface tension and θ = contact angle. θ can be assumed as zero.

Water Transport from Lightweight Aggregate to the Cement Paste

For the concrete with lightweight aggregate an equivalent pore radius for the pore system of the lightweight aggregate pores $R_{eq,LWA}$ can be described like in equation (3) by adapting equation (2) and similar the capillary pressure P_{LWA} in the pores $R_{eq,LWA}$ by equation (4). The number and size of the pores are constant in time.

$$R^2_{eq,LWA} = \frac{\int_{R_{min,LWA}}^{R_{max,LWA}} \frac{dV}{d\log R} \cdot R^2 \cdot d\log R}{\int_{R_{min,LWA}}^{R_{max,LWA}} \frac{dV}{d\log R} \cdot d\log R} \tag{3}$$

$$P_{LWA} = \frac{2 \cdot \sigma \cdot \cos\theta}{R_{eq,LWA}} \tag{4}$$

The different stages of the moisture transport from the LWA to the CP are shown and described in Figure 7 using the model of two cylindrical pores in contact. The cement paste is represented by a capillary pore of radius $r_{eq}(t)$ and the water supply of the lightweight aggregates by their equivalent pore radius $R_{eq,LWA}$. When hydration proceeds the smaller pore exerts suction forces in the greater pore locating the level of fluid at the distance L1. The more the meniscus of the pore of radius $R_{eq,LWA}$ retreats, the higher is the capillary tension in the smaller pore. When this tension exceeds the friction forces in the smaller pore the meniscus change its position. The difference of the capillary pressures $\Delta P = P_K(t) - P_{LWA}$ makes the capillary transport on the distance L_1 possible.

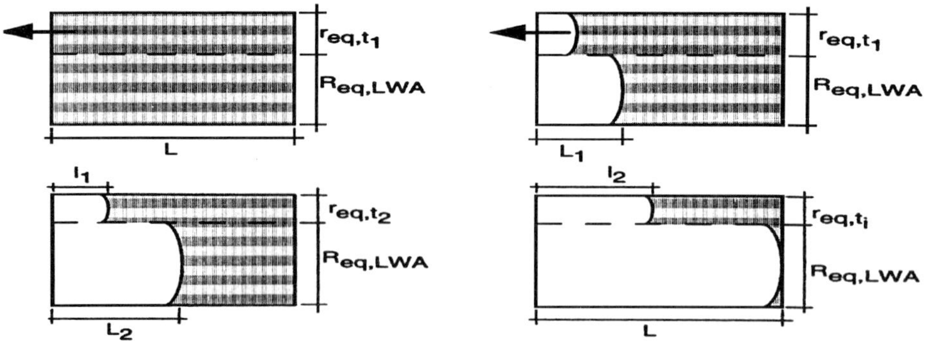

Figure 7 Modelling the water transport from the LWA to the cement paste

A further transport of water is possible only if the meniscus of the smaller pores shifts into position l_1. The capillary transport for this distance is :

$$m \cdot (L_2 - l_1) = \frac{\sigma}{4\eta} \cdot \left(\frac{1}{r_{eq}(t)} - \frac{1}{R_{eq,LWA}} \right) \cdot r_{eq}^2(t) \cdot \pi \cdot \rho_w \tag{5}$$

When the meniscus of the greater pore is in position L, there is no pressure difference ΔP and fluid in the smaller pore is used up either by capillary transport between pores of the cement paste (which can be described similarly) or by diffusion combined with condensation. Between the beginning and the end of hardening this difference is given by

$$\Delta P = \frac{2 \cdot \sigma}{\frac{1}{r_{eq,min}} - \frac{1}{R_{eq,LWA}}} - \frac{2 \cdot \sigma}{\frac{1}{r_{eq,max}} - \frac{1}{R_{eq,LWA}}} \tag{6}$$

For simple calculation the assumption that the pores of the LWA and of the CP have the same length l which is known and the flow of water is laminar through pores is necessary. The distances on which water has to be transported can be modelled starting by the uniform distribution of LWA fraction 4/8 mm through the cement paste. For this the volume of 1m³ concrete is subdivided into a number *n* of identical composite elements each consisting of one aggregate surrounded by a layer of cement paste. The composite elements occupying theoretically the entire volume of the concrete can be spheres, cubes, hexagons. The ratio ζ of the volume sphere / volume cube is 0.52. Using the modelling with spherical composite elements of radius R_{CE}, each aggregate grain is surrounded by a layer of cement paste of constant thickness a. The radius R_{CE} can be calculated with equation (8)

$$n \cdot R_{CE}^3 \cdot \frac{4}{3} \cdot \pi = 1000 \cdot \zeta \qquad (8)$$

An example for this modelling is given in Figure 8 for the LWA grain with R=R $_{LWA}$ and the smallest NWA with radius r_1 and the thickness of the layers of a_{LWA} and a_1 respectively.

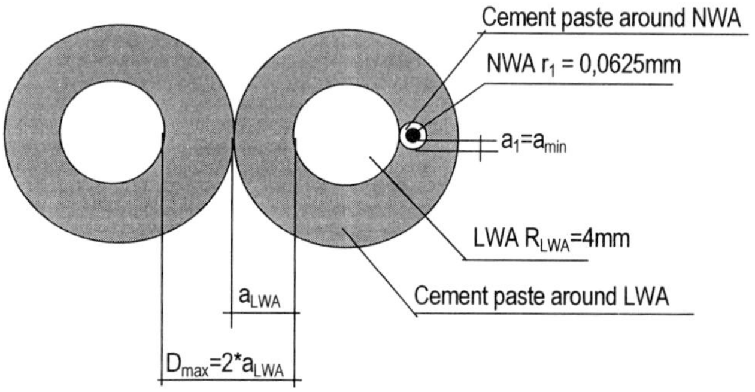

Figure 8 Modelling of composite elements and distances between the aggregates grains

For the LWA grain with a radius of 4 mm, the radius R of the composite element is 5,86 mm. The difference of the radii $R_{CE} - r_a$ gives the thickness of the layer a=1,86mm which surrounds each grain of LWA. The distance between two aggregates of the same size d is two times the thickness of the layer, i.e.= 3,72 mm. Considering that every LWA grain is a water supply for the volume of cement paste surrounding it, the distance on which the water has to be transported can be estimated by considering the limit states of maximum and minimum transport distance. The longest distance D_{max} would be the most convenient case for the concrete, when water transportation on the distance between two lightweight aggregates D_{max}= a_{LWA} =3.72 mm occurs. This would be the longest capillary pore, when assuming their linearity. The most detrimental situation for the concrete would be when a normal weight aggregate is positioned as near as possible to the LWA, being a barrier on the water flow through the cement paste. As the shortest way in this case is the thickness of the layer surrounding the smallest grain, this situation occurs when a aggregate of r_1 =0.0625 mm is in the vicinity of a LWA grain.

DISCUSSION

Independent on curing, a continuous increase in compressive strength with later age of concrete is recorded. The reason lays in the more dense structure of the cement paste indicated by the porosity measurements, which clearly show a decrease of the total porosity with increasing age of concrete. Comparing the pore size distribution of the curves at different ages a refining of the pore system is observed (decrease of the number of pores with great radius and an increase of the pores with smaller radius) leading to the presumption that additional water for hydration was available. The higher degree of hydration at later age of concrete confirms the chemical reaction of this water with unhydrated cement and by X-ray diffraction a higher amount of chemical compounds like calcium hydroxide, ettringite and more C-S-H of the type afwillite at the age of 360 days, compared to the age of 180 days. These aspects are confirming the theoretical assumption of ongoing hydration processes resulting in the predicted more dense structure at later age which is due to the partial filling of the pores by additionally formed hydration products. The peak shifting in the curve of the pore size distribution into the direction of smaller pores stands for creating higher capillary pressure necessary for the water transport from LWA to the CP. From the mass loss of the cubes versus time a humidity gradient is to be expected. From the pore size distribution (Fig.1) and the mass change (Fig.3) the transport conditions for the shown modelling are given (increasing capillary pressure and existence of a humidity gradient). Compared to the conventional HSC the amount of chemically bound water is about 24 l per m^3 concrete higher and represents only the half amount of the water stored in the LWA [5]. Hence this concrete has still water at the disposal which can be transported to the unhydrated cement grains.

The presented concrete shows the enormous advantage of achieving improved macroscopical and microscopical properties in an easy way and with no curing.

ACKNOWLEDGEMENT

The results are part of a research program which was carried out in the framework of AiF (Co-operative Industrial Research) Project AiF -Nr.9816 and co-ordinated by DBV (German Concrete Association) Project Nr.178. The support by Liapor Franken, Elkem, and Heidelberger Zement AG is gratefully acknowledged.

REFERENCES

1. PERSSON, B S M. Self-desiccating high-strength concrete slabs, In Proceedings Third International Symposium Utilisation of High Strength Concrete, Symposium in Lillehammer, Norway, June 20-23, 1993, eds. I. Holland and E. Sellevold, pp 882-889.

2. HAMMER, A. The Maturation of Mechanical Properties of High Strength Concrete Exposed to Different Moisture Conditions. In: Proceedings Third International Symposium on Utilization of High-Strength Concrete, June 20-23, Lillehammer, Norway, 1993, eds. Ivar Holland and Erik Sellevold, Vol 2, pp 1084-1091

3. WEBER, S, REINHARDT, H W. Various curing methods applied to high- performance concrete with natural and blended aggregates. In Proceedings of Fourth International Symposium on the Utilisation of High Strength/High Performance concrete, Volume 3, Paris, May 29-3,1996, eds. F.de Larrard and R.Lacroix, pp 1295-1303

4. GUSE, U. Dauerhaftigkeit von hochfestem Beton. DAfStb-Forschungs-kolloquium. Beiträge zum 32. Forschungskolloquium in Karlsruhe am 21.und 22.März 1996, pp 99-106

5. WEBER, S. Nachbehandlungsunempfindlicher Hochleistungsbeton. Doctoral Dissertation, Stuttgart University, 1996

6. DIN 4226-3. 1983-04. Zuschlag für Beton; Zuschlag mit dichtem Gefüge, Begriffe, Bezeichnungen und Anforderungen- Teil 3: Prüfung von Zuschlag mit dichtem und porigem Gefüge

7. DIN 1164 –1. 1994-10, Zemente- Teil 1: Zusammensetzung, Anforderungen

8. DE LARRARD, F., BOSTVIRONNOIS, J L. Discussion on paper published in Magazine of Concrete Research, 43(1991), no.155, pp. 109-120. Magazine of Concrete Research, 44 (1992) no.159, pp. 143-145

9. CHELOUAH, A. Private communication

10. REINHARDT, H W, GABER, K. Equivalent pore size characterising the pore size distribution of cement mortar. Ceramic Transactions, Vol.16, ed.S. Mindess, Am. Ceram. Soc., Westerville 1991, pp 319-335

THE CHEMISTY AND PERFORMANCE OF PORTLAND CEMENT

C Dungur

B R Heywood K Kendall

Keele University

P Livesey

Castle Cement Ltd

United Kingdom

ABSTRACT. This study involved the characterisation of Ordinary Portland cements obtained from different manufacturing sites, or from the same site at different times. Both qualitative and quantitative data have been collected on a range of samples; the chemical identity of cements was found to vary not only between manufacturing sites but also within sites as a function of ageing.

Keywords: Physical characterisation, Chemical analysis, Ordinary Portland cement, Structure/Performance relationships.

Chanda D Mungur is doing a PhD at Keele University. She graduated (1997) at Keele University with a First Class (BSc) Joint Honours in Chemistry and Mathematics.

Professor Brigid R Heywood is the Chair of Inorganic Chemistry at Keele University. She is Research Director of the Crystal Science Group at Keele which specialises in crystal engineering of inorganics and the microstructure/property/function relationships of solid state materials.

Professor Kevin Kendall is Director of the Birchall Centre for Inorganic and Materials Science at Keele University. He specialises in the chemistry of particulate materials such as polymer latex, zirconium oxide and hydraulic cements.

Paul Livesey is National Technical Services Manager with Castle Cement Ltd, UK, and is a member of Dundee University, CTU, Advisory Committee.

INTRODUCTION

The performance characteristics of cement and hence of concrete are determined by the chemistry of cement during hydration and the developing microstructure of the cement paste system. However, despite continuing research activity in the cement industry, the microstructure/performance relationship of cement hydrates remains poorly defined. This relative lack of success is partly due to a critical absence of good microstructural characterisation [1]. Of the many variables to be considered in establishing the microstructure/performance relationship, the chemistry and performance of the cement powder itself is an essential first step.

In the present study, samples from different manufacturing sites referred to as C1-C8, and samples of different age, (C4A and C4B), were subjected to chemico-physical analysis using a variety of solid state techniques (XRD, XRF, TG/DTA, FTIR, FESEM, spectrophotometry). The chemical sequence of the hydration process was mapped and the data obtained correlated with the performance characteristics of cement and concrete.

This paper presents the preliminary results of a characterisation study of the anhydrous cements. Although characterisation data of an unhydrated oilwell cement has been published by Lota [2], the present study represents the first full, systematic characterisation of UK Portland cements. Moreover since there is little comparative data for mainland UK and continental European sources, the inclusion in this study of such data is also noteworthy. It includes comparisons for inter- and intra-manufacturing sites; work is continuing to correlate the differences reported here with the physical properties of concretes produced from these cements.

MATERIALS AND METHODS

Materials

Ordinary Portland Cement powders, as obtained from Castle Cement Ltd, were investigated.

XRD

A Philips diffractometer with monochromated copper K_α radiation (1.542 Å) and nickel filters, operating at 40 kV and 40 mA, was used to collect the diffracted intensity data using a fixed time step-count procedure over an angular range of $5° \leq 2\theta \leq 60°$ at an interval of 0.05° per step. Samples were placed on a sample holder, which was then front loaded onto the diffractometer. The geometry of the diffractometer and its slit system were kept constant between experiments.

FTIR

A Perkin Elmer 1000 Paragon infrared spectrophotometer was used over a range of 4000-400 cm^{-1} at a resolution of 4 cm^{-1} using the potassium bromide disc technique, whereby 3 mg of unhydrated cement sample was ground very finely with 300 mg of potassium bromide and the mixture was pressed (under a compression of 10 tonnes for a duration of 10 minutes) into a transparent disc. The resulting disc was placed in a suitable holder directly in the infrared beam.

TG/DTA

Simultaneous TG and DTA were recorded from a Rheometric Scientific STA 1500+. The sample (approximately 20mg) was placed in a platinum crucible and heated in a dry nitrogen atmosphere from 25 °C to 1000 °C at a constant heating rate of 10 °C/min and a gas flow of 20 ml/min.

FESEM

A Hitachi S-4500 FESEM machine was used and the samples were mounted on aluminium stubs using adhesive carbon discs and coated with a thin layer (~100 Å) of gold. The accelerating voltage used was 5 kV and the working distance was set at 18 μm.

Reflectance Spectrophotometer

The measurements were taken by an Ihara S-900 Reflectance Spectrophotometer using a 0-45° geometry and measuring over a range of 400 – 700 nm at 10 nm intervals. The instrument was calibrated using a secondary standard white tile which in turn was calibrated using a perfect diffuser and a black trap. The software used to compute the CIE values using D65 was QC++ by Colourware Ltd.

XRF

Studies were carried out on a Philips PW 1606 machine operated at 40 KV and 40 mA.

Specific Surface Area was determined by the air permeability method, setting time by the Vicat needle method and strength tests were performed on EN 196-1 mortar prisms.

RESULTS AND DISCUSSION

XRD

The application of this technique to the study of Portland Cement chemistry is not new and experimental data have continuously been reported in International Symposia since 1938. Here, the XRD diffractograms of the samples revealed the major constituents of cement. Differences in the relative peak intensities within and between samples were observed. The gypsum peak at 11.7° 2θ [3] could only be distinguished in C2 and C8 suggesting that gypsum may be present in a larger proportion in these samples. There was a substantial disparity between the samples in the relative intensity of the alite peak at about 25.3° 2θ [3]. The most intense peak observed was due to the alite, belite and aluminate phases overlapping at about 32.2° 2θ [3], except in C5. Generally, the three major peaks from 32° 2θ, appeared in a descending order with increasing 2θ, apart again from C5 which shows the reverse trend. The XRD patterns of C1 and C2 showed a relatively strong peak at about 44° 2θ [3], which has not, as yet, been identified. The diffractogram of C6 showed no ferrite peak at 12.1° 2θ. The diffractograms of the two samples of different age but from the same manufacturing site revealed that the three major peaks in the middle of the spectrum were in an ascending order of intensity with increasing 2θ for the C4A sample while in the C4B sample, the major peaks appeared in a descending order. This suggests that ageing has an effect on the relative phase composition of the cement and may also indicate that some age-related habit modification of the crystalline phases occurs. Some of these differences were quite marked as illustrated by the two diffractograms presented in Figure 1.

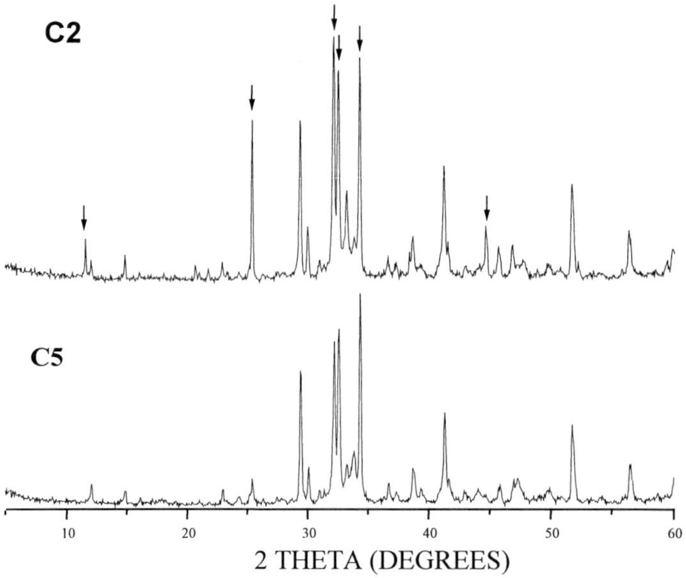

Figure 1 Diffractograms of C2 and C5

FTIR

Interest has been shown in the application of IR spectroscopy to cement chemistry since the 4th International Symposium on the Chemistry of Cement, Washington (1960). In this study it was difficult to identify clearly the belite, aluminate and ferrite phases in the FTIR spectra as they were encompassed in the broad alite bands [4]. However, the spectra of all samples showed the presence of calcium hydroxide, calcium carbonate and gypsum. Calcium hydroxide and calcium carbonate could not be identified in the X-ray diffractograms as they were minor phases and their characteristic d-spacings overlapped with the major alite and belite phases. This shows that the XRD and FTIR techniques complement each other in the information they provide about the samples. An unidentified sharp band at ~1385 cm^{-1} appeared in almost all the samples but was especially pronounced in C1, C2 and C8.

TG/DTA

The TG plots showed three distinct weight losses in the following regions: ~90-200 °C, ~350-450 °C, ~550-650 °C [5]. The first weight loss can be attributed to water egress from gypsum. The decomposition of calcium hydroxide, formed by the reaction of free lime present in the cement samples with moisture, gives rise to a weight loss at ~350-450 °C. The amount of heat associated with this decomposition can be determined by the area under the endotherm at ~350-450 °C from the DTA curve and is shown in Table 1. The large amount of heat evolved in C5 and C7 was attributed to the larger proportion of calcium hydroxide present in these samples; the low heat flow values indicated by C4, C8 and C4B samples reflected their lower calcium hydroxide content. The third weight loss, in the region ~550-650 °C, was ascribed to the decarboxylation of calcium carbonate which may be present as an impurity in gypsum, and, also the carbonation of free lime, rather than the presence of filler. The TG curves have been analysed quantitatively and calcium carbonate was assumed to be the only form of carbonate present in the samples. These quantitative analyses are summarised in Table 2 and Figure 2.

Table 1 DTA analyses

CEMENT SAMPLES	HEAT FLOW (mJ/mg)	MID-POINT TEMP. OF ENDOTHERM (°C)
C1	18.38	426.84
C2	11.63	417.79
C3	21.72	410.15
C4	4.87	425.2
C5	37.49	415.35
C6	17.82	415.96
C7	33.81	415.41
C8	5.19	438.26
C4A	10.54	421.82
C4B	2.84	421.32

Table 2 TG analyses

CEMENT SAMPLES	WEIGHT USED (mg)	WEIGHT LOSS (%)			
		Water	$Ca(OH)_2$	$CaCO_3$	LOI
C1	18.869	0.30	0.30	0.51	1.32
C2	18.752	0.38	0.20	0.44	1.24
C3	17.012	0.10	0.36	0.34	1.01
C4	18.212	0.07	0.17	0.32	0.78
C5	18.956	0.20	0.57	0.92	1.88
C6	18.036	0.14	0.28	0.40	1.14
C7	19.445	0.09	0.51	0.59	1.41
C8	14.941	0.51	0.12	0.42	1.48
C4A	22.068	0.32	0.23	0.32	1.28
C4B	24.050	0.29	0.12	0.25	1.09

The samples showed a range of total loss on ignition, from 0.78 wt % in C4 to 1.88 wt % in C5. This suggests that C4 was a *fresher* sample than C5; we note that in this case the two samples were stored under different conditions with the environment of C4 apparently proving to be optimal. In contrast, it is likely that C5 proved to be sensitive to environmental conditions during transit from Southern Europe to the UK. Calcium carbonate generally seems to make the largest contribution to the loss on ignition, except in sample C8 which had a high water content. Comparing the C4 sample prepared on two different dates (C4A and C4B) also confirmed that the older the sample, the more moisture and carbon dioxide it absorbs from the atmosphere. This would account for the increased water, calcium hydroxide and calcium carbonate content and the larger loss on ignition observed for the C4A sample. Water, calcium hydroxide and calcium carbonate were identified by FTIR and, quantitatively, by the TG/DTA analyses.

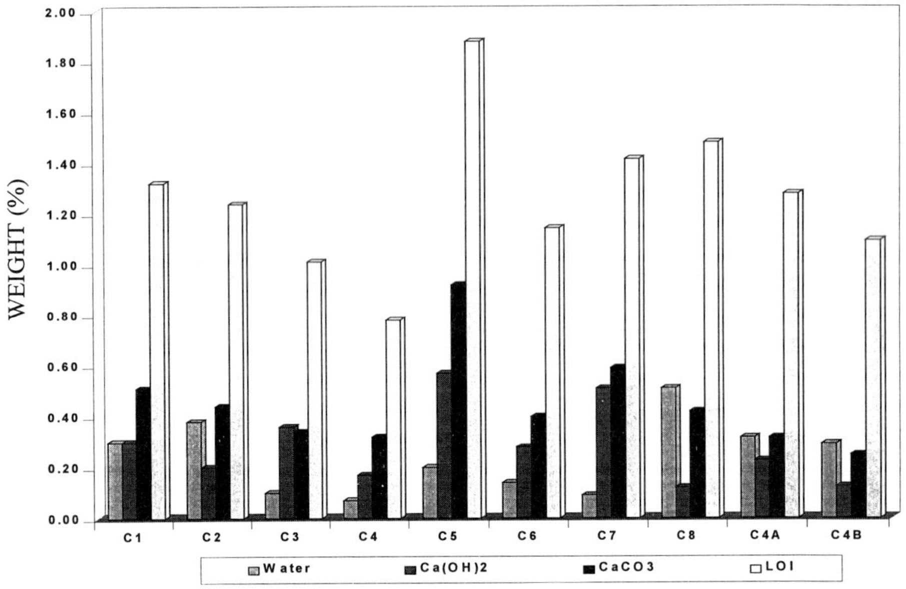

Figure 2 Graphical representation of TG analysis

FESEM

FESEM showed a wide range of particle grain sizes and revealed agglomeration of fine particles on larger grains. At higher magnification, plate-like features of calcium hydroxide could be observed. There was no marked difference in the basic microstucture of any sample.

REFLECTANCE SPECTROPHOTOMETER

The application of this technique to cements for colour measurements is not widespread. The samples show barely any difference in hue and are all yellowish grey although they differed significantly in both *lightness* and *colourfulness* as shown by their reflectance curves in Figure 3. The samples generally reflected more in the red region (at longer wavelength) and less in the blue region (at shorter wavelength) suggesting that they are all yellowish in colour. Some samples reflected more at all wavelengths, indicating that they were lighter than others; as illustrated by C6 and C5 – C6 is lighter than C5.The relative fall off in the lower wavelength region differed significantly between samples; for instance, compared to the rapid tail off displayed by C1, C3 reflected less preferentially in the red and blue regions. C1 is yellower/redder but less blue/green than C3.

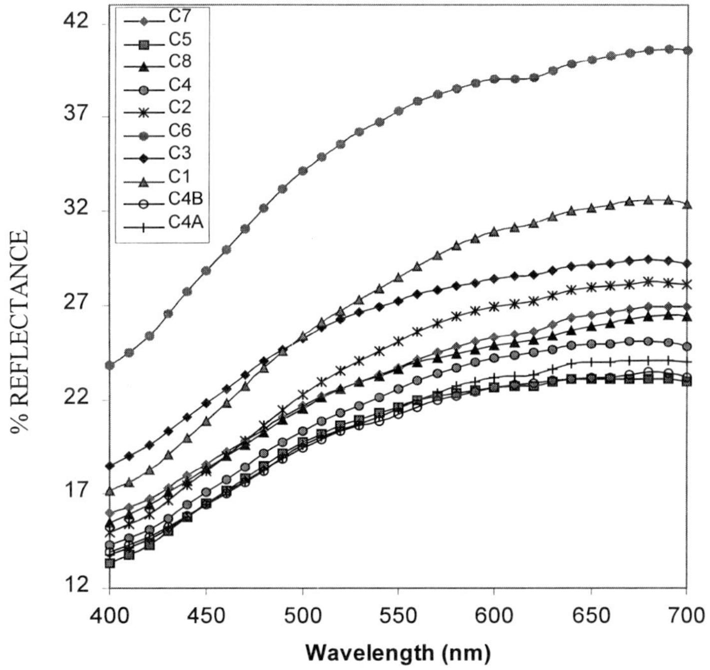

Figure 3 Reflectance curves of the samples

The reflectance curves were transformed into colour coordinates as described by the Commission International d'Eclairage (CIE), using the computing package, Colourware. The results are tabulated in Table 3. Figure 4 and 5 show plots of L vs a* and L vs b* where L is a scale of lightness and 0 represents perfect black and 100 represents pure white, the positive axis of a* shows the degree of redness and the negative axis shows the degree of greeness, the positive axis of b* represents yellow and negative axis blue.

The major variations between the samples lie in their lightness (L-axis) and in their yellowness (b* axis). There was little variation in the redness and greeness of the samples (a* axis).

The grey colour of Portland Cement is attributed essentially to the calcium aluminoferrite phase and possibly the presence of chromium and manganese oxide. The colour intensity of clinker can be reduced significantly and, therefore, the whiteness/lightness increased, by the existence of reducing conditions in the kiln which induce the reduction of iron (III) to iron (II), manganese (III) to manganese (II) and chromium (VI) to chromium (III) or (II)[6]. Incomplete oxidation during cooling has also been reported to affect the colour of cement[7].

There is a substantial amount of iron (II) present in the burning zone in a normal kiln atmosphere and if oxidation is incomplete during cooling, the clinker nodules are likely to be yellower in colour. Therefore, reducing kiln conditions and incomplete oxidation during cooling can account for lighter and yellower clinker nodules and can be extended to explain the varying colour of Ordinary Portland Cements.

Table 3 CIE values of the samples

SAMPLES	L	A	B
C1	58.249	0.019	10.859
C2	52.640	0.026	9.673
C3	57.418	-0.860	8.189
C4	55.418	-0.343	10.020
C5	51.895	-0.990	8.450
C6	64.217	-0.848	10.276
C7	54.619	-0.329	8.603
C8	53.427	-0.420	9.047
C4A	52.627	-0.221	9.278
C4B	52.345	-0.498	8.536

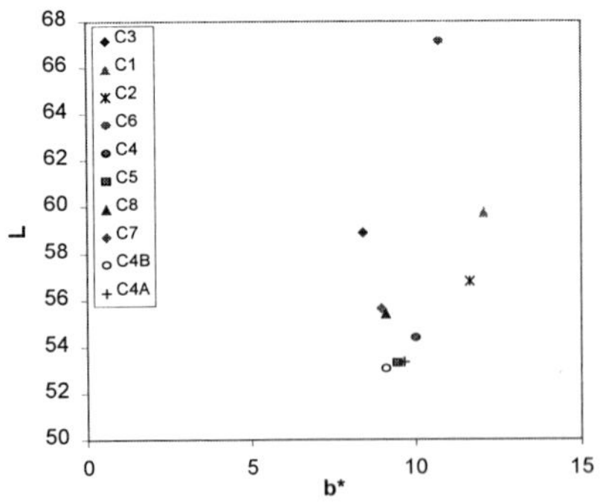

Figure 4 Plot of L versus a^*

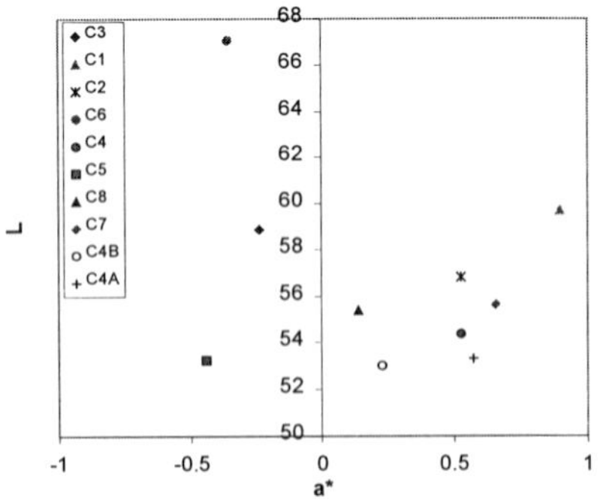

Figure 5 Plot of L versus b^*

XRF

Table 4 summarises the proportion (as a percentage) of iron oxide and manganese oxide present in the samples. The proportion (as a percentage) of the four major phases in the samples were obtained from the modified Bogue calculations and XRF studies and these results coupled with some physical properties are summarised in Table 5. In simple terms, C6 has the lowest iron oxide content, hence its pale appearance; the high iron oxide and manganese oxide contents of C8 explained its dark colouration. Other more subtle differences including the distinction between iron (II) and iron (III) are still to be investigated.

PERFORMANCE DATA

There appears to be a general correlation between alite and belite contents and strength performance and between specific surface area and setting time. As expected the relationship is not linear and other factors influencing the physical properties of cement are under investigation.

Table 4 Iron oxide, chromium oxide and manganese oxide content of the samples

SAMPLES	Fe_2O_3	Cr_2O_3	Mn_3O_4
C1	2.14	<0.01	0.05
C2	2.90	<0.01	0.04
C3	3.22	<0.01	0.10
C4	3.06	0.02	0.04
C5	3.74	0.06	0.04
C6	1.65	0.01	0.05
C7	3.28	<0.01	0.23
C8	3.80	0.01	0.23
C4A	2.95	0.01	0.05
C4B	3.04	<0.01	0.05

Table 5 Chemical and physical properties of the samples

	C1	C2	C3	C4	C5	C6	C7	C8	C4A	C4B
Alite	49.5	53.1	49.8	50.0	56.9	57.3	47.9	61.2	51.1	54.6
Belite	21.2	19.4	20.3	20.6	14	17.7	22.9	11.1	20.7	18.1
Ferrite	6.7	8.4	9.8	9.2	11.7	5.2	9.3	11.8	8.8	9.2
Aluminate	11.5	8.3	7.6	8.5	5.5	9.7	8	5.1	8.3	7.9
SSA (m^2/kg)	350	395	400	380	370	360	380	390	301	305
Set Time (s)	140	100	100	135	190	115	125	150	120	130
2-day strength (N/mm^2)	25.5	28.0	23.0	31.5	24.0	27.5	25.5	29.0	21.0	25.5
7-day strength (N/mm^2)	46.0	44.0	41.0	44.0	41.0	48.5	40.0	48.5	33.0	35.0
28-day strength (N/mm^2)	61.5	59.5	59.0	59.5	58.0	60.5	57.5	60.5	46.0	48.0

CONCLUSIONS

The techniques used have, to date, revealed subtle differences in the relative proportions of some phases present in cements. It is significant to note that the integrated use of a range of solid state techniques was required to effect the complete mapping of composition. The subtle effects of storage and changing environmental conditions on composition of Ordinary Portland Cements can be identified. Moreover, the relationship between chemical/elemental composition and Ordinary Portland Cements colour has been qualified using a new technique. We also note that differences in the chemical profile of the cements do not translate directly into a predictive assay of concrete performance (Table 5). This suggests that the reactivity of the cement may be affected by the aggregate during hydration. Indeed, the results of a parallel study have also highlighted the potential significance of aggregate[8]. No similar analysis has been published before especially regarding colour measurements. This research is ongoing: a study of the hydration process, and the correlation of performance with characterisation data for the anhydrous and hydrated cements continues and will be reported when completed.

ACKNOWLEDGEMENTS

The authors would like to express their thanks to Dr S. Westland for assistance with the colour measurements.

REFERENCES

1. SCRIVENER K L, Materials Science of Concrete, American Ceramic Society, 1989, pp 127-161.

2. LOTA J S, BENSTED J and PRATT PL, L'industria italiana del Cemento, 1998, No. 729.

3. TAYLOR H F W, Cement Chemistry, 2nd Edition, Thomas Telford Services, 1997, pp100.

4. BENSTED J, Cement Technology, 1974, Vol 5, No 4, pp 378-82.

5. POPE M I and JUDD M D, Differential Thermal Analysis: A Guide To The Technique And Its Applications, Heyden and Sons Ltd., 1977, pp149.

6. BENSTED J, White and coloured Portland Cements, World Cement, 1993, pp13-19.

7. SCRIVENER K L, TAYLOR H F W, Clinker nodules with light-coloured centers, Zement-Kalk-Kips, 1995, Vol 48, pp34-39.

8. BUCKLES G, HEYWOOD BR, KENDALL K, MILNES K, see conference proceedings.

FUNDAMENTAL BASES OF FORECAST OF TECHNICAL CHARACTERISTICS OF CONCRETE

L B Svatovskaya **V Y Solovieva**

N N Shangina **P G Komohov**

Railway University

M M Sychov **V G Korsakov**

N A Stepanova

St Petersburg Technological Institute

Russia

ABSTRACT. Work is devoted to the use of the fundamental approaches to the forecast of characteristics of binders and concretes - strength, resistance to cracks, electric conductivity. For the base is taken the thermodynamic aspect of strength, as a reflection of running of chemical reactions in the system, as well as energy and configuration of electronic orbitals, which form the surface of filler, and zone model of solids. It is shown that on the base of such sort of concepts it is possible to forecast characteristics and develop fundamentally new materials – for instance shine concretes.

Keywords: Thermodynamics, Strength, Phosphorescence, Resistance to cracks, Portland cement, Filler, Structure models, Configuration of electronic orbits.

Professor Larisa B Svatovskaya is the Head of Department of Engineering Chemistry and Environment Protection, Railway University, St.Petersburg, Russia. She specializes in the chemistry of the binders.

Professor Valentina Y Solovieva is Lecturer in Department of Engineering Chemistry and Environment Protection, Railway University, St.Petersburg, Russia.

Professor Nina N Shangina is Lecturer in Department of Building Materials, Railway University, St.Petersburg, Russia. Area of scientific interests – binders and concretes.

Professor Pavel G Komohov is Head of Department of Building Materials, Railway University, St.Petersburg, Russia. Area of scientific interests – concretes.

Dr Maxim M Sychov is Lecturer in Theoretical Bases of Materials Technology, St.Petersburg Technological Institute, Russia.

Professor Vladimir G Korsakov is Head of Department of Electronic Engineering Materials, St.Petersburg Technological Institute, Russia. Specialist in the chemistry of solid.

Dr Natalia A Stepanova is Lecturer in St.Petersburg Technological Institute, Russia. Area of scientific interests – electroluminescence.

INTRODUCTION

Thermodynamic consideration brings about the conclusion that depending on nature of cement (silicate, phosphate, aluminate), other things being equal (W/C and others) the strength of material, σ, may be predicted and is a function of energy content of the system ΔH [1...4]. Level of predicted strength σ_i at given time τ_i reflects part of energy of reaction, passed till that time. The difference between the most reached in given reactions σ_0 and σ_i is a thermodynamic reserve of the system TR. Under such setting up a question, knowing the value of energy content of a system ΔH^0_{298}, calculated on reactions, and the value of corresponding strength, it is possible to control strength by means of extracting from the system its energy reserve, with help of additives for instance. In total, due to the law of energy conservation and postulates of thermodynamics it is possible to predict the strength of a material. In this knowledge system is valued durability of stone in environment – the lower is the energy content of stone, or otherwise, the "colder" it is, the more it durable in environment. In this paper technics of corresponding evaluations are proposed.

Structure of solids on the electronic level allows to develop cements with plastic characteristics or material with crack resistance, as well as electroconductive or dielectrical [1]. The main attention is paid to electronic configurations on the interfaces of solid phases. Offered categorization of binders and fillers is based on their energy and electronic nature and allows to predict technical features of materials. Main conclusions of this part of work indicate that directed d-orbitals on the interfaces ensure increased crack-resistance of the material.

EXPERIMENTAL DETAILS

Materials

In work were used Portland cement M 400, sand, rubble; wastes of St.Petersburg area, containing d-metals.

RESULTS AND DISCUSSION

In Table 1 are shown energy calculations of system and system control using additives. Table data indicate that it is possible to control the energy of system – to transform intrinsic energy of the system in its strength. To reach such sort of transformation is possible by the use of the introduced additives of certain nature. The Figure 1 is based on the data of Table 1 and shows the relationship between calculated TR (energy reserve) of the concrete and its measured strength (R^2=0.964).

The problem of development of the electroconductive concretes was solved by the use as fillers and aggregates of fine-grained concrete and carbon-containing natural and artificial materials: carbon black, shungite, shungizite and oxidized graphite. Materials were chosen, the use of which as a filler allows both to reduce a resistance of concrete and to increase its strength. Among these fillers are natural shungite and obtained by its roasting shungizite, as well as oxidized graphite. Characteristics of given materials when changing by them of the part of the cement not to reduce, but in some cases to increase the strength of the concrete is explained by the high intensity of surface active centres of certain nature. Hardening was carried out during 24 hours at 20^{0}C

Table 1 Thermodynamical calculation of model system C_3S-H_2O

ACTIVATOR	MASS LOSS IN DECOMPOSITION OF $CA(OH)_2$, %	DEGREE OF PROCESS PASSING	ΔG_{298}, KJ	TR, KJ	COMPRESSIVE STRENGTH, N/mm^2
$SnCl_4$	2.2	$0.3C_3S+0.65H_2O \rightarrow 0.3Ca(OH)_2+0.3C_2SH_{1.17}$	-14.79	-34.49	24.0
PbO_2	1.9	$0.25C_3S+0.54H_2O \rightarrow 0.25Ca(OH)_2+0.25C_2SH_{1.17}$	-12.33	-36.95	20.0
$KMnO_4$	2.15	$0.28C_3S+0.60H_2O \rightarrow 0.28Ca(OH)_2+0.28C_2SH_{1.17}$	-13.79	-35.49	23.6
Fe	1.8	$0.24C_3S+0.52H_2O \rightarrow 0.24Ca(OH)_2+0.24C_2SH_{1.17}$	-11.83	-37.45	19.2
Fe_2O_3	1.5	$0.20C_3S+0.43H_2O \rightarrow 0.2Ca(OH)_2+0.2C_2SH_{1.17}$	-9.86	-39.42	13.9
$CaSO_4$	1.9	$0.26C_3S+0.56H_2O \rightarrow 0.26Ca(OH)_2+0.26C_2SH_{1.17}$	-12.79	-36.49	20.2

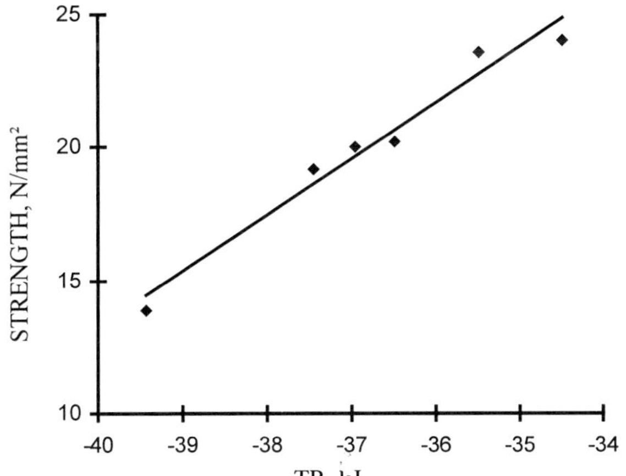

Figure 1 Relationship between the TR of concrete and its strength

The use of shungizite sand (fraction 0...5 mm) instead of building sand in the fine-grained concrete allows to reduce the resistance of material from 15x10^5 Ohmm to 33x10^3 Ohmm. Much more reduces resistance the use of binder, obtained by the joint grind of shungizite or shungite and Portland cement. Binder, filled with shungite (30% from the mass of cement) has resistivity 1400 OhMM and strength of stone 48,4 N/mm². Electroconductive matrix was obtained, which conductivity was enlarged by adding of carbon black, which has defect to enlarge water consumption of the concrete that sharply limits its possible content. The use of three-component binder allows to organize structure of material with increased conductivity.

PRACTICAL APPLICATION

In the Table 2 is shown assortment of developed concrete products, obtained with reduced energy consumption, in the Table 3 are shown examples of product possess increased resistance to cracks. These qualities are providing by additive "DEYA". In the last case Portland cement is used as an activator in the mixtures in the amount 5...7%. Chosen composition of electroconductive concrete allowed to develop on its base the new electrotechnical product, named "shine concrete", representing itself the following. Electroluminescent coating, consisting of consecutively applied over the electroconductive concrete: dielectrical layer, electroluminescent layer and transparent electrode, gives a concrete new characteristic - a phosphorescence under the action of the electrical current. On the base of cement and two types of carbon-containing fillers received electroconductive concrete with the compression strength 42,5 N/mm² and electroresistance 625 OhMM. Electroconductive concrete is used in the new type of technical products - "shine concretes".

Table 2 Assortment of developed concrete products with activated hardening and contracted consumption of energy and cement

PRODUCT	M	CASTING TEMP, °C	ADDITIVE "DEYA", % MASS	W/C	COMPRESSIVE STRENGTH, N/mm² (DAYS) 1	28	90	X,[1] CYCLES	Y,[2] ATM
Column	400	80	–	0.43	32.4	42.1	43.5	100	4
		45	1.5	0.43	33.7	44.1	47.6	150	6
Cross bar	300	80	–	0.48	22.6	29.4	31.8	100	4
		45	1.5	0.48	23.7	30.8	33.3	150	6
Wall panel	200	80	–	0.56	14.2	18.5	20.2	50	2
		45	1.5	0.56	16.6	21.6	23.5	100	4

1 X = Resistance to freezing   2 Y = Degree of waterproofness

The main characteristics of samples of shine concrete: brightness 80...100 cd/m² (220 V, 400 Hz), power input 50...100 W/m². The main advantage of the material is the possibility of getting of durable electroconducting substrate of any form from the monolithic concrete.

Table 2 Assortment of developed building composites

PRODUCT	M	BINDER CONTENT, CONTAINING D-ELEMENT Kg/m^3	W/C	COMPRESSIVE STRENGTH AFTER STEAMING, N/mm^2 3 Months	6 Months	C_{str}/B_{str}*	X, CYCLES
Pavement board	400	460	0.35	48.1	52.4	0.27	200
		500	0.33	45.6	49.2	0.22	200
	300	320	0.39	35.6	38.5	0.26	200
		350	0.38	34.3	36.7	0.21	200
Foundation block	200	245	0.55	24.9	26.4	0.24	100
		270	0.54	23.5	24.8	0.21	100
	150	210	0.58	18.2	19.1	0.23	100
		230	0.56	17.0	17.5	0.2	100
Wall panel	100	180	0.58	12.7	13.0	0.52	50
		190	0.57	11.5	11.7	0.4	50
	75	160	0.60	9.0	9.2	0.5	50
		170	0.59	8.3	8.4	0.4	50

* Compressive Strength / Bending Strength

CONCLUSIONS

1. It is shown that using concepts of the thermodynamic reserve TR of chemical reactions of hardening systems (energy reserve) it is possible, extracting intrinsic energy by means of, for instance, additives, to save heat during steaming and cement and to obtain material with higher quality.

2. It is shown that the use of materials – waste of industry, containing elements with directed d-orbits in high degrees of oxidation it is possible to develop materials with increased resistance to cracks (increase of bending strength to compression strength ratio).

3. It is shown that the use of concepts of electronic configurations, contained, for instance, in the graphite, it is possible to develop the new types of the concretes – "shine concretes".

REFERENCES

1. SVATOVSKAYA L.B. Engineering Chemistry, Part 1. Press of SPb. Railway University, St.Petersburg, 1995, pp 95

2. SVATOVSKAYA L.B. Engineering Chemistry, Part 2. Press of SPb. Railway University, St.Petersburg, 1998, pp 92

3. SVATOVSKAYA L.B. Thermodynamic aspect of strength. Cement, No. 7, 1996, pp 6...8.

4. SVATOVSKAYA L.B. Thermodynamic and electronic level of strength reserves of cement materials. Bulletin of VUZ, Series a construction, №8, 1998, pp 18...22.

THE EFFECT OF WATER-TO-CEMENTITIOUS MATERIAL RATIO ON THE DURABILITY OF CONCRETE CONTAINING FLY ASH

S Amey

H Brown

J Moreau   A Atienza

Master Builders Inc

United States of America

ABSTRACT. The use of fly ash as a replacement of portland cement is an attractive option to the ready-mix concrete industry for reasons such as economics and enhanced engineering properties of hardened concrete. There are also known disadvantages that inhibit the increased use of fly ash (e.g., reduced resistance to deicing salt scaling, set time retardation and slow compressive strength development). The objective of this study was to evaluate the effect of water-to-cementitious material ratio on scaling performance. This ratio was set using either additional water or the addition of an admixture based upon polycarboxalate chemistry and a blend of non-chloride accelerators to achieve a fixed slump. The study compared admixture-treated and non-treated concrete mixtures of different qualities with cementitious material contents of 250 and 305 kg/m^3 and four levels of ASTM C618 class C fly ash (0, 15, 30, 45% by weight). By increasing the cementitious factor, decreasing the permeability, decreasing the initial set and increasing the strength of a mixture, the scaling resistance can be positively affected, however, none of these factors on its own was conclusive. However, concrete that had accelerated set, decreased water-to-cementitious material ratio and increased strength development were not found to scale significantly in the study.

Keywords: Admixture, Fly ash, Water, Scaling, Durability, Permeability, Air-voids.

Stephen Amey is the group technology manager for admixture product development at Master Builders, Inc. He has eight years of experience studying the environmental degradation of materials

Heath Brown is a member of the admixture product development group at Master Builders, Inc. He holds a B.A. in technology from Kent State University. Certified as an

John Moreau is a product marketing manager for admixtures at Master Builders, Inc. He has five years of experience in developing admixtures and the construction industry.

Armand Atienza is a technical specialist for Master Builders, Inc. He has three years of experience in developing admixture products.

INTRODUCTION

The use of fly ash as a replacement for portland cement is an attractive option to the ready-mix concrete industry because of its economic advantages and its improvement in the engineering properties of hardened concrete (e.g, late age strength, permeability) [1,2]. Although these advantages are known throughout the industry, there are also known disadvantages that inhibit the growth of fly ash use. Three of these disadvantages are deicing salt scaling resistance [1,3-10], set time retardation [1,2,4] and slow compressive strength development [1,2,4,7,11].

While the two latter limitations are inconvenient, most regulatory agencies [12,13] limit the use of fly ash in concrete because of its apparent susceptibility to scaling damage. The building code for structural applications in the United States (ACI 318-95 [12]) sets three limitations for concrete containing fly ash that is to be used in applications where freezing and thawing or deicing chemicals are present. First, the fly ash content must be less than 25% of the total cementitious materials. In addition, ACI 318 specifies a minimum compressive strength requirement of 31 MPa and a maximum water-to-cementitious material ratio (w/cm) of 0.45.

The published literature on the subject suggests that there is much agreement that fly ash replacement levels of over 50% result in significant scaling [1,3,4,7,9,10]. However, many studies [5,6,8,10] suggest that no significant scaling occurs when fly ash is used in replacement amounts ranging from 20% to 40%. Some have suggested the range of reported scaling results is due to inadequately developed air void systems [1,10,11,14] or to that system deteriorating over time [8]. However, some of the most severe scaling in the above studies has been observed in concrete with air contents of over 5%, durable air void parameters and good durability factors (by ASTM 666 [15]). Others have suggested curing procedures have an effect upon scaling resistance [1,8] because of its role in strength development. Yet others suggest that the enhanced scaling resistance obtained for mixtures with low values of w/cm is perhaps due to the significant effect of w/cm on strength development and permeability [5,6]. One of these studies [5] also suggests that the pozzolans should be neglected altogether in specifications and water-to-cement (w/c) ratio alone should be controlled to minimize scaling. While defining what is necessary to prevent scaling of concrete containing fly ash is not completely clear, it is evident that durable concrete can be made under conditions that defy the current ACI 318 specification.

Recently, an admixture (PC100), based upon polycarboxylate chemistry and a blend of non-chloride accelerators, has been shown to provide normal setting characteristics to concrete containing up to 30% fly ash [2]. This admixture also provides considerable water reduction and improved strength development in mixtures containing fly ash.

The objective of this study was to evaluate the effect of w/cm on scaling performance of concrete containing fly ash. This was achieved by comparing the PC100-treated and non-treated concrete mixtures of different qualities with four levels of ASTM C618-94a class C [16] fly ash replacement. The value of w/cm was varied by adding additional water to achieve a fixed slump. This strategy produced a range of concrete mixtures which were compared on the basis of w/cm, compressive strength, set time, chloride diffusion coefficient, rapid chloride permeability, freeze-thaw resistance, scaling resistance and fly ash content.

Mechanisms are proposed to explain performance of the mixtures, the fly ash and the admixture under scaling conditions.

EXPERIMENTAL

Materials

The admixtures used in making the concrete are commercially available. In addition, all chloride solutions (for scaling tests) were made using reagent grade chemicals and de-ionized water. The solutions were prepared at least one day prior to use in the testing.

All specimens were made from 0.085 cubic meter laboratory concrete mixers in accordance with ASTM C 192-90a [17] for making, handling and storing laboratory concrete specimens. The mixture proportions for all of the mixtures used are listed in Table 1 with their plastic properties.

Table 1 Concrete mixture proportions

FA %	CF kg/m^3	CEMENT kg/m^3	SAND kg/m^3	STONE kg/m^3	WATER kg/m^3	DENSITY kg/m^3	PC100 mL/100g	AE mL/100g	SLUMP mm	AIR %
0	249	249	821	1087	154	2310	0	110	120	6.4
15	249	211	790	1089	150	2278	0	110	120	6.5
30	249	174	822	1089	150	2278	0	100	120	5.9
45	252	141	811	1089	141	2289	0	105	135	6.5
0	249	249	847	1102	139	2336	1300	60	115	6.4
15	249	211	847	1102	125	2323	1300	50	115	6.3
30	249	174	842	1096	129	2316	1300	40	115	6.2
45	251	138	834	1131	132	2348	1300	40	120	6.4
0	305	305	781	1062	168	2316	0	120	130	6.1
15	307	260	785	1067	162	2321	0	115	160	6.2
30	309	216	790	1074	157	2329	0	110	120	6.0
45	309	169	791	1074	147	2321	0	110	120	6.0
0	304	304	798	1082	152	2336	1300	80	115	6.3
15	306	259	803	1087	147	2342	1300	70	115	5.8
30	304	212	797	1080	137	2316	1300	65	120	6.5
45	309	169	809	1096	134	2348	1300	50	120	5.7

FA = Class C Fly Ash ; CF = Total Cementitious Material Factor ; AE = Air Entraining Admixture

PC100 is a commercially available admixture that is based upon polycarboxalate chemistry and a blend of non-chloride accelerators. It meets ASTM C 494-92 [18] for a type A, C, and E admixture. It has an active solids content of 40% and a typical dosage range of 455 to 1960 mL/100 kg. This admixture is added to concrete mixtures to provide water reduction, to improve setting characteristics (e.g., accelerated set times) and to enhance strength development in mixtures containing fly ash. The dosage of PC100 used for all "treated" mixtures in this study was 1300 mL/100 kg. To gain insight as to the effect of w/cm on scaling performance, two different mixture qualities were used. The first (L) contained a nominal total cementitious content of approximately 250 kg/m^3, while the second (H) contained a nominal total cementitious content of 305 kg/m^3.

Table 2 Primary mineral compositions

	FLY ASH	CEMENT
Class	"C"	Type I
Fineness		
Blaine, m^2/kg	-	333
& ret, 45μm sieve	6.20	-
Specific Gravity	2.62	3.15
SiO_2	36.49	21.23
Al_2O_3	20.25	5.27
Fe_2O_3	4.91	3.17
CaO	25.22	64.04
MgO	6.03	1.39
SO_3	1.16	3.00
K_2O	-	0.57
Na_2O	-	0.14
Total Alkalis	1.21	0.52
Loss on Ignition	0.22	1.75
Free CaO	-	0.72
Insoluble	-	0.52
C_3S		51.0
C_2S		22.0
C_3A		9.0
C_4AF		10.0

Four levels of ASTM C 618-94a Class C [16] fly ash (0,15,30,45% by weight) were used to replace Type I cement and to achieve the desired cementitious factors. Class C fly ash was chosen because of its retardation effect upon setting times and the fact that an accelerating admixture was being used (i.e., PC100). Table 2 lists the primary mineral compositions of the cement and the fly ash used in the evaluations.

The mixing of the concrete took place at ambient temperatures of 20° ± 3°C. The w/cm was set after all materials were batched in the mixer by adding water to achieve a fixed slump of 125 ± 10 mm. One set of mixtures was treated (LT, HT) using a fixed dosage of PC100 (added up front with the mixing water), while another was left untreated (L,H). All of the mixtures had plastic air contents of 6.0 ± 0.5%. These mixture proportions were used because they are typical of residential concrete in the United States. All sand and aggregate weights listed in the tables are saturated surface dry weights. The sand is a natural silica sand (specific gravity: 2.62) and the coarse aggregate (specific gravity: 2.79) used was a crushed limestone fitting a #57 stone gradation. All water contained in the admixtures was accounted for in the mixing water.

Methods

All specimens were stored under conditions of 20°C and 100% relative humidity unless indicated differently by the applicable ASTM methods. Specimens were cast for evaluating compressive strength (three 100x200 mm cylinders per age), air void parameters (one 100x200 mm cylinder), freezing and thawing resistance (three 75x100x400 mm beams), resistance to deicing chemicals (two 75x300x300 mm prisms) and chloride diffusion (three 100x200 mm cylinders).

Air void parameters were determined by ASTM C 457 [19]. The modified point count method was used at 100x magnification. The resistance of all specimens to freezing and thawing was determined by ASTM C666 [15] and the cycling was performed at a rate of 12 cycles per day. In addition to determining durability factors, weight loss was also measured over the 300 cycle period. The resistance of all specimens to scaling by deicing salts was determined by ASTM C 672 [20]. The specimen area exposed to the chloride solution was approximately 0.085 m^2. The specimens were evaluated by a visual rating system (0-5) at five cycle intervals (e.g., "0" would indicate no scaling and "5" indicates severe scaling, as per the test method). The permeability of the concrete to chloride penetration was determined by ASTM C 1202 [21] and by a modified ASTM C 1202 procedure as described by Streicher [22] to gain effective diffusion coefficients. This latter procedure is referred to as a chloride migration technique.

RESULTS

Rate of Hardening Performance

Initial and final setting times are illustrated in Figure 1. The initial set times were found to increase with increasing fly ash content and decreasing cementitious content. Furthermore, the PC100 decreased the initial set time relative to the untreated mixtures.

In all but one mixture (i.e., 45% fly ash, 250 kg/m^3), the treated mixtures allowed 15% more fly ash to be used *without increasing the set time*, when compared to an untreated mixture. For mixtures treated with 1300 mL/100 kg of PC100 and containing no fly ash, the initial and final sets were accelerated by approximately 30 minutes when compared to a comparable untreated mixture. For mixtures containing 15, 30 and 45% fly ash replacements, the initial and final set times were accelerated by approximately one hour when PC100 was present in the mixture.

Water-to-Cementitious Material Ratio

Concrete was produced from all mixtures that was within the target slump (125 ± 10 mm) and air (6.0 ± 0.5%) requirements. The w/cm values for the H and HT (305 kg/m^3) concrete ranged from 0.43 to 0.55. The w/cm values for the L and LT concrete (250 kg/m^3) ranged from 0.50 to 0.62. Note, however, that by neglecting the fly ash, the w/c values were as high as 0.87 and 1.00 for the H/HT and L/LT mixtures, respectively. Only two mixtures (30-HT and 45-HT) in this study were at or below a w/cm value of 0.45. The PC100 admixture provided 7-10% and 6-16% water reduction in the HT and LT mixtures, respectively (Tables 1 and 3). Hence, the treated mixtures had lower w/cm than the comparable untreated mixtures.

Air-Void Evaluation

All of the concrete mixtures exhibited an air void structure typical of durable concrete (Table 3). No air void data was obtained to characterize differences between the surface of the specimens and the bulk concrete.

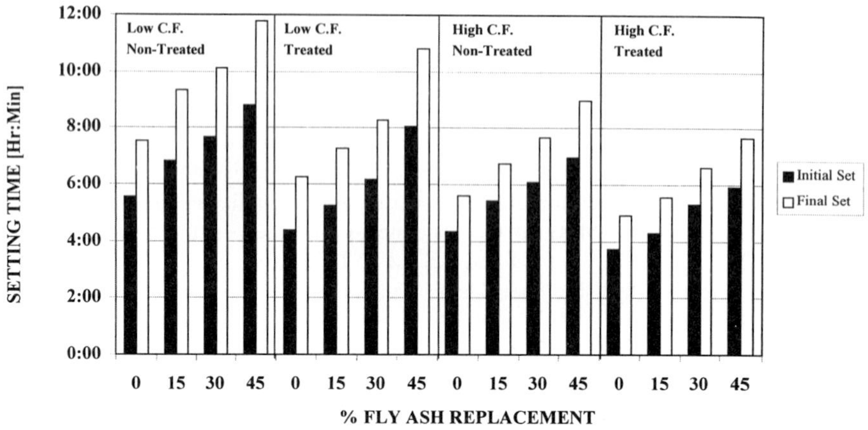

Figure 1 Rate of hardening as a function of mixture

Table 3 Air-void parameters

MIXTURE	W/CM	W/C	PASTE, P (% Vol)	AIR, A (% Vol)	P/A	N (mm-1)	I (mm)	α (mm-1)	<L> (mm)
0-L	0.62	0.62	21.2	8.3	2.57	0.451	0.178	21.9	0.127
15-L	0.60	0.71	21.8	9.2	2.38	0.420	0.229	18.3	0.127
30-L	0.60	0.86	26.4	7.7	3.43	0.440	0.178	22.9	0.152
45-L	0.57	1.00	25.2	8.3	3.05	0.533	0.152	25.9	0.127
0-LT	0.56	0.56	25.0	8.7	2.87	0.499	0.178	22.9	0.127
15-LT	0.50	0.59	26.3	6.4	4.12	0.499	0.127	31.3	0.127
30-LT	0.52	0.74	25.5	6.2	4.11	0.468	0.127	30.2	0.127
45-LT	0.53	1.00	23.7	5.1	4.66	0.444	0.127	34.9	0.127
0-H	0.55	0.55	25.6	6.8	3.79	0.328	0.203	19.4	0.203
15-H	0.53	0.62	27.6	6.2	4.48	0.385	0.152	25.0	0.178
30-H	0.51	0.73	24.2	5.7	4.21	0.666	0.076	46.5	0.102
45-H	0.48	0.87	26.7	5.4	4.93	0.354	0.152	26.1	0.178
0-HT	0.50	0.50	28.8	6.2	4.67	0.300	0.203	19.5	0.229
15-HT	0.48	0.57	31.1	5.6	5.53	0.465	0.127	33.0	0.152
30-HT	0.45	0.65	23.8	7.5	3.19	0.425	0.178	22.8	0.152
45-HT	0.44	0.79	24.5	5.6	4.40	0.377	0.152	27.1	0.152

W/CM: water/(fly ash + cement); W/C: water/cement; P/A: Paste/Air; n: voids per length; I: chord intercept; α: specific surface; <L>: mean spacing; L: 250 kg/m3 cement factor; H: 305 kg/m3 cement factor; LT, HT: mixtures treated with 1300 mL/100 kg of PC100 admixture; #: followed by L,H,LT,HT is the amount of fly ash replacement.

Compressive Strength

Compressive strength data are displayed in Figures 2a and 2b for specimens tested at ages of 3, 7, 28 and 56 days. All PC100 treated mixtures (LT,HT) resulted in greater compressive strength than the corresponding non-treated concrete (L,H). In fact, all PC100 treated mixtures (except one--45% fly ash, 305 kg/m^3), exhibited equal or better strength than the untreated, non-fly ash containing mixture at 3 days. At all other ages, the treated mixtures exhibited equal or greater strength to the non-fly ash reference. Note that only 15-L, 15-H and all of the HT mixtures were above the 31 MPa suggested [12] compressive strength level.

Freezing and Thawing Evaluation

After 300 cycles, all of the concrete had passed ASTM C 666 for freezing and thawing durability (Table 4). The concrete specimens all attained durability factors greater than 95% except one mixture which had 84%.

Due to water penetrating the specimens, some weight loss due to scaling was evident. The mixtures had weight losses less than 1%, except for two mixtures (30-L, 45-LT), during the testing. This is considered low. While no clear trend was observed with fly ash content, there was, in general, less weight loss due to scaling in the higher cementitious factor (H and HT) mixtures. There was an average weight loss of 0.19 ± 0.20% for the higher cementitious factor mixtures, compared to a weight loss of 1.2 ± 1.6% for the lower cementitious factor (L and LT) mixtures. There were no trends evident which distinguish the treated (LT and HT) and untreated (L and H) mixtures.

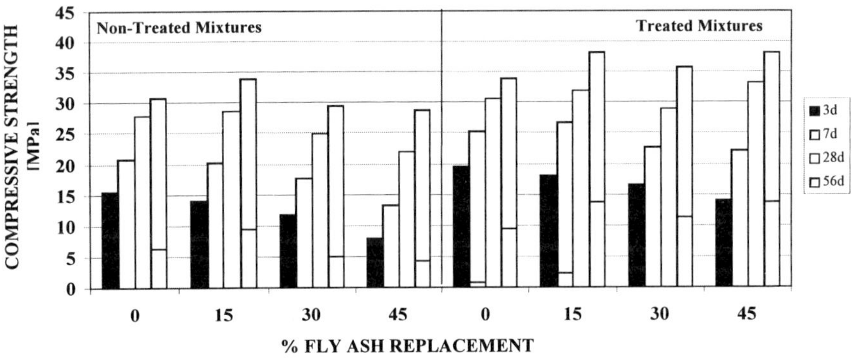

Figure 2a Compressive strengths from low-250 kg/m^3 mixtures

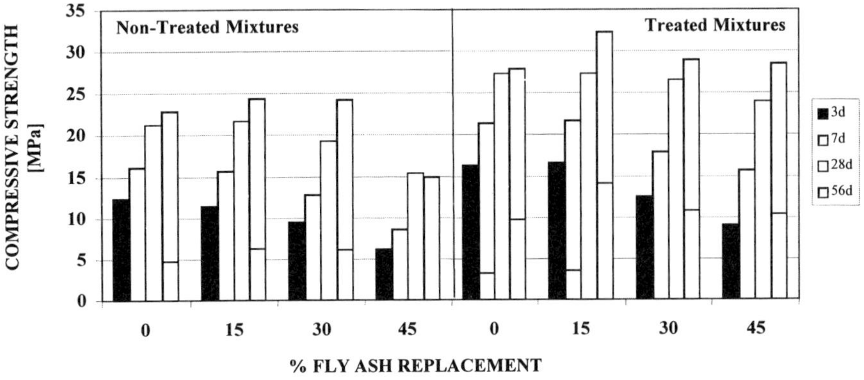

Figure 2b Compressive strengths from high-305 kg/m^3 mixtures

Scaling Resistance Evaluation

Most specimens started showing representative behavior after 30 cycles and maintained that behavior through the end of the test (Table 4). After 50 freeze/thaw cycles, the concrete specimens exhibited scaling ratings from 0 to 3. Note that none of the specimens exhibited severe scaling (i.e., ratings of 4-5).

It should be recognized that the ASTM C 672 ratings are better viewed as "rankings." This explains why some of the specimens that exhibited some weight loss during ASTM C 666 testing, also resulted in ratings of 0 during the ASTM C 672 testing (see LT and HT mixtures in Table 4). Also, it is reasonable to expect differences in scaling due to the different solutions used (water versus chloride) and due to the number of cycles (300 versus 50) used in each test.

Mixtures with increased early strength, accelerated initial set times or higher cement factors, in general, exhibited slightly less scaling damage than the other mixtures under the same conditions. Although lower values of w/cm and lower amounts of scaling were obtained by adding the PC100 to the treated mixtures, reduced amounts of water used in the untreated mixtures did not necessarily result in less scaling damage. Independent of w/cm, the PC100 treated specimens showed less scaling damage than the untreated specimens from the same mixture. Finally, scaling damage appeared independent of fly ash content or ultimate compressive strength.

Table 4 Durability data

MIXTURE	W/CM	W/C	HARDENED AIR (% Vol)	SPACING FACTOR (mm)	ASTM C666 Durability (%)	ASTM C666 Wt.Loss (%)	ASTM C672 Rating (0-5)
0-L	0.62	0.62	8.3	0.127	99	0.08	2.5
15-L	0.60	0.71	9.2	0.127	100	0.41	1.0
30-L	0.60	0.86	7.7	0.152	99	2.15	2.0
45-L	0.57	1.00	8.3	0.127	100	0.08	2.0
0-LT	0.56	0.56	8.7	0.127	100	0.008	0.0
15-LT	0.50	0.59	6.4	0.127	99	0.92	0.0
30-LT	0.52	0.74	6.2	0.127	100	0.93	0.0
45-LT	0.53	1.00	5.1	0.127	84	4.81	0.0
0-H	0.55	0.55	6.8	0.203	97	0.27	1.0
15-H	0.53	0.62	6.2	0.178	97	0.08	1.0
30-H	0.51	0.73	5.7	0.102	97	0.00	2.0
45-H	0.48	0.87	5.4	0.178	97	0.14	2.0
0-HT	0.50	0.50	6.2	0.229	99	0.13	1.0
15-HT	0.48	0.57	5.6	0.152	97	0.08	1.0
30-HT	0.45	0.65	7.5	0.152	99	0.13	0.0
45-HT	0.44	0.79	5.6	0.152	95	0.66	0.0

L: 250 kg/m3 cement factor; H: 305 kg/m3 cement factor; LT, HT: mixtures treated with 1300 ml/100kg of PC100 admixture. ASTM C666 and C672 results are the average of 3 beams and 2 prisms, respectively.

Chloride Permeability

All specimens showed permeabilities of between 2400 and 4800 coulombs at an age of 101 days (Figure 3). Four specimens (0-L, 30-L, 0-LT and 0-H) were indicative of "high" permeability (i.e., >4000 coulombs). The remaining specimens were typical of "moderate" permeability concrete. These results are consistent with the strength development measured (Figure 2a,2b).

The effective chloride diffusion coefficients, D, as determined by the migration technique indicated the H and HT mixtures had a slightly reduced permeability when compared to L and LT mixtures at an age of 90 days. Finally, the PC100 treated mixtures (HT,LT) exhibited slightly lower chloride diffusivities than non-treated mixtures (H,L).

With both these techniques it is difficult to determine with any degree of accuracy the actual permeability of the concrete during durability testing. This is true because the concrete containing fly ash is still hydrating and the permeability of the surface concrete may be different than the bulk. Hence, 90 days and 101 days were chosen to give characteristic values for the "type" of concrete being tested.

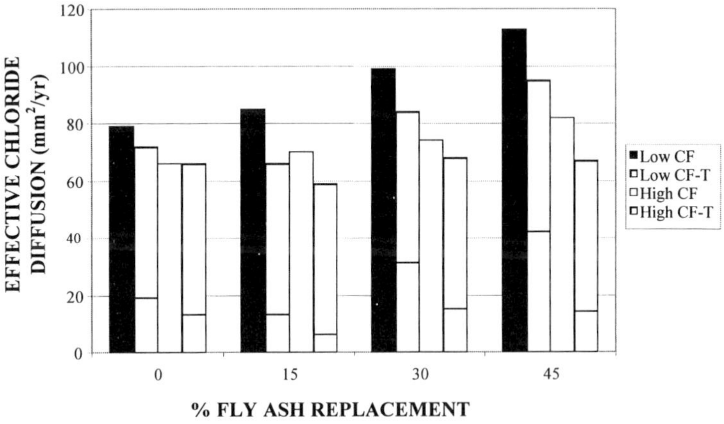

Figure 3 Chloride diffusion coefficients as a function of mixture quality at 90 days

DISCUSSION

Performance based specifications are a good idea, however, they should focus on *performance*, rather than mixture proportioning (e.g., w/cm, fly ash content). This is supported by the variability in the reported results in the literature [1,3-10,14] and makes it difficult to make sweeping conclusions. Several studies [3,5,6] conclude that scaling resistance is inversely proportional to w/cm, decreases with fly ash content [6,7] and the key to good scaling resistance is a good air void structure [1,5,6,7,11,14]. Despite these trends, the situation appears to be more cloudy than clear. The latter conclusion is well supported by this paper.

The scaling ratings for the concrete tested in this study ranged from 0-3 (Table 4). The ratings of none to moderate scaling levels obtained are consistent with what others have reported [6,14], and yet are inconsistent with current specifications [12,13]. Interestingly enough, there was not a clear trend with respect to fly ash content, strength level, nor with respect to a minimum w/cm in this paper. In fact, very durable concrete was made containing elevated amounts of fly ash (30-45%) and w/cm above 0.45. This finding supports the hypothesis that parameters besides fly ash content and w/cm are equally, and perhaps, more important to the scaling resistance of concrete.

Another interesting result is the clear trend of improved scaling resistance when PC100 was used in the concrete. The treated mixtures were always one rating value less than or equal to the non-treated mixtures, independent of mixture quality. While the admixture increased the strength development (Figures 2a, 2b), decreased the set time (Figure 1) and decreased the permeability (Figure 3) of the mixtures, each of these properties on their own does not explain the scaling performance. If they could, then these parameters should also hold true with respect to scaling resistance in the non-treated mixtures -- which they do not in all cases. Hence, the mechanism of the improved scaling resistance may be due to something else or is a synergistic effect among all three effects.

Adding the PC100 to concrete reduces the amount of water necessary to achieve a given workability. Note that water reduction also plays a role in the above three properties (strength development, set time, permeability). The PC100 admixture reduced the w/cm in the treated mixtures relative to the untreated mixtures. In the lower cement factor mixtures, the treated mixtures had ratings of 0 versus ratings of 1-2.5 for the non-treated mixtures. The treated mixtures reduced the w/cm by 0.08 ± 0.06 when compared with the non-treated mixtures. In the higher cement factor mixtures, the treated mixtures had ratings of 0-1 versus ratings of 1-2 for the non-treated mixtures. The treated mixtures reduced the w/cm by 0.07 ± 0.02 when compared with the non-treated mixtures.

By reducing the w/cm, there are several ramifications. First, decreasing the w/cm, decreases the permeability of the concrete (Figure 3). This makes it more difficult for chlorides to be transported into the concrete and cause damage. Second, by decreasing the w/cm, this typically increases strength development. Thus, more chlorides may be necessary to create damage as strength is increased. Finally, by decreasing the w/cm and maintaining the same slump/workability, it is possible that the finishing of the specimens might be easier and, thus, the resulting air void structure at the surface would be less affected.

Attiogbe [23] concluded that using a fixed spacing factor to determine durable concrete was not accurate (e.g., <0.2 mm). He found that the paste volume, the air volume and the w/cm were also important. He concluded that by decreasing the w/cm, the allowable spacing factor for durable concrete could be increased and that increasing the volume of air for a given volume of paste had a similar effect. He was concerned with freezing and thawing of bulk concrete, but the concepts may be applicable to scaling studies, as well.

If during finishing, the surface is worked in such a manner that increases the paste volume or decreases the air volume, the scaling resistance could be affected (i.e., the maximum allowable spacing factor for durable concrete would be decreased). This might be true despite having seemingly "durable" air contents. Lowering the w/cm and maintaining workability may make the surface concrete's air-void system more forgiving to such damaging features as "over-finishing." To prove such a concept would require air-void parameters to be measured on a cross section from the surface to the bulk of a concrete specimen. During this study, no surface air-voids were measured. Hence, more research is necessary. While there are not enough data in this study to conclusively prove this later hypothesis, the concept appears reasonable and warrants further study.

SUMMARY AND CONCLUSIONS

1. Concrete durable to deicing salt scaling can be made using fly ash contents as high as 45% and w/cm as high as 0.60.

2. All mixtures in this study were freeze-thaw durable, had good air-void parameters and ranged from 0 to 3 in scaling resistance by ASTM C 672.

3. Decreasing the w/cm, increasing the cementitious factor, decreasing the permeability, decreasing the initial set, and increasing the strength of a mixture to play a role in increasing scaling resistance, but none on its own was conclusive.

4. Higher levels of fly ash replacement did not necessarily increase scaling.

5. The PC100 admixture had the following effect on the concrete mixtures:

- scaling was reduced in treated mixtures over non-treated mixtures
- reduced w/cm significantly at similar slumps to comparable, non-treated mixtures
- set times of treated mixtures containing 15% more fly ash were the same as non-treated mixtures
- strength development of all treated mixtures was higher than non-treated mixtures
- permeability was lower in treated than non-treated mixtures

ACKNOWLEDGEMENTS

The authors wish to express their appreciation to Dr. Emmanuel Attiogbe for his participation in discussions surrounding air void parameters.

REFERENCES

1. MALHOTRA, V M AND RAMEZANIANPOUR, A A. Fly Ash in Concrete, 2nd edition, CANMET, Ontario, 1994, pp.21-25, 44-50; pp. 73-81; pp. 153-154.

2. NMAI, C K, SCHLAGBAUM, T, AND VIOLETTA, B, "A History of Mid-Range Water-Reducing Admixtures," *Concrete International*, April, 1998, pp. 45-50.

3. BILODEAU, A, SIVASUNDARAM, V, PAINTER, K E, MALHOTRA, V M, "Durability of Concrete Incorporating High Volumes of Fly Ash from Sources in the United States," *ACI Materials Journal*, Jan-Feb, 1994, pp. 3-12.

4. BILODEAU, A AND MALHOTRA, V M. "Concrete Incorporating High Volumes of ASTM Class F Fly Ash : Mechanical Properties and Resistance to Deicing Salt Scaling and to Chloride Ion Penetration," *Proc. from Fly Ash, Silica Fume, Slag, and Natural Pozzolans in Concrete*, 4th Intl Conf., Istanbul, Turkey, 1992, SP 132, ed. V.M. Malhotra, American Concrete Institute, Detroit, MI, 1993, pp. 319-349.

5. JOHNSTON, C D, "W/Cm Code Requirements Inappropriate for Resistance to Deicer Salt Scaling," *Durability of Concrete*, 3rd Intl. Conf., Nice, France, 1994, SP-145, ed. V.M. Malhotra, American Concrete Institute, Detroit, MI, 1994, pp. 85-105.

6. BILODEAU, A AND MALHOTRA, V M. "Deicing Salt Scaling Resistance of Concrete Incorporating Supplementary Cementing Materials:CANMET Research,"*Freeze-Thw Dur. of Conc.*, eds. J.Marchand, M.Pigeon & M.Setzer,E. & FN Spon, London,UK, 1997, pp.121-156.

7. MALHOTRA, V M AND PAINTER, K E. "Early Age Strength Properties and Freeze-Thaw Resistance of Concrete Incorporating High Volumes of Fly Ash," *Intl. J. of Cement Composites and Lightweight Concrete*, vol. 11, #1, 1989, pp. 37-46.

8. KLIEGER, P AND GEBLER, S, "Fly Ash and Concrete Durability," *Concrete Durability*, SP-100, ed. J. Scanlon, American Concrete Institute, Detroit, MI, 1987, pp. 1043-1046.

9. MALHOTRA, V M, CARETTE, G G, BILODEAU, A, AND SIVASUNDARAM, V. "Some Aspects of Durability of High Volume ASTM Class F Fly Ash Concrete," *Dur. of Conc.*, 2nd Intl., Montreal,Canada,1991,SP-126,ed. V.M. Malhotra, American Concrete Institute, Detroit, MI, 1991, pp. 65-82.

10. NAIK, T R, AND SINGH, S S, "Use of High Calcium Fly Ash in Cement Based Construction Materials," *Proc. from Fly Ash, Silica Fume, Slag, and Natural Pozzolans in Concrete*, 5th Intl Conf., Supplementary Papers, American Concrete Institute, Milwaukee, WI, 1995, pp. 1-44.

11. ARCHULETA, L G, TIKALSKY, P J, AND CARRASQUILLO, R L. "Production of Concrete Containing Fly Ash for Structural Applications," Research Report 364-1, Transportation Research Board, May, 1986, pp.1-141.

12. ACI 318-89. "Building Code Requirements for Reinforced Concrete," ACI, 1989.

13. CAN/CSA A23.1-M90. "Concrete Materials and Methods of Concrete Construction," Canadian Standards Association, 1990.

14. TIKALSKY, P J, CARRASQUILLO, P M, CARRASQUILLO, R L. "Strength and Durability Considerations Affecting Mix Proportioning of Concrete Containing Fly Ash," *ACI Materials Journal,* Nov-Dec. 1988, pp.505-511.

15. ASTM C666-92. "Standard Test Method for Resistance of Concrete to Rapid Freezing and Thawing," Annual Book of ASTM Standards, *V. 04.02, Concrete and Aggregates*, American Society for Testing and Materials, Philadelphia, 1997, pp.320-325.

16. ASTM C618-94a. "Standard Specification for Coal Fly Ash and Raw or Calcined Natural Pozzolan for Use as a Mineral Admixture in Portland Cement Concrete," Annual Book of ASTM Standards, *V. 04.02, Concrete and Aggregates*, American Society for Testing and Materials, Philadelphia, 1997, pp. 304-306.

17. ASTM C192-90a. "Standard Practice for Making and Curing Concrete Test Specimens in the Laboratory," Annual Book of ASTM Standards, *V. 04.02, Concrete and Aggregates*, American Society for Testing and Materials, Philadelphia, 1997, pp.116-122.

18. ASTM C494-92. "Standard Specification for Chemical Admixtures for Concrete," Annual Book of ASTM Standards, *V. 04.02, Concrete and Aggregates*, American Society for Testing and Materials, Philadelphia, 1997, pp. 254-261.

19. ASTM C457-90. "Standard Test Method for Microscopical Determination of Parameters of the Air-Void System in Hardened Concrete," Annual Book of ASTM Standards, *V. 04.02, Conc. and Agg.*, American Society for Testing & Materials, Philadelphia, 1997, pp. 228-240.

20. ASTM C672-92. "Standard Test Method for Scaling Resistance of Concrete Surfaces Exposed to Deicing Chemicals," Annual Book of ASTM Standards, *V. 04.02, Concrete and Aggregates*, American Society for Testing and Materials, Philadelphia, 1997, pp.339-341.

21. ASTM 1202-94. "Standard Test Method for Electrical Indication of Concrete's Ability to Resist Chloride Ion Penetration," Annual Book of ASTM Standards, *V. 04.02, Concrete and Aggregates*, American Society for Testing and Materials, Philadelphia, 1997, pp. 624-629.

22. STREICHER, P E, AND ALEXANDER, M G. "A Chloride Conduction Test for Concrete, *Cement and Concrete Research*, vol. 25, no. 6, 1995, pp. 1284-1294.

23. ATTIOGBE, E K. "Predicting Freeze-Thaw Durability of Concrete—A New Approach," *ACI Materials Journal*, Sept.-Oct. 1996, pp. 457-464.

ENHANCING THE FRACTURE TOUGHNESS OF CEMENT-BASED MATERIALS

H W Chandler

D E MacPhee

I Atkinson

R J Henderson I J Merchant

University of Aberdeen

United Kingdom

ABSTRACT. A mechanism for the toughening of ceramic materials has been successfully developed in the refractories industry using residual stresses at the microstructural scale. This paper addresses the application of this approach to cement-based systems. In the cement case, the growth of expansive phases has been controlled, along with the set characteristics of the paste, to establish internal stresses which, while significant, have not been sufficient to cause expansive failure. Macro-crack propagation, initiated by external loading, has been suppressed either by the release of residual stresses which act to close the crack or by grain bridging. Because such stresses are dispersed throughout the matrix, this toughening mechanism would add to the value of steel reinforcement in the strengthening of concrete structures. This paper describes some early work concerning the factors which control internal stresses in cement microstructures and the experimentally measured stress-strain behaviour of pastes and mortars which confirm the dramatic effect of microstructural engineering.

Keywords: Cement, Microstructure, Structural applications, Mechanical properties.

Professor Howard W Chandler leads the Mechanical Engineering Group in the Engineering Department of Aberdeen University, Scotland, UK. His research interests are many and varied and include mathematical modelling of ceramic microstructures.

Dr Donald E MacPhee is a Lecturer in the Chemistry Department of Aberdeen University, Scotland, UK. His primary area of research is in cement chemistry and microstructure.

Mr Ian Atkinson is currently a Postgraduate Student in the Chemistry Department of Aberdeen University, Scotland, UK.

Dr Robin J Henderson is a Postdoctoral Research Fellow in the Engineering Department of Aberdeen University, Scotland, UK.

Dr Ian J Merchant is a Postdoctoral Research Fellow working jointly in Departments of Engineering and Chemistry in Aberdeen University, Scotland, UK. Ian is currently devloping the work described in this paper and is funded by EPSRC.

INTRODUCTION

Dual phase refractories, in particular magnesia-spinel composites, have been recently developed specifically to withstand severe thermal shock without the propagation of major cracks. This has been achieved by the subtle manipulation of residual stresses within the microstructure such that the applied tensile stresses trigger microcracking. This has two effects.

1. It releases these residual stresses producing a permanent overall volume expansion [1,2] which in turn suppresses the growth of major cracks in a similar way to transformation toughening in zirconia containing ceramics[3].

2. It also produces a complex network of microcracks ahead of the main crack which can lead to grain bridging across the major crack.

Figure 1 shows the stress-strain properties of magnesia and magnesia-spinel refractories and highlights the enhanced energy absorption (represented by the area under the stress-strain curve) of the dual phase refractory. Whilst the residual strains may seem small, the magnesia-spinel material exhibits increasing toughness as the crack propagates (R-Curve behaviour) compared to the magnesia material which has a constant value of toughness. Mathematical modelling of the refractory materials indicates that microstructures where the included phase is in residual tension, and the matrix phase in residual compression offers the best opportunities for improving the mechanical behaviour of these systems. Can a similar beneficial effect be tailored in cement mortars? The key to transferring this mechanism of toughening from fired ceramic to cement based materials is the development and control of residual stresses within the mortar.

Development of Residual Stresses in Portland Cement Pastes, Mortars and Concretes

Under normal conditions, the formation of hydration products during the hydration of Portland cement results in a net shrinkage, as water is lost from the pore structure into products or to the atmosphere. However, addition of certain additives can cause components of the paste to expand. By appropriate combination of rate of expansion and set time, internal stresses on the microstructure can be generated by the growth of expansive phases within the confined porosity (see Figure 2). Expansion causing internal stresses is said to be ‘active’ expansion, whereas expansion occurring while the paste is plastic or is accommodated for in the available porosity after the paste has set, is considered to be ‘passive’ [4] . By manipulation of mix design and set times, expansion characteristics can be engineered to optimise the distribution and level of internal stresses.

Expansive cements are not new. Existing products aimed at shrinkage compensation and self-stressing cements all rely on the same chemistry. The expansion is associated with the rapid development of ettringite, a calcium aluminosulphate hydrate of approximate composition $Ca_6Al_2(SO_4)_3.(OH)_{12}.26H_2O$. There has been much discussion on the chemistry that causes ettringite expansion. Two main theories have been proposed [5]. According to the crystal growth theory, expansion is caused by the growth of ettringite crystals on solid surfaces or from solution, resulting in a crystallisation pressure. Alternatively, the incorporation of water into an ettringite gel, causing an overall expansion, is the basis of the swelling theory.

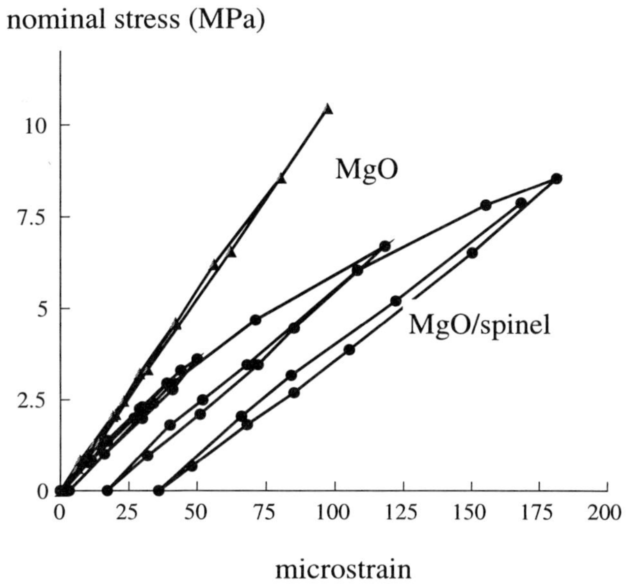

Figure 1 Stress-strain properties of MgO and MgO-spinel refractories highlighting the influence of residual stresses. The energy absorption is represented by the area inside the hysteresis loop

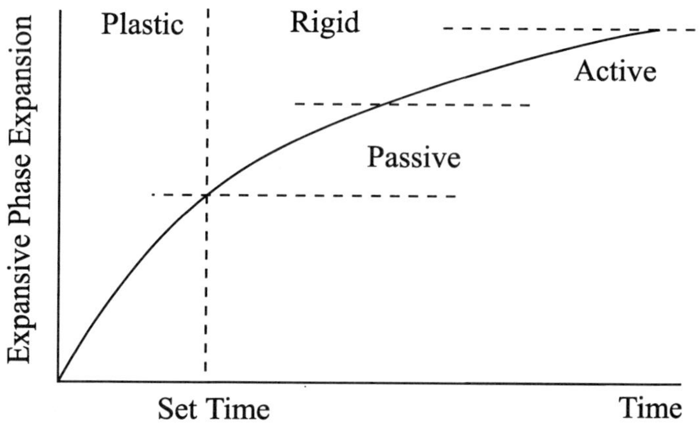

Figure 2 Influence of phase expansion on the development of internal stresses

A number of factors influence both models, not least the chemistry of the solution, in which the concentration of calcium ions is thought to play an important role in defining the size and aspect ratio of ettringite crystallites [6-8].

Portland cement pastes develop ettringite quickly due to the presence of gypsum, deliberately added to retard the tricalcium aluminate hydration reactions. While the sulphate concentration in solution remains high, ettringite remains the stable phase. This may be achieved by the addition of a soluble sulphate salt during mixing. At very high sulphate concentrations, gypsum ($CaSO_4.2H_2O$) will precipitate. This is also an expansive reaction contributing to pore filling and paste expansion. Sulphate additions not only stabilise and promote ettringite formation (assuming an adequate supply of aluminium), but they also can significantly shorten set times. This paper describes the effect of sulphate additions on Portland cement paste set times and dimensional properties, and on the mechanical performance of the associated mortars.

METHODOLOGY

Sample Preparation, Curing and Set Time Measurement

Portland cement pastes were made from ordinary Portland cement (OPC)(Supplied by Blue Cicle Industries plc, Dunbar Works, East Lothian, Scotland.), of composition given in Table 1, and various additions of sodium sulphate (see Table 2 for mix compositions). These were mixed with water to a water/cement ratio (w/c) of 0.35. Sodium sulphate was used in this study because it is a convenient source of soluble sulphate but, in practice, an alternative such as calcium sulphoaluminate cement would be used to minimise the addition of alkalis to the cement.

Table 1 Portland cement (OPC) composition

OXIDE	Wt.%
SiO_2	20.1
Al_2O_3	5.07
Fe_2O_3	3.3
CaO	64.21
MgO	2.27
SO_3	3.14
Eq. Na	0.43

Table 2 Mix proportions for the preparation of the cement binder

CEMENT/g	Na_2SO_4/g	WATER/g
292	8	105
285	15	105
278	22	105
270	30	105

Pastes were thoroughly mixed using a domestic food blender and cast into perspex moulds of dimensions 10 x 10 x 100 mm. Five bars were cast for each composition. A portion of the paste was retained for set time determination using the Vicat needle method (BS 4550). After approximately 15 minutes, the top surface of the moulded bars were screeded to produce smooth paste bar surfaces and the moulds, with bars, were placed on a tray above water in a sealed desiccator for 24 hours at 23°C. The bars were then demoulded and Demec studs were fitted to the ends using fast-setting Araldite. The bars were then weighed and their dimensions recorded using a digital comparator (Mitutoyo Digimatic Indicator) and a measuring jig, to ensure reproducable positioning of the bar each time. The bars were then returned to the above curing regime to await further measurements.

Mortar bars, made with a sand: binder ratio of 1:1, were also prepared. The preparation technique was similar to that for paste bars except that river sand (<1 mm) was saturated with water prior to mixing with the binder (OPC) so that the w/c of 0.35 would be preserved on mixing.

Mechanical Property Measurements

The stress-strain behaviour of an OPC mortar with (7.5%) and without sodium sulphate additions was investigated using 3 point flexure tests on the fabricated bars. The strain was measured directly on the specimen by strain gauges attached to the tensile face of the specimen. The nominal stress was calculated from the applied moment and dimensions of the specimen assuming linear elasticity. The specimens were loaded and unloaded, the maximum load being increased every cycle, until failure occurred.

Electron Microscopy

Scanning electron microscopy was carried out on fracture surfaces generated during mechanical testing. The fractured end of a sample bar was removed using a hacksaw and fixed to a sample holder using a conducting carbon glue. The samples were then coated with a thin layer of gold to prevent specimen charging or overheating before being mounted onto the sample stage of a ISI-SS40 Scanning Electron Microscope for examination.

RESULTS AND DISCUSSION

Figures 3 and 4 show the dimensional and weight changes, respectively, on ageing of cement paste and mortar bars as a function of time. The exhibited trends are derived from averages of the readings from five bars per composition at each measurement time. All dimensional and weight changes reported are referenced against the initial dimensions and weights recorded on demoulding. The data show an expansion and weight increase with time for all specimens, but these changes are exaggerated as the sulphate additions are increased, at least up to 7.5%. The combination of weight increase and expansion suggests that moisture is absorbed from the curing environment to enable the continuing development of ettringite or gypsum. This would be consistent with expectations. The higher addition (10%) actually demonstrates a lower expansion than for the 7.5% addition at longer ages.

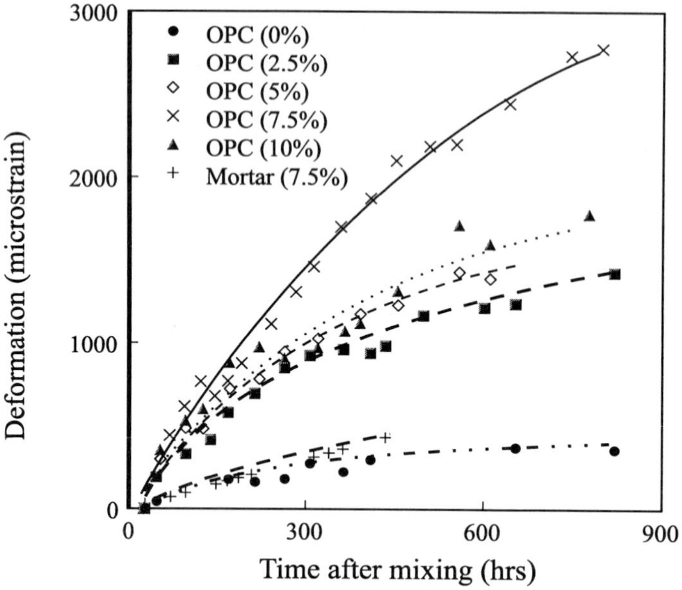

Figure 3 Deformation vs. time for OPC paste and mortar bars aged in a high humidity environment

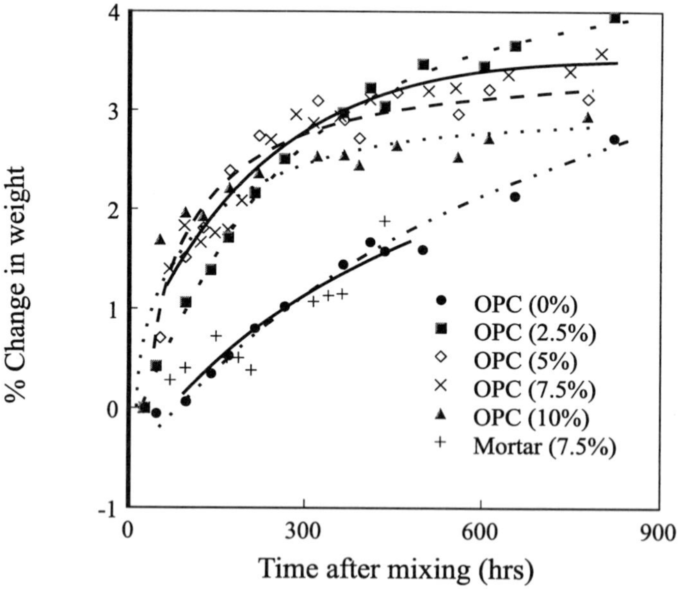

Figure 4 Weight change vs. time for OPC paste and mortar bars aged in a high humidity environment

As expected, there is less expansion in the mortar than in the corresponding OPC paste (i.e. both containing 7.5% sulphate addition), due to the dilution of cement by the aggregate.

Figure 5 shows the accelerating effect of sulphate additions. This acceleration may account for the reduced expansion measured for the sample containing 10% relative to the 7.5% sulphate additions.

Assuming it has attained a final set, the paste is no longer able to yield to expansion unless expansion is accommodated for in available porosity. While this occurs, macroscopic expansion stops. Further phase expansion internally places the microstructure under stress until its tensile strength is exceeded and cracking is observed. Indeed, macroscopic cracking was observed on the bars containing the highest sulphate loading suggesting that on-going expansion was sufficiently 'active' to crack the matrix.

This is an example where the mix composition was unsuitable although the generation of internal stresses is quite evident.

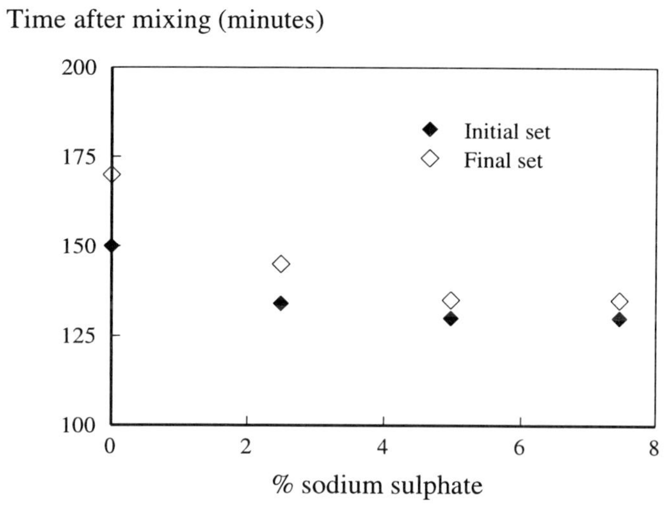

Figure 5 Set time characteristics of OPC pastes with and without sulphate additions

Further evidence of released internal stresses can be obtained from the electron micrographs (Figure 6) which show radial cracks emerging from pores containing ettringite needles and/or gypsum platelets.

These crack patterns are consistent with expansion stresses and the presence of massive crystallites, and the concentration and orientation of ettringite needles in these circumstances, are strong indications of 'active' expansion in this paste.

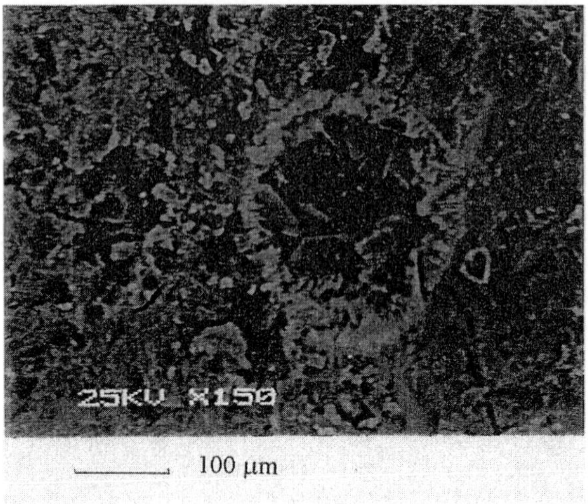

Figure 6 SEM of a fracture surface of an OPC paste showing radial cracking around an ettringite/gypsum-filled pore

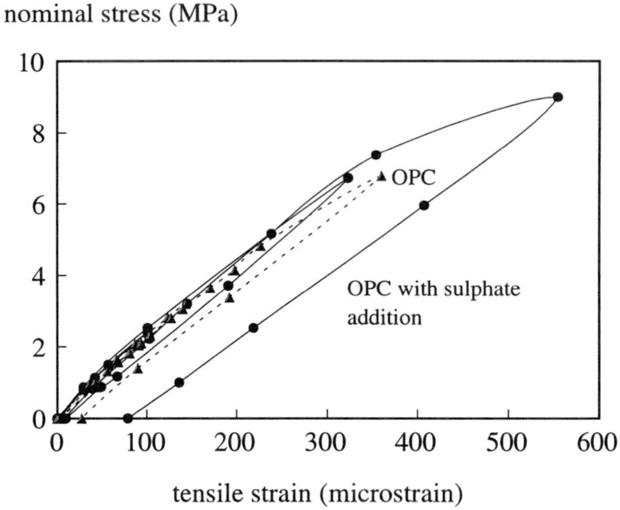

Figure 7 Influence of sodium sulphate additions on the stress-strain properties of Portland cement pastes (w/c = 0.35), aged 42 days under high humidity conditions

But the final, and most convincing argument for the existence and effectiveness of internal stresses in these cement pastes and mortars comes from the mechanical performance measurements which show non-linear stress-strain characteristics consistent with those observed for the MgO-spinel ceramic system in Figure 1. Figure 7 shows the characteristics of mortars made from OPC and OPC with added sulphate (7.5\%).

The Portland cement mortar without sulphate addition exhibits some non-linear behaviour but this is greatly enhanced by the addition of sulphate. After the peak load was reached, these samples remained intact and, on reloading, were capable of withstanding 50% of the initial maximum stress. This indicates extensive microcracking rather than the propagation of one large crack and consequently a toughened matrix.

Most failures are a result of tensile loading, either arising from excessive expansion of phases within the concrete or as a result of impact. In either case, concretes with better resistance to catastrophic cracking would provide enhanced protection against surface deterioration and spalling. In addition, a consequence of the toughening mechanism is a lower porosity microstructure. This will improve durability because the matrix modifications, whilst reducing porosity overall, will shift the emphasis from interconnected porosity, which permits material transport throughout the interior of the microstructure, to isolated porosity. The latter cannot contribute to internal degradation of the matrix so that the exposed surfaces are primarily restricted to the geometric surface of the concrete structure and deterioration rates will be reduced. However, this would require to be confirmed by experimentation.

CONCLUSIONS

A few simple experiments with conventional cement mortars have shown that residual stresses can also be manipulated in hydraulically bonded systems and the tensile mechanical properties dramatically enhanced (see Figure 7). Thus, by suitable mix design, products with enhanced toughness can be manufactured without the use of reinforcing fibres. This is an important observation especially relevant to high strength concrete where the low toughness is particularly noticeable.

REFERENCES

1. HENDERSON, R J AND CHANDLER, H W. A model for non-linear mechanical behaviour of refractories. Brit.Ceram.Trans., Vol 96,1997, pp 85-96.

2. HENDERSON, R J. Thermomechanics of Dual-Phase Refractories, PhD thesis, University of Aberdeen, 1997.

3. SWAIN, M V. R-Curve behaviour and thermal shock resistance of ceramics. J. Amer.Ceram.Soc., Vol.73 No.3, 1991, pp 621-628.

4. PANCHENKO, A I. Control of expansion and structure formation of expansive cement. Cem.Concr.Res., Vol.20 No.3, 1990, pp 602-609.

5. COHEN, M D. Theories of expansion in sulfoaluminate-type expansive cements-schools of thought. Cem.Concr.Res., Vol.13, 1993, pp 809-818.

6. OGAWA, K AND ROY, D M. C_4A_3S hydration, ettringite formation, and its expansion mechanism. 1. Expansion - ettringite stability. Cem.Concr.Res., Vol.11, 1981, pp 741.

7. OGAWA, K AND ROY, D M. C_4A_3S hydration, ettringite formation, and its expansion mechanism. 2. Microstructural observation of expansion. Cem.Concr.Res., Vol.12, 1982, pp 101-109.

8. OGAWA, K AND ROY, D M. C_4A_3S hydration, ettringite formation, and its expansion mechanism. 3. Effect of CaO, NaOH and NaCl - Conclusions. Cem.Concr.Res., Vol.12, 1982, pp 247-256.

SELECTION OF THE OPTIMAL MICROSTRUCTURE FOR A FROST RESISTANT CEMENTITIOUS COMPOSITE

A J Klemm

Glasgow Caledonian University

United Kingdom

P Klemm **M Jablonski**

University of Lodz

W Marks

Institute of Fundamental Technical Research

Poland

ABSTRACT. The aim of the present study was to select the optimal pore size distribution of pores which would ensure significant improvement of resistance to alternate freezing and thawing cycles. The Mercury Intrusion Porosimetry (MIP) and Differential Scanning Calorimetry (DSC) techniques have been used to identify microstructural features and to determine the phase transition parameters for the optimisation procedure respectively. In the multicriteria optimisation the mathematical model considered is formulated by the selection of functions describing with sufficient accuracy sets, of experimental data. The set of values of objective functions corresponding to certain values of decisive variables is approximated by functions of these variables. The solution of the problem involves the determination of the values of the design variables which extremize the objective functions.

Keywords: Frost resistant concrete, Durability, Microstructure, Multicryterion optimisation

Dr Agnieszka J Klemm is a Lecturer in Construction Materials, Glasgow Caledonian University, UK. Her research interests are in the field of durability and protection of polymer modified concrete, with particular reference to frost resistance.

Professor Piotr Klemm is Head of Department of Building Physics and Building Materials, Technical University of Lodz, Poland. His research interests include thermomechanics and thermodifussion in composite materials, physical, chemical and mechanical problems in building materials.

Marek Jablonski is a Lecturer in Building Physics, Technical University of Lodz, Poland. His main research interests include computational methods and experimental measurements of phase transition process 'water to ice' in cementitious composites.

Professor Wojciech Marks is Head of Division of Multicryterion optimisation, Institute of Fundamental Technical Research, Warsaw, Poland. He specialises in multicryterion optimisation of structures and materials.

$$V_i (p_w - p_i) = \Delta S_f (T - T_o)$$

where : V_i is the mole volume of ice;
ΔS_f is the entropy of phase transition.

The migration process of water from capillary to pore lasts until $\mu_{ice\ in\ pore}$ becomes equal to $\mu_{ice\ in\ capillary}$.

Unfortunately, the above model is not free from problems, particularly the ice plasticity seems to be too small to maintain thermodynamic equilibrium over a period of time in the total volume of ice [7]. Nevertheless it offers an approximate picture of the process which is suitable for most comparative purposes.

EXPERIMENTAL PROCEDURE

The Mercury Intrusion Porosimetry (MIP) and Differential Scanning Calorimetry (DSC) techniques have been used to identify microstructural features and determine the phase transition parameters for the optimisation procedure respectively. [9], [10] Two polymeric admixtures - methyl hydroxy ethyl cellulose, MHEC, and polyvinyl acetate, PVA, have been used to modify the composites microstructure.

In order to analyse the results of the porosimetric measurements (Mercury Intrusion Porosimeter, model: AUTOPORE II 9220 V3.00 produced by Micromeritics) three ranges of pore diameters have been considered: small pores with diameters below 300nm, medium sized - between 300 and 3000nm and the large ones with diameters in excess of 3000nm.

Calorimetric measurements were carried out using the automatic mode of the UNIPAN 605M system linked with an IBM PC computer, allowing the processing and analysis of the calorimetric measurements. Amounts of heat evolved during three stages of phase transition (nucleus creation, spontaneous growth of nucleus, ice crystal growth) and critical temperatures below which spontaneous growth of ice nucleus takes place, have been determined. Since the second stage of transition i.e. spontaneous growth of ice nucleus, is believed to be the most dangerous stage for the material (mainly because of comparatively high increase of ice volume takes place during a very short period of time), the attempt was made to relate amount of heat emmited during this stage to the the amount of heat evolved during the third stage of transition..

MULTICRITERIA OPTIMISATION

The optimal pore size distribution of pores which would ensure significant improvement of resistance to alternate freezing and thawing cycles has been investigated. Cementitious material modified by two admixtures - methyl hydroxy ethyl cellulose, MHEC, and polyvinyl acetate, PVA, was the subject of multicriteria optimisation approach.

Formulation of the Objective Functions - $F_i(x_j)$

For the purpose of this investigation the following criteria of optimisation have been assumed:

- minimum latent heat of the phase transition ΔQ [J/g],
- minimum ratio of the heat associated with spontaneous nucleus creation to the heat associated with crystal growth $\vartheta = \Delta Q_2/\Delta Q_3$ [-],
- maximum critical temperature ΔT [K],
- minimum water vapour absorption $\varphi(S)$ [-].
- maximum flexural strength f_z [MPa],
- maximum compressive strength f_c [MPa].

The volumes of pores, [mL/g], from three size ranges (I - below 300nm (x_1), II - between 300 -3000nm (x_2), III - exeeding 3000nm (x_3)) have been defined as decisive variables.

These decisive variables have to satisfied following boundary conditions:

$$0.02 \le x_1 \le 0.10,$$
$$0{,}03 \le x_2 \le 0.15,$$
$$0.005 \le x_3 \le 0.17,$$
$$0.10 \le x_1 + x_2 + x_3 \le 0.30.$$

Relationships between objective functions and decisive variables have been experimentaly determined. (Table 1) Analytical forms of the objective functions have been found by approximation of the experimental data by multinomials of first or second degree:

$$F1 = 85.6117 - 1888.74x1 + 13758.5x1^2 - 580.736x2 + 4202.18x2^2 + 260.085x3 - 966.753x3^2$$

$$F2 = 0.413509 - 2.36686x1 + 0.378854x2 - 0.408484x3$$

$$F3 = 243.343 + 852.617x1 - 6634.33x1^2 + 145.224x2 + 1839.91x1x2 - 327.105x2^2 \quad 268.098x3 + 1542.63x1x3 - 126.239x2x3 + 1043.46x3^2$$

$$F4 = 0.425473 + 5.16206x1 - 4.59241x1^2 + 0.34519x2 - 30.7719x1x2 + 9.72467x2^2 - 3.22523x3 - 8.44693x1x3 - 4.89628x2x3 + 16.4012x3^2$$

$$F5 = 7.86876 - 26.4987x1 - 13.8286x2 - 9.29686x3$$

$$F6 = 42.9371 - 135.958x1 + 147.937x2 - 95.2775x3$$

Coefficients at variables x_i, i = 1,2,3 were determined by using the least square method (MATHEMATICA program).

Microstructure itself is described by the set of factors including total porosity and pore size distribution, pore shape etc. It is believed that the pores within a certain range of sizes may facilitate ice nucleus initiation and crystal growth thus leading to accelerated deterioration of the composite. They naturally require a lower value of supercooling (understood as a temperature difference) and make the process of phase transition more likely to occur.

The preferable solution in multicriteria optimisation is usually determined by either the utility function method or the matrix function method. The utility function method requires the estimation of the appropriate weights for all objective functions which in many cases may be impossible or very difficult. The final results of optimisation are very sensitive to the variations of such coefficients.

The second method seems to be a suitable option when the weights coefficient are difficult to specify. The preferable solution can be found as a point of the compromise set which is closest to the ideal solution. The following volumes of small, medium and large pores - 0.06591 mL/g, 0.03 mL/g and 0.1119mL/g, have been found to satisfy the above condition. This corresponds to approximately 32%, 14% and 54% of pores in three specified zones. The results obtained are indeed very close to those estimated from a number of industrial experiments and correlates with the previous research findings on frost resistance of ceramics [3], although the dangerous zone of the pore diameters has been narrowed to 250 - 1400 nm. The microstructure characterised by the high percentage (>40%) of small and large pores is likely to provide frost resistance. Pores having smaller diameters (because of significant effect of the friction forces) and bigger diameters (because of significant effect of the gravitation) are supposed to affect the capillary flow to a lesser degree thus reducing potential frost damage.

An appropriate formulation of the composite model is absolutely crucial to obtain reliable results. Since the number of optimisation criteria is usually reduced by considering only the most relevant properties of the composite, the solution is automatically simplified. In such cases the optimisation process is only a part of the design procedure for a composite and the final decision as to the composition is left to the designer, who can take into account other properties and other issues not contained in the mathematical optimisation process.

REFERENCES

1. KLEMM A J, KLEMM P. “Ice formation in pores in polymer modified concrete I. The influence of the admixture on the water to ice transition” Building and Environment, Vol.32, No 3, pp. 195-198, 1997

2. KLEMM, A J., KLEMM, P., “Ice formation in pores in polymer modified concrete II. The influence of the admixtures on the water to ice transition in the cementitious composites subjected to freezing/thawing cycles” Building and Environment, Vol.32, No.3, pp. 199-202, 1997

3. RAVAGLIOLI, A., "Evaluation of the Frost Resistance of Pressed Ceramic Products", Trans.Brit.Ceram.Soc., Vol.75, No5, pp 92-95, 1976

4. HILLING, W B, TURNBULL, D, "Theory of Crystal Growth in Undercooled Pore Liquids", J.Chem.Phys.24 (1956)

5. G. PILTZ, "Porengrossenverteillung im Zigelscherben und ihre Bedeutung fur die Frostfestigkeit", Ziegelindustrie, nr15/16 (1970) pp.327 – 336

6. D.H. EVERETT, "Thermodynamics of Frost Damage to Porous Solids", Trans.Farad.Soc. Vol.56 (1961) pp. 1541 –1551

7. D.H. EVERETT, J.M. HYNES, "Capillary Properties of Some Model Pore Systems with Special Reference to Frost Damage" Rilem Bull. No27 (1965) pp. 31 –38

8. SKAPSKI, R. BILLUPS, R. ROONEY, "Capillary-Cone Method for Determination of Surface Tension of Solids", J.Chem.Phys. 26, No5 (1967) pp.1350 – 1351

9. M. JABLONSKI, P. KLEMM, A.J. KLEMM, "Calorimetric Experiment in Cementitious Composites" Proc. IV Conference "Building Physics in Theory and Practice" Lodz, Poland (1993)

10. KLEMM A.J., "The Influence of Admixtures on the Microstructural Features and Mechanical Properties of Cementitious Materials Subjected to Freezing and Thawing Cycles" PhD Thesis, Strathclyde University, Glasgow 1994

11. ASTM Standard C666-75 "Standard method of test for resistance of concrete to rapid freezing and thawing" Part 14, pp 371-375, 1975

INTRODUCTION

In general, hydraulic materials are composite materials constituted of granulates inert walnuts in an active matrix that constitutes the Portland cement paste. The knowledge of the behavior of the active phase allows to focus on the evolution during hardening of some parameters up to there poorly known . A correct prediction of the physico-chemicals performances of hydraulic materials has to be based on the degree of hydration of the binder, the active mix phase. In what follows, first experimental results are presented then a the mathematical model of the speed of fixing of the combined water in Portland cement pastes is shown. Finally, that it is possible to represent the speed of fixing of the combined water in hydraulic binder by a mathematical model and that physic-chemical evolution of hydraulic materials are closely linked to the degree of hydration of the binder.

MATERIALS

During this work refer to experimental results achieved by other laboratories. These results obtained from an experimental study of the speed of fixing the water by the Portland cement paste [1]. In these tests an ordinary Portland cement (OPC) is used. The mineralogical and chemical compositions are shown in Table 1.

Table 1 Chemical analysis of ordinary Portland cement

CHEMICAL COMPOSITION (%)		COMPOUND COMPOSITION (USING BOGUE FORMULA)	
Insoluble Residue	1.08	C_3S	55%
SiO_2	21.40	C_2S	24%
Al_2O_3	5.30	C_3A	10%
Fe_2O_3	2.30	C_4AF	7%
CaO	64.00		
MgO	1.40		
K_2O	0.49		
Na_2O	0.04		
SO_3	2.98		
Loss on Ignition	1.02		

EXPERIMENTAL RESULTS

The technique proposed in this study for the measurement of the quantity of combined water in Portland cement paste , is a technique of extraction by solvent. The quantity of combined water has been determined by burning at 900° C, during 1 hour, a sample is dried to a temperature of 48° C during 20 to 24 hours. Before the drying, the hydration reactions have been stopped by means of the acetone. The combined water percentage is given, by the following relationship:

$$\text{Combined water (\%)} = \frac{\text{Quantity of water chemically linked}}{\text{Quantity of anhydrous cement}} \quad (1)$$

Tests have been done at various ages, from the first hours of conservation. Figure 1 shows the influence of the w/c ratio on the evolution of the content in combined water. Notice, that for these samples, tests have been conserved of water-tight conditions.

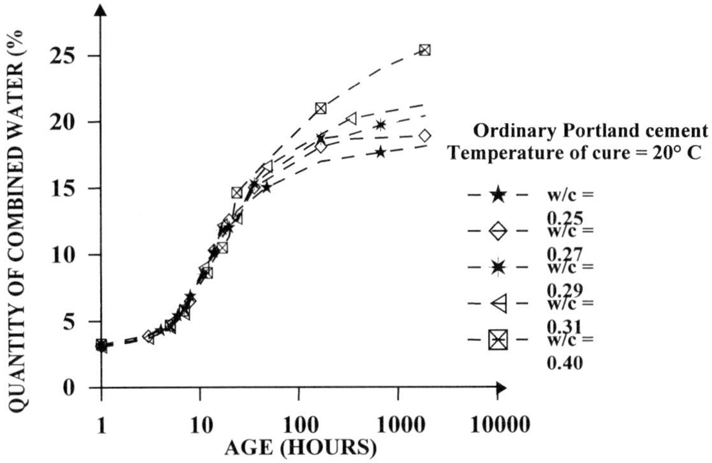

Figure 1 Variation of the quantity of combined water for various w/c ratios

Figure 2 illustrates the influence of the temperature of cure. Measures have been done since the first hours of conservation. It appears clearly on the last figure that the temperature influence strongly the speed of water fixing by the cement paste. Nevertheless the final values of the total quantity of combined water is approximately same since the samples have been conserved in water-tight conditions.

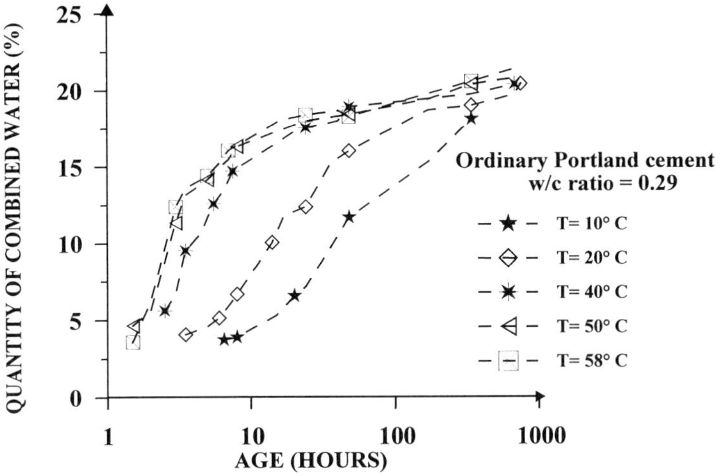

Figure 2 Variation of the quantity of combined water for different temperatures

The degree of hydration can be appreciated by different manners Note here, that the different definitions of the degree of hydration give the same results. In our work, we retain the equation (4) to dcfine the degree of hydration, in which the quantity of water combined at time « t » given by the ratio of the quantity of combined water in the sample to the quantity of anhydrous cement. The total quantity of combined water is estimated from formulae of Bogue

$$\alpha = \frac{\text{Combined water at time (t)}}{\text{Combined water at } (t = \alpha)} \tag{4}$$

The comparison between experimental results and these anticipated by the model is shown in the figure 6 and 7. These curves show the good concordance that exists between the experimental results and the predicted by our model. Nevertheless a light difference is observed at early age and more precisely before the setting (2 to 6 hours in our case).

The period before setting is complex and does not concern the increase of performances of physic-chemical of hydraulic materials.

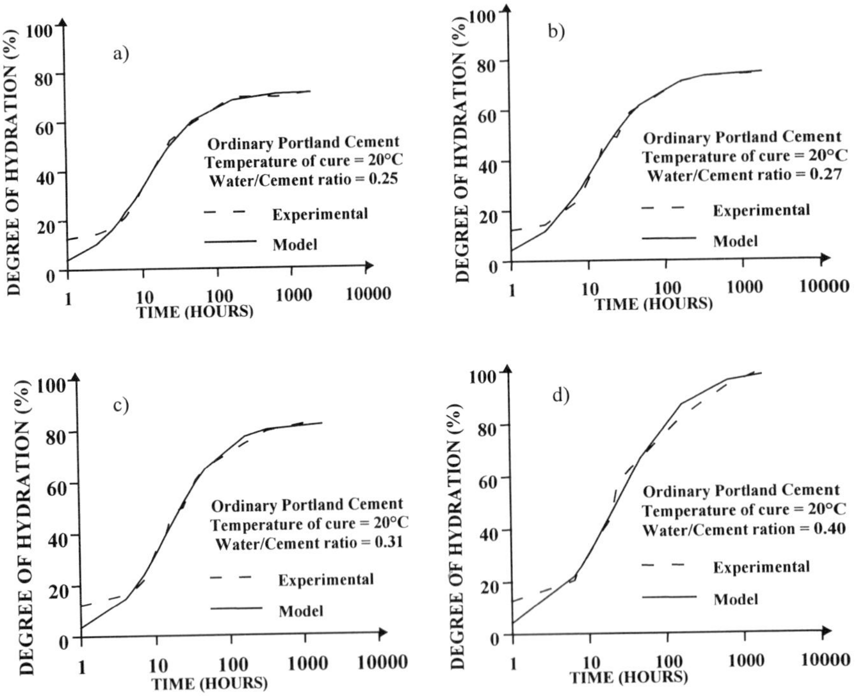

Figure 5 Evolution of degree of hydration for various w/c ratios

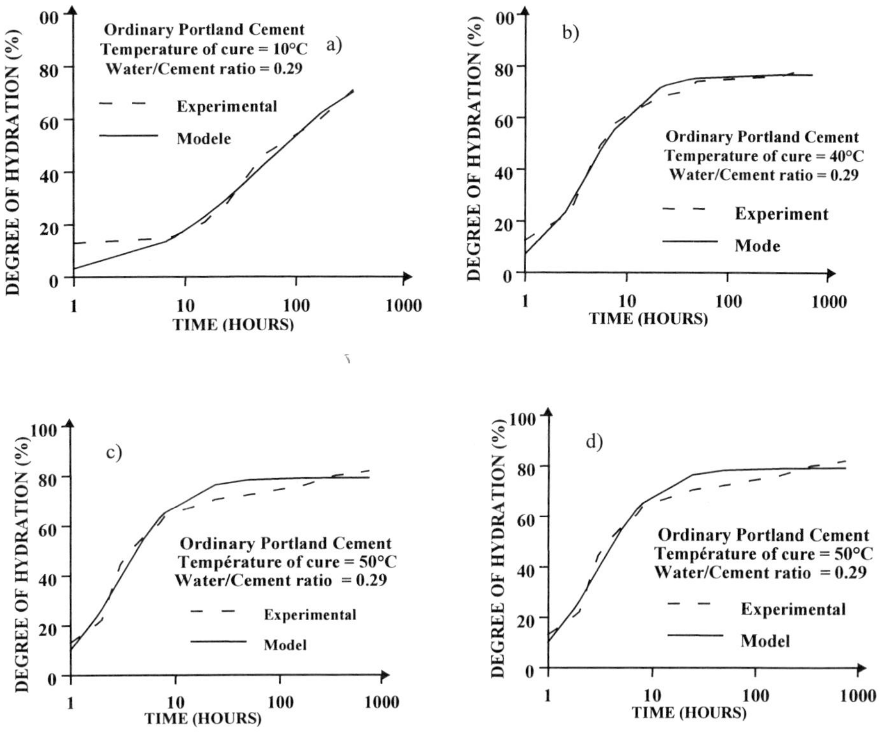

Figure 6 Evolution of degree hydration for different temperatures of cure

PREDICTION OF STRENGTH

The compressive strength is often used to characterize the concrete. In most cases, the binder constitutes the only active phase of hydraulic materials (concrete in our case). The strength to anticipate for concrete can then be easily estimated from the equation (7) in which the binder is represented by its degree of hydration at the temperature of cure.

$$Rc(T,t) = \frac{\alpha(T,t)}{\alpha(T_{ref},t)} Rc(T_{ref},t) \quad (5)$$

Where, $Rc\ (T)$ = concrete compressive strength at a temperature T
$Rc\ (T_{ref})$ = compressive strength of the concrete at 28 days and a temperature of 20° C
$\propto(T)$ = degree of hydration at a temperature T
$\propto(T_{ref})$ = degree of hydration of temperature of 20° C
t = days

Knowing the curve of reference giving the evolution of the resistance in isothermal cure conditions (generally 20° C); the predicted strength for a hydraulic material can therefore be obtained by an affinity of the $\propto (T, t)/ \propto (Tref, t)$ ratio.

Application

An example of development of mechanical performances of concrete by proposed model (in absent of other influencing factors) is given hereafter. Figure 7 shows strength evolution versus the age for various temperatures.

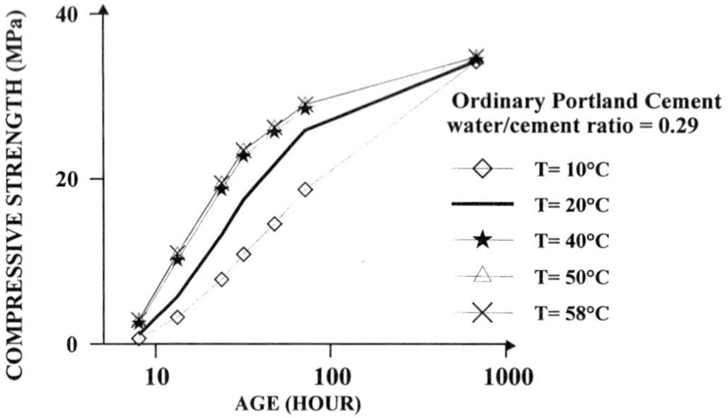

Figure 7 Evolution of strength for different temperatures

This figure shown that the final strengths are the same under different cure temperatures. This can de explained by the fact that, without water exchange , at any temperature, the degree of hydration tends to 1.

CONCLUSIONS

We can tell, here, that the model that we have developed is very satisfactory since the difference between the experimental results and these predicted by the model are minimal. In all cases the evolution of the hydration of the cement is the main factor for the physic-chemical developments of hydraulic materials, which shows well the importance of the model that we have developed. The introduction of this model in predicting of physic-chemical performances of materials allows to illuminate some factors until there poorly known .

REFERENCES

1. PAPAMICHALI, E. Etude de la vitesse de fixation de l'eau dans les pâtes de ciment Portland . Mémoire de DEA, 1980, INSA de Toulouse, 49p.

2. TORRENTI, J M. La résistance du béton au très jeune âge. Bull. Liaison L.C.P.C. No 179. mai-juin 1992. pp. 31-41.

3. LEBERTON, J D, MILLIER, C, Modèles dynamiques déterministes en biologie . Ed. Masson, 1982.

4. BOUMAHRAT, M, GOURDIN, A, Méthodes numériques appliquées . O.P.U. 1980

MEETING COMPRESSIVE STRENGTH REQUIREMENTS AT EARLY AGE BY USING NUMERICAL TOOLS – DETERMINATION OF APPARENT ACTIVATION ENERGY OF CONCRETE

L D'Aloia

J L Clement

Laboratoire Central des Ponts et Chaussées

France

ABSTRACT. Meeting compressive strength requirements at early age is of the utmost importance in civil engineering as in the case of avoidance of thermal cracking. Numerical tools are useful to find a good compromise between mix design, structural design and phasing of concreting. The TEXO and MEXO modules of the finite element program CESAR-LCPC developed by the Laboratoire Central des Ponts et Chaussées enable temperature and induced thermal stress fields in concrete structures to be simulated. Besides, the "method of equivalent age" makes possible the compressive strength of concrete at early age to be estimated from actual or simulated temperature histories of the in-place concrete. It is based on the Arrhenius' law, that the method has shown itself to be the most accurate to describe the kinetic influence of the temperature. The "apparent activation energy" : Ea, characterizes the concrete mixture and is the only parameter which can be found in the expression of the Arrhenius' law. The experimental determination of Ea could be questionable. A proper estimation of concrete strength at early age is directly related to a correct determination of Ea. This paper gives a panel describing an extensive study of concrete at early age by using numerical tools. Its major aims are firstly, to show the influence of apparent activation energy of concrete on estimated temperature and compressive strength, and secondly, to propose a new method of determining Ea from calorimetric tests performed under semi-adiabatic conditions at different initial temperatures.

Keywords : Concrete, Compressive strength, Thermal stress, Early age, Equivalent age, Numerical tools, Apparent activation energy, Calorimetric test

Laetitia D'Aloia is a Civil Engineer. Doctoral dissertation in 1998 on the mechanical behavior of concrete at early age at the Ecole Nationale des Travaux Publics de l'Etat, Lyon. Now she is part of the research group on Mix Design and Casting of Concrete at the Laboratoire Central des Ponts et Chaussées, Paris.

Jean Luc Clement is the head of the research group on Mechanical Behavior of Concrete and Modeling at the Laboratoire Central des Ponts et Chaussées, Paris. He is also associate Professor at the Ecole Normale Supérieure de Cachan, Paris.

INTRODUCTION

Nowadays, a proper estimation of the compressive strength of concrete at early age and the occurrence of thermal cracking are essential in civil engineering. The possibility of estimating the mechanical properties in every sensitive part of a concrete structure is of great interest to building firms regarding formwork removal, early safe prestressing and even avoidance of thermal cracking. The control of all these operations is directly related to a proper understanding of the concrete hardening process at early age. Moreover, the computing tools and a correct knowledge of the physico-chemical phenomena involved in cement hydration make it easier to reconsider the mathematical models of the mechanical behavior of concrete, by taking into account several phenomena as the heat of hydration, the shrinkage and the creep.

A rise of temperature due to the cement hydration reaction is intensified by the mass effect in large structures and can induce thermal gradients ranging from 5°C up to 50°C between the inside and the outside of a concrete block. This thermal consideration affects concrete durability and its mechanical properties. Two points can be underlined. First comes the kinetic influence of temperature on concrete strength development. The higher the temperature, the faster the concrete strength development occurs. Second concerns the risk of thermal cracking due to thermal gradients during cooling of concrete and to restrained conditions.

Numerical tools enable an extensive study of concrete at early age. They can be useful to find a good compromise between mix design, structural design and phasing of concreting. The Finite Element (FE) Program CESAR-LCPC has been developed by the Laboratoire Central des Ponts et Chaussées (LCPC, Paris) for more than 15 years [1,2]. The TEXO and MEXO modules enable a simulation of the temperature evolution and induced thermal stress in every point of the structure [3,4]. The reliability of the temperature field and induced thermal stress calculation depends on the proper description of the boundary and initial conditions, and on the choice of appropriate input data.

This paper gives a panel describing an extensive study of concrete at early age by using CESAR-LCPC (TEXO and MEXO modules) and the method referred to as "method of equivalent age". This method is based on the Arrhenius' law which has turned out to be the most accurate in order to describe the kinetic influence of the temperature [5,6,7]. The major aims of this paper are firstly, to underline and illustrate the usefulness of numerical tools on studying concrete at early age. Secondly, to show the influence of "apparent activation energy" of concrete : Ea (input data) on estimated compressive strength. And finally, to propose a new method of determining Ea from calorimetric tests performed under semi-adiabatic conditions at different initial temperatures.

EXTENSIVE STUDY OF CONCRETE AT EARLY AGE

The Problems Linked to Early Age

The thermal effects are predominant at early age. The rise of temperature in concrete structures is due to the hydration heat of cement. This phenomenon is liable to be intensified by the mass effect in large structures, and partly modified by ambient temperatures and diffusion conditions to the surroundings.

Moreover, the compressive strength development of concrete is influenced by temperature. At early ages, it is a kinetic-type influence. A rise of temperature leads to an acceleration of the chemical reactions between water and cement and hence, the strength development occurs at a faster rate.

Moreover, a risk of thermal cracking can appear during cooling of concrete. It is due to the combination of several factors : high rigidity of the material, low dilatation and contraction capacities, large thermal gradients and restrained conditions.
The two questions of paramount importance are :

- *how to estimate the in-place concrete strength at early age in order to ensure, for example, formwork removal or early safe prestressing,*
- *how to estimate the induced thermal stress in order to avoid thermal cracking and to ensure the durability of the structure.*

The Use of Numerical Tools : CESAR-LCPC

Solutions to these two problems can be found by using numerical tools as the FE Program CESAR-LCPC.

The TEXO and MEXO modules of CESAR-LCPC enable the simulation of the temperature and induced thermal stress field in concrete structures.

The mesh of the structure can be easily obtained with the interactive pre-processor MAX. 2D (plane or axi-symetric) and 3D calculations can be made. CESAR-LCPC makes it possible for the initial and boundary conditions to be taken into account (Initial temperature of fresh concrete, variable ambient or imposed temperature for TEXO. Initial strains and stresses for MEXO). After the calculations, the interactive post-processor PEGGY enables the graphical representation of results to be obtained.

The formwork removal can be modeled with appropriate coefficients of thermal exchange. Successive calculations and declaration of "inactive" areas in the structure can be done to simulate the phasing of casting. Such characteristics of the studied concrete mixture as the specific heat c, the thermal conductivity λ, the final autogeneous shrinkage β_{final}, the final elastic modulus E_{final} and the thermal dilatation coefficient α_{th} must be incorporated in the data file of CESAR-LCPC.

The results of the quasi-adiabatic test (QAB developed by the LCPC [8]) and the apparent activation energy : Ea, makes possible the concrete's hydration heat development to be described and taken into account.

The "method of equivalent age" is then required to obtain compressive strength fields from the temperature ones. Besides, a supplementary analysis of creep (which is not included in the program at the present stage) can be necessary to refine the estimation of the risk of thermal cracking. It is to be more detailed in the analysis and interpretation of the simulation results.

TEXO module :

- **Solution of the diffusion heat equation with a heat source :**

$$\lambda \Delta T + s = \rho c \frac{\partial T}{\partial t}$$

λ : thermal conductivity,
c : specific heat,
s : heat source.

- **Experimental Arrhenius' law :**

$$\frac{\partial \alpha}{\partial t} = f(\alpha)\exp\left(-\frac{Ea}{RT}\right)$$

α : hydration degree,
T : absolute temperature, K,
R : perfect gas constant, 8.314 J/mol K,
Ea : apparent activation energy, J/mol.

- **Material characteristics and input data**

λ : thermal conductivity,
c : specific heat,
QAB test results,
Ea : apparent activation energy of concrete.

- **Initial and boundary conditions :**

T_{ini} : initial temperature of fresh concrete,
K : thermal exchange coefficient,
T_{amb} : ambient temperature (variable)
T_{imp}, ϕ_{imp} : imposed temperature and flow.

- **Results :**

Temperature field : $T(t)$
Hydration degree field : $\alpha(t) = q(t) / q_{final}$

MEXO module :

- **Aging thermo-elastic law :**

$$\dot{\varepsilon}_{ij} = \dot{\varepsilon}_{ij}^{el} + \dot{\varepsilon}_{ij}^{r} + \dot{\varepsilon}_{ij}^{th} = \left(\alpha_{th}\dot{T} - \beta\dot{\alpha}\right)\delta_{ij} + \frac{1+\nu}{E(\alpha)}\dot{\sigma}_{ij} - \frac{\nu}{E(\alpha)}\dot{\sigma}_{kk}\delta_{ij}$$

Elastic strain : $\varepsilon_{ij}^{el} = \frac{1+\nu}{E(\alpha)}\sigma_{ij} - \frac{\nu}{E(\alpha)}\sigma_{kk}\delta_{ij}$

Autogeneous strain : $\varepsilon_{ij}^{r} = -\alpha\beta_{fin}\delta_{ij}$

Thermal strain : $\varepsilon_{ij}^{th} = \alpha_{th}\Delta T\delta_{ij}$

- **Evolution laws :**

Autogeneous strain : $\beta(\alpha) = -\alpha\beta_{fin}$

Elastic modulus : $\mathrm{E}(\alpha) = \mathrm{E}_{final} h(\alpha)$

$$h(\alpha) = \frac{1 + 1.37 Sc_{final}^{\ 2.204}}{1 + 1.37 Sc(\alpha)^{2.204}} \left(\frac{Sc(\alpha)}{Sc_{final}} \right)^{2.675}$$

$$Sc_{final} = \left(\frac{\mathrm{E}_{final}}{7250} \right)^{\frac{1}{0.471}}$$

$$\begin{cases} \alpha \le \alpha_0,\ Sc(\alpha) = \dfrac{\alpha}{\alpha r_0} Sc_0 \\ \alpha \ge \alpha_0,\ Sc(\alpha) = \left(Sc_{final} - Sc_0\right)\left(\dfrac{\alpha - \alpha_0}{1 - \alpha_0}\right) + Sc_0 \\ Sc_0 = \alpha_0 \dfrac{Sc_{final}}{10} \end{cases}$$

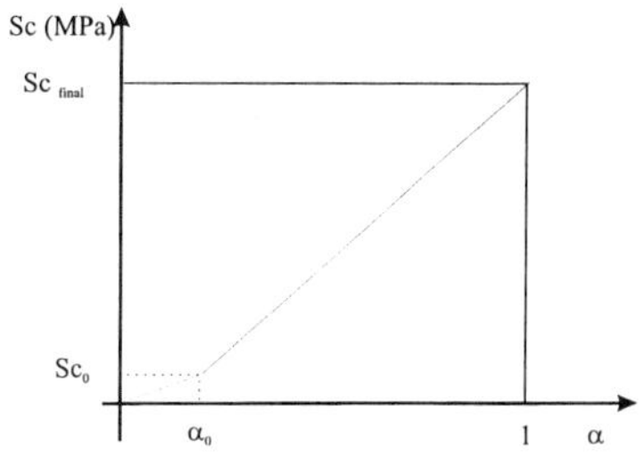

α_0 : hardening threshold : 0.05 for ordinary concrete, 0.1 for high performance concrete.

- **Material characteristics and input data :**

$\nu = 0.2$: Poisson ratio,

α_{th} : thermal dilatation coefficient,

β_{final} : final autogeneous shrinkage,

E_{final} : final elastic modulus.

- **Initial and boundary conditions :**

Initial stresses,
Imposed displacements.

TEXO Results : $T(t), \alpha(t)$

- **Results :**

Induced thermal stress fields : $\sigma(t)$

Interpretation and analysis of TEXO and MEXO results

- *how to estimate the in-place concrete strength at early age in order to ensure, for example, formwork removal or early safe prestressing.*

The kinetic influence of temperature on compressive strength development is already underlined. Meeting strength requirements at early age implies estimating compressive strength from actual or simulated temperature histories with the "method of equivalent age". The usage of equivalent age based on the Arrhenius' law, with the purpose of estimating concrete strength at early age is gaining acceptance [5,6,7]. The equivalent age represents the age of curing at the reference temperature, generally 20°C. It would result in the same strength as would result from curing at other temperatures.

According to the Arrhenius law, the expression of equivalent age t_{eq} is :

$$t_{eq}(20°C) = \int_0^{t0} \exp\left(-\frac{Ea}{R}\left(\frac{1}{T(s)} - \frac{1}{293}\right)\right) ds \qquad (1)$$

$T(s)$: actual or simulated absolute temperature of concrete, in Kelvin,

The method of equivalent age is summarized in Figure 1.

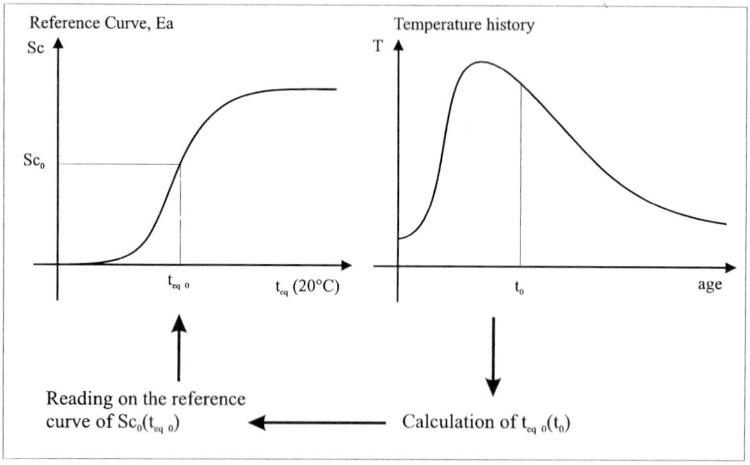

Figure 1 Scheme of the Application of the "Method of Equivalent Age"

The method of equivalent age can be analyzed in 3 stages :

- the relationship between strength and equivalent age is established from laboratory compressive strength tests on concrete specimens, and is called the "reference curve at 20°C",
- the equivalent age is computed from the actual or simulated temperature history of the in-place concrete in accordance with (1),
- the strength is estimated from the reference curve according to the previously computed equivalent age value.

Remark : the final strength of the in-place concrete is supposed to be independent of the temperature history.

- *how to estimate induced thermal stresses in order to avoid thermal cracking and to ensure the durability of the structure.*

The induced thermal stress fields can be simulated by CESAR-LCPC. The autogeneous shrinkage is taken into account. The risk of thermal cracking can be estimated by comparing induced thermal stress with mechanical strength of concrete.

The tensile strength can be measured or estimated from compressive strength fields which are obtained by using the method of equivalent age.

If a more refined analysis is required, a supplementary study of stress relaxation due to creep (which is not included in the program) can be followed. Stress relaxation due to creep is favorable to avoidance of thermal cracking by reducing the thermal stress level.

An extensive study at early age with numerical tools is summarized in Figure 2.

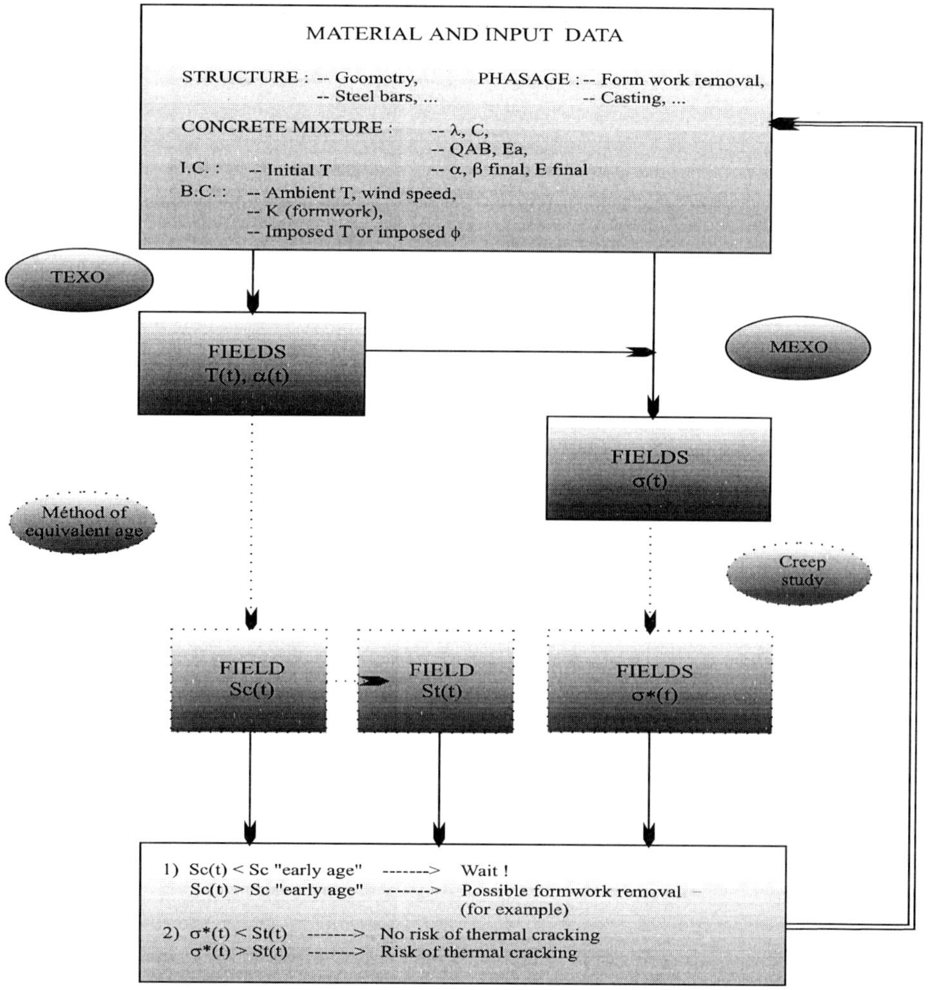

λ, C: thermal conductivity and specific heat
QAB : stands for Quasi Adiabatic tests resulst
Ea : apparent activation energy of concrete
α_{th} : thermal dilatation coefficient
β final : final autogeous shrinkage
E final : final elastic modulus

K : coefficient of thermal exchange
ϕ : thermal flow
α : hydration degree
Sc, St : compressive and tensile strengths
σ : induced thermal stresses
σ* : induced thermal stresses (including creep)
Sc "early age" : required strength at early age

Figure 2 Scheme of an Extensive Study at Early Age by Using Numerical Tools

As far as the interpretation and the analyses of the simulation results have been done, it is possible to modify the type of cement (lower heat of hydration), the age of formwork removal, the curing conditions, even the geometry of the structure, in order to meet strength requirements at early age and to limit the risk of thermal cracking.

It is why numerical tools are essential to find a good compromise between mix design, structural design and phasing of casting.

Proper estimation of temperature fields, compressive strength and risk of thermal cracking are directly related to a correct modeling of the structure. This also implies an accurate modeling of boundary conditions and appropriate values of input data.

Now, we should like to focus on input data and particularly on the choice of the apparent activation energy of concrete : Ea.

INFLUENCE OF APPARENT ACTIVATION ENERGY : Ea, ON CONCRETE STRENGTH ESTIMATION

Ea is introduced in the data file of CESAR-LCPC to obtain temperature fields (TEXO module). The estimation of its influence on compressive strength with the method of equivalent age is to be illustrated hereafter.

Ea Significance

Remark : the Arrhenius' law allows the description of the temperature influence on a "simple" chemical reaction kinetic. Its application to the case of cement hydration is an approximation in that the hydration of the different anhydrous components of cement corresponds to several simultaneous and inter-dependent chemical reactions. It is why the activation energy is called "apparent".

Ea is the only parameter which can be found in the expression of the equivalent age and has to be experimentally determined. Ea represents the sensitivity of the concrete mixture to a rise of temperature. The higher the apparent activation energy, the more sensitive the concrete mixture. It characterizes the concrete mixture and depends on several parameters :

- mix design : type of cement, fineness, water to cement ratio : w/c, mineral additions, admixtures, aggregates,
- physical variables : hydration degree, temperature.

Illustration

The influence of the adopted Ea value on the reliability of the estimated compressive strength can be easily illustrated.

Putting forth the problem

Materials used and temperature cycle : A typical ordinary concrete is chosen (equivalent type I cement, w/c=0.45). The reference curve is based on compressive strength tests performed on 11 × 22 cm cylinders stored at 20°C (Figure 3a). Test results are interpolated with an appropriate mathematical model. The thermal evolution of a point located at the surface of a 50 × 50 cm cross-section is obtained from CESAR-LCPC simulations (TEXO' module) and is shown on Figure 3b. Concrete is cast in a wood formwork. Initial and ambient temperatures are equal to 20°C.

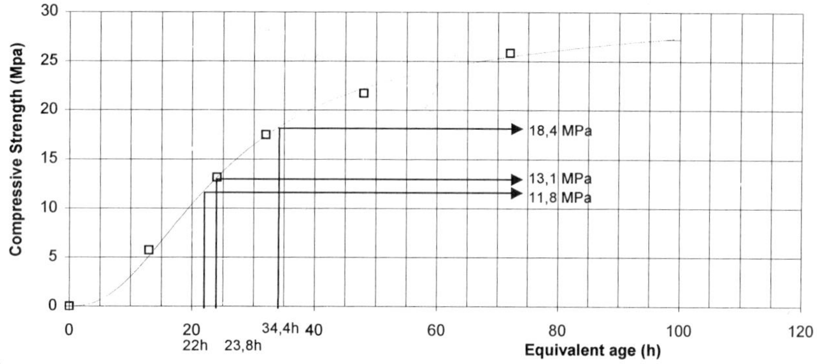

Figure 3a Influence of the selected Ea value on the estimated concrete strength, reference curve at 20°c.

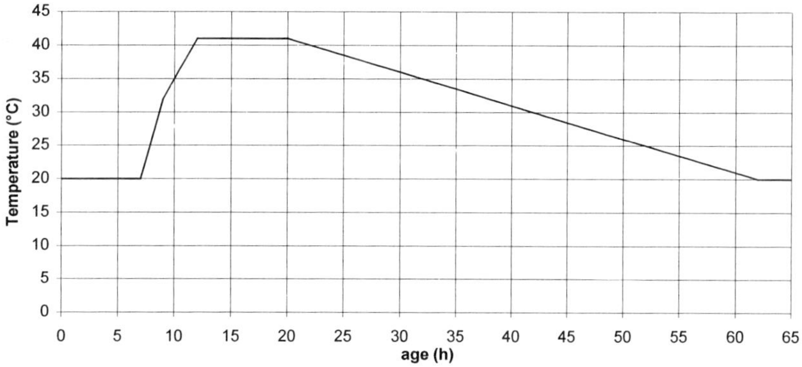

Figure 3b Influence of the Selected Ea value on the Estimated Concrete Strength, temperature evolution in concrete (TEXO simulation)

Selected Ea Values : Several Ea values are selected in order to describe fully the range of the Ea values which have been published in various research papers. Typical values are presented in a database [8]. The minimum and maximum values are : 29.4 kJ/mol and 55.7 kJ/mol, according to the type of cement. The value of 34.5 kJ/mol which corresponds to the studied cement is also selected in the previously mentioned database.

How to use the method of equivalent age

The equivalent ages calculated with the temperature cycle and in accordance with (1) for the three different Ea values, are plotted against the actual age of curing (Figure 3c). For example, a 15-hours curing leads to a difference of nearly 6 MPa in the corresponding compressive strength, according to the adopted Ea value (Figures 3c and 3a). Table 1 summarizes the results obtained for the three different Ea values.

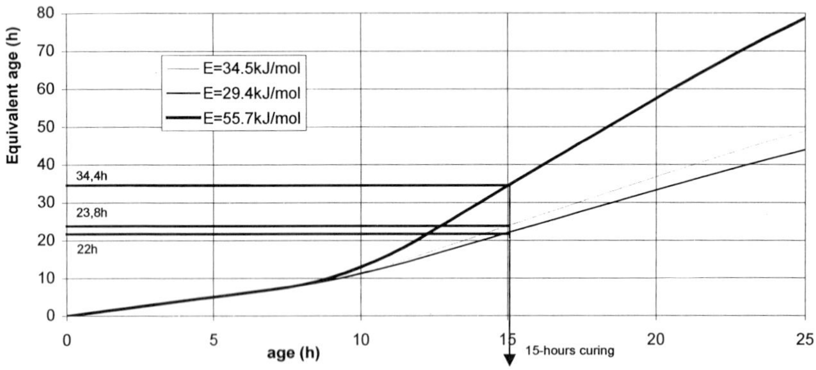

Figure 3c Influence of the selected Ea value on the estimated concrete strength, equivalent age vs. actual age of curing.

Table 1 Comparison between equivalent age and actual age of curing

Ea (kJ/mol)	29.4	34.5	55.7
Compressive strength after 15-hours curing (MPa)	11.8	13.1	18.4

Concluding remarks about the influence of Ea

The Ea value cannot be arbitrarily chosen. An approximate Ea value chosen according to the type of cement can be satisfactory only in the case of limited variations of temperature, when neither admixture nor mineral addition is incorporated. The apparent activation energy characterizes the overall mixture and is liable to depend on any parameter modifying the kinetic of cement hydration. Hence, the dependence of Ea on the degree of hydration and possibly on temperature can be explained [9,10,11].

Remark : the degree of hydration is defined as the relative evolution of a mechanical or physical property (compressive strength or heat of hydration) [12].

EXPERIMENTAL DETERMINATION OF Ea : CALORIMETRIC TESTS

Why Using Calorimetric Tests

The proper correlation obtained between the development of mechanical and physical properties during cement hydration made it possible to determine Ea from calorimetric tests [6,12]. The proposed method is based on heat of hydration tests under semi-adiabatic conditions (possibly adiabatic) at different initial temperatures [10,11].

The main advantages of this method consist in more reliable continuous measurements of temperature and in easy test procedures. It must be pointed out that semi-adiabatic conditions lead to temperature histories similar to the actual thermal evolution of in-place concrete in large structures. A new method of determining Ea from semi-adiabatic tests at different initial temperatures has been implemented and is briefly presented below.

Experimental Procedure

Two points must be underlined with regard to the experimental procedure :

- *possible formulation of a mortar thermally equivalent to the concrete,*
- *thermal equilibrium of calorimeters and fresh concrete at the beginning of test.*

Testing concrete

The semi-adiabatic apparatus (the QAB calorimeter) used for testing concrete, was developed by the Laboratoire des Ponts et Chaussées (LCPC, Paris) [13]. The core temperature of an insulated 16×32 cm concrete cylinder is continuously recorded at an appropriate frequency by using a temperature probe connected to a temperature recording/read out device. The coefficient of thermal losses and the thermal characteristics of the apparatus enable the recorded temperature to be corrected. The heat of hydration is then obtained by multiplying the fictitious rise in corrected temperature by the specific heat calculated from the mixture proportions [13,14] :

$$q(t) = \frac{C}{m_c}\left(T_c^*(t) - T_0\right) \qquad (2)$$

C : specific heat of the specimen, J/°C,
m_c : mass of cement in the specimen, g,
$T_c^*(t)$: fictitious temperature corrected from thermal losses,
T_0 : initial temperature of the concrete mixture.

Testing equivalent mortars

The semi-adiabatic apparatus used for testing mortars is the French standard Langavant calorimeter [14]. It operates in the same way as the QAB calorimeter, but is more reliable. The experimental conditions differ from the standard ones. Mortars proportioned from the same concrete mixture are used, instead of the standard mortar.

Two mortars have been proportioned with the same fine fractions as concrete. w/c is kept as a constant parameter in the two mortars, so that the chemical reactions between cement and water are not modified compared with the concrete mixture. The two ideas supporting the two mortar mixtures, namely : Mortar 1 and Mortar 2, are the following :

Mortar 1 : the specific surface of aggregates divided by the weight of cement is kept constant in both Mortar 1 and concrete. The assumption that thermal phenomena be mainly controlled by the specific surface of heat exchange between aggregates and cement paste has already been suggested and developed by Italcimenti - Ciments Français [15].

Mortar 2 : The second mortar has been proportioned on the basis of a new concept developed at the Ecole Nationale des Travaux Publics de l'Etat (ENTPE) in the framework of proportioning sand concrete [16]. According to this concept, a granular extension factor is evaluated. It quantifies the cement paste thickness surrounding each aggregate assimilated to a spherical grain. As far as the porosity of the packing of aggregates is known, the paste volume can be calculated by summing the volume required to fill this porosity and the volume required to keep constant this extension factor in both Mortar 2 and concrete.

Thermal equilibrium at the beginning of test

In order to avoid thermal problems caused by imbalances when placing concrete and mortar specimens in the calorimeter, experiments are carried out in a temperature regulated chamber. The ambient temperature is equal ($\pm 1°C$) to the planned initial temperature. Devices and materials are stored in the regulated chamber. All the batches are prepared in the chamber. The temperature is registered during at least 48 hours depending on the reactivity and mass of cement. Moreover, thermal equilibrium at the beginning of test enables to reduce thermal losses. Hence, the influence of the coefficients of thermal losses of calorimeters which are experimentally determined, is minimized.

How to Analyze the Results

The Arrhenius' law leads to the following expression of the rate of reaction :

$$\frac{d\alpha}{dt} = g(\alpha)K(T) = f(\alpha)\exp\left(-\frac{Ea}{RT}\right) \qquad (3)$$

α : hydration degree,
$K(T)$: rate constant at temperature T,
g, f : functions of the hydration degree.

The overall hydration process is assumed to be described by two separable variables. Considering two different temperature histories obtained by choosing two different initial temperatures under semi-adiabatic conditions : $T_1(t)$ and $T_2(t)$, one can write :

$$For\,all\,\alpha_0, \begin{cases} \left(\frac{d\alpha}{dt}\right)_{T_1}(\alpha_0) = f(\alpha_0)\exp\left(-\frac{Ea}{RT_1(\alpha_0)}\right) \\ \left(\frac{d\alpha}{dt}\right)_{T_2}(\alpha_0) = f(\alpha_0)\exp\left(-\frac{Ea}{RT_2(\alpha_0)}\right) \end{cases} \qquad (4)$$

For each fixed value of the degree of hydration : α_0, a corresponding value of the apparent activation energy : Ea is obtained :

$$Ea(\alpha_0) = -\frac{R}{\dfrac{1}{T_2(\alpha_0)} - \dfrac{1}{T_1(\alpha_0)}} \ln\left(\frac{\left(\dfrac{d\alpha}{dt}\right)_{T_2}(\alpha_0)}{\left(\dfrac{d\alpha}{dt}\right)_{T_1}(\alpha_0)} \right) \qquad (5)$$

With the definition of the degree of hydration as the relative heat released during hydration :

$$\alpha(t) = \frac{q(t)}{q_{final}} \qquad (6)$$

$q(t), q_{final}$: heat of hydration expressed in J/g of cement (at time t and final value).

And with the assumption that the final heat of hydration do not depend on the curing temperature during the hydration process, the preceding equation can be alternatively written by replacing α by q and α_0 by q_0. The right-hand side of (5) can also be expressed in terms of heat of hydration.

Considering that Ea vs. q curves are obtained, defining a period corresponding to the early age according to the concrete mixture becomes easier. An average value can be computed, and a field of definition can be given. It must be pointed out that by considering the rates of released heat, no assumption of Ea's dependence on the heat of hydration is made. Even if Ea depends on q, the Ea values computed according to equation (5) remain valid. Only (3) must be replaced by (7), for the sake of rigor :

$$\frac{d\alpha}{dt} = g(\alpha)K'(T,\alpha) \qquad (7)$$

The existed classical analyses generally and implicitly assume that Ea do not depend on q. Ea is calculated after integration and according to the following equation :

$$Ea = -\frac{R}{\dfrac{1}{T_2(\alpha_0)} - \dfrac{1}{T_1(\alpha_0)}} \ln\left(\frac{t_2(\alpha_0)}{t_1(\alpha_0)} \right) \qquad (8)$$

If Ea depends on q, any conclusion about its variation with q can be drawn. Moreover, the setting time is different with regard to the temperature, but this has not to be taken into account when considering differences or rates of released heat.

First Results

Several results can be found in a previously published paper [11]. Typical Ea versus hydration heat curves are shown in Figures 4. The main results of this study are the following :

- All the curves Ea versus hydration heat show similar trends. Ea cannot be considered as a constant independent of the heat of hydration, except during the beginning of the hydration process : at early ages when the extent of hydration is less than 25 to 45%. An average value independent of temperature can be defined at early ages. However, the range of definition seems to be strongly affected by temperature. The higher the temperature, the shorter the definition range.

- Similar trends have been observed in Ea vs. heat of hydration curves, by testing concrete or mortars. The influence of the granular skeleton can be considered as a secondary one. The parameter which appears to be of the utmost importance to proportion a mortar thermally equivalent to the concrete mixture, is w/c. This point is to lead to a simplified procedure of Ea determination by testing mortar instead of concrete.

- The thermal equilibrium at the beginning of test enables a reliable determination of Ea to be ensured.

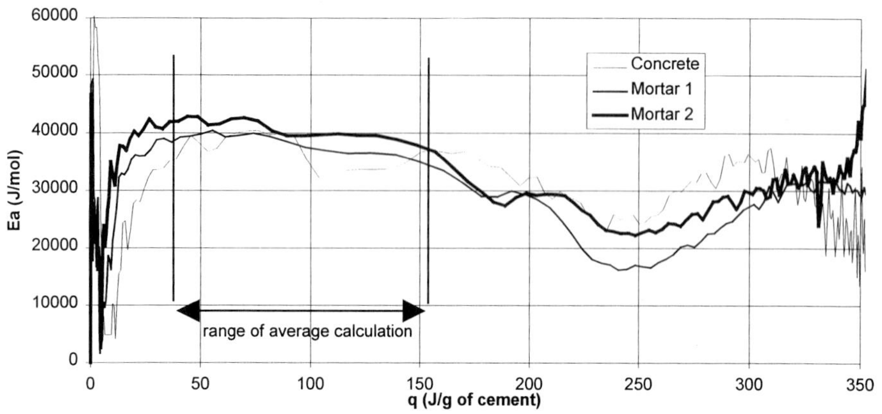

Figure 4 Ea vs. Hydration heat (equivalent type I cement), initial temperatures of tests : 10 and 20°C

CONCLUSIONS

Firstly, the use of the TEXO and MEXO modules of the FE program CESAR-LCPC enables the link between concrete hydration heat and induced strains and stresses. Nowadays, several numerical simulations are performed at the LCPC in order to help civil engineers to chose appropriate concrete mixture and building process [3,4].

The most important parameter which is required for thermo-mechanical simulations is the apparent activation energy : Ea. At present, Ea is experimentally determined from calorimetric tests performed under semi-adiabatic conditions instead of mechanical ones. Several aspects have been underlined and led to a more reliable experimental procedure, that is to say, definition of a mortar thermally equivalent to the concrete mixture and thermal equilibrium at the beginning of the test.

The secondary influence of aggregates has been pointed out. Moreover, the analysis of test results in terms of heat flow, enables a Ea versus hydration degree curve to be obtained. The dependence of Ea on hydration degree (or hydration heat) is underlined. The dependence on temperature is much more questionable.

Experimental results and phenomenological aspects led to the conclusion that Ea can be considered as a constant only within a limited range of hydration degree, depending on temperature. We now propose to implement this dependence into the numerical tools to get more realistic results.

REFERENCES

1. HUMBERT, P, CESAR-LCPC . Un code général de calcul par éléments finis. Bulletin de liaison des Laboratoires des Ponts et Chaussées, March-April 1989, n°160, pp 112-115.

2. Logiciels LPC, CESAR-LCPC. Edited by the Laboratoire Central des Ponts et Chaussées, France, 1994, ISBN 2-7208-8520-0, pp 58-65.

3. TORENTI, J M, ARISTAGHES, P, DOMBERNOWSKI, K. et al. La simulation des effets thermiques dans le béton au jeune âge : exemples d'application. Annales de l'ITBTP, July-August 1995, n°535, serie : béton 322, pp 1-27.

4. TORRENTI, J M, GUENOT, I, LAPLANTE, P, ACKER, P, DE LARRARD, F. Numerical simulation of temperatures and stresses in concrete at early ages. Computational Modeling of Concrete Structures. Proceedings International Conference, EURO-C 1994. H. Mang, N. Bicanic, and R. De Borst, eds., Pineridge Press, Swansea, Wales, pp 559-568.

5. FREIESLEBEN HANSEN, P, PEDERSEN, E J. Maleinstrument til Kontrol af betons haerdning. Nordisk betong, 1977, pp 21-25.

6. REGOURD, M, GAUTIER, E. Comportement des ciments soumis au durcissement accéléré. Durcissement accéléré des bétons, Annales de l'ITBTP, October 1980, n°387, pp 83-96.

7. TANK, R C, CARINO, N J. Rate constant functions for strength development of concrete. ACI Materials Journal, Vol. 88, N°1, January-February 1991, pp 74-83.

8. BRESSON, J. La prévision des résistances des produits en béton. RILEM International Conference on Concrete at Early Ages, ENPC Paris, April 1982, Vol. 1, pp 111-115.

9. D'ALOIA, L. Détermination de l'énergie d'activation apparente du béton dans le cadre de l'application de la méthode du temps équivalent à la prévision de la résistance en compression au jeune âge : approches mécanique et calorimétrique, simulations numériques, Thèse de l'INSA de Lyon, France, 3 mars 1998.

10. D'ALOIA, L, CHANVILLARD, G. Determination of the apparent activation energy of concrete : Ea. Semi-adiabatic tests of heat development. Proceedings of the 4th CANMET/ACI/JCI International Symposium on Recent Advances in Concrete Technology, Japan, Tokushima 7-11 June 1998. Edited par V. M. Malhotra. pp 561-586.

11. KJELLSEN, K O, DETWILER, R J. Reaction kinetics of Portland cement mortars hydrated at different temperatures. Cement and Concrete Research, Vol. 22, N°1, 1992, pp 112-120.

12. TORRENTI, J M. Comportement mécanique du béton : bilan de 6 années de recherche. OA23, LCPC Presses, Paris, October 1996, Chapter 1 : Comportement du béton au jeune âge, pp 7-50.

13. ACKER, P, CHAUVIN, J J. Essais calorimétriques sur béton : banque de données. Paris : LCPC, Bordeaux : CETE du sud-ouest, report, 1990, 11p.

14. NFP 15-436 Binders - Measuring the hydration heat of cements by means of semi-adiabatic calorimetry (Langavant method). September 1988, 21p.

15. LAPLANTE, P, ROUSSEL, S, LECRUX, S. Technique maturométrique : La loi d'Arrhenius au service des chantiers. RILEM Symposium on Concrete : From material to structure, Arles, France, 11-12 September 1996, 20p.

16. CHANVILLARD, G, BASUYAUX, O. Une méthode de formulation des bétons de sable à maniabilité et résistance fixées. Bulletin des laboratoires des Ponts et Chaussées, N°205, September-October 1996, pp 49-63.

INFLUENCE OF CHEMICAL ADMIXTURES AND MIXING ON THE MIX PROPORTION OF GENERAL PURPOSE SELF-COMPACTING CONCRETE

K Takada

Kajima Technical Research Institute

Japan

G I Pelova J C Walraven

Delft University of Technology

Netherlands

ABSTRACT. In this paper, the influences of the chemical admixtures and the mixing intensity on the mixture proportioning of general purpose Self-Compacting Concrete (SCC), as a internal and a external governing factor respectively, were tried to be understood qualitatively. From the experimental results, it was found that the adequate value of the water to powder volume ratio (V_w/V_p) was varied due to the type of superplasticizer applied. The polycarboxylate type of superplasticizer used made the mortars less viscous than the melamine sulfonate and/or naphthalene sulfonate type of superplasticizer used when V_w/V_p was constant under the same mixing condition applied. The welan gum, a kind of natural polysaccharide-based viscosity agent, raised the value of adequate V_w/V_p due to its characteristics to make the mixture viscous. Moreover, it was also clarified that the adequate V_w/V_p was strongly influenced by the mixer type and the mixing time. It could be found in a different value for each mixing way even though the same materials were used. This phenomena was tried to be explained by the three effects of mixing efficiency on the property of the paste phase in fresh concrete.

Keywords: Self-compacting concrete, Superplasticizer, Viscosity agent, Mixing intensity Flowability, Viscosity, Particle dispersion, Initial hydration product, Air bubble

Mr Kazunori Takada is a research engineer of Kajima Technical Research Institute, Japan, and a guest researcher (1996-98) of Delft University of Technology, the Netherlands. His main research interest is in the development of new value concrete including self-compacting concrete and its scientific fundamentals. He is a committee member of RILEM TC-SCC.

Ms Galia I Pelova is a research associate of Delft University of Technology, The Netherlands. Her research topic is specialized on the rheology of fluid concrete.

Professor Joost C Walraven is the dean of the department of concrete structure in the faculty of civil engineering, Delft University of Technology, The Netherlands. He has published widely in the both fields of structure and material. He has joined many international committees and has taken the chairmanships, especially in *fib* and former CEB.

mixed in 6 different mixing ways shown in Figure 4. The F5.0 and F3.5 are the mixings produced by a forced pan mixer. The G7.5, G5.5, G3.5 and G2.5 are the mixings produced by a gravity tilting mixer. Both mixers have a capacity of 50 liters. In this test, the coarse aggregate volume was fixed at G/G_{lim}=55% (percentage to the solid volume of the coarse aggregate)[4][6]. The fine aggregate volume in the mortar was fixed at 40% as well as for the mortar test. Only the powder composition of BSC60%&FA40% was applied. Chemical admixtures were varied in 3 types, (a) polycarboxylic ether type SP (SP-A and B combined 1:2.5 in weight), (b) polycarboxylic ether type SP (ditto) and welan gum, (c) naphthalene sulfonate type SP (SP-D).

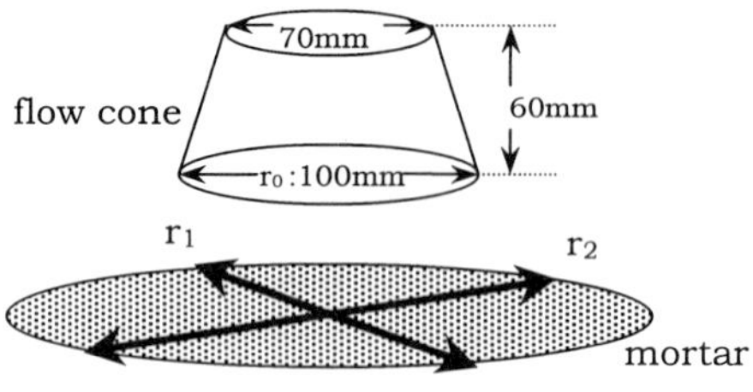

$$r = \frac{(r_1 + r_2)}{2}$$

$$\Gamma_m = \left(\frac{r}{r_0}\right)^2 - 1$$

Figure 1 Mortar flow test

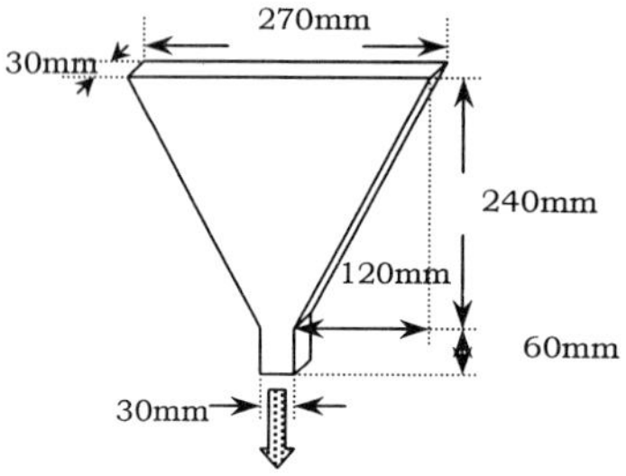

$$R_m = \frac{10}{t}$$

Figure 2 Mortar funnel test

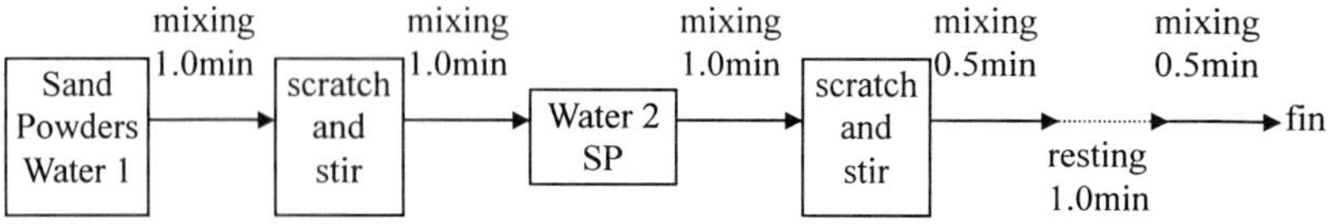

Figure 3 Mixing procedure of mortar

For all mixtures, the slump flow test and the V funnel test[4][6] were carried out. The dosage of superplasticizer (Sp/P, percentage to powder weight) was adjusted to obtain a proper deformability. In this study, a slump flow value=650±30mm and a V funnel time=11±2sec are considered as the adequate values for a workable SCC[6].

When a concrete mix resulted in an inadequate value of V funnel time in spite of a proper result of the slump flow, V_w/V_p was adjusted so that the V funnel time resulted in the control value range under the applied mixing conditions.

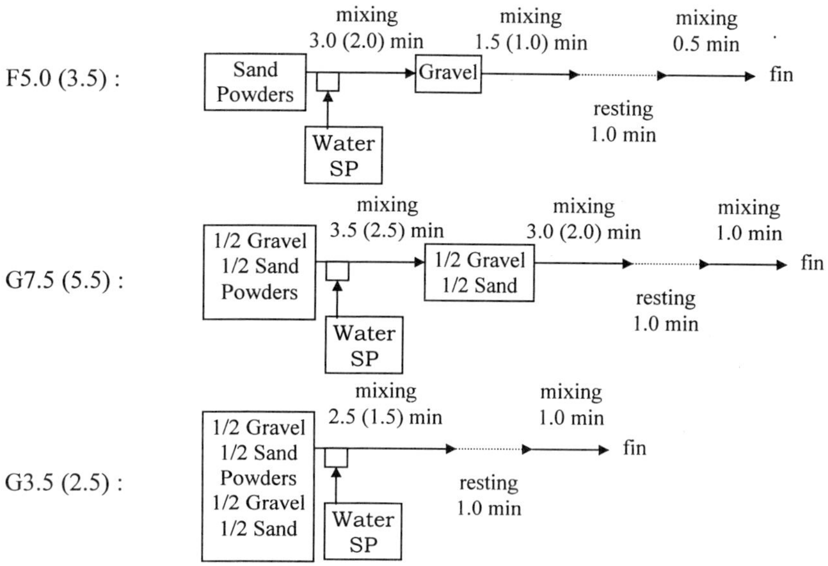

Figure 4 Mixing procedures of concrete

RESULTS AND DISCUSSION

Series A: Tests on Mortar

Dispersibility of superplasticizer

Table 2 shows the results of the mortar tests. The adequate ratio V_w/V_p determined by the method of the general purpose approach varies due to the combination of the powder composition and the chemical admixture. In the mortars 1-4, the type of powder composition was fixed as BSC100%. However, adequate values of V_w/V_p were dependent on the type of the superplasticizer. To realize a proper viscosity of the mixture, each superplasticizer requires a different ratio V_w/V_p. The adequate values are smaller in the order SP-A(polycar.)<SP-B(polycar.)<SP-D(naphth.)<SP-C(naphth.&melam.). SP-A resulted in an 11% smaller value than SP-C.

This result can be considered to be related to the dispersibility of each superplasticizer under the same mixing condition. It has been explained that naphthalene sulfonate or melamine sulfonate type of superplasticizers have a dispersion mechanism based on the electrostatic repulsion effect. On the other hand, the dispersion mechanism of polycarboxylic ether type is based on the steric hindrance effect[8]. Furthermore, it has been also explained that the dispersibility and its retainability of polycarboxylic superplasticizers can be controlled by varying a miner part of the chemical structure[9]. This dispersion mechanism naturally reflects on the results observed. The result mentioned above can be interpreted in such a way that a mixture with a polycarboxylate SP becomes less viscous than a mixture with a naphthalene SP when the ratio V_w/V_p is constant. The viscosity, or in other words the deforming resistance, of the paste phase in the mixture is considered to be governed by the degree of agglomeration of powder particles when the ratio V_w/V_p is constant. Even in the mixture composed with a superplasticizer, the powder particles cannot disperse perfectly individually and the degree of agglomeration controls the deforming resistance of the mixture. From this point of view, it can be said that a naphthalene or melamine type SP could not disperse the powder particles better than a polycarboxylate SP under the mixing condition applied, and it resulted in a larger V_w/V_p to obtain the proper viscosity for SCC.

Table 2 Results of mortar tests

MIX NO	POWDER COMPOSITION (% in volume)	β_p	E_p	SUPERPLASTICIZER	WELAN GUM	ADEQUATE Vw/Vp
1	BSC1 100%	0.951	0.0900	SP-A	-	0.745
2				SP-B	-	0.810
3				SP-C	-	0.835
4				SP-D	-	0.815
5				SP-B	0.05%*	0.850
6	BSC2:60% + FA1:40%	0.864	0.0623	SP-A&B(1:2.5)	-	0.765
7				SP-A&B(1:2.5)	0.05%*	0.790
8	BSC2:60% + FA2:40%	0.874	0.0727	SP-A&B(1:2.5)	0.05%*	0.805
9				SP-D	-	0.830

* percentage to water content

Influence of welan gum

In the mortars 5 and 7, welan gum was used and those mixes resulted in larger values of V_w/V_p than the mortars 2 and 6 respectively. For both cases, the polycarboxylate SP was applied. According to a recent study on the basic property of the welan gum[10], it is explained that welan gum increases the viscosity of the free water in the fresh concrete by virtue of the welan gum's polymer's characteristics to associate each other in water. Moreover, it is also considered that the polymer in the welan gum and the polymer in the superplasticizer restrain each other and this phenomena results in a higher segregation resistance and a larger dosage of SP for a particular deformability. The test result shown in Table 2 agrees well with this explanation. When the welan gum is dosed, the mixture tends to be more viscous and requires a higher value of V_w/V_p to obtain the appropriate viscosity for SCC than the mixture without the welan gum.

Series B: Tests on Concrete

Influence of Mixing Intensity on Proper Proportion of SCC

Table 3 shows the test result for the material combination of BSC2:60%&FA1:40% and chemical admixture type (a) polycarboxylic ether type SP (SP-A and B combined in 1:2.5 weight). α is the premium factor to the original value of the adequate V_w/V_p for general purpose SCC estimated by the mortar test. Figure 5 shows the relation among the V funnel time, V_w/V_p and the mixing intensity when the slump flow was controlled.

From this result, the following tendencies can be clearly observed:

(1) The original value of V_w/V_p was quite adequate for the mixing procedure F5.0 but not for other mixing ways. A shorter mixing by the same forced mixer, F3.5, resulted in a longer V funnel time when V_w/V_p was maintained as the original value. A 5% increase of V_w/V_p from 0.765 to 0.803 was sufficient to obtain nearly the same funnel time as for the shorter mixing F3.5.

(2) In the case that a gravity tilting mixer was applied, the mixture became extremely viscous and heavy to handle when it was mixed with the V_w/V_p ratio which was adequate for the mixing F3.5. For the G7.5 mixing, a 12% increased value from the original V_w/V_p , from 0.765 to 0.858, was adequate to arrive at the appropriate funnel time.

(3) The shorter mixings by the gravity tilting mixer require higher values of V_w/V_p than the mixing G7.5 to realize the adequate V funnel time. However, the mixing G3.5 and G2.5 show almost the same results and this could imply that the highest limit of V_w/V_p exists for the material combination independently of the mixing efficiency.

Table 3 Results of concrete test: admixture type (a)

MIX NO	MIXING	V_w/V_p	α	Sp/P (%)	SLUMP FLOW (after mix) (mm)	SLUMP FLOW (after tests) (mm)	V FUNNEL TIME (sec)
a1	F5.0	0.765	1.000	0.570	657.5	662.5	11.05
a2	F5.0	0.803	1.050	0.540	670.0	-	9.80
a3	F3.5	0.765	1.000	0.565	622.5	665.0	16.00
a4	F3.5	0.803	1.050	0.520	655.0	672.5	10.55
a5	G7.5	0.803	1.050	0.410	650.0	630.0	35.50
a6	G7.5	0.858	1.120	0.368	652.5	640.0	11.55
a7	G7.5	0.861	1.125	0.365	645.0	647.5	9.75
a8	G5.5	0.861	1.125	0.390	635.0	630.0	13.60
a9	G5.5	0.872	1.140	0.360	647.5	640.0	11.35
a10	G5.5	0.880	1.150	0.360	652.5	655.0	9.75
a11	G3.5	0.899	1.175	0.360	622.5	630.0	9.80
a12	G2.5	0.899	1.175	0.375	657.5	660.0	9.05

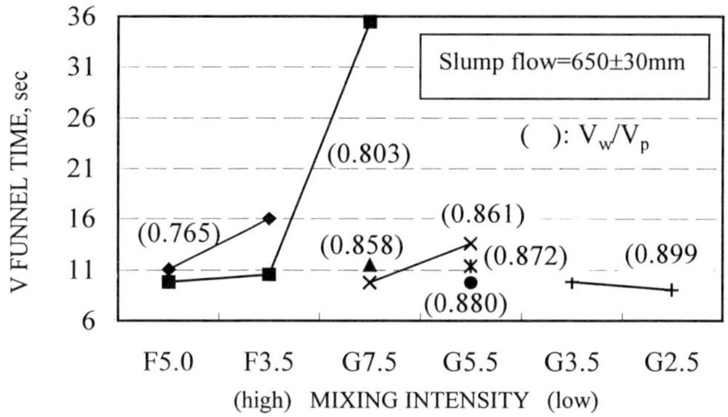

Figure 5 Influence of mixing on V funnel time

Figure 6 shows the relation between the mixing intensity and the appropriate dosage of superplasticizer to obtain the proper result of the slump flow. Regarding these results, the following observations are made:

(1) The forced pan mixer used in this study needs a larger amount of superplasticizer than the gravity tilting mixer in order to obtain the particular deformability. Mixing F5.0 requires a slightly larger amount of SP than the shorter mixing F3.5 and a much more SP than G7.5 mixing by the gravity mixer even when the water to powder volume ratio is constant.

(2) Among the mixing G7.5-2.5, adequate dosages of SP to obtain the proper slump flow and the V funnel time are nearly constant or even slightly increased when the mixing time becomes shorter. This result is contrary to the result of the forced mixer.

From the results presented before, adequate values of V_w/V_p and Sp/P can be given for each mixing procedure when applying the material combination mentioned above. Figure 7 shows the relation between optimized values of V_w/V_p ,Sp/P and the mixing procedures. In order to produce a concrete with the constant slump flow and the V funnel time, the different mixing intensities require different values of V_w/V_p and Sp/P even when materials of the same qualities are used.

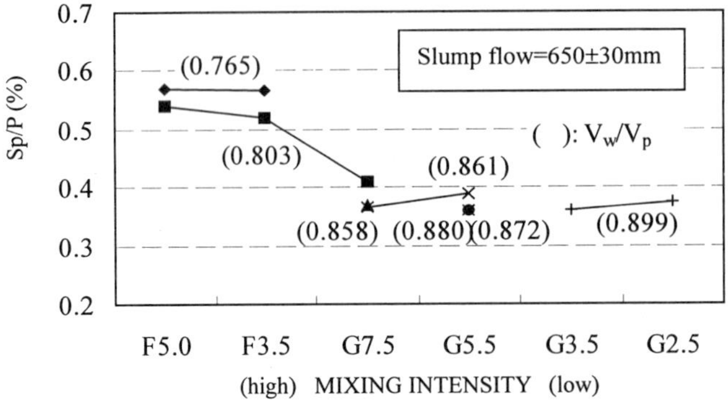

Figure 6 Influence of mixing on superplasticizer dosage

Relation between the Mixing Influence and the Type of Chemical Admixtures

Tables 4 shows the test results for the material combination BSC2:60%&FA2:40% and chemical admixture type (b) polycarboxylic ether type SP (SP-A and B combined in 1:2.5 weight)+welan gum. From this result, it can be said that a concrete including welan gum has a tendency to be more viscous than a concrete without welan gum if the mixing is not sufficient. In other words, welan gum concrete needs a larger increase of V_w/V_p than non-welan gum concrete to realize the same workability when the mixing becomes less intensive. In the case of mixture b5 and b6 which were mixed by the gravity tilting mixer, the initial flowability was not sufficient and it grew in dependence of the time. This behavior implies that the welan gum concrete requires to be mixed more intensively to perform stable flowability from the initial stage.

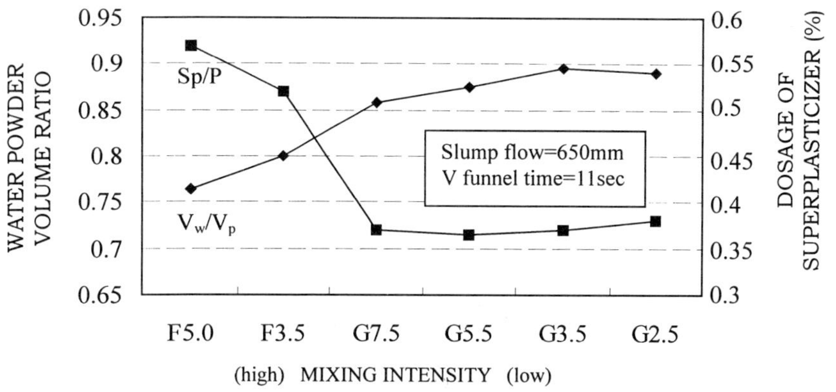

Figure 7 Optimized value of V_w/V_p and Sp/P due to mixing

Table 5 shows the test results for the material combination of BSC2:60%&FA2:40% and the chemical admixture type (c) naphthalene sulfonate type SP (SP-D). Similar to the results for the admixture (a), the gravity mixer needed higher values of V_w/V_p than the forced mixer in order to obtain a proper viscosity of the fresh concrete. However, the premium factor α, due to the mixing procedure, tends to be smaller than for the case of admixture (a). This result implies that the naphthalene based superplasticizer requires a smaller increase of V_w/V_p in order to adjust the property of concrete according to less intensive mixing than the polycarboxylic ether based superplasticizer.

Table 4 Results of Concrete Test: admixture type (b)

MIX NO	MIXING	V_w/V_p	α	Sp/P (%)	SLUMP FLOW (after mix) (mm)	SLUMP FLOW (after tests) (mm)	V FUNNEL TIME (sec)
b1	F5.0	0.805	1.000	0.670	660.0	675.0	11.70
b2	F3.5	0.805	1.000	0.650	607.5	637.5	21.15
b3	F3.5	0.845	1.050	0.550	632.5	640.0	13.10
b4	F3.5	0.865	1.070	0.530	635.0	647.5	10.70
b5	G7.5	0.906	1.125	0.500	615.0	-	15.45
b6	G3.5	0.946	1.175	0.550	570.0	637.5	13.15

Hypothesis for Mixing Effect

From the above mentioned experimental results, the following tendencies of the influence of mixing efficiency on the properties of fresh concrete can be established in general.

(a) More intensive mixing makes the concrete less viscous when the water to powder volume ratio is constant.

(b) More intensive mixing requires a larger amount of superplasticizer in order to obtain the same deformability of the fresh concrete when the water to powder volume ratio is constant.

(c) In order to explain this phenomenon, the following three effects of mixing efficiency expected in the paste phase of the concrete are considered.

Table 5 Results of concrete test: admixture type (c)

MIX NO	MIXING	V_w/V_p	α	Sp/P (%)	SLUMP FLOW (after mix) (mm)	V FUNNEL TIME (sec)
c1	F5.0	0.830	1.000	1.250	635.0	14.45
c2	F5.0	0.872	1.050	1.150	655.0	8.80
c3	F3.5	0.830	1.000	1.250	722.5	16.25
c4	F3.5	0.830	1.000	1.170	685.0	18.05
c5	F3.5	0.872	1.050	1.050	665.0	12.10
c6	G7.5	0.872	1.050	1.050	677.5	68.35
c7	G7.5	0.913	1.100	0.990	627.5	9.90
c8	G3.5	0.913	1.100	0.990	660.0	13.35

Mixing Effect 1: Effect on Dispersion of Powder Particles

An intensive mixing probably makes the powder particles better dispersed than a less intensive mixing. Agglomerated particles make the paste phase of the concrete more viscous (higher deforming resistance) than well dispersed particles.

Mixing Effect 2: Effect on Surface Procession of Powder Particles

When cement (binder) particles are mixed with water and superplasticizer, the initial hydration product is generated on their surface and polymers of superplasticizer adsorb on it. Intensive mixing probably breaks and tears off the initial hydration product from the surface of the cement (binder) particles and the process described above repeats again and again during the mixing. In consequence, the intensive mixing requires more superplasticizer than less intensive mixing to obtain the same slump flow value (deforming capacity), because it consumes a larger amount of the superplasticizer and the water as well during the mixing. Furthermore, the fine pieces of hydration product possibly act as a lubricant among the binder particles and they make the concrete mixture less viscous (lower deforming resistance).

Mixing Effect 3: Effect on Air Bubble Generation in the Paste Phase of Fresh Concrete

More intensively mixed concrete mixtures contain a larger amount of air in it in the cases of this experiment. The entrained air bubbles might be very small and acting as a good lubricant among the aggregate particles. If then, they would help to make the concrete mixture less viscous.

CONCLUSIONS

1. The adequate value of water to powder volume ratio estimated by the mortar test described in the general purpose designing system is dependent not only on the properties of the powder composition but also on the type of chemical admixture.

2. Naphthalene sulfonate and melamine sulfonate types of superplasticizers require a higher V_w/V_p than polycarboxylate type of superplasticizer for the applied mixing method. This tendency is considered to be related to the dispersion mechanisms of each superplasticizer.

3. Welan gum raises the adequate value of V_w/V_p compared to the mixtures without welan gum.

4. The adequate value of V_w/V_p for SCC with proper workability must be estimated considering the mixing efficiency. It is varied by the mixer type and mixing time and can be adjusted due to the mixing intensity in order to obtain a proper quality of SCC.

5. The degree of the influence of the mixing efficiency on the proper proportioning of SCC is dependent on the type of chemical admixtures even when the same powder composition is applied. Welan gum concrete shows a bigger tendency of variation due to the mixing way than non-welan gum concrete. Concretes with a naphthalene sulfonate superplasticizer show a smaller tendency than concretes with a polycarboxylate superplasticizer.

REFERENCES

1. NANAYAKKARA, A., OZAWA, K. and MAEKAWA, K., "Flow and Segregation of Fresh Concrete in Tapered Pipes", Proc. of 3rd International Symposium on Liquid-Solid Flows, ASME, 1988

2. OZAWA, K., TANGTERMSIRIKUL, S. and MAEKAWA, K., "Role of Powder Materials on the Filling Capacity of Fresh Concrete", 4th Canmet/ACI International Conference on Fly Ash, Silica Fume, Slag and Natural Pozzolans in Concrete, 1992

3. YAMAGUCHI, S., EDAMATSU, Y. and OKAMURA, H., "Characterization of Sand in View of Mortar Flow (written in Japanese)", Proc. of JCI, Vol.16, No.1, 1994

4. OKAMURA, H., MAEKAWA, K. and OZAWA, K., "High Performance Concrete (written in Japanese)", *Gihodo Shuppan*, Tokyo, 1993

5. OKAMURA, H. and OZAWA, K., "Self-Compacting High Performance Concrete", Proc. of Cairo First Conference on Concrete Structures, Vol.1, 1996

6. TAKADA, K., PELOVA, G.I. and WALRAVEN, J.C., "Self-Compacting Concrete Produced by Japanese Method with Dutch Materials", European Ready Mixed Concrete Organization Congress ERMCO98, Lisbon, Portugal, June 23-26, 1998

7. YURUGI, M., SAKAI, G. and SAKATA, N., "Viscosity Agent and Mineral Admixtures for Highly Fluidized Concrete", Proc. of International Conference on Concrete Under Severe Conditions, Sapporo, 1995

8. SAKAI, E. and DAIMON, M., "Dispersion Mechanism of Alite Stabilized by Superplasticizers Containing Polyethylene Oxide Graft Chains, 5th Canmet/ACI International Conference on Superplasticizers and Other Chemical Admixtures in Concrete, 1997

9. OHTA, A., SUGIYAMA, T. and TANAKA, Y., "Fluidizing Mechanism and Application of Polycarboxylate-Based Superplasticizer, 5th Canmet/ACI International Conference on Superplasticizers and Other Chemical Admixtures in Concrete, 1997

10. SAKATA, N., MARUYAMA, K. and MINAMI, M., "Basic Properties of Viscosity Agent Welan Gum and Effects on Self-Consolidating Property of Fresh Concrete (written in Japanese)", Proc. of JSCE No.538/V-31, 57-68, 1996.5

UNDERSTANDING MECHANICAL PROPERTIES AND MIX PROPORTIONING OF MORTARS MADE WITH 53 GRADE CEMENT

G A Rao

Indian Institute of Science

P Purushotham

M G Reddy

Government of Andhra Pradesh

India

ABSTRACT. This paper deals with some experimental investigations on cement mortars. In this study an attempt is made to understand some mechanical properties and mix proportioning of mortars prepared with 53 grade ordinary portland cement. In this attempt, three important variables namely water-cement ratio, cement-sand ratio and age of mortar were adopted.. The water-cement ratio was varied from 0.30 to 0.65 at a constant increment of 0.05. The cement-sand ratios of 1:2, 1:2.5 and 1:3 were adopted. At ages of 3, 7 and 28 days, the behaviour and mechanical properties were investigated. The mechanical properties were (i) compressive strength (ii) split tensile strength (iii) load-deformation variations and (iv) initial tangent of stress-strain curves. From the experimental results it was clear that the compressive strength as well as split tensile strength increases as the water-cement ratio decreased. Further, brittle failure was observed on strong mixes in which crack formed in the hardened cement paste propagate across the sand grains. Where as some mixes exhibited ductile failure in which, cracks crossed the grains and some propagated around the aggregate resulting in interfacial bond failure. The shape of the ascending part of the load-deformation curve is more linear and steeper in relatively stronger mortar mixes. Emphirical equations were derived for predicting (i) compressive strength (ii) split tensile strength and (iii) initial tangent of load-deformation curves for mortar at 28 days as functions of 3 days or 7 days compressive strength and water-cement ratio.

Keywords: Mechanical properties, Mix proportioning, Mortar, 53 grade cement, Compressive strength, Split tensile strength, Load-deformation.

G Appa Rao is an Assistant Professor in Civil Engineering, Sri Venkateswara University, Tirupathi, A.P., India. He specialises in the use of cementitious materials, fracture mechanics of concrete and finite element analysis. He has more than twenty publications in his credit. Presently, he is a Research Scholar at the Indian Institute of Science, Bangalore under Quality Improvement Programme sponsored by MHRD, India.

P Purushotham is an Assistant Executive Engineer, Roads and Buildings, Government of Andhra Pradesh, Chittoor.

M Gopal Reddy is an Assistant Executive Engineer, Roads and Buildings, Government of Andhra Pradesh, Cuddapah.

INTRODUCTION

Since long time, in the art of civil engineering certain finely divided silicious materials have been combined with lime to produce hydraulic cementitious products. Dates back to the days of ancient Rome and India, cementitious concept was extensively employed in the masonry construction of aqueducts, arch bridges, retaining structures and buildings. In fact, the proportioning of materials was arbitrarily adopted.

In general, grading of concretes was adopted to select suitable concrete mix depending on the type of loading, type of structure and other environmental impacts. In India, prior to 1987 only one grade of OPC often known to be 33 grade OPC, which conforms to IS:269-1976, was generally available for all types of construction activities. In more recent times, use of higher grades of ordinary portland cement are being widely encouraged in most of the advanced countries to suit the conditions for producing high strength and high performance concretes. In fact, lot of research work in the world is concentrated on mortars and their behaviour under various loading and environmental conditions. In India, introduction of high grade cements is a very recent development. In the present construction practices, 33, 43 and 53 grade OPC are being produced on regular basis and are easily available to the consumers.

It is a fact that the concrete and mortar are heterogeneous materials with three phase and two phase systems respectively and their mechanical properties are significantly influenced by the interaction of various constitutents. Generally it was observed that the physical properties of cement based materials are primarily affected by the water-cement ratio, the chemical composition, micro structure and pore geometry of the cementitious materials, properties of aggregates, cement-aggragate ratio, and properties of cement - aggregate interfacial zone.

Ingenious efforts have revealed that the performance of concrete is directly related with the characteristics of mortar matrix. In this study, efforts were made to understand the mortar as medium for high strength and high performance concretes. It is also tried to standardize the mix proportions for mortars.

EXPERIMENTAL PROGRAMME

Materials

Cement

A 53 grade ordinary portland cement, which conforms to IS:12269 - 1987, was used. The properties of the cement are listed in Table 1.

Fine aggregate

The natural river sand available nearby Tirupati was used. The sand fraction passing through 4.75 mm sieve and 600 um sieve was used in the mixes. In order to avoid the presence of inorganic and silt particles the above fraction was adopted. The fineness modulus was 2.98.

Water

Potable tap water available in the laboratory with pH value of 7.85 ± 0.1 was used for mixing mortar and curing the mortar specimens

Table 1 Properties of Cement and Specified Values of 53 Grade Cement

SL.NO.	PROPERTY	RESULT	IS: 12267-1987 RECOMMENDATIONS FOR 53 GRADE CEMENT
1.	Normal Consistency	27.75	-
2.	Setting times		
	a) Initial (Minutes)	100	> 30
	b) Final (Minutes)	275	< 600
3.	Specific gravity	3.12	-
4.	Compressive strength (N/mm^2)		
	a) 3 days	29.00	27.00
	b) 7 days	38.50	37.00
	c) 28 days	53.50	53.00

Mix proportions

In this study, the cement-sand ratios of 1:2 (CM-A), 1:2.5 (CM-B) and 1:3 (CM-C) were adopted. For each cement - sand ratio, the water-cement ratio was varied from 0.30 to 0.65 at constant value of 0.05. Table 2 shows the details of various mix proportions.

Preparation of Test Specimens

Compressive strength

100 mm metalic cubes were used for the preparation of specimen for compressive strength. Table vibrator was used for compacting the mortar filled cubes. The specimens were demoulded after 24 hours and immersed in water till the time of testing.

Split tensile strength and load-deformation variations

Cylinders measuring 100 mm diameter and 200 mm height were adopted for preparation of test specimens for split tensile strength and load-deformation variations. The compacted specimens were demoulded after 24 hours and immersed in water till the time of testing.

Testing of Specimens

The cube specimens were tested for compressive strength at the age of 3, 7, and 28 days using 200 tonne compression testing machine. The cylindrical specimens were splitted using a splitting frame under 40 tonnes UTM.

The split tensile strength is calculated as below:

$$\sigma_t = 2P/\pi LD$$

Where σ_t = split tensile strength
P = load on the splitting frame
L = Length of cylinder
D = Diameter of the cylinder

The load - deformations were measured over a distance of 100 mm in the middle portion of the specimens using demec point and dial gauge system.

Table 2 Mix proportioning

SL.NO.	MIX DESIGNATION	MIX PROPORTIONS
		C : W : S:
1.	CM A1	1 : 0.30 : 2.0
2.	CM A2	1 : 0.35 : 2.0
3.	CM A3	1 : 0.40 : 2.0
4.	CM A4	1 : 0.45 : 2.0
5.	CM A5	1 : 0.50 : 2.0
6.	CM A6	1 : 0.55 : 2.0
7.	CM A7	1 : 0.60 : 2.0
8.	CM A8	1 : 0.65 : 2.0
9.	CM B1	1 : 0.30 : 2.5
10.	CM B2	1 : 0.35 : 2.5
11.	CM B3	1 : 0.40 : 2.5
12.	CM B4	1 : 0.45 : 2.5
13.	CM B5	1 : 0.50 : 2.5
14.	CM B6	1 : 0.55 : 2.5
15.	CM B7	1 : 0.60 : 2.5
16.	CM B8	1 : 0.65 : 2.5
17.	CM C1	1 : 0.30 : 3.0
18.	CM C2	1 : 0.35 : 3.0
19.	CM C3	1 : 0.40 : 3.0
20.	CM C4	1 : 0.45 : 3.0
21.	CM C5	1 : 0.50 : 3.0
22.	CM C6	1 : 0.55 : 3.0
23.	CM C7	1 : 0.60 : 3.0
24.	CM C8	1 : 0.65 : 3.0

TEST RESULTS AND DISCUSSION

Various results that were evaluated are

1. Variation of comporessive strength with water-cement ratio for different cement mortar mixes at ages of 3,7, and 28 days.

2. Variation of split tensile strength with water-cement ratio and compressive strength of mortar.

3. Variation of Load-deformation in various cement mortar mixes.

Empirical equations are developed to estimate the compressive strength and split tensile strength and initial tangent as functions of water-cement ratio, and compressive strength at 3, 7 and/or 28 days.

Variation of Compressive Strength with Water-Cement Ratio

CM-A, CM-B and CM-C are the three mortar mix groups in which cement-sand ratios adopted are 1: 2, 1:2.5 and 1:3 respectively. From the test results it is observed that CM-A mortar mix group are showing higher values of compressive strength than that in the other two mix groups. i.e. CM-B and CM-C. It is generally observed that the compressive strength at any age of mortar, i.e. 3, 7 and 28 days, is increasing as the water-cement ratio decreases except at w/c = 0.30. At w/c = 0.30, the mortar was in a semi-dry state due to insufficient amount of water available. As a result, full compaction was not possible and also the hydration of cement was incomplete.

Figures. 1 to 3 are drawn to show the variation of compressive strength with water-cement ratio at three ages i.e. 3, 7 and 28 days for mixes CM-A, CM-B and CM-C respectively. In engineering practice the strength of fully compacted concrete or mortar processed with a specific type and size of aggregates at a given age and curing at a prescribed temperature, is assumed to depend primarily on the water-cement ratio. But the compressive strength at a particular water-cement ratio also depends on chemical composition and fineness of cement. The recent literature shows that the gel-space ratio has much increased influence on the compressive strength of cementitious materials. In proportioning of mortar/concrete mixes, the first and fore most thing is to arrive at water-cement ratio which primarily dictates the desired strength at different ages. Critical review of the existing literature [2,3] reveals that as the free water-cement ratio increases the compressive strength decreases. The results in this study are conforming to the existing data on variation of compressive strength with water-cement ratio.

Figure 4 shows the variation of compressive strength of different mortar mixes at the age of 3 days. From the figure it is clearly showing that the strength in mortar mix CM-A at any water-cement ratio is higher than that in CM-B and CM-C mortar mixes. Similarly, Figures 5 and 6 show the compressive strength variation with water-cement ratio in different mortar mixes at 7 and 28 days respectively. It is generally observed that CM-A mortar mixes are stronger than CM-B and CM-C mortar mixes at any age. A general variation of compressive strength with water-cement ratio is also shown in Figure 9, 10 and 11 at 3, 7, and 28 days

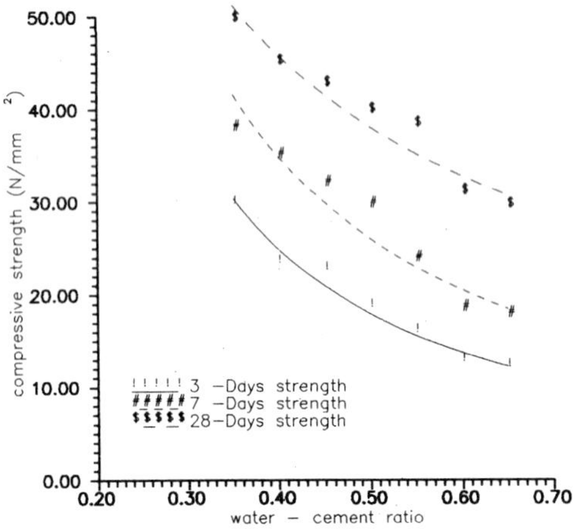

Figure 1 Variation of compressive strength with w/c ratio for 1:2 cement mortar

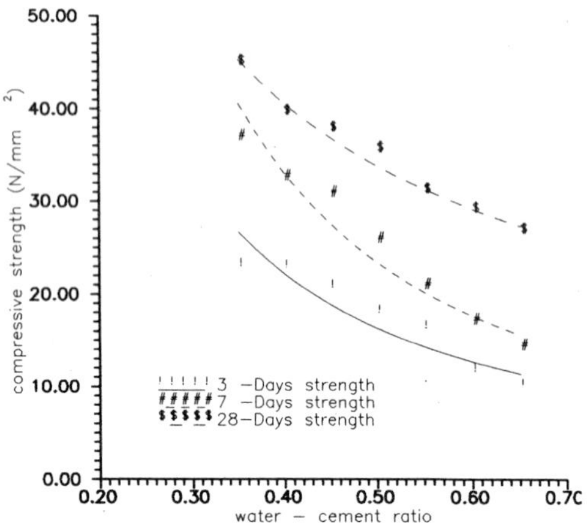

Figure 2 Variation of compressive strength with w/c ratio for 1:2:5 cement mortar

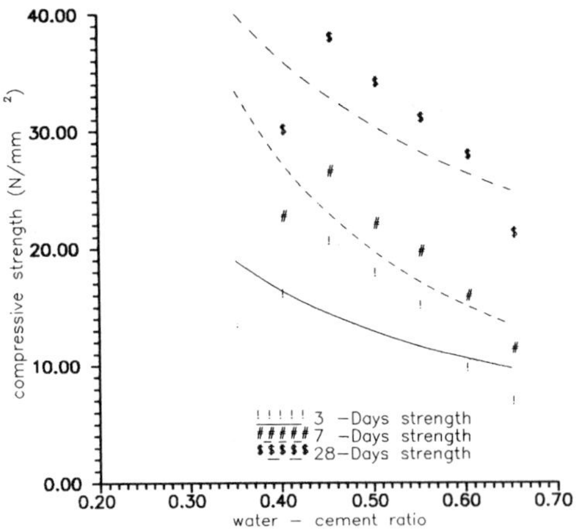

Figure 3 Variation of compressive strength with w/c ratio for 1:3 cement mortar

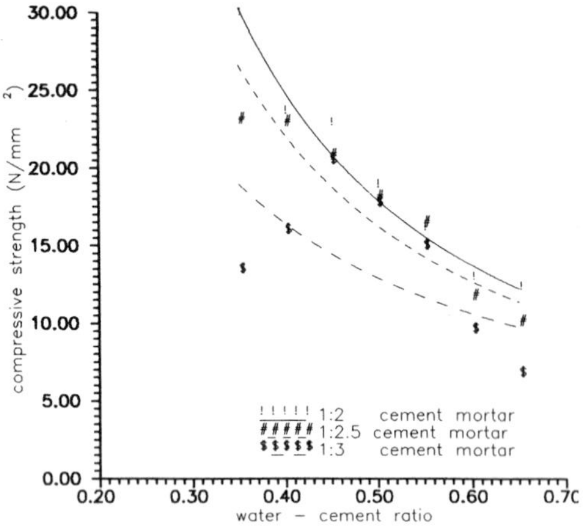

Figure 4 Variation of 3 day compressive strength with w/c ratio for various mixes

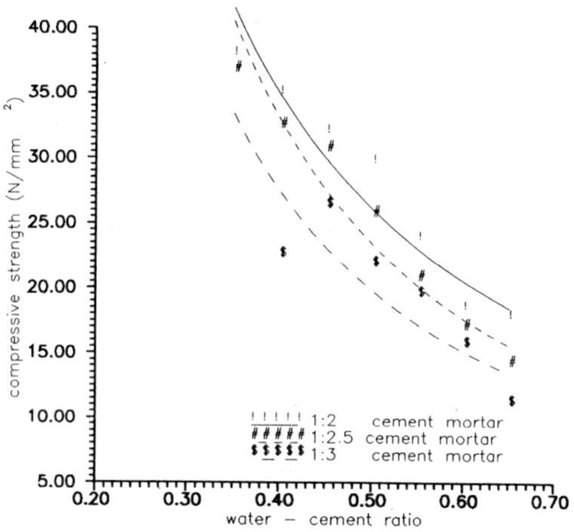

Figure 5 Variation of 7 day compressive strength with w/c ratio for various mixes

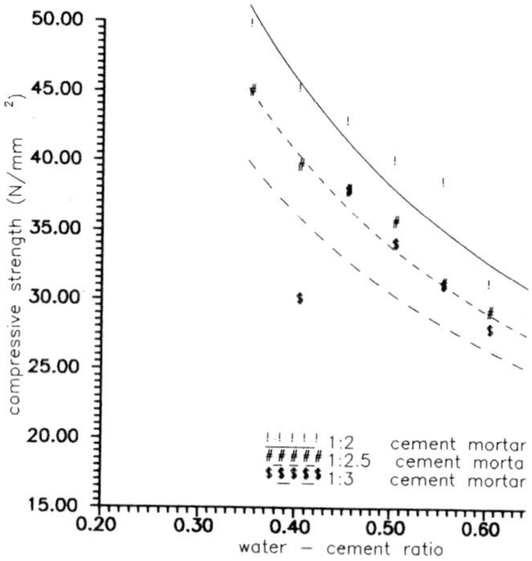

Figure 6 Variation of 28 day compressive strength with w/c ratio for various mixes

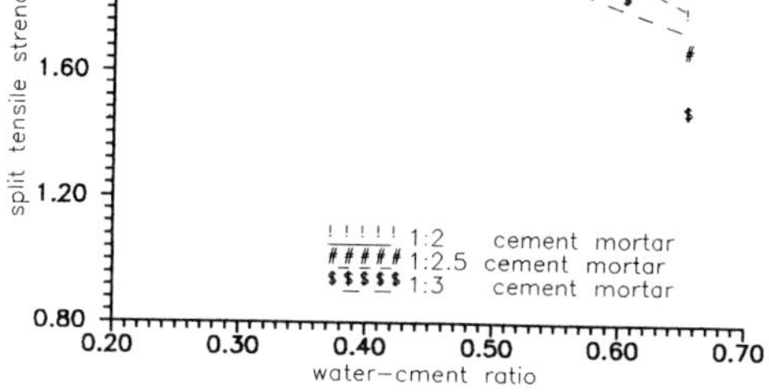

Figure 7 Variation of 28 day split tensile strength with w/c ratio for various mixes

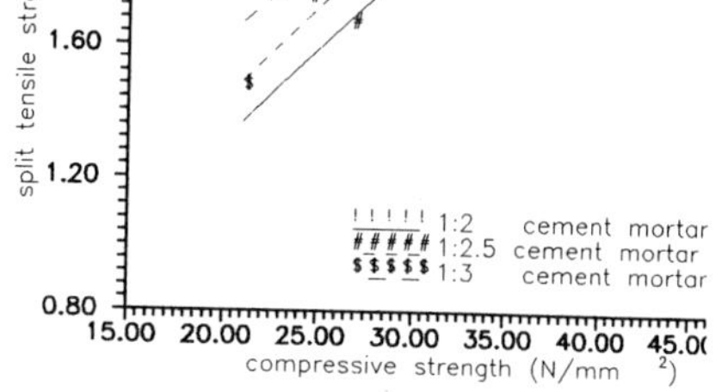

Figure 8 Variation of split tensile strength with 28 day compressive strength

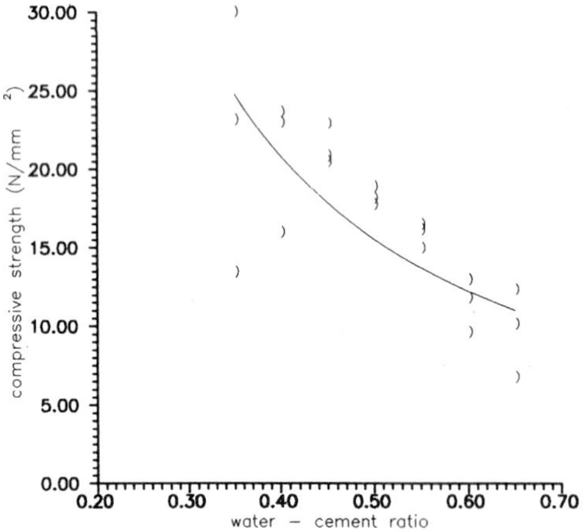

Figure 9 Variation of 3 day compressive strength with w/c ratio

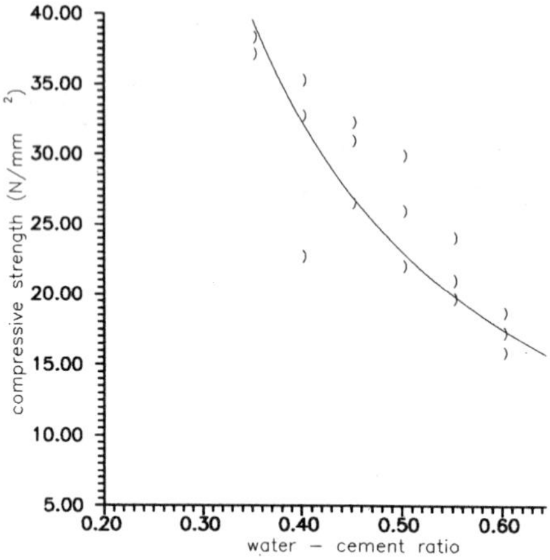

Figure 10 Variation of 7 day compressive strength with w/c ratio

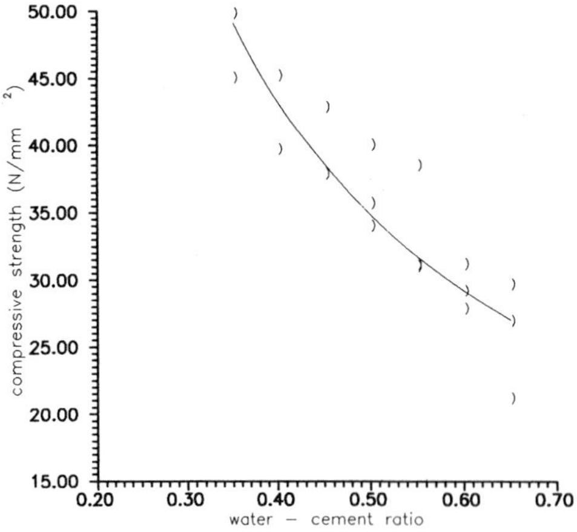

Figure 11 Variation of 28 day compressive strength with w/c ratio

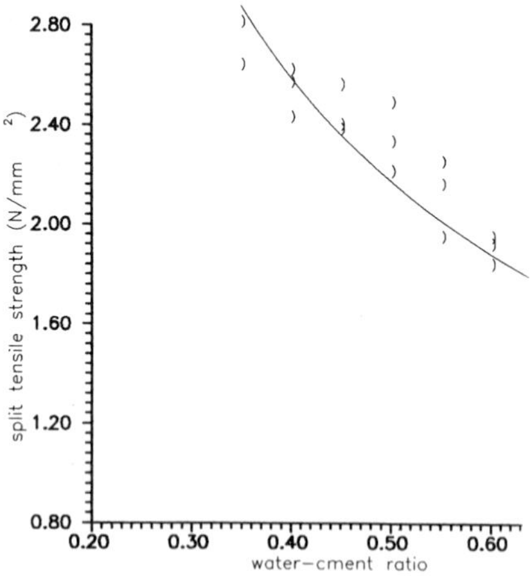

Figure 12 Variation of split tensile strength with w/c ratio

respectively irrespective of the mix designation. In all the curves the generalised compressive strength with water-cement ratio law is perfectly observed.

The empirical formulae derived for predicting the 28 days compressive strength as a function of water-cement ratio are as follows:

For CM-A mortar mixes

$$\sigma_{c28} = 29.00 + (w/c)\,[87.26 - 127.87\,(w/c)]$$

For CM-B mortar mixes

$$\sigma_{c28} = 22.35 + (w/c)\,[92.62 - 130.58\,(w/c)]$$

For CM-C mortar mixes

$$\sigma_{c28} = 38.69 + (w/c)\,[23.23\,(w/c) - 29.11]$$

The generalised equation for 28 days compressive strength of mortar as a function of water-cement ratio is

$$\sigma_{c28} = 47.68 - (w/c)\,[26.89 + 0.48\,(w/c)]$$

Variation of Split Tensile Strength with Water-Cement Ratio and Compressive Strength

Figure 7 shows the variation of 28 days split tensile strength for different mortar mixes as a function of water-cement ratio. In this case also the trend of the split tensile strength is much similar to the compressive strength. CM-A is much stronger than CM-B and CM-C mixes.

The generalized equation for predicting the 28 days split tensile strength as a function of water-cement ratio is

$$\sigma_{t28} = 3.04 + (w/c)\,[0.24\,(w/c) - 1.86]$$

The relationship between compressive strength and split tensile strength at 28 days is also derived, which is as follows:

$$\sigma_{c28} = \sigma_{t28}\,[4.33 + 6.95\,\sigma_{t28}] - 3.62$$

The variation is shown in Figure 13 and it is a linear variation.

Variation of Load-Deformations in Various Mortar Mixes

Figures 14 to 16 shows the variation of compressive load versus deformation. From the curves it is clear that the slope of ascending part decreasing as the water-cement ratio increases. This is observed in all the mortar mixes. The shape of the ascending part of the load-deformation curve is more linear and steeper for high strength mortars. Brittle failure

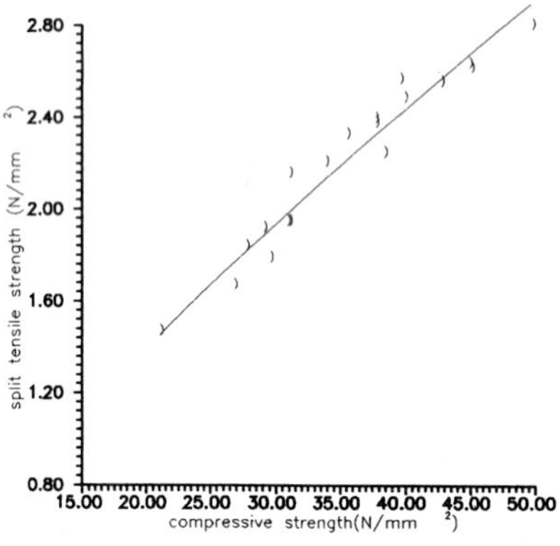

Figure 13 Variation of split tensile strength with 28 day compressive strength

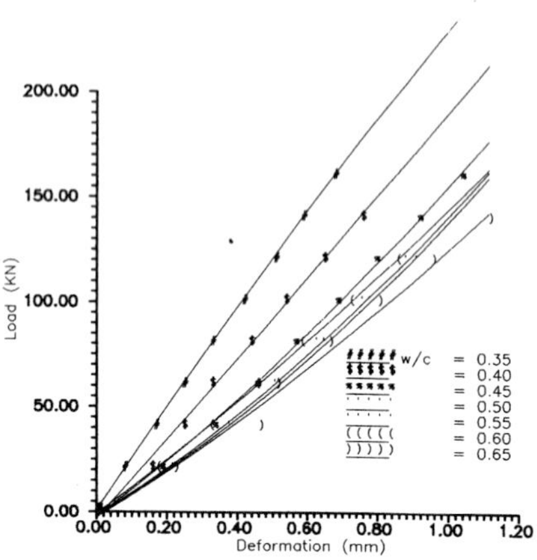

Figure 14 Load-deformation curves for 1:2 cement mortar

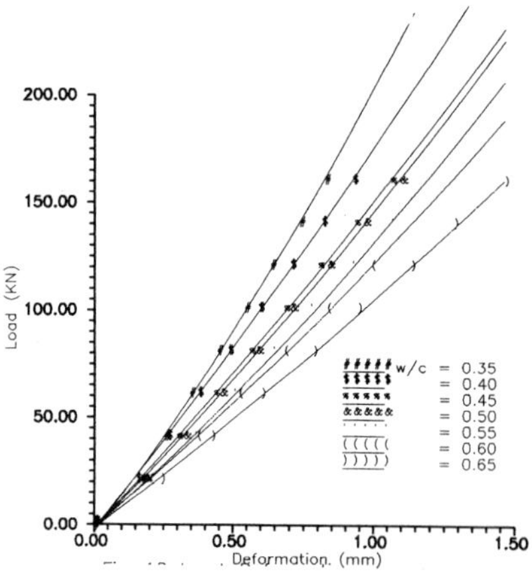

Figure 15 Load-deformation curves for 1:2:5 cement mortar

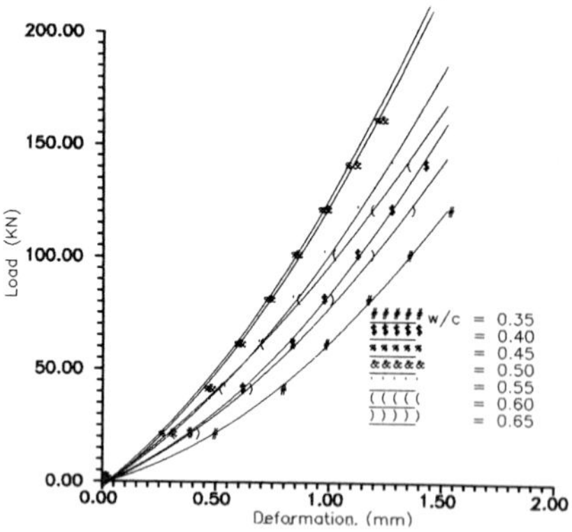

Figure 16 Load deformation curves for 1:3 cement mortar

was observed in mortars at low water-cement ratios. The failure was sudden. At higher water-cement ratios, decrease in slope of the ascending part and reduction in brittleness of mortar was observed. The bulging of lateral dimensions was also lesser in high strength mortars in compression.

Initial-Tangent of Load-Deformation /Stress-Strain Curves

Figure 17 shows the variation of initial tangent with the compressive strength for various mixes. From the Figure 17, it is clear that as the compressive strength increases the tangent is increasing. A general equation is derived to show the relationship between compressive strength at 28 days and initial tangent. This is given by

$$\sigma_{c28} = 32.2 + E_m\,[1.66\,E_m - 6.17]$$

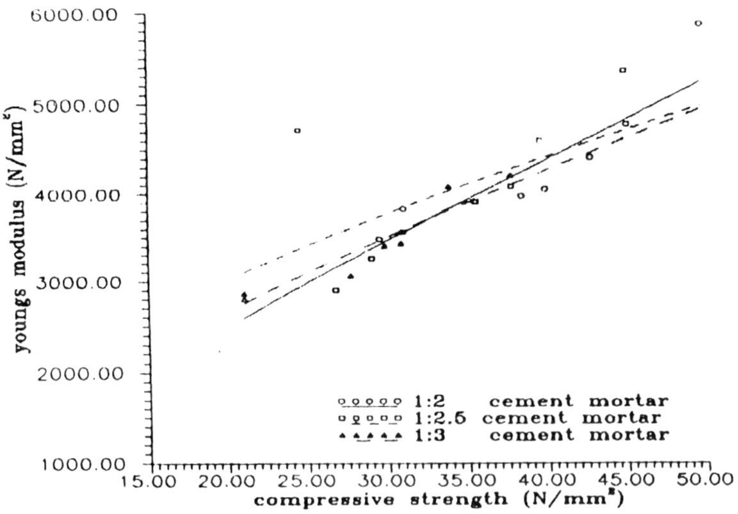

Figure 17 Variation of Youngs modulus with compressive strength for mixes at 28 days

CONCLUSIONS

From the test results the following conclusions may be drawn.

1. All the cement mortars are perfectly obeying the strength water-cement ratio law.

2. CM-A mortar mixes exhibit relatively higher strength than the other two mixes i.e. CM-B and CM-C.

3. The ascending part of load-deformation curve is more steeper in high strength mortars.

4. Brittle failure was observed in high strength mortars and the failure was sudden.

5. Split tensile strength of mortar also exhibit the trend similar to that for compressive strength.

6. The initial tangent of stress-strain curves is higher at lower water-cement ratios and lower at higher water-cement ratios.

7. Among the mortar mixes, CM-A always show better performance in all respects.

REFERENCES

1. PURUSHOTHAM, P, Understanding the mix proportioning and behaviour of cement mortars using 53 grade cement. M.Tech. thesis under the guidance of G. Appa Rao, S.V. University, Tirupati, August, 1997.

2. NAGARAJ, T G AND ZAHIDA BANU, Generalization of Abram's law, Cement and Concrete Research, Vol.26, No.6, 1996, pp.933-942.

3. ACI Committee 363, State-of-the Art Report on High-strength Concrete, ACI Journal, July-August, 1984, pp.364-411.

INDEX OF AUTHORS

SUBJECT INDEX

This index has been compiled from the keywords assigned to the papers, edited and extended as appropriate. The page references are to the first page of the relevant paper.